Crop Improvement Strategies in *Brassica* species: Applied Science

Adesh Kumar • Bhupendra Koul
Editors

Crop Improvement Strategies in *Brassica* species: Applied Science

Volume 1

Springer

Editors
Adesh Kumar
Department of Plant Pathology, School of Agriculture
Lovely Professional University
Phagwara, Punjab, India

Bhupendra Koul
Department of Biotechnology, School of Bioengineering and Biosciences
Lovely Professional University
Phagwara, Punjab, India

ISBN 978-981-95-3860-7 ISBN 978-981-95-3861-4 (eBook)
https://doi.org/10.1007/978-981-95-3861-4

This Springer imprint is published by the registered company Springer Nature Singapore Pte Ltd.
The registered company address is: 152 Beach Road, #21-01/04 Gateway East, Singapore 189721, Singapore

Preface

Brassica species—encompassing vital crops like mustard, cabbage, cauliflower, broccoli, and rapeseed—hold exceptional economic, nutritional, and agricultural significance across the globe. As the demands of food security, climate resilience, and sustainable agriculture continue to grow, the scientific community is called upon to innovate and modernize crop improvement strategies. This book, *Crop Improvement Strategies in Brassica species: Applied Science (Volume 1)*, emerges as a response to that call.

This volume offers an extensive compilation of cutting-edge research and practical advancements spanning diverse disciplines such as plant breeding, biotechnology, genetics, phytochemistry, pathology, agronomy, and seed technology. Through 17 comprehensive chapters, renowned scholars and researchers from around the world delve into the economic importance, genetic enhancement techniques, biofortification, phytochemical potential, and the role of *Brassica* crops in climate-smart agriculture. Special attention is given to the integration of modern technologies such as robotics, machine learning, and microbial applications in enhancing crop resilience and productivity.

The content of this volume reflects a harmonious blend of foundational knowledge and innovative methodologies aimed at transforming *Brassica* crop production into a more sustainable and resilient system. We believe that this book will serve as a valuable resource for researchers, academicians, students, and policymakers working in the areas of plant science, agricultural biotechnology, and crop improvement.

We express our sincere gratitude to all the contributing authors for their scholarly insights and commitment to excellence. We are also thankful to our institutions and colleagues for their support throughout the editorial process.

As editors, we are pleased to present this volume as the first in a two-part series. While Volume 1 emphasizes applied strategies, the forthcoming Volume 2 will focus specifically on stress resistance strategies in *Brassica* crops—further deepening the reader's understanding of the challenges and opportunities in this vital crop group.

We hope this volume sparks further research, collaboration, and innovation in the field of Brassica improvement.

Phagwara, Punjab, India — Adesh Kumar
Phagwara, Punjab, India — Bhupendra Koul

Contents

Editors and Contributors

About the Editors

Adesh Kumar is currently working as Associate Professor and Head in the Department of Plant Pathology at Lovely Professional University, Phagwara, Punjab (India). He obtained his M.Sc. (Ag.) from Department of Plant Pathology, CSA University, Kanpur (Uttar Pradesh) and Ph.D. (Ag.) from Department of Plant Pathology, CSA University, Meerut (Uttar Pradesh), during 2008 and 2012, respectively. He was appointed and worked as Assistant Professor, School of Agriculture, Selaqui, Dehradun (Uttrakhand), India, from Oct 2013 to Dec 2014. His areas of research interest are Fungal Pathology, Biological Control of Phytopathogens, Mycology, Seed Pathology, Plant Microbe Interaction, and Fungal Decomposition. He has to his credit more than 48 research papers published in reputed national and international journals, 1 lab manual, 4 books and 10+ book chapters, 20+ patents, and 10+ copyrights. He has been awarded Excellence in Research Award. He has served as a member of scientific societies such as Indian Phytopathological Society and Indian Science Congress Association.

Bhupendra Koul is an Associate Professor in the Department of Biotechnology, School of Bioengineering and Biosciences, Lovely Professional University (LPU), Punjab, India. During his Ph.D. at the Plant Transgenics Lab, CSIR-National Botanical Research Institute (CSIR-NBRI), Lucknow, he worked on the optimization, introduction, and expression of modified full-length and truncated versions of Bt-*cry1Ab* and *1Ac* genes in tomato for developing non-chimeric and stable transgenic lines resistant to two lepidopteran insects (*Helicoverpa armigera* and *Spodoptera litura*) and evaluated the performance of both the versions of *cry1Ab* and *1Ac* genes for the stability and efficacy of insecticidal toxin in transgenic plants. He also evaluated the performance and role of various cis-motifs of synthetic promoters for overexpression of genes in tomato and performed comparative in silico analyses of several *cry1A* genes for toxicity to target insects. He has also optimized the regeneration and *Agrobacterium*-mediated transformation of Stevia (*Stevia rebaudiana* Bertoni) and has developed herbicide-resistant transgenic Stevia for effective weed management in Stevia cultivation. He was awarded CSIR-Senior Research Fellowship (SRF) in the year 2013. He has 9.5 years of teaching experience and has received the Teacher Appreciation Award 2016 from LPU in the discipline of biotechnology, through the MHRD Minister, Government of India. He has designed the full-length synthetic *cry1Ac* gene (GenBank: KP195020.1) and has authored 45 + research papers in national and international journals as well as 21 book chapters, 3 books with Springer Nature, and 1 international patent (granted).

Contributors

Kanik Kumar Bansal Faculty of Agriculture, Guru Kashi University, Talwandi Sabo, Bathinda, Punjab, India

Bharti School of Agriculture Sciences & Engineering, Maharaja Ranjit Singh Punjab Technical University, Bathinda, Punjab, India

Monika Chaudhary School of Biotechnology Engineering & Life Sciences, Shobhit Institute of Engineering & Technology (NAAC Accredited Grade 'A', Deemed to-be-University), Meerut, Uttar Pradesh, India

Vladislav Čurn Faculty of Agriculture and Technology, Department of Genetics and Biotechnology, University of South Bohemia in České Budějovice, České Budějovice, Czech Republic

Pankaj Dahiya Department of Agronomy, School of Agriculture, Lovely Professional University, Phagwara, India

Sinouvassane Djearamane Department of Biomedical Science, Faculty of Science, Universiti Tunku Abdul Rahman (UTAR), Kampar, Perak, Malaysia

A. Ganesh Kumar Department of Microbiology, Saveetha Medical College and Hospital, Saveetha Institute of Medical and Technical Sciences (SIMATS), Saveetha University, Chennai, Tamil Nadu, India

Etalesh Goutam Department of Horticulture, Guru Kashi University, Bathinda, Punjab, India

Jyoti Gupta Department of Education, Lovely Professional University, Phagwara, Punjab, India

Ondřej Hejna Faculty of Agriculture and Technology, Department of Genetics and Biotechnology, University of South Bohemia in České Budějovice, České Budějovice, Czech Republic

Himanshu Department of Horticulture, School of Agriculture, Lovely Professional University, Phagwara, Punjab, India

Dilfuza Jabborova Institute of Genetics and Plant Experimental Biology, Kibray, Uzbekistan

Eva Jozová Faculty of Agriculture and Technology, Department of Genetics and Biotechnology, University of South Bohemia in České Budějovice, České Budějovice, Czech Republic

Lokender Kashyap Faculty of Agriculture, Guru Kashi University, Talwandi Sabo, Bathinda, Punjab, India

Khushboo Kathayat School of Agriculture, Lovely Professional University, Phagwara, Punjab, India

Jasjit Kaur Department of Food Technology and Nutrition, Lovely Professional University, Phagwara, Punjab, India

Deepika Kaushik Department of Biotechnology, Faculty of Applied Sciences and Biotechnology, Shoolini University, Solan, India

Rahmatullah Kazimi Department of Horticulture, Daikundi of Higher Education Institution, Daikundi, Afghanistan

Khalid Ali Khan Center of Bee Research and its Products (CBRP), King Khalid University, Abha, Saudi Arabia

Miroslav Klíma Laboratory of Plant Stress Biology and Biotechnology, Department of Plant Genetics and Crop Breeding, Crop Research Institute, Prague, Czechia

Bhupendra Koul Department of Biotechnology, School of Bioengineering & Biosciences, Lovely Professional University, Phagwara, Punjab, India

Adesh Kumar Department of Plant Pathology, School of Agriculture, Lovely Professional University, Phagwara, Punjab, India

Akshay Kumar Agriculture and Environment Research Unit, School of Life and Medical Sciences, University of Hertfordshire, Hatfield, England, UK

Mukul Kumar Department of Food Technology and Nutrition, Lovely Professional University, Phagwara, Punjab, India

Roushan Kumar Department of Pharmaceutical Sciences and Technology, Maharaja Ranjit Singh Punjab Technical University, Bathinda, Punjab, India

Vishal Kumar Department of Food Technology and Nutrition, Lovely Professional University, Phagwara, Punjab, India

Sudhir Kumar Upadhyay Research and Development Cell, Lovely Professional University, Phagwara, Punjab, India

Anjana Kumari Department of Applied Sciences and Humanities, Ambala College of Engineering and Applied Research, Haryana, India

Shivanshu Ladohia Department of Agronomy, Chaudhary Sarwan Kumar Himachal Pradesh Krishi Vishvavidyalaya (CSK HPKV), Palampur, HP, India

Mohamed A. Mattar Department of Agricultural Engineering, College of Food and Agricultural Sciences, King Saud University, Riyadh, Saudi Arabia

Agricultural Engineering Research Institute (AEnRI), Agricultural Research Centre, Giza, Egypt

Swati Mehta Department of Agronomy, School of Agriculture, Lovely Professional University, Phagwara, Punjab, India

Fatih Oz Department of Food Engineering, Faculty of Agricultur, Ataturk University, Erzurum, Türkiye

Satya Prakash Department of Genetics and Plant Breeding, Lovely Professional University, Phagwara, Punjab, India

Research and Development Cell, Lovely Professional University, Phagwara, Punjab, India

Baveesh Pudhuvai Faculty of Agriculture and Technology, Department of Plant Production, University of South Bohemia in České Budějovice, České Budějovice, Czech Republic

Ranjithkumar Rajamani Department of Pharmacology, Saveetha Medical College and Hospital, Saveetha Institute of Medical and Technical Sciences (SIMATS), Saveetha University, Chennai, Tamil Nadu, India

Navjot Rana Department of Agronomy, School of Agriculture, Lovely Professional University, Phagwara, India

Prasad Rasane Department of Food Technology and Nutrition, Lovely Professional University, Phagwara, Punjab, India

Monisha Rawat School of Agriculture, Lovely Professional University, Phagwara, Punjab, India

Robin Rijal Natural Resources Institute of University of Greenwich, Kent, UK

Department of Plant Pathology, School of Agriculture, Lovely Professional University, Phagwara, Punjab, India

Ramesh Kumar Sadawarti Department of Horticulture, School of Agriculture, Lovely Professional University, Phagwara, Punjab, India

Rakesh Roshan Satapathy Faculty of Agricultural Sciences, Siksha O Anusandhan University, Bhubaneswar, Odisha, India

Deepika Saxena Department of Horticulture, School of Agriculture, Lovely Professional University, Phagwara, Punjab, India

Althaf Sayyad Department of Food Technology and Nutrition, Lovely Professional University, Phagwara, Punjab, India

Aashish Kumar Sharm Department of Plant Pathology, School of Agriculture, Lovely Professional University, Phagwara, Punjab, India

Aditi Sharma College of Horticulture & Forestry, Dr Y S Parmar University of Horticulture & Forestry, Nauni, Solan, Himachal Pradesh, India

Anchal Sharma Department of Mathematics and Humanities, Maharishi Markandeshwar (Deemed to be) University, Mullana, Ambala, India

Geeta Sharma School of Computer Application, Lovely Professional University, Phagwara, Punjab, India

Mamta Sharma School of Bioengineering and Biosciences, Lovely Professional University, Phagwara, Punjab, India

Shobna Department of Horticulture, School of Agriculture, Lovely Professional University, Phagwara, Punjab, India

Lakhwinder Singh Faculty of Agriculture, Guru Kashi University, Talwandi Sabo, Bathinda, Punjab, India

Souravdeep Singh School of Computer Application, Lovely Professional University, Phagwara, Punjab, India

K. Sunitha kumari Department of Botany, PSGR Krishnammal College for Women, Peelamedu, Coimbatore, Tamil Nadu, India

S. T. Gopukumar Center for Global Health Research, Saveetha Medical College and Hospital, Saveetha Institute of Medical and Technical Sciences (SIMATS), Saveetha University, Chennai, Tamil Nadu, India

Akshay Kumar Vats Department of Plant Breeding and Genetics, Dr. Kalam Agricultural College, Arrabari, Bihar Agricultural University, Bhagalpur, Kishanganj, Bihar, India

Abbreviations

4-HPEA	4-Hydroxyphenethyl Alcohol
6-MSITC	6-Methylsulfinylhexyl Isothiocyanate
ABTS	2,2′-Azino-bis(3-ethylbenzothiazoline-6-sulfonic acid)
ACE2	Angiotensin-Converting Enzyme 2
AFLP	Amplified Fragment Length Polymorphism
AI	Artificial Intelligence
ALA	Alpha-Linolenic Acid
AMF	Arbuscular Mycorrhizal Fungi
AMPK	AMP-Activated Protein Kinase
ANFIS	Adaptive Neuro-Fuzzy Inference System
ARE	Antioxidant Response Element
AT	Associative Transcriptomics
B. juncea	*Brassica juncea*
B. napus	*Brassica napus*
B. oleracea	*Brassica oleracea*
B. rapa	*Brassica rapa*
B:C	Benefit-to-Cost Ratio
Bt	*Bacillus thuringiensis*
CAGR	Compound Annual Growth Rate
CANFIS	Coactive Adaptive Neuro-Fuzzy Inference System
CMS	Cytoplasmic Male Sterility
CoCo	Genome from *Brassica oleracea*
CODEX	Codex Alimentarius (Food standards)
cpSSR	Chloroplast Simple Sequence Repeats
CRISPR	Clustered Regularly Interspaced Short Palindromic Repeats
CRISPR-Cas9	CRISPR-associated protein 9
DAT	Days After Transplanting
DH	Double Haploid
DHA	Docosahexaenoic Acid
DHAP	2,6-Dihydroxyacetophenone
DNMT	DNA Methyltransferase
DPA	Docosapentaenoic Acid
DPPH	2,2-Diphenyl-1-picrylhydrazyl
EFSA	European Food Safety Authority

EPA	Eicosapentaenoic Acid
FAO	Food and Agriculture Organization
FAOSTAT	Food and Agriculture Organization Statistical Database
FDA	Food and Drug Administration
FFNN	Feed-Forward Neural Network
FYM	Farmyard Manure
GAE	Gallic Acid Equivalent
GEMs	Gene Expression Markers
GMOs	Genetically Modified Organisms
GS	Genomic Selection
GSL	Glucosinolate
GWAS	Genome-Wide Association Studies
ha	hectare
HBA	4-Hydroxybenzaldehyde
HD	Heterosis Breeding
HDAC	Histone Deacetylase
HOLL	High Oleic and Low Linolenic
INR	Indian Rupees
INS	International Numbering System for Food Additives
IPM	Integrated Pest Management
ISR	Induced Systemic Resistance
ITC(s)	Isothiocyanate(s)
LAB	Lactic Acid Bacteria
MAS	Marker-Assisted Selection
ML	Machine Learning
MLPNN	Multi-Layer Perceptron Neural Network
MLR	Multiple Linear Regression
MOFA	Ministry of Foreign Affairs
MRA-WT-LSTM	Multi-resolution Analysis-Wavelet Transform–Long Short-Term Memory
MSL	Male Sterile Lembke
MT	Metric Tons
MUFA	Monounsaturated Fatty Acids
MY	Marketing Year
NGS	Next Generation Sequencing
NRF2	Nuclear Factor Erythroid 2–Related Factor 2
OCL	*Oleagineux, Corps Gras, Lipides* (Journal Abbreviation)
OEC	Observatory of Economic Complexity
OP-ELM	Optimally Pruned Extreme Learning Machine
PGPR	Plant Growth-Promoting Rhizobacteria
PPARα	Peroxisome Proliferator-Activated Receptor Alpha
PSA	Prostate-Specific Antigen
PUFA	Polyunsaturated Fatty Acids
q/ha	Quintal per Hectare
QE	Quercetin Equivalent

QTL	Quantitative Trait Loci
RAMP	Random Amplified Microsatellite Polymorphism
RAPD	Random Amplified Polymorphic DNA
RF	Random Forest
Rf genes	Fertility Restoration Genes
RFLP	Restriction Fragment Length Polymorphism
RGA	Resistance Gene Analog
RNAi	RNA Interference
ROS	Reactive Oxygen Species
RSM	Response Surface Methodology
SAR	Systemic Acquired Resistance
SDA	Stearidonic Acid
SFA	Saturated Fatty Acids
SFE	Supercritical Fluid Extraction
SFN	Sulforaphane
SG	Single Guide (RNA)
SNP	Single Nucleotide Polymorphism
SPEI	Standardized Precipitation Evapotranspiration Index
SPI	Standardized Precipitation Index
spp.	Species (plural)
SSR	Simple Sequence Repeats
subsp.	Subspecies
SVM	Support Vector Machine
TCM	Traditional Chinese Medicine
TuMV	Turnip Mosaic Virus
UPOV	International Union for the Protection of New Varieties of Plants
USDA	United States Department of Agriculture
var.	Variety
WA	Wavelet Analysis
WA-ANN	Wavelet-Transformed Artificial Neural Network
WAM	Weighted Arithmetic Mean
WPART	Weighted PART (a type of rule-based ML algorithm)
WT	Wavelet Transform
XGBoost	Extreme Gradient Boosting

Economic Importance of *Brassica* Crops

Robin Rijal, Aashish Kumar Sharm, and Adesh Kumar

Abstract

Brassicaceae crops hold significant economic importance in the global food supply as oilseeds, vegetables, and condiments, and are also used as fodder, organic fertilisers, and value-added products such as fermented foods. Popular *Brassica* species, including cabbage, broccoli, kale, and mustard, are widely consumed for their nutritional benefits, which include antioxidants, vitamins, and other bioactive compounds. These vegetables are consumed in numerous forms, including heads, flower buds, stems, roots, and curds. *Brassica*-based fermented foods such as kimchi and sauerkraut, made from Chinese cabbage, are commercialised globally and are recognised for their considerable health benefits, including antioxidant, anti-inflammatory, and anticancer properties. They also serve as functional foods due to their prebiotic, probiotic, and postbiotic content. Moreover, Brassicaceae condiments, such as mustard, contribute to the global spice market and provide health benefits as antioxidants and anti-inflammatory agents. In addition, oil cakes from mustard and rapeseed provide cost-effective, protein-rich livestock feed and serve as sustainable natural fertilisers that improve soil productivity and reduce farming costs. As a result, the Brassicaceae family has attracted growing interest for its diverse applications across the food, pharmaceutical, and medicinal industries. Its health-promoting properties support well-

R. Rijal
Natural Resources Institute of University of Greenwich, Kent, UK

Department of Plant Pathology, School of Agriculture, Lovely Professional University, Phagwara, Punjab, India

A. K. Sharm · A. Kumar (✉)
Department of Plant Pathology, School of Agriculture, Lovely Professional University, Phagwara, Punjab, India
e-mail: adesh.19078@lpu.co.in

A. Kumar, B. Koul (eds.), *Crop Improvement Strategies in Brassica species: Applied Science*, https://doi.org/10.1007/978-981-95-3861-4_1

being and disease prevention, while its economic value underpins global trade, agriculture, and industry.

Keywords

Brassicaceae · Condiments · Functional foods · Nutritional benefits · Oilseeds

1 Introduction

The Brassicaceae family, also known as Cruciferae, is one of the largest angiospermic families in the order Brassicales, comprising more than 360 genera and 4000 species (Jat et al. 2019; Raza et al. 2020), and is cultivated and diversified almost all over the world (Subramanian et al. 2023). *Brassica* species play an important role in both the Indian and global economies due to their wide range of uses, which include oilseeds, vegetables, condiments, fermented foods, animal feed, and organic fertilisers (Fig. 1). They are a major source of essential vitamins A, B6, C, E, K, dietary fibre, and minerals like iron, magnesium, calcium, potassium, phosphorus, and folates (Jabeen 2020; Raza et al. 2020). *Brassica* crops are the most extensively grown crop worldwide, cultivated in more than 124 countries. They are highly polymorphic, encompassing oilseed crops, root crops, and vegetables such as Chinese cabbage, Brussels sprouts, and broccoli, while some are also cultivated for use as salads and condiments (Shree et al. 2022).

Mustard oil has been a major source of edible vegetable oil in the Indian subcontinent since ancient times. The primary *Brassica* species cultivated for oilseeds are *Brassica rapa* (syn. *B. campestris*), *Brassica napus*, and *Brassica juncea*. Among these, *B. juncea* is predominantly cultivated across Asia, particularly in China, India, Nepal, Pakistan, and Bangladesh, whereas *B. rapa* and *B. napus* are largely grown in countries such as Germany, Sweden, France, Australia, and Canada (Jat et al. 2019; Rai et al. 2022). In India, nine oilseed crops are cultivated under diverse agro-ecosystems, among which oilseed *Brassica* rank second after soybean. The major *Brassica* species cultivated in India include *Brassica juncea, B. oleracea, B. rapa (*syn. *campestris), B. carinata*, and *B. napus* for edible oil, *and B. nigra* and *B. oleracea* for seed condiments (Jat et al. 2019). Among these, *B. juncea* (Indian mustard) is the predominant species cultivated for oilseeds, accounting for nearly 90% of the total area under rapeseed-mustard cultivation in India (Daun 2011).

Brassica crops, particularly oilseed types, are economically significant due to their high oil and protein content. For instance, the average oil content in *B. napus*, *B. rapa*, and *B. juncea* ranges between 38 and 50%. The seed meal of *Brassica* spp. is widely used as animal feed, serving as a primary protein source for livestock in many parts of the world (Adegbeye et al. 2020). It is also applied in sustainable and organic farming to improve soil health due to its richness in minerals and nutrients. In addition to oil, its seed meal is widely used in sustainable and organic farming to enhance soil health, owing to its richness in minerals and nutrients. For example, it can supply up to 5% organic nitrogen, 3% phosphorus, and 1% potassium. Thus, the

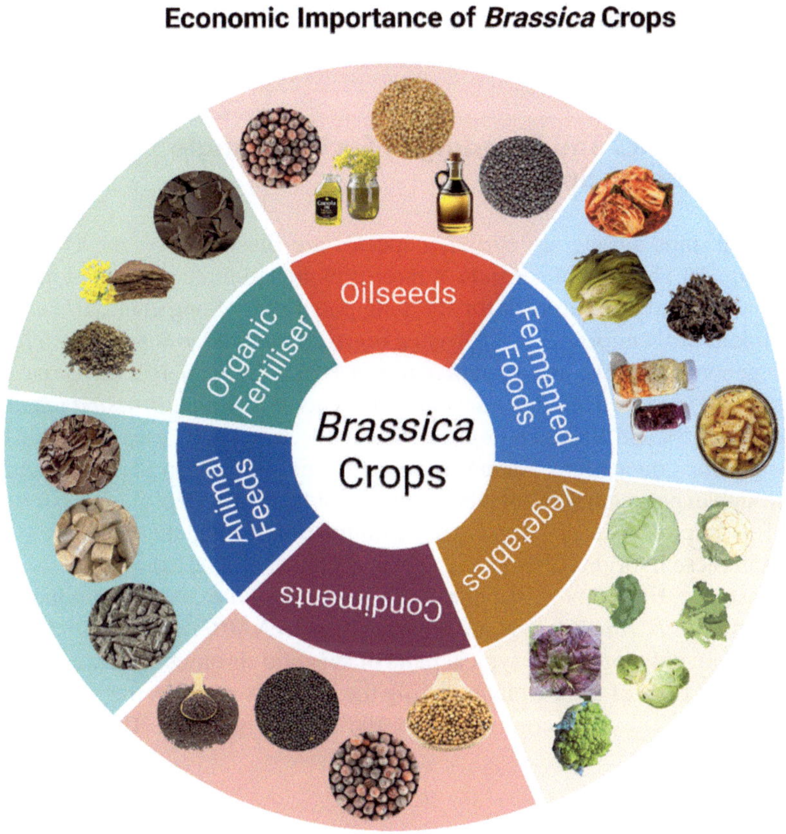

Fig. 1 Economic importance of *Brassica* crops across multiple sectors (created in BioRender; Rijal (2024) BioRender.com/m26r800)

cultivation and utilisation of *Brassica* species provide multiple economic benefits (Temmen et al. 2022).

The most diverse group of *B. rapa* belongs to the pekinensis type, also known as Napa cabbage or Chinese cabbage. It originated near the Beijing region of China and is widely used in East Asian and European cuisines (Cartea et al. 2021; Kim et al. 2022). For instance, Napa cabbage is the main ingredient in kimchi, a popular fermented dish in Korea and Japan, while sauerkraut, a famous fermented dish made from white cabbage (*Brassica oleracea* var. *capitata*), and Napa cabbage are predominantly consumed in European countries like the United Kingdom, Germany, Austria, Switzerland, and Belgium (Ashaolu and Reale 2020; Surya and Lee 2022). Additionally, Brassicaceae seeds, from species such as *Brassica juncea* (brown mustard) and *Brassica nigra* (black mustard), are widely used in Chinese and South Asian cuisines to prepare hot mustard (Karmakar et al. 2022; Singh and Singh 2025). Other notable *Brassica* crops include the *perviridis* group, Pak choi or bok choy (*chinensis* group), and the *npposinica* group, which are valued for their leaves and can be consumed either cooked or raw (Cartea et al. 2021).

2 Genomics and Breeding of Six U's Triangle of *Brassica* Species

The relationships among six *Brassica* species are illustrated by the U's Triangle model (Fig. 2). This model consists of three diploid and three amphidiploid species and is a widely used framework for studying polyploidy (Song et al. 2021a, b). The three diploid progenitor species, *Brassica nigra* (B genome, $n = 8$), *Brassica oleracea* (*C* genome, $n = 9$), and *Brassica rapa* (*A* genome, $n = 10$), gave rise to the amphidiploid species: *Brassica carinata* (*BC* genome, $n = 17$), *Brassica juncea* (*AB* genome, $n = 18$), and *Brassica napus* (*AC* genome, $n = 19$) (Cartea et al. 2021).

The genetic connections among *Brassica* species have been identified and verified through experimental crosses between diploid and/or tetraploid *Brassica* crops, as well as microscopic or karyotyping examination during the synapsis stage of meiosis in these hybrids (Cheng et al. 2014; Zhang et al. 2021a).

For instance, F1 plants with a chromosome count of $n = 29$ were produced from a cross between *B. rapa* (AA, n = 10) and *B. napus* (AACC, $n = 19$). Under microscopic examination, only nine typical pairs of synaptic chromosomes were observed in the meiotic cells of the F1 plants during the synapsis stage, corresponding to the two sets of chromosomes contributed by the *B. rapa* genome. The remaining 10

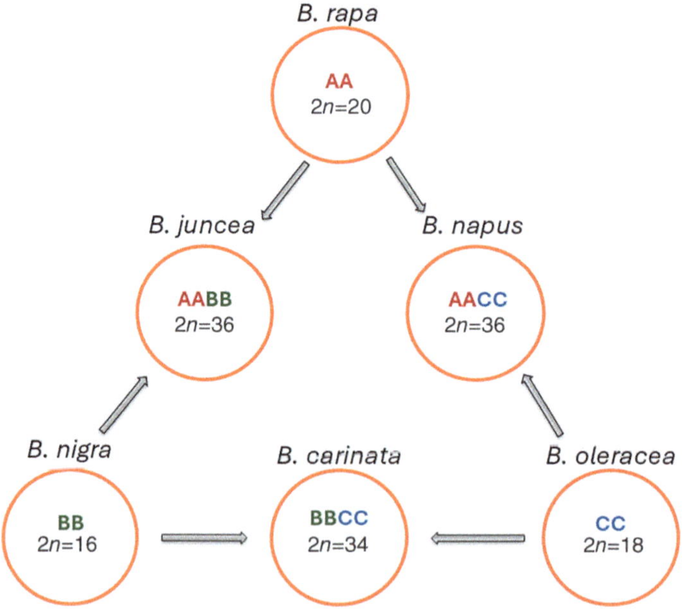

Fig. 2 Genetic relationships among six *Brassica* species (U's Triangle)

chromosomes of *B. napus*, which were unable to form normal pairs of synaptic chromosomes in the F1 generation, were influenced by the chromosomes from the A genome of *B. rapa*. This provides credence to the hypothesis that these 10 chromosomes in *B. napus* are homologous to the corresponding 10 chromosomes in *B. rapa* (Morinaga 1929; Cheng et al. 2014). Furthermore, similar experiments involving crosses between *B. oleracea* and *B. napus* demonstrated that the other nine chromosomes of *B. napus* are homologous to the nine chromosomes of *B. oleracea*. Collectively, these findings support the conclusion that B. napus is a tetraploid resulting from interspecific hybridisation between *B. oleracea* and *B. rapa* (Cheng et al. 2014; Rong et al. 2019).

Based on experimental findings, the relationships among six *Brassica* species are succinctly described by the U's Triangle model. In this model, the three diploid species, *B. rapa*, *B. nigra* and *B. oleracea*, represent the fundamental A, B, and C genomes, respectively, that are positioned at the three vertices (Fig. 2). The three allotetraploid species, *B. juncea* (AB), *B. napus* (AC), and *B. carinata* (BC), which resulted from hybridisation between these diploids, occupy the midpoints of the triangle's edges. Thus, the U's Triangle model has been successfully applied to elucidate phylogenetic relationships within the *Brassica* species and has significantly contributed to their genetic studies (Cheng et al. 2014; Koh et al. 2017).

3 Oilseeds

Globally, rapeseed-mustard is the third most important oilseed crop after soybean and palm oil, with over 90% of the total production concentrated in Asia, with China alone contributing approximately 62.4% of Asia's total production. In North America, Canada accounts for about 93.8% of rapeseed-mustard production, while in Europe, France produced approximately 21.0% of the crop in 2019. Globally, Canada remained the leading producer of rapeseed in 2024/2025, with 19.24 million metric tons, accounting for 22% of global output. The European Union followed with 16.86 million metric tons (20%), led by France and Germany. China ranked third with 15.8 million metric tons (18%), and India ranked fourth with 11.5 million metric tons (13%) (USDA 2025; Shahbandeh 2025).

In India, breeding programmes are primarily focused on *Brassica juncea*, which occupies nearly 75% of the total area under rapeseed-mustard cultivation. In contrast, global programs predominantly focused on *B. rapa* and *B. napus* (Kapusta-Duch et al., 2012; Yadava et al., 2022). Oilseeds constitute the second most important sector of India's agricultural economy after cereals, with an average annual growth rate of 4.1% over the past three decades. Within this sector, *Brassica* oilseeds account for about 23.5% of the cultivated area and contribute approximately 24.2% of the total oilseed production in the country (Jat et al. 2019; Chand et al. 2021). India is one of the largest agrarian subcontinents and represents the fifth-largest vegetable oil economy in the world, accounting for 7.4% of global oilseed production. Among domestic oilseeds, *Brassica* spp. alone contribute 31% of edible oil production, followed by soybean (26%) and groundnut (25%). During 2024–2025, the global production of rapeseed-mustard was 85.73 million metric tons (USDA 2025), cultivated on 36.6 million hectares with an average productivity of 19.8 q/ha (Chand et al. 2021).

According to the USDA reports on oilseeds and products update, the total area under rapeseed-mustard cultivation in India is 9.30 million hectares. Based on 2022/2023 data, Rajasthan is the leading state for rapeseed-mustard cultivation, producing 44.9% of national output from 40.5% of the cropped area, followed by Madhya Pradesh (11.3% from 14.29% of the area) and Uttar Pradesh (10.6% from 13.57% of the area) (USDA 2023a). According to the report published by USDA on production-rapeseed (2025), global rapeseed and mustard production declined by 5% in 2024/2025 compared to 2023/2024. In India, production dropped by 0.69%, primarily due to early-season frost damage, elevated late-season temperatures, soil erosion, and increased disease severity, which adversely affected crop quality (Chand et al. 2021; USDA 2025).

According to Das and Brown (2025), the MY import and export quantity of 2024/2025 in India was about 2000 MT and 975 MT, respectively, compared to 2000 MT (remained unchanged) and 1000 MT of import and export, respectively, in 2023/2024. In addition, according to the report of World Integrated Trade Solution (2019), the top exporters of vegetable oils, including rape, colza, or mustard oil and their fractions in 2019, were Canada (approximately $1.48 billion, 1,651,200 metric tons), Germany ($0.39 billion, 436,484 metric tons), Belgium ($0.20 billion, 211,206 metric tons), European Union ($0.173 billion, 167,357 metric tons), and Netherlands ($123 billion, 130,386 metric tons). Similarly, according to the Observatory of Economic Complexity (OEC) (2021a), report on the trade data of canola, rape, and colza or mustard oil revealed that the combined trade for rape, colza, or mustard, and canola oil ranked them as the world's 632nd most traded product, with a total value of $5.75 billion. For instance, the export of these commodities grew by 54.3%, from $3.73 billion to $5.75 billion between 2020 and 2021, which represent about 0.027% of the total world trade. In addition, the top exporters of rape, colza, or mustard oil, and canola were Canada ($2600 million), Germany ($977 million), Belgium ($443 million), Netherlands ($186 million), and Czechia ($180 million) in 2021. On the other hand, the major importers of these commodities in 2021 were USA ($2320 million), Netherlands ($636 million), Mexico ($295 million), China ($235 million), and France ($224 million).

The recent report of USDA revealed that the estimated production of rapeseed and mustard (rapeseed-mustard) in India is forecasted to be about 11.4 million metric tons, and 3.9 million MT of rapeseed oil is forecasted in MY 2022/2023. Due to larger stocks of rapeseed-mustards, it will lead to increased rapeseed crush in the current MY, resulting in advantageous pricing of domestic rapeseed oil meals compared to other origins, which can assist to sustain export growth to regional markets (USDA 2023a). Hence, *Brassica* crops have become economically significant in the global food supply chain due to their significant contribution to the oilseed industry.

4 Edible Vegetables

Apart from oilseeds, Brassicaceae vegetables are one of the largest, and most widely consumed plants around the world today due to its several health benefits. Some of the examples of Brassicaceae vegetables are cabbage, cauliflower, broccoli, Brussels Sprouts, collards, kale, kohlrabi, turnips, rutabagas, and radish (Kalia and Singh 2023). Brassicaceae vegetables, such as *Brassica rapa* crops, including *rapifera* and leafy types, are widely consumed in Europe, Asia, and all over the world as raw or steamed vegetables, as they are a significant source of nutrients and bioactive compounds in the daily diet (Kapusta-Duch et al. 2012; Yadava et al. 2022). The most common and widely consumed Brassicaceae vegetables subspecies/varieties are *botrytis* (cauliflower and Romanesco broccoli), *alboglabra* (Chinese kale and broccoli), *acephela* (kale and collard greens), capitata (cabbage), *gongylodes* (kohlrabi), *gemmifera* (Brussels sprouts), and *italica* (broccoli) (FAO 2023; Greer et al. 2023). These vegetables are rich in phytochemicals that can prevent oxidative stress, stimulate the immune system, induce detoxification enzymes, decrease the risk of cancers, and inhibit malignant transformation, as well as fight against free radicals. Furthermore, the vegetables from this species are a significant source of vitamin C, vitamin E, carotenoids, and antioxidants enzymes such as catalase, superoxide dismutase, and peroxidase. As a result, the demand for *Brassica* vegetables has been experiencing significant growth in the market (Kapusta-Duch et al. 2012; Ramirez et al. 2020; Zhang and Jing 2022).

In 2020, the cabbage and other Brassicaceae vegetables were produced around 96.4 million tonnes globally with a 10.6% increase over the past decades (Greer et al. 2023; FAO 2023). According to FAO (2022), the global production of vegetables increased by up to 68% from 2000 to 2021, reaching 1.15 billion tonnes in 2021, with cabbage ranking as the fourth most produced vegetable at 71 million tonnes. According to the report of Mordor Intelligence (2023), the Brassicaceae vegetables market size is estimated at approximately $40 billion in 2023 and is expected to reach up to $45.70 billion by 2028, growing at a CAGR of about 3.0% from 2023 to 2028. In terms of production, Asia-Pacific produces the highest amount of cabbage and other Brassicaceae vegetables, among which China being the leading country in both production and export. China is the leading producer of cabbage in the world, followed by India and Russia (Cao et al. 2023; FAO 2023). In 2021, China exported around 34% of cabbage and other Brassicaceae vegetables of global exports, in which about 97% were exported to the European Union countries, such as Germany (famous for Sauerkraut), Asia, and America (Mordor Intelligence 2023; Ren et al. 2023). The combined gross production value of cabbage, cauliflower, broccoli, mustard, and rapeseeds was tipping point to around $67.5 billion during the year 2020 (Subramanian et al. 2023).

On the other hand, cauliflower and broccoli are one of the most popular vegetables that comes under the family Brassicaceae that have a global production of approximately 26 million tons annually in which China is the leading producer of accounting for about 40% of the global production, and India is in the second-largest producer of accounting about nearly 35% of the global production (FAO

2022; Majumder and Annegowda 2021). According to OEC (2021b), the cauliflowers and headed broccoli including both fresh and chilled were the world's 1700th most traded commodity in the world in 2021, out of 4641. The top importers of cauliflower and broccoli were Spain ($582 million), the USA ($159 million), Italy ($129 million), Mexico ($109 million), and France ($107 million). On the other hand, its top importers were United Kingdom ($218 million), Germany ($198 million), the USA ($158 million), Canada ($145 million), and Netherlands ($90.8 million). Furthermore, Brussels sprouts is one of the popular vegetables consumed in the world, with a total trade of $264 million in 2021, in which exports grew by 12.8% between 2020 and 2021, from $234 M to $264 million, and represents about 0.0013% of the total world trade. The major exporters of both fresh and chilled Brussels sprouts were Mexico ($99.3 million), Netherlands ($73.8 million), and the USA ($29.3 million). On the other hand, the major importers were the USA ($102 million), Germany ($33.9 million), Canada ($28 million), and UK ($14.1 million) (OEC 2021c). A non-exhaustive list of some important vegetable crops under the *Brassica* crops is presented in Table 1.

5 Kimchi

Kimchi is one of the most widely consumed traditional Korean fermented side dishes, and mostly consumed in East Asia, such as China, Korea, Japan, and Taiwan (Yun et al. 2021a). It is a symbol of South Korea's identity, cross-cultural exchange, economic growth, and health promotion (Jung et al. 2022). Kimchi is prepared via spontaneous fermentation, which is primarily made from Chinese cabbage or Korean cabbage, radish, and garlic. Its taste and flavour mainly depends on the raw materials used, such as the freshness of cabbage, quantity of salt used, and the use of red peppers (Lee et al. 2020). Kimchi is well recognised as a healthy food due to its essential physiological nutrients, phytochemicals, as well as rich in probiotics such as *Leuconostoc, Lactobacillus*, and *Weissella*, (Lee et al. 2021). Furthermore, Kimchi cabbage, a synonym for Chinese cabbage and napa cabbage, was added to the CODEX standard in 2012 (MOFA 2013); the CODEX standard for kimchi was officially recognised in 2001 (FAO 2001). In addition, UNESCO proudly included Kimjang, a traditional method for preparing and preserving kimchi, on its list of the intangible cultural heritage of humankind in 2013 (MOFA 2013).

The preparation of kimchi is simple (Fig. 3); however, good-quality raw materials are crucial for producing high-quality kimchi. For instance, the cabbage and radish are cut into appropriate sizes, and then about 10–15% of salt is added, either directly or indirectly (Jung et al. 2022). The direct method involves adding salt directly to the raw material, a process also known as salting. On the other hand, the indirect method includes dipping raw materials in saltwater. Subsequently, excess water should be drained, and the ingredients should be well mixed. At this stage, seasoning for kimchi can be done using mixed spices, salted fish, red pepper powder, garlic, and ginger (Kim 2021; Kwon et al. 2021). Generally, a salt concentration of 2–3% is considered adequate at this stage, allowing for adjustments after

Table 1 A non-exhaustive list of economically significant *Brassica* vegetables, including their edible parts

Vegetable name	Botanical name	Edible parts
Broccoli	*Brassica oleracea* var. *italica*	Curds, stems, outer leaves, seeds, and florets
Cauliflower	*Brassica oleracea* var. *botrytis*	Curds, leaves, stems, florets, and seeds
Brussels sprouts	*Brassica oleracea* var. *gemmifera*	Its sprouts, top leaves, and stems
Cabbage	*Brassica oleracea* var. *capitata*	Heads, stems, seeds, and leaves
Red cabbage	*Brassica oleracea* var. *capitata f. rubra*	Heads, stems, and leaves
Kale	*Brassica oleracea* var. *sabellica*	Leaves and stems
Siberian kale	*Brassica napus* var. *pabularia*	Leaves and stems
Turnip	*Brassica rapa* subsp. *rapa*	Bulbous roots and leaves
Turnip rape	*Brassica rapa* subsp. *rapifera*	Leaves
Radish	*Raphanus sativus*	Roots, leaves, and seeds
Red-tailed radish	*Raphanus sativus* var. *caudatus*	Roots, leaves, and seeds
Bok choy	*Brassica rapa* subsp. *chinensis*	White stalks and green leaves
Collard greens	*Brassica oleracea* var. *viridis*	Leaves
Mustard greens	*Brassica juncea*	Leaves, seeds, flowers, and stems
Rutabaga	*Brassica napus* var. *napobrassica*	Roots and leaves
Kohlrabi	*Brassica olearcea* var. *gongylodes*	Enlarged stem (bulb), and leaves
Chinese broccoli	*Brassica oleracea* var. *alboglabra*	Leaves, stems, and flower buds
Napa cabbage	*Brassica rapa* subsp. *pekinensis*	Head and stem
Savoy cabbage	*Brassica oleracea* var. *sabauda*	Head of crinkled leaves and stem
Watercress	*Nasturtium officinale*	Leaves, flowers, and stalks
Rapini	*Brassica rapa* var. *ruvo* / *Brassica rapa* subsp. *sylvestris* var. *esculenta.*	Stem, leaves, buds, and flowers
Mizuna	*Brassica rapa* var. *nipposinica*	Leaves, stems, and flowers
Tatsoi	*Brassica rapa* subsp. *narinosa*	Leaves and stems
Arugula (some varieties)	*Eruca vesicaria* subsp. *sativa*	Leaves, stems, and flowers
Broccolini	*Brassica oleracea* italica × alboglabra	Leaves, tender stems, and florets
Daikon/ white radish	*Raphanus sativus* var. *longipinnatus*	Roots and leaves
Komatsuna Japanese mustard green	*Brassica rapa* var. *perviridis*	Leaves and stems
Romanesco broccoli	*Brassica oleracea* convar. *Botrytis* var. *botrytis*	Flowering head, stem, and stalks

(continued)

Table 1 (continued)

Vegetable name	Botanical name	Edible parts
Field mustard	*Brassica rapa* var. *rapa*	Leaves, flowers, pods, and roots
Mizuna mustard greens	*Brassica rapa* var. *nipposinica*	Leaves and tender stems
Snow mustard	*Brassica juncea* subsp. *integrifolia* var. *subintegrifolia*	Leaves, stem, and flowers
Mustard red Giant	*Brassica juncea*	Leaves and stem
Kalettes	*Brassica oleracea* var. *viridis* x *gemmifera*	Leaves, stem, and florets

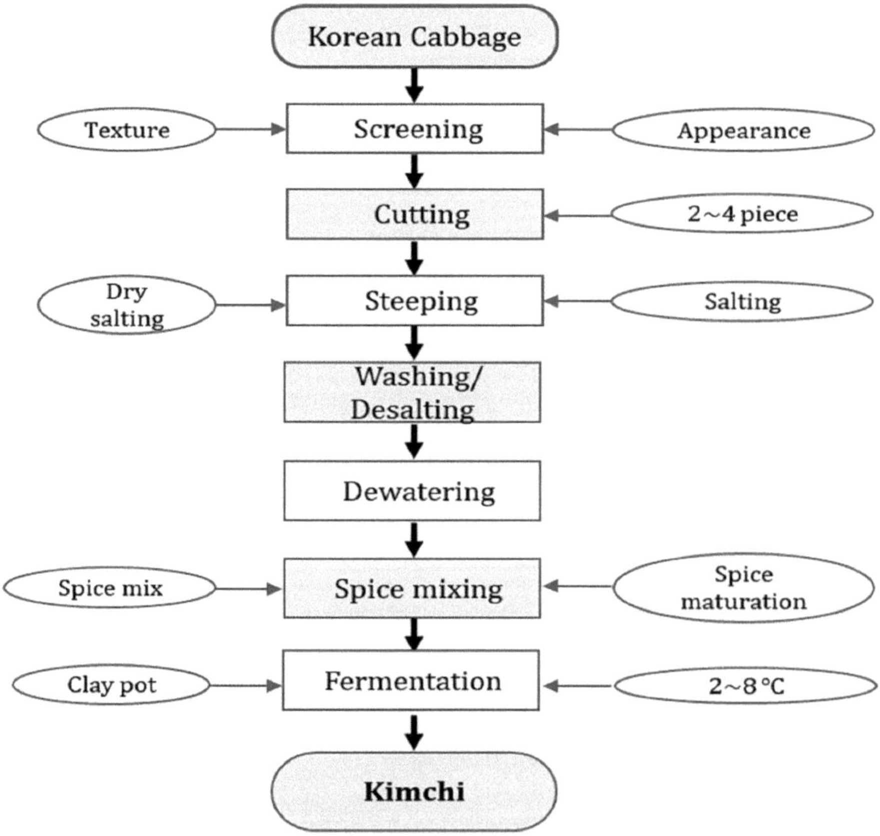

Fig. 3 A schematic diagram for the simple manufacturing process of kimchi

seasoning. In the subsequent stage, the raw materials are mixed and stored in containers at chilling temperatures for short-term (5–7 days) or long-term (3–4 weeks) fermentation. In contrast, sauerkraut and Chinese Pao Cai, also made from cabbage, do not require special seasoning (Jung et al. 2022).

Nowadays, Kimchi has been widely industrialised due to globalisation so that it is practically available at the market all year long. According to Surya and Nugroho (2023), the annual size of kimchi market in Korea was about \$1.2 million in 2017, whereas at the global scale the kimchi market is anticipated to attain \$2.39 billion between 2020 and 2024. According to the global kimchi market projection report published by Intellect Insights Journal (2023), the global kimchi market was valued at \$3.53 billion in 2022 and is projected to reach up to \$6.06 billion by 2030, growing at a compound annual growth rate (CAGR) of 5.32% from 2023 to 2030. South Korea is the largest producer and exporter of kimchi globally. The exported value reached a tipping point to over \$144.5 million in 2020, making about 37.6% increase from the previous year (Nihayati et al. 2022). In this context, Japan stands out as the main importer of kimchi, with a value of \$71.1 million, followed by the United States (\$23.06 million), Hongkong, Taiwan, and Australia, each worth under \$ ten million (Nihayati et al. 2022; Surya and Nugroho 2023). In fact, South Korea has exported kimchi to nearly 80 countries, and kimchi exports have been ongoing since 2015.

In 2021, the Korean Culture and Information Service published an article estimating that the average Korean consumes approximately 37 kg of kimchi annually. About 43.0% of Korean households still make kimchi at home, whereas around 52% Koreans purchase at stores or get it from their acquaintances. In 2008, Korean food scientists created a special kimchi named "Space Kimchi" for So-yeon Yi, the first Korean astronaut to take the shuttle to space to eat aboard the shuttle. "If a Korean goes to space, kimchi must go there, too," Kim Sung Soo, a Korea Food Research Institute scientist, stated to the New York Times (Sang-Hun 2008). Similarly, the country will seek to achieve approximately \$300 million in kimchi exports by 2027. This goal is inspired by a dedication to innovation-driven supremacy over other countries that produce kimchi, especially China and Japan (Lee 2023). Moreover, Nihayati et al. (2022) reported that there has been an escalation in kimchi export in the last 10 years and found that there was an increase in kimchi exports even during the Covid pandemic. Bousquet et al. (2021) revealed that fermented cabbage has the capacity to inhibit the production angiotensin-converting enzyme 2 (ACE 2), a key binding site for the coronavirus that assists in its crossing of the virus's cell membrane, as a result, preventing COVID-19. Moreover, kimchi enriched with Jeju citrus concentrates exhibits anti-obesity effects in both in vitro and in vivo by lowering triglyceride levels and suppressing the expression of obesity-related genes (Yun et al. 2019; Yun et al. 2021b).

6 Sauerkraut

Sauerkraut is finely cut cabbage that has been naturally fermented and consumed as a food for hundreds of years. It is produced through spontaneous fermentation in the presence of heterofermentative lactic acid bacteria (LAB), which are naturally found in cabbage (Siddeeg et al. 2022). The process flow diagram for making sauerkraut is illustrated in Fig. 4. The most commonly found natural LAB in cabbage

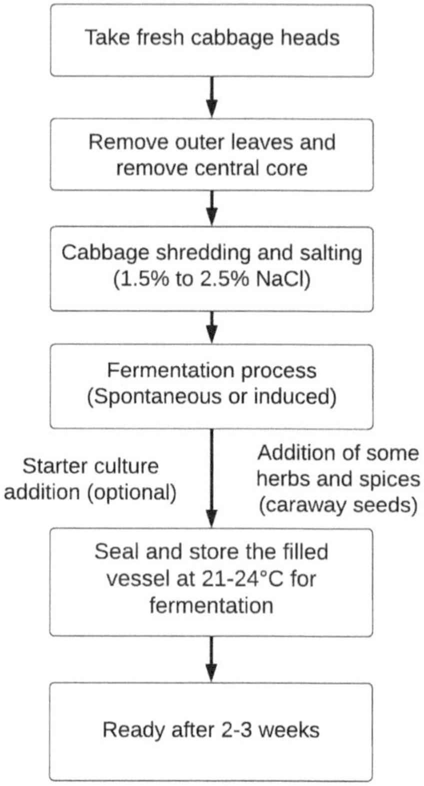

Fig. 4 Process flow chart for making sauerkraut

are *Leuconostoc mesenteroides*, *Pediococcus cerevisiae*, and *Lactobacillus plantarum*. These LAB break down fermentable sugars into byproducts such as lactic acid, acetic acids, alcohol, and carbon dioxide gas. Furthermore, the addition of a starter culture not only accelerates the fermentation of sauerkraut but also enhances its quality and flavour (Ashaolu and Reale 2020). It is a low-calorie food that has a high phytochemical and nutritional profile (Siddeeg et al. 2022). Traditionally, sauerkraut was used to treat scurvy and various other illnesses due to its high vitamin content. As a result, it was regarded as one of the most important foods for sailors in the past (Short 2020; Siddeeg et al. 2022).

Sauerkraut is one of the most important fermented foods, consumed not only in Europe but also in different parts of the world, such as in East Asia, Australia, and the USA (Tlais et al. 2022). The global market for Sauerkraut valued at $10.90 billion and $11.48 billion in 2022 and 2023, respectively, which is anticipated to reach about $16.3 billion by 2023, growing as a CAGR of 5.5% from 2023 to 2032 (Polaris Market Research 2023). The current average per capita consumption of sauerkraut is approximately 4 kg in Eastern Europe. In Germany even today around 200,000 tonnes of cabbages are processed into sauerkraut. Similarly, Serbia also used most of the fresh cabbage processed into sauerkraut at small family farms and households. About half of the sauerkraut is distributed through retail channels, while the remaining half being sold at green markets or directly at the farm gate (Raak et al. 2014). The biggest consumers of sauerkraut are Germany, France, and the USA. Besides its domestic use in foods, sauerkraut is also employed by small restaurants and multinational chains, such as KFC, McDonald's, and Subway, in burgers, sandwiches, and baguettes. Additionally, sauerkraut exhibits antiproliferative, antioxidant, anti-inflammatory, and anticancer activities. As a result, it can be included in our daily diets due to its potential as a functional food (Zubaidah et al. 2020; Amin and Borchgrevink 2022).

7 Condiments

Apart from spices, condiments are an important part of human history due to their economic, cultural, and nutritional value worldwide, owing to their potential to enhance taste and aroma in foods (García-Casal et al. 2016). Among various condiments, the genus *Brassica* is widely used as condiments contributing to the world's condiment's economic market, along with oilseeds, which were also mentioned as condiments in Sumerian and Sanskrit texts as early as 3000 BC (Kang et al. 2021). Generally, condiments do not provide high nutritional value for our body; however, they can certainly make foods more flavourful. In addition, Brassicaceae condiments offer several health benefits as they act as excellent antioxidants, antibiotics, anticoagulants, anticarcinogens, and anti-inflammatory agents, thereby preventing diseases and improving health (García-Casal et al. 2016).

Rapid economic growth and rising per capita income in the Asia-Pacific region have significantly influenced food production. The use and consumption of condiments are not limited only to the Asia-Pacific region due to cross-cultural interaction and increase in foreign travel owing to the food consumption pattern. As a result, the demand for condiments has significantly increased not only in the Asia-Pacific region but also in other parts of the world (Stankus 2014; García-Casal et al. 2016). The *Brassica* family dominates the condiment industry in the Asia-Pacific region, given the widespread consumption of meat in countries like Japan, China, Nepal, and India. The Asia-Pacific condiments market is anticipated to grow due to rising consumer demand for these products. The increasing demand for meat and seafood is contributing to the expansion of the condiments market in Asia-Pacific. Some of the common *Brassica* species seeds used for condiment purposes are brown mustard seeds, rapeseed seeds, cabbage seeds, Ethiopian rape seeds, field mustard seeds, yellow mustard seeds, black mustard seeds, and white mustard seeds (Agrawal et al. 2019; Šamec and Salopek-Sondi 2019; Kang et al. 2021; Salehi et al. 2021).

The worldwide condiment market has witnessed an increase in the demand for *Brassica* condiments due to their multiple uses in the food sector. For example, USDA report on US Condiments and Sauces exports in 2022 revealed that the compound average growth of condiments increased by 4.8% from 2013 to 2022. The three-year average export value is approximately $1.95 billion, with Canada being the top export market at $775.46 million, followed by Mexico ($390.64 million) and the European Union ($96.17 million) in 2022, where mustard is also one of the significant contributors. According to the 2018 report published by Allied Market Research on Asia-Pacific Condiments Market size, the size of condiments market in Asia-Pacific was $45,762.75 million in 2017; however, this figure is projected to reach up to $72,000 million by 2025, growing at a CAGR of approximately 6.0% from 2018 to 2025.

8 Animal Feed

Mustard cakes after extraction of oil are considered a potential source of plant-based protein, used as animal feed due to its rich nutrient content (Greer et al. 2023). The crude protein in mustard oil cake is between 31% and 38%, including both black and yellow mustards. Mustard oil cake offers a good balance of essential amino acids such as methionine, lysine, isoleucine, and threonine that are present in relatively high concentration in black mustard oil cake (Sarker et al. 2015). Similarly, the crude fibre, crude fat, and carbohydrate content in mustard seed cake are approximately 16.0%, 7.5%, and 40.0%, respectively (Rani and Badwaik 2021). Mustard cakes are cheaper than soybean meal and peanut cake, therefore, used more in the feeding of buffaloes, cattle, sheep, goats, fisheries, and pigs. For instance, a study conducted by Kumar et al. (2002) revealed that lambs fed at 50% and 100% levels with mustard cake on a protein basis for 120 days showed a significant increase in weight. The total weight gain was 14.50 kg in Treatment 1 (50% mustard oil cake and 50% peanut oil cake on a protein basis), followed by 11.43 kg in Treatment 2 (100% mustard oil cake on a protein basis), and 11.30 kg in Treatment 3 (100% peanut cake). The study found that the average daily body weight gain values were 120.83 g, 95.26 g, and 94.16 g per day in Treatment 1, Treatment 2, and Treatment 3, respectively. The results indicate no significant differences between Treatment 2 and Treatment 3; however, the higher gain was noted with Treatment 1, possibly due to the combined effect of two oil cakes. The study suggests that mustard cake may completely replace peanut cake without affecting feed intake, efficiency, mineral balance, nitrogen balance, and the growth performance of growing lambs. This is more economical as mustard oil cake is more affordable than peanut cake.

In addition, Itskovich et al. (2021) found that the average milk yield in Holstein breed cows increased by 1.39 kg (11.43%) compared to control animals. Additionally, the average fat and protein content in the milk of cows fed with mustard protein increased by 1.62 kg (11.98%) and 0.05%, respectively, compared to control animals that were fed a basic diet with insufficient digestible protein content. This result is also supported by Sirohi et al. (2013) in lactating Karan Fries crossbred cows and by Sohail et al. (2017) in Achai cattle. On the other hand, the application of mustard oil cake in aquaculture not only provides a natural source of feed to fish but also minimises the incidence of pathogenic microorganisms in fish ponds, unlike the use of cattle and pork manure. Mustard oil cake applied as an organic fertiliser exhibits minimal chances of food-borne serious pathogens and chemical contamination in the aquaculture system. For example, the total bacterial count, total coliform count, *Streptococcus* spp., and *Salmonella* spp. counts were significantly high in cattle manure than mustard oil cake treated pond water. Thus, it can be used in pond fertiliser over cattle manure to produce safe pond fishes. Hence, mustard cake serves as a natural protein supplement for domestic animals, enhancing milk production and body weight. Additionally, it can replace animal manures in fish pond systems, contributing to sustainable and safe fish production (Mredul et al. 2022).

9 Organic Fertiliser in Organic Agriculture

Mustard cake serves as a valuable resource for plant nutrition in natural farming, being abundant in essential macronutrients essential for the growth and development of plants, such as nitrogen, phosphorus, and potassium. Besides, it is a rich source of micronutrients, such as iron, copper, manganese, sulphur, and zinc (Islam et al. 2020). In the agricultural practices of South Asian countries such as India, Nepal, and Pakistan, Mustard seed oil cake holds a pivotal role, being an integral component of traditional farming (Punetha et al. 2018). Indeed, the slow-release nature of nutrients from mustard cakes into the soil contributes to the improvement of soil microbiomes and enhances overall soil fertility. This characteristic makes mustard cake a natural and beneficial plant food, particularly conducive to promoting vegetative growth and fostering the blooming of flowers in plants (Mredul et al. 2022; Rathor et al. 2023).

Furthermore, the oil cakes contain a high amount of organic material, ensuring both the long-term fertility and productivity of the soil. Additionally, they enhance the physico-chemical characteristics of the soil. Consequently, integrating oil cakes into agricultural soil not only boosts soil productivity but also lowers overall agricultural production costs. For instance, mustard oil cake contains 5.2% nitrogen, 1.8% potassium, and 1.2% potassium (Rathor et al. 2023). The combined application of mustard cake oil, either as solid or liquid formulation, along with other manures such as farmyard manure and vermicompost, led to improved growth, flower yield, and vase life in the commercial cultivation of French marigold. This approach also resulted in a net income of 249,248 INR/ha and a benefits-to-cost ratio (B:C) of 1:1.96 (Rathor et al. 2023). In addition, Chakraborty (2021) found that the three applications of mustard oil cake at 75 kg/ha (1 DAT, 20 DAT, and 35 DAT) as a basal dose up to 35 DAT showed significant results in the growth and development attributes of rice plants. For instance, the plant height, tiller height, panicle length, filled grains per panicle, 1000 grains weight, grain yield, and straw yield were 81.67 cm, 21.42 cm, 22.84 cm, 132.24, 29.35 g, 7.6 tonnes/ha, and 5.15 tonnes/ha, respectively.

Moreover, mustard cake oil is also popular for controlling soil-borne pathogens in plants, including soil-borne fungal pathogens and nematodes. For instance, Goswami et al. (2006) demonstrated that the application of mustard oil cake, incorporated with two fungal bioagents, *Paecilomyces lilacinus* and *Trichoderma viride*, in soil cultivated with tomatoes effectively controlled root-knot nematodes caused by *Meloidogyne incognita*. This combination not only promoted plant growth but also reduced the number of galls per plant. It demonstrates that mustard oil cakes can be incorporated with bioagents as nematicides to control nematodes in several plants. This is due to the presence of antimicrobial activity of glucosinolates and isothiocyanates in mustard cakes, which also contribute to the pungency in mustards. This result is also supported by Devi et al. (2019), who utilised mustard oil cakes for the management of *Meloidogyne graminicola* in rice nurseries. They reported that the use of mustard cakes (4.5 galls/seedlings) in rice beds resulted in a threefold reduction in the number of galls per seedling compared to control seedlings (15.2 galls/seedlings). From numerous studies, it has been proven that mustard oil cakes have greater potential for use in both natural farming and organic agriculture to preserve agro-diversity without adversely affecting soil health and fertility. As a result, there are innumerable economic importance of the family Brassicaceae, as well as various species of Brassicaceae explored for their versatile uses in different industries (Raza et al. 2020).

Although oilseed *Brassica* has grown significantly in India in the past, large-scale rain-fed production, biotic and abiotic stressors, and resource scarcity continue to result in poor productivity levels. In addition, its cultivation is also facing major challenges, such as climate change, low genotypic potential, and price fluctuation (Siomos et al. 2022; Duarte-Aké et al. 2023). Despite challenges, there is a significant opportunity to enhance the production both conventionally and unconventionally all over the world through advanced breeding techniques, technological advancements, and effective policy framework (Jat et al. 2019). Moreover, some of the value-added products made from *Brassica* crops that are being commercialised under various brands are listed in Table 2.

10 Conclusion

Brassica crops hold significant economic importance across various industries, including the food industry, pharmaceuticals, and the medicinal industry due to their richness in beneficial phytochemicals, vitamins, minerals, and nutrients. The *Brassica* genus includes many crop plants that provide edible roots, leaves, stems, buds, flowers, and seeds that possess economic importance before or after processing. The most common economically significant value-added products include vegetables, edible oils, fermented food products, and condiments, which not only possess economic significance but also improve health and well-being of the people. In addition, the diversification of *Brassica* spp. can offer a promising option for both small and marginal farmers, as well as small and big businesses due to high-value crops in this family, particularly during the current economic crisis. The primary benefit of this crop is its ability to grow in diverse climates, weather and soil conditions, thereby contributing to the development of rural areas. *Brassica* crop farming is a very promising activity and is gaining economic momentum globally due to its increasing market demand. However, the existing inconsistencies in production, as well as throughout the marketing chain, result in a failure to meet the demand for

Table 2 A non-exhaustive list of value-added products made from *Brassica* spp.

Products	Ingredients	Compounds	Health benefits	Make
Fermented products				
Pickled cabbage/ Pickled red cabbage (*Brassica oleracea* var. *capitata/ B. oleracea* var. *capitata* f. *rubra*)	(red) cabbage, thyme, barley malt salt, vinegar, water, firming agent (calcium acetate), flavouring, sweetener (sodium saccharin), preservative (sodium Metabisulphite)	FRAP (ferric reducing antioxidant power) antioxidant compound, phenolic compounds, flavonoids, DPPH (2,2-Diphenyl-1-picrylhydrazyl) radical-scavenging activity, ABTS (2,2′-azino-bis(3-ethylbenzothiazoline-6-sulfonic acid)) (Kaşıkçı and Bağdatlıoğlu 2022; Wieczorek and Drabińska 2022)	Beneficial for gastrointestinal microbiome, inflammatory illnesses, and good for gut bacteria Reduction of diseases and illness by releasing free radicles due to its radicle-scavenging activity *Lactobacillus plantarum, Leuconostoc mesenteroides, Lactobacillus paracasei, Weissella cibaria,* and *Pediococcus* spp. (Kaşıkçı and Bağdatlıoğlu 2022; Wieczorek and Drabińska 2022)	Shu Xin yuan, China; ASDA, UK; Haywards, UK; De Mână, Romania
Gundruk (a fermented food prepared from leaves of *brassica Juncea*, *Raphanus sativus*, and *Brassica oleracea* var. *botrytis*)	Green mustard leaves, radish leaves, cauliflower leaves, yeasts, and wheat flour	Rich in ascorbic acid, carotene, lactic acid, vitamin K, dietary fibre, and high levels of antioxidants	Improves gut health due to rich in lactic acid bacteria, such as *Lactobacilli*, *Leuconostoc*, and *Pediococcus*. It also acts as an anti-carcinogenic agent, and its soup serves as an appetiser (Jhamb and Swaminathan 2023)	NAU Aamako achar, Nepal; Gorkha food products, Nepal; Nepal foods, Nepal
Suancai (also known as Chinese sauerkraut, made from Napa cabbage)	Fresh fish, Chinese mustard greens, peppers, sea salt, Sichuan peppercorn seeds, dried pepper flakes, green onions, root ginger, sliced garlic cloves (sliced/ minced), and water	Rich in organic acids (acetate, citrate, lactate, malate, succinate, tartrate), sodium, potassium, and dietary fibre. Act as a good source of antioxidant properties, such as immunomodulatory properties, anti-inflammatory properties, and radical-scavenging activities (Song et al. 2021a, b; Xu et al. 2022)	Improves gut health as this food is rich in lactic acid bacteria, such as *Lactobacillus plantarum, L. brevis, L. fermentum, L. pentosus, L. acetotolerans, L. buchneri, Leuconostoc mesenteroides, Leuconostoc gelidum, Leuconostoc citreum, Pediococcus pentosaceus,* and *P. ethanolidurans* (Song et al. 2021a, b; Xu et al. 2022).	Sing-Kee, China; Caili send food (Shanghai) co., ltd., China

(continued)

Table 2 (continued)

Products	Ingredients	Compounds	Health benefits	Make
Pickled radish	Radish, sugar, salt, perilla spice, cumin seeds, vinegar, water, yeast extract, sweeteners (E950, E951, E954), acids (E260, E330), colour (E101), sulphites (E223)	It is a good source of antimicrobial phenolic compounds, such as 4-hydroxyphenethyl alcohol (4-HPEA), 2,6-dihydroxyacetophenone (DHAP), and 4-hydroxybenzaldehyde (HBA) (Li et al. 2020)	It has good free radical-scavenging capacity, antimicrobial activity, and high ferric reducing ability. For instance, DHAP has good antimicrobial activity against *Bacillus subtilis, Escherichia coli*, and *Candida albicans* (Li et al. 2020)	Yutaka, Japan; S&B, Japan; Blue Ville, South Korea; Sung, China
Mulako Achar (Nepali radish pickle, made from *Raphanus sativus* L.)	Fresh radish dices, mustard seeds, cumin seeds, coriander seeds, salt, fenugreek, fennel seed, mustard oil, turmeric, and green chilli, red chilli powder, ginger, garlic	Total volatiles (ascorbic acid, ethyl alcohol, methyl propanol, butyl alcohol, pentyl alcohol), ascorbic acid, flavonoids, terpenes and glucosinolates (Joshi and Sharma 2009; Gamba et al. 2021)	Effective appetiser, relief for diarrhoea, and stomach pain. Improve gut heath due to the presence of LAB, such as *Pediococcus pentosaceus, lactobacillus plantarum, lactobacillus brevis*, and *Leuconostoc mesenteroides* (Behera et al. 2020)	NAU Aamako achar, Nepal; Gharko Achar, Nepal; Farm star, India
Sichuan pao cai (Chinese kimchi) It is produced by pickling in brine or vinegar	Chinese cabbage, brine solution (5–8%), coarse salt, garlic, ginger, caster sugar, Sichuan pepper, chili, cilantro, green onions, bay leaves, and star anises	Rich in vitamin C, dietary fibre, amino acids, phenolic acids, probiotics, alkaloids, terpenoids, flavonoids, and antioxidant activity. The hepatoprotective properties of pao cai are associated with an increase in the expression and activity of major liver antioxidant enzymes, including glutathione reductase, glutathione S-transferase, glutathione peroxidase, catalase, and superoxide dismutase (Wang and Shao 2018; Nugroho et al. 2023)	An excellent source of LAB, including *Lentilactobacillus buchneri, lactobacillus acetotolerans, Lactiplantibacillus paraplantarum, Pediococcus metallidurans, Pediococcus damnosus, Secundilactobacillus aparacollinoides*, is good for our gut health (Jiang et al. 2021); as a result, improve the functionality of liver health. Regular consumption of pao cai can reduce cholesterol levels and prevent constipation issues. Therefore, it acts as an excellent functional food beneficial for human health (Nugroho et al. 2023)	Yingpeng, China

Zhacai (made from pickled mustard plant stem)	Mustard plant stem, sesame oil, soy sauce, salt, water, sugar, yeast extract, monosodium L-glutamate, glucono-delta-lactone, onion powder, citric acid, disodium 5-guanylate, and disodium 5-inosinate.	A good source of organic acids (acetic and lactic acid), free amino acids (alanine, glutamic, and lysine), and volatiles (alcohols, aldehydes, esters, and ketones) (Zhang et al. 2021b)	Contribute to improving immune function, digestive and heart health. In addition, abundance in Lactobacillus-related species, such as *Companilactobacillus alimentarius*, *Halomonas* spp., *Idiomarina loihiensis*, *Weissella* spp., and yeast *Debaryomyces hansenii* (Zhang et al. 2021b)	Momoya, Japan; Wujiang Fuling, China
Sunki (made from stems and leaves of turnip, *Brassica rapa* subsp. *rapa*)	Daikon mooli radish (88%), mixed vegetables (3%), cucumber, ginger, lotus root, purple perilla, salt, water, sugar, soy sauce, soybeans, vinegar, perilla spice, sesame seeds, yeast extract, and monascus red	Lactic acid, acetic acid, malic acid, amino acids, and volatile compounds (Endo et al. 2008)	A good source of gut bacteria, such as *Lactobacillus delbrueckii*, *L. fermentum*, *L. plantarum*, *L. brevis*, and *Bacillus coagulans*, is essential for supplying nutrients, as well as aiding in the digestion of cellulose, synthesising vitamin K, and promoting enteric nerve function (Endo et al., 2008; Zhang et al. 2015)	Yutaka, Japan
Yan-tsai-shin (made from fermented broccoli stems)	Contains crunchy florets of broccoli preserved in a light brine of vinegar, water, salt, sugar, and Italian herbs	Antimicrobial compounds, bioactive compounds (such as organic acids, enzymes, peptides, and polysaccharides), as well as bacteriocins produced by LAB, exhibit inhibitory activities against food-borne pathogens. Additionally, they produce antifungal compounds that can inhibit mycotoxins and bind to, thereby reducing the content of mycotoxins in the fermented foods (Abdul Hakim et al. 2023)	An excellent source of health beneficial heterofermentive LAB, including *Weissella paramesenteroides, Weissella cibaria, Weissella minor, Weissella Hellenica, Leuconostoc mesenteroides,* and *Lactobacillus plantarum* (Chen et al. 2013; Abdul Hakim et al. 2023)	Home-made

(continued)

Table 2 (continued)

Products	Ingredients	Compounds	Health benefits	Make
Suan-tsai (Chinese cabbage, cabbage, mustard leaves)	Cabbage (98%), green mustard, salt, water, acidity regulator (citric acid, acetic acid), preservative (sodium benzoate, and sodium Metabisulphite), flavour enhancer (monosodium glutamate), and colour (Tartrazine)	Rich in dietary fibre, vitamins, and probiotics. The bacteriocin Plantaricin JLA-9, produced by *L. plantarum*, is effective against *Bacillus* spp. in the food industry (Zhao et al. 2016)	*Lactobacillus futsaii* sp. nov., *Lactobacillus alimentarius, L. brevis, L. coryniformis, L. farciminis, L. plantarum, L. versmoldensis, L. crustorum, L. mindensis, L. mesenteroides, L. pseudomesenteroides, Weissella cibaria,* and *W. paramesenteroides* (Chao et al. 2009; Zhao et al. 2016)	Ma ling, China; Penta, Thailand
Kimchi	Salted Napa cabbage, salt, radish, red pepper, seasoned paste (water, radish, red pepper powder, garlic, onion, ginger), flavour enhancer (monosodium glutamate), thickener (xanthan gum), sweetener (sorbitol), flavour enhancer (monosodium glutamate), fructose, green onion, starter culture (*lactobacillus* fermented powder), and acidity regulator (calcium lactate, Glucono-delta-lactone)	Kimchi has revealed several health benefits due to its high antioxidative potential, attributed to flavonoids, vitamin C, carotenoids, and LAB. In addition, bioactive compounds such as quercetin, capsaicin, ascorbic acid, quercetin, 3-(4-hydroxy-3-methoxyphenyl) propionic acid, and polyphenolic compounds present in kimchi inhibit oxidation by scavenging free radicals by suppressing reactive oxygen species production (Woo et al. 2018; Choi et al. 2024). It has prebiotic, probiotic, and postbiotic properties (Cha et al. 2023)	A good source of LAB including, but not limited to *Leuconostoc* spp., *Lactobacillus* spp., *Weissella* spp., *Pediococcus pentosaceus*, and *Levilactobacillus brevis* contribute to several health and nutritional benefits (Woo et al. 2018; Choi et al. 2024)	Bibigo, South Korea; Jongga, South Korea; Yutaka, Japan

Sauerkraut	White cabbage, sea salt, caraway seeds, peppercorns, juniper berries (1%), white wine, salt, and antioxidant (ascorbic acid)	Bioactive compounds, including indole-3-carbinol, ascorbigen, sulforaphane, and allyl isothiocyanate, offer several health benefits. Moreover, rich in the antioxidant, such as vitamin C, vitamin E, beta-carotene, folic acid, polyphenols, flavonoids, and glucosinolates (Zabat et al. 2018; Siddeeg et al. 2022)	A good source of LAB including, but not limited to *Leuconostoc mesenteroides, L. plantarum, L. brevis, P. pentosaceus, P. cerevisiae*, and *Enterococcus faecalis* that not only enhances innate and adaptive immunity but also attenuates inflammation by modifying the gut microbiota (Zabat et al. 2018; Siddeeg et al. 2022)	Biona organic, the Netherlands; Hengstenberg, Germany; Raw, United Kingdom; Loving foods, UK
Pickled mustard greens (primarily made from *B. juncea*)	Mustard 83%, salt 5%, citric acid E330, water 9%, sugar, emulsifier E472a, colour E102, and preservative E211 and E223	Rich in LAB, dietary fibres, and essential micronutrients, including vitamins A, C, and K; therefore, it offers benefits for eye and heart health, as well as possessing anticancer and immune-boosting properties (Rodzi and Lee 2021)	*Lactobacillus fermentum* strain MBUL69, *Lactobacillus fermentum* KLB12, *Lactobacillus panis*, *Lactobacillus plantarum, Weisella cibaria*, *Staphylococcus carnosus, Enterococcus faecium, Lactobacillus brevis*, and *Leuconostoc mesenteroides* (Rodzi and Lee 2021)	Chang, China; RSSHOPS, Philippines; Kambal's, Philippines
Som Pak (made from pickled cabbage)	Cabbage, water, salt, green onions, condiments (garlic, ginger, chilli, turmeric), and preservatives (sodium benzoate)	A rich source of LAB and other antimicrobial substances, Nam pak, can inhibit undesirable pathogenic bacteria. Some Lactobacilli can also hinder microbial pathogens in the gastrointestinal tract. Therefore, Nam pak could be considered a potential food for promoting human health due to its well-known functional and therapeutic properties (Pakwan et al. 2020)	*Lactococcus* spp., *Pediococcus* spp., *Micrococcus* spp., *Leuconostoc mesenteroides, Lactobacillus brevis, Bacillus subtilis, Bacillus pumilus,* and *Bacillus megaterium* (Pakwan et al. 2020)	LinLin, Thailand; Murata, Romania

(continued)

Table 2 (continued)

Products	Ingredients	Compounds	Health benefits	Make
Dua cai (primarily made from fermented cabbage)	Cabbage, green mustard, salt (5%), water, citric acid (E330), sugar, preservative (E211), sodium Metabisulphite (E223), acidity regulator (citric acid, and acetic acid), flavour enhancers (monosodium glutamate), preservative (sodium benzoate, sodium Metabisulphite), and colour (E102)	An excellent source of essential nutrients, such as vitamins A, C, and K, folate, and minerals like calcium and potassium that help combat oxidative stress and inflammation in the body; as a result, they help the body defend against infections and illnesses (Phan et al. 2017)	*Lactobacillus plantarum, Lactobacillus fermentum, Lactobacillus delbrueckii, Lactobacillus pontis, Lactobacillus helveticus, Lactobacillus acidophilus,* and *Lactobacillus casei* (Phan et al. 2017)	NẤM LÝ TƯỞNG, Vietnam; LinLin, Thailand
Sayur asin (made from mustard cabbage leaf)	Mustard green 65%, soy sauce 20%, sugar 14%, iodised salt 1%, tamarind, pepper, turmeric, shrimp paste, acidity regulator (INS 330), and flavour enhancer (monosodium glutamate)	Its functional properties, attributed to probiotics, include hypocholesterolaemia activity by lowering plasma cholesterol, preventing and treating diarrhoea, and boosting the immune system. Other beneficial effects of these probiotics include competing with pathogens for adhesion sites, secreting antimicrobial substances, and inactivating toxins (Mangunwardoyo et al. 2016)	This food contributes to excellent acid and bile tolerance, as well as exhibits antimicrobial activity against pathogenic bacteria. Some of the beneficial LAB essential to improving gut health are *lactobacillus fermentum, L. plantarum, L. farciminis, L. namurensis, L. paralimentarius, L. parafarraginis, L. composti, L. confusus, L. curvatus, Leuconostoc mesenteroides,* and *Pediococcus pentosaceus* (Mangunwardoyo et al. 2016)	Lotus, Indonesia; Prima rasa, Indonesia; Pigeon, Thailand

Spicy pickled cauliflower	Fresh cauliflower, water, sugar, spirit vinegar, dried red pepper, cumin seeds, coriander, turmeric oleoresin, ginger oleoresin, fenugreek powder, garlic and tamarind, and preservative (sodium Metabisulphite)	High in vitamins, minerals, proteins, and fibre. Pickled cauliflower is also a source of glucosinolates, a cancer-fighting compound, and is rich in antioxidants that help reduce inflammation	Great for fighting cancer, detoxifying our digestive system, boosting heart and brain health. In addition, the LAB (*Lactobacillus plantarum, Leuconostoc mesenteroides, Enterococcus faecium,* and *Enterococcus faecalis)* present in pickled cauliflower help relieve stomach bloat, combat harmful bacteria, support the immune system, and aid in detoxifying the digestive system (Paramithiotis et al. 2010; Thompson et al. 2020)	Garner's, Scotland; Mrs. Klein's, USA; Amoga's, India
Oilseeds				
Rapeseed (*Brassica napus*) oil	Rapeseed oil, and anti-foaming agent: Dimethyl polysiloxane (E900)	The major functional components found in the oil include vitamin E, flavonoids, carotenoids, squalene, indole-3-carbinol, glucoraphanin, phospholipids, tocopherols, sterols, alpha-linolenic acid, and ferulic acid (Zhang et al. 2022; Shen et al. 2023)	Possess several health-benefiting properties, contributing to its potential for antimicrobial, anti-inflammatory, anticancer, anti-diabetic, neuroprotective, and cardioprotective effects. Moreover, rapeseed oil is popularly used in the dairy industry as a blend to make butter (Chew 2020; Shen et al. 2023)	KTC, United Kingdom; Tropical sun, United Kingdom; Mazola, USA; fortune, India; Kujawski, Poland

(continued)

Table 2 (continued)

Products	Ingredients	Compounds	Health benefits	Make
Black mustard (*Brassica nigra*) oil	100% cold-pressed mustard oil	Vitamin A and E, 3,4-di-hydroxybenzoic acid, omega-3 alpha-linolenic acid, omega-6 linoleic acid, ferulic acid, sinapic acid, rutin, saponins, alkaloids, flavonoids, glycosides, reducing sugar, phlobatannin (Rasera et al. 2019)	The oil can reduce low-density lipoprotein cholesterol (bad cholesterol), thereby lowering your cardiovascular risk. Additionally, it can be used for hair care, body massage, baby massage, and cooking. In addition, mustard oil controls microbial infection, reduces inflammation, and helps treat cold symptoms (Danlami et al. 2016; Rasera et al. 2019; Tian and Deng 2020)	KTC, United Kingdom; TRS, United Kingdom; Dhara, Nepal; fortune, India; Dabur, India; Crisco, USA; Niharti, India; Assav, India; Candidorga, India; Heera, UK
White mustard (*Sinapis alba*) oil	100% pressed cold mustard seed oil	3,4-di-hydroxybenzoic acid, omega-3 alpha-linolenic acid, omega-6 linoleic acid, ferulic acid, and sinapic acid (Rasera et al. 2019)		
Indian mustard/ brown mustard (*brassica junceae*) oil	100% pressed cold mustard seed oil	Vitamins (C and K), omega-3 alpha-linolenic acid, omega-6 linoleic acid trace minerals (Ca, Cu, Fe, Zn, Mn, and Mg), electrolytes (Na and K), and is rich in numerous phytochemicals, such as glucosinolates (sinigrin and sinalbin), isothiocyanates, polyphenols, phytosterols, and volatile components (allylisothiocyanate and 3-butyl isothiocyanate) (Rasera et al. 2019; Tian and Deng 2020)		

Radish (*Raphanus sativus*) oil	100% pressed cold mustard seed oil	This oil is renowned for having high concentrations of vitamins, minerals, and antioxidants. For example, the major components of the oil are pentyl hexyl, dimethyl disulfide, 1-methylthio-3-pentanone, 4-methylpentyl isothiocyanate, methyl methanethiol sulfinate, and alkenyl glucosinolates. Additionally, erucic, palmitic, oleic, linoleic, and linolenic acids are the main fatty acids found in the oil (Gutiérrez and Perez 2004)	The oil has been used to treat respiratory ailments, digestive issues, and skin conditions, in addition to promoting blood circulation and reducing inflammation. As a result, it is gaining popularity in the food, medicinal, and pharmaceutical industries for its benefits to the skin, hair, and overall health	Hemani, Pakistan; Home genie, India; Deve Herbes, India; Guzarish, India
Broccoli seed (*Brassica oleracea* var. *italica)* oil	100% pressed cold mustard seed oil	Vitamins A, C, and K, omega-6 fatty acid, omega-9 fatty acid, glucosinolates, linoleic acid, and high in erucic acid (Li et al. 2022)	A versatile oil used in hair care, skincare, and lip balms to protect, moisturise, and add shine. It guards against thermal damage and dryness in hair, smoothing the cuticle for radiant shine	Naturally balmy, UK; Soil association, UK; Wooden spoon, Bulgaria
Cauliflower oil	100% pressed cold mustard seed oil	It contains antioxidants like glucosinolates, carotenoids, and flavonoids, which protect cells from oxidative stress and may possess anti-inflammatory properties	Cauliflower seed oil may offer potential health benefits, including improved immune function and digestion, as well as effectiveness for hair health	RV essential, India; Cloud bud, India; DEVINEZ, India
Cabbage oil	100% pressed cold mustard seed oil	Omega-3 fatty acids, glucosinolates, minerals (calcium, magnesium, and potassium), vitamin C and K, and folate	It can help to improve digestion, keep the heart healthy, treat constipation, and prevent stomach ulcers. Moreover, the oil keeps the skin, eyes, and hair healthy	Biopurus, UK; Savage cabbage, UK; Hemani, UAE
Medicinal/pharmaceutical uses				

(continued)

Table 2 (continued)

Products	Ingredients	Compounds	Health benefits	Make
Wasabia japonica	Rhizomes, stems, and leaves	Anti-carcinogenic, anti-inflammatory, and free radical-scavenging activities due to the presence of bioactive compounds including, but not limited to 6-(methylsulfinyl)hexyl isothiocyanate, trans-feruric acid, vanillic acid, p-coumaric acid, 2″-O-trans sinapoyl isovitexin, astragalin, and isorhamnetin-3-glucoside (Depree et al. 1998; Uto et al. 2012; Kurata et al. 2019)	Extensively used in Japanese cuisine, which is used in the preparation of sauces and condiments. It acts as a powerful antibacterial and antifungal agent against yeast, bacteria, and moulds (Kurata et al. 2019)	Shizuoka, Japan; WASABI-O, Japan
Wasabi (*Eutrema japonica)* extract tincture	Essential oil from 100% wasabi dried root, vegetable glycerine, organic cane alcohol, and water	6-Methylsulfinylhexyl isothiocyanate (6-MSITC)	6-MSITC have anti-inflammatory and antioxidant effects, as well as improve neuroinflammation and memory. Moreover, it is beneficial for treating myalgic encephalomyelitis/chronic fatigue syndrome (Oka et al. 2022)	Montana Farmacy, USA; Garden organics, USA
Wasabi (*Eutrema japonicum*) powder	100% pure wasabi	–	It is beneficial for digestive health, heart health, reduces the risk of cancer, and boosts our immune system	Buy whole foods online, UK
Erysimum herb *(Erysimum kotschyanum, Erysimum cheiranthoides)*	100% crushed *Erysimum* herb essential oil, flower, steam	Vanilic acid, 3,4-hydroxybenzoic acid, gallic acid, chlorogenic acid, ferulic acid, and cinnamic acid (Kılınçarslan and Mammadov 2018)	Used against heart problems, constipation, liver disease, and gallbladder disease. In addition, it has long been used to treat dry bronchitis, heart diseases, fevers, eye injuries, and has shown potential as a cancer therapy (Kılınçarslan and Mammadov 2018)	Plantas, Spain

Red cabbage cream	Red cabbage, water, glycerin, cetearyl alcohol, shea butter, coconut oil, palm oil, glyceryl monostearate, sunflower seed oil, beeswax, phenoxyethanol, dimethicone, carbomer, vitamin E, and sodium citrate			Bianca Rosa, Canada
Broccoli detoxifier	Broccoli extract, 10% Sulforaphane (50 mg)	Sulforaphane	Natural inducer of phase 2 detox enzymes, improves estrogen metabolite ration, free of contaminants, and has no side effects. It can be used against cancer, inflammation, balance estrogen, and breakdown estrogen into friendly metabolites	Life tropics, UK
Broccoli sprouts dietary supplement	Organic fermented broccoli sprout, organic broccoli (whole plant), sulforaphane, organic rice hull concentrate, myrosinase, and glucoraphanin	Sulforaphane, glucoraphanin, and myrosinase	It offers potent support for the detoxification process in your body, promotes cellular health, protects cells, tissues, and DNA from free radical oxidative damage, maintains cholesterol levels, supports gut health, improves gut microbial balance, and supports prostate health	Dr. Mercola, USA

(continued)

Table 2 (continued)

Products	Ingredients	Compounds	Health benefits	Make
Sterigmostemum incanum	Flower, leaves and leaves.	Antioxidant compounds, and phenolic compounds	It possesses medicinal properties and is popularly known for its antibacterial, anti-inflammatory, and antifungal properties (Sheidai and Sonboli 2011; Duran et al. 2015).	–
Aethionema sancakense, Aethionema dumanii, Aethionema umbellatum, Aethionema alpinum, and *Aethionema zagricum*	Flower and essential oil	Glucosinolates phytochemicals, such as 3-methylsulfonylpropyl, (4-hydroxy-indolyl-3-methyl, 7-methylsulfinylheptyl glucosinolate, 8-Methylsulfinyl-n-octyl glucosinolate, methylthioalkyl, methylsulfinylalkyl, and methylsulfonylalkyl (Mohammadin et al. 2017; Moazzeni et al. 2023)	It can be used against typhoid, bacterial infections, and meningitis due to its richness in phenolics, flavonoids, radical-scavenging properties, and antioxidant activity. Consequently, it is suitable as a raw material in the food, pharmaceutical, and perfume industries, owing to its high antioxidant and antimicrobial capacity (Moazzeni et al. 2023)	–

oilseeds, vegetables, and animal supplements, not only at the domestic level but also on an industrial scale. This is due to low productivity of existing cultivars, insufficient pest and disease control, extreme weather conditions, and a lack of improved cultivars, all of which greatly affect their production. Hence, there is a significant opportunity to enhance their production through advance breeding techniques, technological advancements, and effective policy framework. In this respect, *Brassica* crops have the potential to boost the agricultural sector and create new economic opportunities for the growth of the rural economy as well as the gross domestic product of nations. As a result, there are encouraging signs in the market due to the increasing number of value-added products made from *Brassica* crops, as their global market is expected to witness substantial growth during the next decade. However, the overexploitation of plants in the name of breeding should be limited. It is imperative to conserve valuable species, including wild and indigenous varieties, and their cultivation should be promoted for future use.

Authorship Contribution Statement: Robin Rijal Writing—review and editing, writing—original draft, visualisation, conceptualisation, software, methodology, investigation, formal analysis, and data curation. **Aashish Kumar Sharma:** writing—original draft, editing, formal analysis. **Adesh Kumar:** review & editing, supervision, project administration, formal analysis, conceptualisation.

Declaration of Competing Interest The authors declare that they have no known competing financial interests or personal relationships that could have appeared to influence the work reported in this paper.

References

Abdul Hakim BN, Xuan NJ, Oslan SNH (2023) A comprehensive review of bioactive compounds from lactic acid bacteria: potential functions as functional food in dietetics and the food industry. Foods 12:2850. https://doi.org/10.3390/foods12152850

Adegbeye MJ, Oloruntola OD, Asaniyan EK et al (2020) Pawpaw, black cumin, and mustard seed meals dietary supplementation in broiler chickens: effect on performance, gut microflora, and gut morphology. J Agric Sci 22:1235–1246

Agrawal S, Yallatikar T, Gurjar P (2019) *Brassica nigra*: Ethopharmacological review of a routinely used condiment. Curr Drug Discov Technol 16:40–47. https://doi.org/10.2174/1570163815666180308143400

Allied Market Research (2018) Asia-Pacific Condiments Market Size, Share, Competitive Landscape and Trend Analysis Report by Type, Application, and Form: Opportunity Analysis and Industry Forecast, 2018–2025. Available via https://www.alliedmarketresearch.com/asia-pacific-condiments-market. Accessed 25 Dec 2023

Amin S, Borchgrevink CP (2022) A Culinology® perspective of dry beans and other pulses. In: Siddiq M, Uebersax MA (eds) Dry beans and pulses: production, processing, and nutrition, 2nd edn. Wiley, New York. https://doi.org/10.1002/9781119776802.ch18

Ashaolu TJ, Reale A (2020) A holistic review on euro-Asian lactic acid bacteria fermented cereals and vegetables. Microorganisms 8:1176

Behera SS, El Sheikha AF, Hammami R et al (2020) Traditionally fermented pickles: how the microbial diversity associated with their nutritional and health benefits? J Funct Foods 70:103971. https://doi.org/10.1016/j.jff.2020.103971

Bousquet J, Anto JM, Czarlewski W et al (2021) Cabbage and fermented vegetables: from death rate heterogeneity in countries to candidates for mitigation strategies of severe COVID-19. Allergy: Eur J Allergy Clin Immunol 76:735–750. https://doi.org/10.1111/all.14549

Cao H, Zhang F, Fu J et al (2023) Optimized nitrogen fertilizer rate can increase yield and nitrogen use efficiency for open-field Chinese cabbage in Southwest China. Agronomy 13:1578. https://doi.org/10.3390/agronomy13061578

Cartea ME, Cámara-Martos F, Obregón S et al (2021) Advances in breeding in vegetable *Brassica rapa* crops. In: Aminul Islam AKM, Hossain MA, Mominul Islam AKM (eds) Brassica breeding and biotechnology. IntechOpen, London. https://doi.org/10.5772/intechopen.95769

Cha J, Kim YB, Park SE et al (2023) Does kimchi deserve the status of a probiotic food? Crit Rev Food Sci Nutr 64:6512–6525. https://doi.org/10.1080/10408398.2023.2170319

Chakraborty S (2021) Effect of different doses and time of application of mustard oil cake on growth and yield of Rice (*Oryza sativa* L.) Thesis, Sher-e-Bangla Agricultural University. Available via http://archive.saulibrary.edu.bd:8080/xmlui/handle/123456789/4951

Chand S, Patidar OP, Chaudhary R et al (2021) Rapeseed-mustard breeding in India: scenario, achievements and research needs. In: Aminul Islam AKM, Hossain MA, Mominul Islam AKM (eds) *Brassica* breeding and biotechnology. IntechOpen, London. https://doi.org/10.5772/intechopen.96319

Chao SH, Wu RJ, Watanabe K et al (2009) Diversity of lactic acid bacteria in suan-tsai and fu-tsai, traditional fermented mustard products of Taiwan. Int J Food Microbiol 135:203–210. https://doi.org/10.1016/j.ijfoodmicro.2009.07.032

Chen YS, Liou MS, Ji SH et al (2013) Isolation and characterization of lactic acid bacteria from Yan-tsai-shin (fermented broccoli stems), a traditional fermented food in Taiwan. J Appl Microbiol 115:125–132. https://doi.org/10.1111/jam.12199

Cheng F, Wu J, Wang X (2014) Genome triplication drove the diversification of *brassica* plants. Hortic Res 1:14024. https://doi.org/10.1038/hortres.2014.24

Chew SC (2020) Cold-pressed rapeseed (*Brassica napus*) oil: chemistry and functionality. Food Res Int 131:108997. https://doi.org/10.1016/j.foodres.2020.108997

Choi HW, Park SE, Kim EJ et al (2024) Selective influence of garlic as a key ingredient in kimchi on lactic acid bacteria in a fermentation model system. Heliyon 10:e24503. https://doi.org/10.1016/j.heliyon.2024.e24503

Danlami U, Orishadipe AT, Lawal DR (2016) Phytochemical, nutritional and antimicrobial evaluations of the aqueous extract of *Brassica nigra* (Brassicaceae) seeds. Am J Appl Chem 4:161–163. https://doi.org/10.11648/j.ajac.20160404.17

Das S, Brown J (2025) Oilseeds and products annual. USDA report on oilseeds and products annual report number: IN2025–0021 9. Available via https://appsfasusdagov/newgainapi/api/Report/DownloadReportByFileName?fileName=Oilseeds%20and%20Products%20Annual_New%20Delhi_India_IN2025-0021pdf. Accessed 4 Oct 2025

Daun JK (2011) Origin, distribution, and production. In: Daun LK, Michael Eskin NA, Hickling D (eds) Canola chemistry, production, processing, and utilization, 1st edn. AOCS Press. https://doi.org/10.1016/B978-0-9818936-5-5.50005-X

Depree JA, Howard TM, Savage GP (1998) Flavour and pharmaceutical properties of the volatile Sulphur compounds of wasabi (*Wasabia japonica*). Food Res Int 31:329–337. https://doi.org/10.1016/S0963-9969(98)00105-7

Devi P, Kanwar RS, Patil JA (2019) Effect of oil cakes for the management of *Meloidogyne graminicola* in rice nursery. J Entomol Zool 7:180–182

Duarte-Aké F, Us-Camas R, De-la-Peña C (2023) Epigenetic regulation in Heterosis and environmental stress: the challenge of producing hybrid Epigenomes to face climate change. Epigenomes 7:14. https://doi.org/10.3390/epigenomes7030014

Duran A, Uslu N, Dogan B et al (2015) Antioxidant activity and phenolic contents of 30 selected medicinal plants. J Agroaliment Process Technol 21:136–141

Endo A, Mizuno H, Okada S (2008) Monitoring the bacterial community during fermentation of sunki, an unsalted, fermented vegetable traditional to the Kiso area of Japan. Lett Appl Microbiol 47:221–226. https://doi.org/10.1111/j.1472-765X.2008.02404.x

FAO (2001) Standard for kimchi. Available via http://wwwfaoorg/fao-who codexalimentarius/shproxy/en//?lnk=1&url=https%253A%252F%252Fworkspacefaoorg%252Fsites%252Fcodex%252FStandards%252FCXS%2B223-2001%252FCXS_223epdf. Accessed 1 Jan 2024

FAO (2022) Agricultural production statistics. 2000–2021. FAOSTAT analytical brief series No. 60. Rome. doi:https://doi.org/10.4060/cc3751en

FAO (2023) FAOSTAT—worldwide production and yield of cabbage, cauliflower, broccoli and rape or colza seed. https://www.fao.org/faostat/en/#data/QCL. Accessed 17 Nov 2023

Gamba M, Asllanaj E, Raguindin PF et al (2021) Nutritional and phytochemical characterization of radish (*Raphanus sativus*): a systematic review. Trends Food Sci Technol 113:205–218. https://doi.org/10.1016/j.tifs.2021.04.045

García-Casal MN, Peña-Rosas JP, Malavé HG (2016) Sauces, spices, and condiments: definitions, potential benefits, consumption patterns, and global markets. Ann N Y Acad Sci 1379:3–16. https://doi.org/10.1111/nyas.13045

Goswami BK, Pandey RK, Rathour KS et al (2006) Integrated application of some compatible biocontrol agents along with mustard oil seed cake and furadan on *Meloidogyne incognita* infecting tomato plants. J Zhejiang Univ Sci B 7:873–875. https://doi.org/10.1631/jzus.2006.B0873

Greer SF, Surendran A, Grant M et al (2023) The current status, challenges, and future perspectives for managing diseases of brassicas. Front Microbiol 14:1209258. https://doi.org/10.3389/fmicb.2023.1209258

Gutiérrez RMP, Perez RL (2004) *Raphanus sativus* (radish): their chemistry and biology. Sci World J 4:811. https://doi.org/10.1100/tsw.2004.131

Intellect Insights Journal (2023) Kimchi Market 2023 Size, Share Industry Analysis by Future Demand, Top Players, Opportunities, Revenue and Growth Rate Through 2030 Available via https://wwwlinkedincom/pulse/kimchi-market-2023-size-share-industry-analysis/. Accessed 17 Nov 2023

Islam S, Shelly NJ, Ahmed KU et al (2020) Study on the nutritional composition of oil cakes of different released and line cultivars of mustard and rapeseed (*Brassica* spp.). Int J Stress Manag 11:437–444. https://doi.org/10.23910/1.2020.2144

Itskovich AY, Varakin AT, Ryadnov AA (2021) Use of Sarepta-5 mustard protein for Milk production. Bull Samara State Agric Acad 6:76–81

Jabeen N (2020) Agricultural, economic and societal importance of Brassicaceae plants. In: Hasanuzzaman M (ed) The plant family Brassicaceae. Springer, Singapore. https://doi.org/10.1007/978-981-15-6345-4_2

Jat RS, Singh VV, Sharma P et al (2019) Oilseed brassica in India: demand, supply, policy perspective and future potential. OCL – Oilseeds Fats Crops Lipids 26:8. https://doi.org/10.1051/ocl/2019005

Jhamb V, Swaminathan P (2023) Role and importance of lactic acid bacteria in different Indian fermented foods. Biologia 78:3609–3623. https://doi.org/10.1007/s11756-023-01512-5

Jiang L, Xian S, Liu X et al (2021) Metagenomic study on Chinese homemade paocai: the effects of raw materials and fermentation periods on the microbial ecology and volatile components. Foods 11:62. https://doi.org/10.3390/foods11010062

Joshi VK, Sharma S (2009) Lactic acid fermentation of radish for shelf-stability and pickling. Nat Prod Radiance 8:19–24

Jung SJ, Chae SW, Shin DH (2022) Fermented foods of Korea and their functionalities. Fermentation 8:645. https://doi.org/10.3390/fermentation8110645

Kalia P, Singh S (2023) Nutritional enhancement of vegetable crops (with major emphasis on broccoli: a new Cole crop in India). In: Singh B, Kalia P (eds) Vegetables for nutrition and entrepreneurship. Springer, Singapore. https://doi.org/10.1007/978-981-19-9016-8_1

Kang L, Qian L, Zheng M et al (2021) Genomic insights into the origin, domestication and diversification of *Brassica juncea*. Nat Genet 53:1392–1402. https://doi.org/10.1038/s41588-021-00922-y

Kapusta-Duch J, Kopec A, Piatkowska E et al (2012) The beneficial effects of *brassica* vegetables on human health. Rocz Panstw Zakl Hig 63:389–395

Karmakar M, Jana D, Manna T et al (2022) Immunostimulant properties of some commonly used Indian spices and herbs with special reference to region-specific cuisines. In: Sangwan NS, Farag MA, Modolo LV (eds) Plants and Phytomolecules for immunomodulation. Springer, Singapore. https://doi.org/10.1007/978-981-16-8117-2_8

Kaşıkçı MB, Bağdatlıoğlu N (2022) Assessment of the bioaccessibility of phenolic compounds and antioxidant activity in raw and pickled white cabbage and gherkins. Ital J Food Sci 34:1–10. https://doi.org/10.15586/ijfs.v34i4.2267

Kılınçarslan O, Mammadov R (2018) HPLC analysis and antioxidant, antibacterial and cytotoxicity activities of various solvent extracts of *Erysimum kotschyanum* gay.(Brassicaceae). J Chem Soc Pak 40:707–714

Kim HR (2021) The background and truth of the controversy over kimchi suzerainty, and countermeasures-focusing on the birth, process, and uniqueness of kimchi. J Korean Soc Food Cult 36:274–283

Kim S, Rho HY, Kim S (2022) The effects of climate change on heading type Chinese cabbage (*Brassica rapa* L. Ssp Pekinensis) Economic Production in South Korea. Agronomy 12:3172. https://doi.org/10.3390/agronomy12123172

Koh JC, Barbulescu DM, Norton S et al (2017) A multiplex PCR for rapid identification of *Brassica* species in the triangle of U. Plant Methods 13:1–8. https://doi.org/10.1186/s13007-017-0200-8

Korean Culture and Information Service Korea (2021) Here and now, Ask Us about Korea: Kimchi Available via https://wwwkocisgokr/eng/fpcBoard/viewdo?seq=1039491. Accessed 23 Dec 2024

Kumar GA, Panwar VS, Yadav KR et al (2002) Mustard cake as a source of dietary protein for growing lambs. Small Rumin Res 44:47–51. https://doi.org/10.1016/S0921-4488(02)00034-2

Kurata T, Misawa N, Hosoya T et al (2019) Isolation and identification of components from wasabi (*Wasabia japonica* Matsumura) flowers and investigation of their antioxidant and anti-inflammatory activities. Food Sci Technol Res 25:449–457. https://doi.org/10.3136/fstr.25.449

Kwon LS, Park HY, Lee SR et al (2021) Current status and future Prospect of the kimchi industry. J Wellbeing Manag Appl Psychol 4:1–13. https://doi.org/10.13106/jwmap.2021.Vol4.no2.1

Lee K (2023) Kimchi exports up 20 percent in H1 on robust demand from US, EU, The Korea Times, 06 August. Available via https://wwwkoreatimescokr/www/tech/2023/12/129_356436html Accessed 2 Jan 2024

Lee SH, Whon TW, Roh SW et al (2020) Unraveling microbial fermentation features in kimchi: from classical to meta-omics approaches. Appl Microbiol Biotechnol 104:7731–7744. https://doi.org/10.1007/s00253-020-10804-8

Lee SJ, Jeon HS, Yoo JY et al (2021) Some important metabolites produced by lactic acid bacteria originated from kimchi. Foods 10:2148

Li J, Huang SY, Deng Q et al (2020) Extraction and characterization of phenolic compounds with antioxidant and antimicrobial activities from pickled radish. Food Chem Toxicol 136:111050. https://doi.org/10.1016/j.fct.2019.111050

Li H, Xia Y, Liu HY et al (2022) Nutritional values, beneficial effects, and food applications of broccoli (*Brassica oleracea* var. *italica* Plenck). Trends Food Sci Technol 119:288–308. https://doi.org/10.1016/j.tifs.2021.12.015

Majumder P, Annegowda HV (2021) Fruit and vegetable by-products: novel ingredients for a sustainable society. In: Bhat R (ed) Valorization of Agri-food wastes and by-products, 1st edn. Academic Press. https://doi.org/10.1016/B978-0-12-824044-1.00006-4

Mangunwardoyo W, Salamah A, Sukara E et al (2016) Diversity and distribution of culturable lactic acid bacterial species in Indonesian Sayur Asin. Iran J Microbiol 8:274

Moazzeni H, Mahmoodi M, Jafari M et al (2023) Underestimated diversity in high elevations of a global biodiversity hotspot: two new endemic species of *Aethionema* (Brassicaceae) from the alpine zone of Iran. Front Plant Sci 14:1182073. https://doi.org/10.3389/fpls.2023.1182073

MOFA (2013) Kimjang: making and sharing kimchi" makes UNESCO'S list of the intangible cultural heritage of humanity. Available via https://wwwmofagokr/eng/brd/m_5676/viewdo?seq=313150. Accessed 1 Jan 2024

Mohammadin S, Nguyen TP, van Weij MS et al (2017) Flowering locus C (FLC) is a potential major regulator of glucosinolate content across developmental stages of *Aethionema arabicum* (Brassicaceae). Front Plant Sci 8:876. https://doi.org/10.3389/fpls.2017.00876

Mordor Intelligence (2023) Cabbages and other brassicas market size & share analysis – Growth Trends & Forecasts (2023–2028). Available via https://wwwmordorintelligencecom/industry-reports/cabbages-and-other-brassicas-market. Accessed 17 Nov 2023

Morinaga T (1929) Interspecific hybridization in *brassica* I. The cytology of F 1 hybrids of B. napella and various other species with 10 chromosomes. Cytologia 1:16–27. https://doi.org/10.1508/cytologia.1.16

Mredul MMH, Akkas AB, ElSaidy N et al (2022) Using of mustard oil cake in safe organic aquaculture through increasing pond primary productivity. Aquac Rep 23:101073. https://doi.org/10.1016/j.aqrep.2022.101073

Nihayati A, Pertiwi EA, Haripa TI (2022) The analysis of south Korean Gastrodiplomacy towards the increases of south Korean kimchi export. J World Trade Stud 6:53–66. https://doi.org/10.22146/jwts.v6i2.3607

Nugroho D, Surya R, Nurkolis F et al (2023) Hepatoprotective effects of ethnic cabbage dishes: a comparison study on kimchi and pao cai. J Ethn Foods 10:31. https://doi.org/10.1186/s42779-023-00201-7

OEC (2021a) The latest trade data of Canola, rape, colza or mustard oil, fractions, refine. Available via https://oec.world/en/profile/hs/canola-rape-colza-or-mustard-oil-fractions-refine#trade. Accessed 01 Jan 2024

OEC (2021b) Cauliflowers and headed broccoli, fresh or chilled. Available via https://oec.world/en/profile/hs/cauliflowers-and-headed-broccoli-fresh-or-chilled. Accessed 23 Dec 2023

OEC (2021c) Brussels sprouts, fresh or chilled. Available via https://oec.world/en/profile/hs/brussels-sprouts-fresh-or-chilled. Accessed 23 Dec 2023

Oka T, Yamada Y, Lkhagvasuren B (2022) Clinical effects of wasabi extract containing 6-MSITC on myalgic encephalomyelitis/chronic fatigue syndrome: an open-label trial. Biopsychosoc Med 16:1–12. https://doi.org/10.1186/s13030-022-00255-0

Pakwan C, Chitov T, Chantawannakul P et al (2020) Bacterial compositions of indigenous Lanna (northern Thai) fermented foods and their potential functional properties. PLoS One 15:e0242560. https://doi.org/10.1371/journal.pone.0242560

Paramithiotis S, Hondrodimou OL, Drosinos EH (2010) Development of the microbial community during spontaneous cauliflower fermentation. Food Res Int 43:1098–1103. https://doi.org/10.1016/j.foodres.2010.01.023

Phan YTN, Tang MT, Tran TTM et al (2017) Diversity of lactic acid bacteria in vegetable-based and meat-based fermented foods produced in the central region of Vietnam. AIMS Microbiol 3:61. dio:https://doi.org/10.3934/microbiol.2017.1.61

Polaris Market Research (2023) Sauerkrauts market share, size, trends, industry analysis report, by form (Solid and Liquid); by packaging; by distribution channel; by region; segment forecast, 2023–2032. Available via https://wwwpolarismarketresearchcom/industry-analysis/sauerkrauts-market Accessed 17 Nov 2023

Punetha H, Sajwan N, Pandey D (2018) Phytochemical evaluation and assessment of potential nutrients from defatted meal of promising genotypes of *Brassica juncea* (Indian mustard). Biotech Today: An Int J Biol Sci 8:33–36

Raak C, Ostermann T, Boehm K et al (2014) Regular consumption of sauerkraut and its effect on human health: a bibliometric analysis. Glob Adv Health Med 3:12–18. https://doi.org/10.7453/gahmj.2014.038

Rai PK, Yadav P, Kumar A et al (2022) *Brassica juncea*: a crop for food and health. In: Kole C, Mohapatra T (eds) The *Brassica juncea* genome. Compendium of plant genomes. Springer, Cham. https://doi.org/10.1007/978-3-030-91507-0_1

Ramirez D, Abellán-Victorio A, Beretta V et al (2020) Functional ingredients from Brassicaceae species: overview and perspectives. Int J Mol Sci 21:1998. https://doi.org/10.3390/ijms21061998

Rani R, Badwaik LS (2021) Functional properties of oilseed cakes and defatted meals of mustard, soybean and flaxseed. Waste Biomass Valorization 12:5639–5647. https://doi.org/10.1007/s12649-021-01407-z

Rasera GB, Hilkner MH, de Alencar SM et al (2019) Biologically active compounds from white and black mustard grains: an optimization study for recovery and identification of phenolic antioxidants. Ind Crop Prod 135:294–300. https://doi.org/10.1016/j.indcrop.2019.04.059

Rathor PS, Singh SK, Adhikary K et al (2023) Improvement in flowering, yield and vase life of French Marigold (red brocade) through application of mustard oil cake and Vermicompost extracts. Ann Agri Bio Res 28:114–120

Raza A, Hafeez MB, Zahra N et al (2020) The plant family Brassicaceae: introduction, biology, and importance. In: Hasanuzzaman M (ed) The plant family Brassicaceae: biology and physiological responses to environmental stresses. Springer, Singapore. https://doi.org/10.1007/978-981-15-6345-4_1

Ren H, Xu D, Xiao W et al (2023) Chromosome-level genome assembly and annotation of Zicaitai (*Brassica rapa* var. *purpuraria*). Sci Data 10:759. https://doi.org/10.1038/s41597-023-02668-0

Rodzi NARM, Lee LK (2021) Traditional fermented foods as vehicle of non-dairy probiotics: perspectives in South East Asia countries. Food Res Int 150:110814. https://doi.org/10.1016/j.foodres.2021.110814

Rong H, Zhu S, Xie T et al (2019) Phenotypic and seed structural comparison in hybrids of different *Brassica rapa* and *B. oleracea*. Plant Breed 138:925–936. https://doi.org/10.1111/pbr.12748

Salehi B, Quispe C, Butnariu M et al (2021) Phytotherapy and food applications from brassica genus. Phytother Res 35:3590–3609. https://doi.org/10.1002/ptr.7048

Šamec D, Salopek-Sondi B (2019) Cruciferous (brassicaceae) vegetables. In: Mohammad Nabavi S, Silva AS (eds) Nonvitamin and nonmineral nutritional supplements. Academic Press, Amsterdam. https://doi.org/10.1016/B978-0-12-812491-8.00027-8

Sang-Hun C (2008) Kimchi goes to space, along with first Korean astronaut. The New York Times. Available via https://wwwnytimescom/2008/02/22/world/asia/22iht-kimchi110302283html. Accessed 23 Dec 2023

Sarker AK, Saha D, Begum H et al (2015) Comparison of cake compositions, pepsin digestibility and amino acids concentration of proteins isolated from black mustard and yellow mustard cakes. AMB Express 5:1–6. https://doi.org/10.1186/s13568-015-0110-y

Shahbandeh M (2025). Rapeseed production volume worldwide 2024/2025, by country. Available via https://www.statista.com/statistics/263930/worldwide-production-of-rapeseed-by-country/?. Accessed 4 Oct 2025

Sheidai M, Sonboli A (2011) Cytotaxonomy of four species of *Sterigmostemum* (Brassicaceae) in Iran. Cytologia 76:33–39

Shen J, Liu Y, Wang X et al (2023) A comprehensive review of health-benefiting components in rapeseed oil. Nutrients 15:999. https://doi.org/10.3390/nu15040999

Short BH (2020) Cook's Endeavour, ship of discovery or ship of distemper: an assessment after 250 years. Intern Med J 50:495–498. https://doi.org/10.1111/imj.14795

Shree B, Kumar S, Sharma S et al (2022) Functional significance of underutilized high value cruciferous vegetables-an exotic gleam in the gloomy guise of their functional importance. S Afr J Bot 145:420–437. https://doi.org/10.1016/j.sajb.2022.02.028

Siddeeg A, Afzaal M, Saeed F et al (2022) Recent updates and perspectives of fermented healthy super food sauerkraut: a review. Int J Food Prop 25:2320–2331. https://doi.org/10.1080/10942912.2022.2135531

Singh A, Singh VK (2025) Chemical and pharmacological activities of rape seed mustard *Brassica nigra* (black mustard): a magic herb. Asian J Biol Sci 18:43–54. https://doi.org/10.3923/ajbs.2025.43.54

Siomos AS, Koularmanis K, Tsouvaltzis P (2022) The impacts of the emerging climate change on broccoli (*Brassica oleracea* L. var *italica* Plenck) Crop. Horticulturae 8:1032. https://doi.org/10.3390/horticulturae8111032

Sirohi SK, Walli TK, Garg MR (2013) Effect of formaldehyde treated mustard cake on nutrient utilization and milk production performance in crossbred cows fed wheat straw based diet. Indian J Animal Nut 30:5–11

Sohail M, Nauman-ul-Islam M, Hayazuddin Shah IA et al (2017) Effect of supplementing different levels of mustard seed cake on Milk yield and Milk composition of Achai cattle. Meat Sci Vet Public Health 2:5–8

Song G, He Z, Wang X et al (2021a) Improving the quality of Suancai by inoculating with *lactobacillus plantarum* and *Pediococcus pentosaceus*. Food Res Int 148:110581. https://doi.org/10.1016/j.foodres.2021.110581

Song X, Wei Y, Xiao D et al (2021b) *Brassica carinata* genome characterization clarifies u's triangle model of evolution and polyploidy in *brassica*. Plant Physiol 186:388–406. https://doi.org/10.1093/plphys/kiab048

Stankus T (2014) Pickled vegetable condiments: a global industry and its literature. J Agric Food Inf 15:3–18. https://doi.org/10.1080/10496505.2013.858048

Subramanian P, Kim SH, Hahn BS (2023) *Brassica* biodiversity conservation: prevailing constraints and future avenues for sustainable distribution of plant genetic resources. Front Plant Sci 14:1220134. https://doi.org/10.3389/fpls.2023.1220134

Surya R, Lee AGY (2022) Exploring the philosophical values of kimchi and kimjang culture. J Ethn Foods 9:1–14. https://doi.org/10.1186/s42779-022-00136-5

Surya R, Nugroho D (2023) Kimchi throughout millennia: a narrative review on the early and modern history of kimchi. J Ethn Foods 10:5. https://doi.org/10.1186/s42779-023-00171-w

Temmen D, Randall J, Popova I (2022) The effect of Sinapis alba mustard seed meal extract on potato tuber quality in organic potato production. Agronomy 12:2782. https://doi.org/10.3390/agronomy12112782

Thompson HO, Önning G, Holmgren K et al (2020) Fermentation of cauliflower and white beans with *lactobacillus plantarum*–impact on levels of riboflavin, folate, vitamin B 12, and amino acid composition. Plant Foods Hum Nutr 75:236–242. https://doi.org/10.1007/11130-020-00806-2

Tian Y, Deng F (2020) Phytochemistry and biological activity of mustard (*Brassica juncea*): a review. CyTA-J Food 18:704–718. https://doi.org/10.1080/19476337.2020.1833988

Tlais AZA, Lemos Junior WJF, Filannino P et al (2022) How microbiome composition correlates with biochemical changes during sauerkraut fermentation: a focus on neglected bacterial players and functionalities. Microbiol Spectr 10:e00168–22. https://doi.org/10.1128/spectrum.00168-22

USDA (2022) U.S. Condiments & Sauces Exports in 2022. Available via https://fas.usda.gov/data/commodities/condiments-sauces. Accessed 25 Dec 2023

USDA (2023a) India: Oilseeds and Products Update. Available online: https://fas.usda.gov/data/india-oilseeds-and-products-update-25. Accessed 21 Dec 2023

USDA (2023b) Rapeseed 2023World production. Available via https://ipad.fas.usda.gov/cropexplorer/cropview/commodityView.aspx?cropid=2226000. Accessed 21 Dec 2023

USDA (2025). Production – Rapeseed. 2024/2025 Rapeseed production. Available via https://www.fas.usda.gov/data/production/commodity/2226000. Accessed 4 Oct 2025

Uto T, Hou DX, Morinaga O et al (2012) Molecular mechanisms underlying anti-inflammatory actions of 6-(methylsulfinyl) hexyl isothiocyanate derived from wasabi (*Wasabia japonica*). Adv Pharmacol Pharm Sci 2012:614046. https://doi.org/10.1155/2012/614046

Wang Z, Shao Y (2018) Effects of microbial diversity on nitrite concentration in pao cai, a naturally fermented cabbage product from China. Food Microbiol 72:185–192. https://doi.org/10.1016/j.fm.2017.12.003

Wieczorek MN, Drabińska N (2022) Flavour generation during lactic acid fermentation of *brassica* vegetables—literature review. Appl Sci 12:5598. https://doi.org/10.3390/app12115598

Woo M, Kim MJ, Song YO (2018) Bioactive compounds in kimchi improve the cognitive and memory functions impaired by amyloid beta. Nutrients 10:1554. https://doi.org/10.3390/nu10101554

World Integrated Trade Solution (2019). https://wits.worldbank.org/trade/comtrade/en/country/ALL/year/2019/tradeflow/Exports/partner/WLD/product/151490. Accessed 19 Dec 2023

Xu X, Qiao Y, Peng Q et al (2022) Antioxidant and immunomodulatory properties of partially purified exopolysaccharide from *lactobacillus casei* isolated from Chinese northeast sauerkraut. Immunol Investig 51:748–765. https://doi.org/10.1080/08820139.2020.1869777

Yadava DK, Yashpal Saini N, Nanjundan J et al (2022) *Brassica* breeding. In: Yadava DK, Dikshit HK, Mishra GP, Tripathi S (eds) Fundamentals of field crop breeding. Springer, Singapore. https://doi.org/10.1007/978-981-16-9257-4_15

Yun YR, Park SH, Kim IH (2019) Antioxidant effect of kimchi supplemented with Jeju citrus concentrate and its antiobesity effect on 3T3-L1 adipocytes. Food Sci Nutr 7:2740–2746. https://doi.org/10.1002/fsn3.1138

Yun YR, Lee JJ, Lee HJ et al (2021a) Comparison of quality characteristics of commercial kimchi manufactured in Korea, China, and the United States. Foods 10:2488. https://doi.org/10.3390/foods10102488

Yun YR, Kim HC, Seo HY (2021b) Antiobesity effects of kimchi added with Jeju citrus concentrate on high-fat diet-induced obese mice. Nutr Res 86:50–59. https://doi.org/10.1016/j.nutres.2020.11.007

Zabat MA, Sano WH, Wurster JI et al (2018) Microbial community analysis of sauerkraut fermentation reveals a stable and rapidly established community. Foods 7:77. https://doi.org/10.3390/foods7050077

Zhang N, Jing P (2022) Anthocyanins in Brassicaceae: composition, stability, bioavailability, and potential health benefits. Crit Rev Food Sci Nutr 62:2205–2220. https://doi.org/10.1080/10408398.2020.1852170

Zhang YJ, Li S, Gan RY et al (2015) Impacts of gut bacteria on human health and diseases. Int J Mol Sci 16:7493–7519. https://doi.org/10.3390/ijms16047493

Zhang K, Mason AS, Farooq MA et al (2021a) Challenges and prospects for a potential allohexaploid *brassica* crop. Theor Appl Genet 134:2711–2726. https://doi.org/10.1007/s00122-021-03845-8

Zhang C, Zhang J, Liu D (2021b) Biochemical changes and microbial community dynamics during spontaneous fermentation of Zhacai, a traditional pickled mustard tuber from China. Int J Food Microbiol 347:109199. https://doi.org/10.1016/j.ijfoodmicro.2021.109199

Zhang Y, Qi X, Wang X et al (2022) Contribution of tocopherols in commonly consumed foods to estimated tocopherol intake in the Chinese diet. Front Nutr 9:829091. https://doi.org/10.3389/fnut.2022.829091

Zhao S, Han J, Bie X et al (2016) Purification and characterization of plantaricin JLA-9: a novel bacteriocin against *bacillus* spp. produced by *lactobacillus plantarum* JLA-9 from Suan-Tsai, a traditional Chinese fermented cabbage. J Agric Food Chem 64:2754–2764. https://doi.org/10.1021/acs.jafc.5b05717

Zubaidah E, Susanti I, Yuwono SS et al (2020) Effect of *lactobacillus plantarum* and *Leuconostoc mesenteroides* starter cultures in lower salt concentration fermentation on the sauerkraut quality. Food Res 4:1038–1044. https://doi.org/10.26656/fr.2017.4(4).029

Significance of *Brassica* Crops in Oilseed Industry

Vishal Kumar, Deepika Kaushik, Jasjit Kaur, Prasad Rasane, Althaf Sayyad, Fatih Oz, and Mukul Kumar

Abstract

This chapter reviews the importance of oilseeds, particularly *Brassica napus*, *Brassica juncea*, *Brassica rapa*, and *Brassica carinata* in the global agricultural economy, as they are the channels of vegetable oil and protein sources popular with people and livestock. Their contributions toward food security have, therefore, increased and continue to increase as demand increases worldwide because of the improved yields through breeding and agronomic innovations. This chapter gives information on the nutritional composition of each species regarding some important minerals, beneficial fatty acids, and bioactive compounds, including flavonoids, tocopherols, and glucosinolates, toward their health-promoting properties. *Brassica napus* would offer the best oil content along with a profile favorable in fatty acids, while *Brassica rapa* is rich in omega-3 fatty acids. Antioxidant and anti-inflammatory properties are shown by *Brassica juncea*, which is attributed to the significant bioactive constituents present in its oil. Another key advancement is seen in *Brassica carinata* as a prospect for biofuels. In addition, global oilseed market trends are discussed, along with factors such as population increase and changing consumer preferences that contribute to the growing demand for plant-based oils. Emerging economies are expected to drive a substantial portion of the growth in this market. Therefore, it is essential to

V. Kumar · J. Kaur · P. Rasane · A. Sayyad · M. Kumar (✉)
Department of Food Technology and Nutrition, Lovely Professional University, Phagwara, Punjab, India

D. Kaushik (✉)
Department of Biotechnology, Faculty of Applied Sciences and Biotechnology, Shoolini University, Solan, India

F. Oz
Department of Food Engineering, Faculty of Agriculture, Ataturk University, Erzurum, Türkiye

A. Kumar, B. Koul (eds.), *Crop Improvement Strategies in Brassica species: Applied Science*, https://doi.org/10.1007/978-981-95-3861-4_2

know these trends and their relationship with the nutritional quality of *Brassica* oilseeds, given that they offer opportunities that can be exploited for achieving food security and improving agricultural growth in the rapidly evolving oilseed sector.

Keywords

Brassica · *Brassica* oils · Extraction · Genetic improvement · Global trend

1 Introduction

Brassica crops represent a significant oilseed source that serves the purposes of edible oils, biofuel, and industrial uses. *Brassica napus* is known for edible vegetable oil production across the globe, alongside *Brassica juncea, Brassica rapa,* and *Brassica carinata* (Weselake et al. 2024; Licata et al. 2024). It is also crucial for food security across the world, being an important source of vegetable oil and protein for both human and livestock consumption (Borges et al. 2023). Demand for oilseed from *Brassica* is increasing globally, and the past 50 years have seen significant improvements in yield potential. The continuous agronomic and breeding efforts have brought about tremendous increases in *Brassica napus* production to satisfy cooking oil demand, with yield increases of over 450 times since the 1970s (Zandberg et al. 2022). Oilseed *Brassica*, grown on 35 million hectares, produced 72 million tonnes worldwide. Canada leads in production, with China second and India third in production and area (Jat et al. 2019). *Brassica napus*, a major oilseed crop in the Brassicaceae family, is highly prized for its nutritional value and oil content, along with essential minerals, including magnesium, calcium, potassium, and phosphorus. The phytochemical profile includes flavonoids and total phenolics contributing to its health benefits, such as anti-inflammatory and antioxidant effects (Gagour et al. 2022). Another economically significant oilseed crop with a great nutritional profile is *Brassica rapa*, or field mustard. It contains bioactive compounds, including glycosaponins, glucosinolates, glucosamine, gluconasturtiin, and neoglucobrassin, which are associated with antioxidant and antihypertensive effects (Serrano et al. 2024). The phytosterols, tocopherols, and omega-3 fatty acids present in the crop improve heart health and provide antioxidant effects. *Brassica juncea* is mainly cultivated for extraction as an oilseed, yielding an oil content of 30–40% by weight, therefore serving in both industry and cooking purposes. Together with sinigrin, its other phytochemicals, such as phenolics and flavonoids, help confer antioxidant, anticancer, and anti-inflammatory effects (Kwon et al. 2020; Batool et al. 2024). Due to its high oil content and capacity to grow on marginal lands, *Brassica carinata* is becoming more and more popular as a biofuel crop. It contains sulfur components, as well as flavonoids and phenolic compounds, are linked to several health advantages. The crop seeds yield roughly 35–40% oil by weight, with oil extracted containing high concentrations of linoleic and oleic acids, and increasingly being used for biodiesel manufacture

(Yadav et al. 2022; Bashyal et al. 2021). Soybeans currently dominate oilseed production globally, with the United States, Brazil, China, India, Argentina, and the European Union being the main oilseed producers. Changing consumer preferences also favor oils with nutritional benefits like soybean and canola oils for their heart-healthy properties and demand products with sustainable sources, which force producers to embrace sustainable agricultural methods and certifications alongside the expansion of the plant-based movement, raising demand for plant-based oils as substitutes for animal fats (Yaheliuk et al. 2024). Trade policies, climate change, and regional conflict all define a global oilseed picture; shifting weather patterns and geopolitical uncertainty directly or indirectly affect crop yield and trade; changing crop choices for farmers tend to alter income determination through changes in commodity prices and access to financial and technological resources (Kephe et al. 2020; Yaheliuk et al. 2024); thus, it requires proper understanding from policymakers and agricultural stakeholders on these dynamics toward enhancing resilience and ensuring of food security in this complex world.

2 Types of *Brassica* Crops Used in the Oilseed Industry

2.1 *Brassica napus*

One of the major oilseed crops in the Brassicaceae family, *Brassica napus*, known as canola or rapeseed, is valued for its nutritional and oil contents. The crop's prominence in the market is largely due to its favorable proximate and phytochemical composition. Seeds of *Brassica napus* contain approximately 20–25% protein, 40–45% oil, and 10–15% carbohydrates, along with significant minerals, such as magnesium (Mg), calcium (Ca), potassium (K), and phosphorus (P) (Gagour et al. 2022; Carré et al. 2016). The importance of oil content for industrial as well as edible purposes is emphasized. Distinctly, the minerals recorded were potassium—approximately 7936.53 mg/kg and phosphorus—about 4713.95 mg/kg. Whereas the concentration of Fe, Cu, Mn, and Zn as trace elements is comparatively lower in these seeds. Bioactive compounds, such as flavonoids and total phenolics, constitute the phytochemical profile of *Brassica napus*. The total flavonoid content of rapeseed is approximately 22.84 mg QE/g, whereas the total phenolic content is around 38.49 mg GAE/g (Gagour et al. 2022). These compounds have been associated with various health benefits, including antimicrobial, anticancer, cardioprotective, and anti-inflammatory activities. *Brassica napus* is one of the highest oilseed-yielding crops in the world, producing oil at the rate of about 20–30 million metric tonnes annually (Borges et al. 2023). Major unsaturated fatty acids, mainly oleic acid (C18:1) at 62.19%, a heart-healthy fatty acid, dominate the oil obtained from *Brassica napus* seeds. The oil's nutritional value is further enhanced by two other significant fatty acids: alpha-linolenic acid (C18:3) and linoleic acid (C18:2) (Gagour et al. 2022). Bioactive constituents in *Brassica napus* oil, such as tocopherols (vitamin E), carotenoids, and phytosterols, boost its nutritional value. Canola oils are more appealing for human consumption when compared to oils from other *Brassica* species because of their lower erucic acid and glucosinolate contents (Tileuberdi et al. 2022).

2.2 *Brassica rapa*

Field mustard or turnip is the other term associated with *Brassica rapa,* which is among the other principal oilseed crops. The proximate composition suffices to demonstrate the nutritional richness of *Brassica rapa*. Leaves contain high concentrations of the most important constituents, such as 11.90 mg/g of protein, 3.48 mg/g of fat, and 57.45 mg/g of carbohydrates, with about moisture by 9.10% and total ash at 18.10% (Abid et al. 2022). Majorly, the seeds are consumed and become an important nutrient as good sources of major essential minerals, such as potassium (K), phosphorus (P), calcium (Ca), and magnesium (Mg) (Shekhawat et al. 2023). Regarding the phytochemical profile, plant methanolic extracts have shown a high content of flavonoids (259.13 mg/g) and polyphenols (93.63 mg/g) (Abid et al. 2022). The glycosaponins (70 mg/g) found in water extracts of this crop further indicate its health-enhancing potential. *Brassica rapa* is associated with several health benefits, such as antioxidants and antihypertensive activities because of the important glucosinolates, glucosamine, gluconasturtiin, and neoglucobrassin found in this species (Serrano et al. 2024). *Brassica rapa* is predominantly grown for its oil-rich seeds that are used to produce an edible oil. This oil contains omega-3 fatty acids, tocopherols, and phytosterols, among other beneficial fatty acids that come from these seeds. These compounds add to the health benefits of oil through the reduction of cholesterol levels and antioxidant properties (Tileuberdi et al. 2022). It shows richness in bioactive compounds that make improved nutritional profiles as per the oil extracted from *Brassica rapa*. Besides, oleic acid is very important in this aspect as it is known to lower blood pressure. At the same time, the phytosterols and tocopherols (vitamin E) present in the oil also improve antioxidant properties while promoting heart health (Poudel et al. 2023).

2.3 *Brassica juncea*

Brassica juncea (leaf or Indian mustard) is an oilseed crop, grown primarily for oil extraction for several industrial and culinary uses. *Brassica juncea* has a varied proximate composition; moisture can range from 91.96% to 93.49%, crude protein values can vary from 1.66% to 1.93%, while that for crude lipids is between 0.40% and 0.61%, important in oil extraction; crude ash is between 1.11% and 1.55% and gives an indication of the presence of some important minerals (Kwon et al. 2020). *Brassica juncea* is also a significant source of vitamin A and carbohydrates. These constituents point to *Brassica juncea* being a crop with good nutritional value, giving a good balance of protein and fat suitable for oil production. This crop abounds with phytochemicals that impart health benefits. Total phenols have been reported to vary from 22.5 mg to 49.17 mg GAE/100 g (Batool et al. 2024), while flavonoids can be present in the range of 34.81 mg QE/g. Sinigrin, a glucosinolate present in considerable concentration (16.16 mg/g), also contributes to antioxidant activity (Kwon et al. 2020). These phytochemicals are said to be responsible for a number of health-giving properties, including anticancer activities and anti-inflammatory properties. Leaf mustard also contains carotenoids, ascorbic acid, and chlorophylls (Batool et al. 2024). *Brassica juncea* gives about 30–40% oil by weight. The oil derived from the seed is well-known for its favorable fatty acid profiles, high in linoleic acid and oleic acid content, making it suitable for food processing and cooking (McVetty et al. 2016). The oil's nutritional status is further enhanced by the presence of phytosterols and tocopherols (vitamin E), which are also beneficial for cardiovascular health. *Brassica juncea* oil has many bioactive components, such as linoleic acid and oleic acid, phytosterols, and tocopherols that increase stability and shelf life of the oil while providing health benefits (Poudel et al. 2023).

2.4 *Brassica carinata*

Brassica carinata, popularly known as Ethiopian mustard, is coming out as a popular biofuel crop owing to its high oil content and adaptability to low-input marginal areas. The proximate composition of *Brassica carinata* includes moisture (11.5–12.5%), crude protein (39.2%) (Yadav et al. 2022), and high values of crude fat (30–40%) suited for oil extraction. The seeds also contain other minerals, such as calcium, potassium, and magnesium, and dietary fiber (Bashyal et al. 2021). Phytochemicals, especially glucosinolates, which are sulfur-containing constituents related to several health benefits, are abundant in *Brassica carinata*. The property of *Brassica carinata* contributing to its antioxidant activity is a high concentration of glucosinolate, which can rise to 100 μmol/g (Yadav et al. 2022). Other phytochemicals, such as flavonoids and phenolic compounds, improve the net nutritional content of the plant with possible health benefits. *Brassica carinata* seeds yield about 35–40% oil by weight, making it a direct competitor in the oilseed market. Oil extracted from these seeds has a favorable fatty acid composition, including relatively high concentrations of linoleic and oleic acids that favor good heart health. This oil is now being used more for biodiesel production as well as cooking because of its favorable characteristics (Bashyal et al. 2021; Redda et al. 2024). Linoleic acid, oleic acid, tocopherols, and phytosterols are some important bioactive constituents found in *Brassica carinata* oil that enhance its nutritional profile. These constituents offer health benefits, such as antioxidant properties and lowering of cholesterol, as well as improve the shelf life and stability of the oil (Poudel et al. 2023). Table 1 represents the composition of *Brassica* seeds.

Table 1 Composition of *Brassica* seeds

Crop	Oil yield (%)	Protein content (%)	Fiber content	Ash content (%)	Carbohydrates content (%)	Key fatty acids	Tocopherols	Phytosterols	Glucosinolates (μmol/g)	Other compounds	Reference
Brassica napus	40–45	20–25	Approx. 10–15	4–5	10–20	Oleic acid, linoleic acid, alpha-linolenic acid, and erucic acid	α, β, γ tocopherol	Beta-sitosterol, sitosterol, campesterol, stigmasterol, brassicasterol,	Low (under 30)	Minerals, vitamins, phenolics, flavonoids, hydroxycinnamic acid derivative	Poudel et al. (2023), McVetty et al.(2016), Tileuberdi et al. (2022), and Gagour et al. (2022)
Brassica rapa	35–42	20–30		3–4	20–30				Moderate (around 30–60)		
Brassica juncea	30–40	25–35		4–5					High (around 60–150)		
Brassica carinata	35–40	25–30		4–5					High (around 60–150)		

3 Global Oilseed Market Trends

The tremendous growth of the global oilseed market is influencing the demand for vegetable oils. Vegetable oils are mainly used in cooking, food processing, and numerous industrial applications. The market is anticipated to register a compound annual growth rate (CAGR) of 9.6% for a market value of $326.04 billion in 2024 to $357.49 billion in 2025. Thereafter, it is expected to reach approximately $519.39 billion with a CAGR of 9.8% in 2029. This rising demand has many factors contributing to it, like the birth rate, shifting eating habits, and increasingly large uses of oilseeds in animal feed (Wang 2016). Therefore, with high levels of protein, vitamins, and minerals, oilseeds are gaining popularity among the nutritional supplements launched into the market, further driving growth. It is also expected that favorable governmental activities to support domestic oilseed production in key growing areas will assist in market advancement. The growing biodiesel demand, motivated by the increasing realization of the need to go easy on conventional fuels, puts more demand for vegetable oils and consequently, oilseeds. Nearly two-thirds of the expansion of the global oilseed industry is likely to take place in the hands of nations, such as China, India, Brazil, Malaysia, and Indonesia. The USA, Brazil, China, India, Argentina, and the European Union are the major producers of oilseeds. Because of rising demand for the needs of the increased population, soybeans currently dominate oilseed production worldwide, and they are projected to continue increasing at the fastest rate (Willer et al. 2020; Vinnichek et al. 2019). Figure 1 depicts the global production of mustard seed and rapeseed or canola oil, mustard seed yield with its harvested area across the years (FAOSTAT 2025). Consumer preferences are evolving markedly across the world oilseed market, in areas like sustainability and health. These transformations are causing shifts in the production techniques used throughout the sector, while simultaneously changing demand patterns. Customers' health concerns are on the rise, and with that, oils boasting nutritional advantages are becoming popular. As high unsaturation fatty acid and low saturated fatty acid oils, oils like canola, soybean, and sunflower oils are preferred for heart health. This trend is reflected in the increasing demand for functional foods enriched with these healthier oils, as people look to dietary changes to boost their overall well-being (Zhang et al. 2025). Sustainability is the second significant factor shaping consumer preferences. Demand is increasing for products with sustainable sourcing, including vegetable oils produced with reduced environmental impact. Consumers are increasingly questioning the climate change effects of agricultural methods, and oilseeds grown with environmentally friendly practices are gaining attention. This shift is driving producers to adopt sustainable farming practices and certifications to attract environmentally conscious consumers (Braun 2018). The plant-based movement is also contributing to increased demand for oilseeds. Individuals following vegetarian or vegan diets are increasingly demanding plant-based oils as possible substitutes for animal fats. Oilseeds play a crucial role in this trend as they supply the essential fatty acids and minerals necessary to maintain a balanced diet. Diversification and innovation within the industry have been further promoted by the emergence of market opportunities for oilseed-derived ingredients

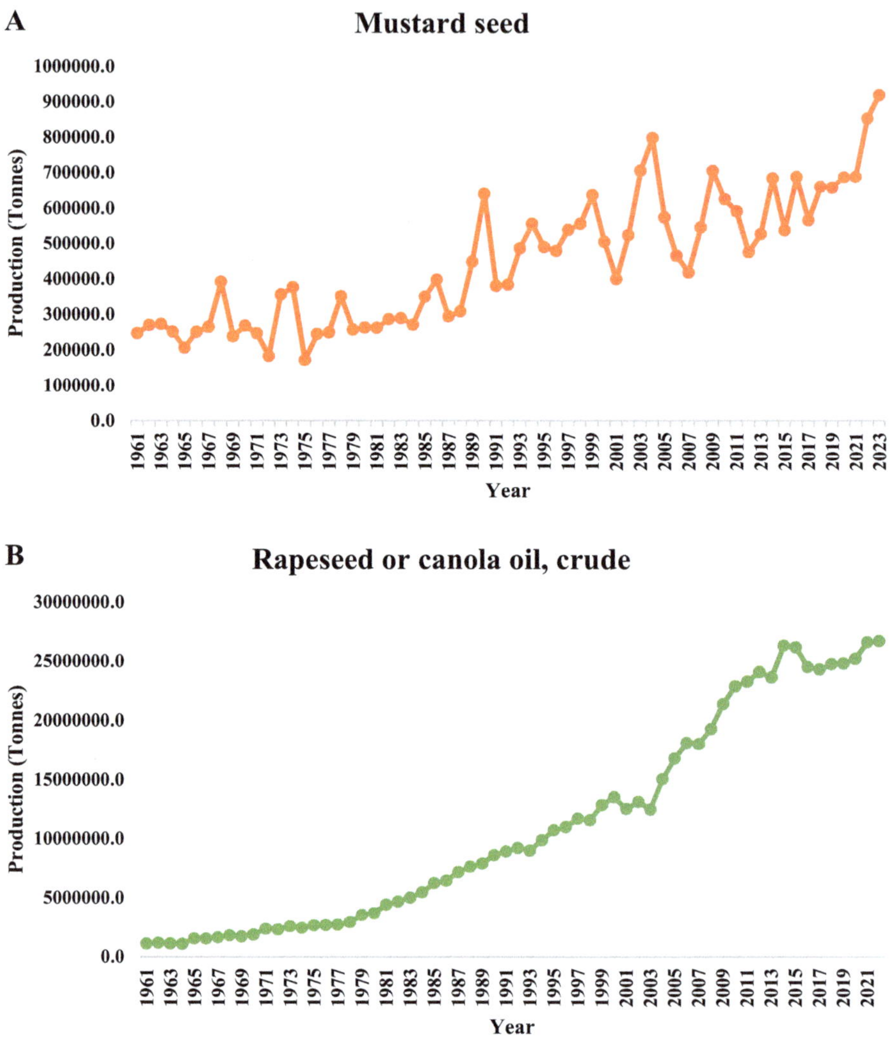

Fig. 1 (**a**) Mustard seed global production in tonnes. (**b**) Rapeseed or canola oil global production in tonnes. (**c**) Yield of mustard seed in kilograms/hectare (**d**) Harvested area of mustard seed in hectares

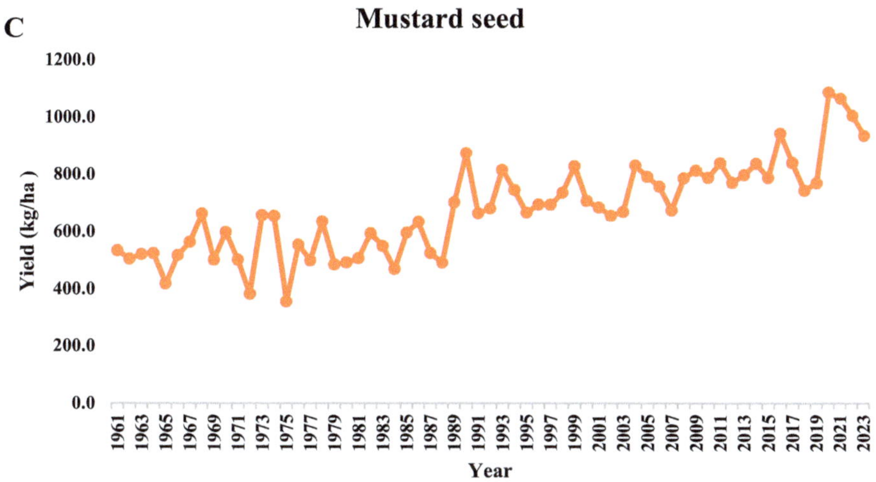

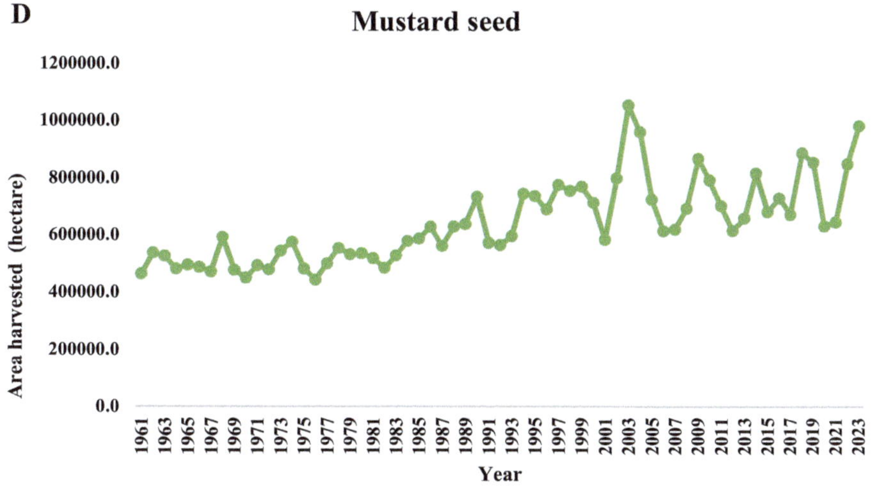

Fig. 1 (continued)

(Yaheliuk et al. 2024). Geopolitical factors influence oilseed production from the selection of crops to trade dynamics. The interplay of international trade policies, climate change, and regional conflicts shapes the global oilseed scenario. Climate change is drastically challenging the growth of oilseed production. Crop yield changes, together with the variability in weather patterns (increased incidence of droughts and floods), influence farmers' choices on the oilseeds to grow. For instance, the study in Limpopo Province of South Africa established that temperature and rainfall influence oilseed production in a significant way. Floods were shown to have the most significant effect on crop selection, indicating that a flexible approach to farming will be required to mitigate climate risks (Kephe et al. 2020). Oilseed trade policies are another tool that shapes the sector. Changes in global supply chains of soybean that stem from trade negotiations currently underway between major producers, such as the USA and China, may also play a factor. Brazil now stands firmly as a player in soybean exports, but that is contingent on the outcome of these negotiations. Conversely, countries like India are ramping up domestic oilseed production by way of government programs intended to reduce reliance on imports with an increasing demand for vegetable oils (Vinnichek et al. 2019). Supply chains and oilseed production may also be influenced by geopolitical conflicts. Among two major oilseed production countries, the war between Russia and Ukraine considerably disrupts the world's grain and oilseed markets. Disruptions in agricultural production in those regions will result in global supply crunches and price hikes. In addition, interspersed agricultural growth within this geopolitical area would mean greater instability in oilseed output (Tass et al. 2024). The economics also matter greatly in oilseed production. The fluctuations in the prices of commodities will dictate the farmers in deciding crop choices. For instance, profitable soybean prices could lead them to turn away from other possible crops. On the other hand, the expansion of oilseed agricultural production will depend on access to finance and technology. To devise policies for the realization of resilience and ensure food security in an increasingly complex world, agricultural stakeholders and policymakers need an in-depth understanding of these dynamics (Yaheliuk et al. 2024). Changing consumer preferences, a rise in demand for vegetable oil, and geopolitical causes are important drivers for the major transformation of the global oilseed industry. Growing rapidly in developing countries, the market is turning increasingly important. Important trends like the plant-based movement and integrated technology, which open rooms for innovative and sustainable growth, currently shape the future of the oilseed industry.

4 Oil Extraction and Processing from *Brassica* Crops

Oil extraction and processing from *Brassica* crops, such as rapeseed and mustard, have been changing over the years. The transition for this whole process has gone from highly traditional practices to quite modern technological innovations, which have made the development in itself very bright from both ends. Traditional extraction methods, although they give reasonably good results, are labor-intensive and involve human effort, which could have been used to raise yields and quality production (Beszterda and Nogala-Kałucka 2019). Current extraction methods, on the other hand, adopt new developments in technology in view of modern needs and standards to improve production efficiency, processing time, and oil quality. Various oilseed crops, such as *Brassica* crops, rapeseed, canola, and mustard, have gained commercial importance due to crude oil extraction and oil-related processing variables to maximize oil quality and quantity. Traditional oil extraction methods include cold pressing, expeller pressing, and solvent extraction (Rani et al. 2021). Cold-press extraction is the mechanical extraction of oil from seeds at low temperatures to prevent change, rancidity, or destruction of their flavor and nutritional value. Although this method is often utilized for its ability to maintain beneficial compounds, its relative oil output may be lower than that of other techniques (Wandhekar et al. 2023). Although it may generate heat, expeller pressing employs a mechanical press to extract oil by applying high pressure, making it still quite a natural extraction method. Commonly using hexane, solvent extraction is a more industrialized technique that helps to maximize oil recovery rates, but it may reduce some tastes and nutrients from the oil (Cravotto et al. 2022). The process begins with seed cleaning and classification to remove impurities; it then proceeds with conditioning to improve oil extraction by varying moisture levels or heating application. The chosen extraction method is then used after grinding the seeds into a paste. After extraction, the oil is filtered to remove particles and pollutants; it may then be refined further using techniques like degumming, bleaching, and deodorizing to raise its quality for use in food. While lowering saturated fat content, these traditional approaches improve the nutritional composition of the oil and match consumer inclinations for healthier cooking oils abundant in unsaturated fatty acids, such as oleic and linoleic acids (Cravotto et al. 2023; Kaur et al. 2022; Czwartkowski et al. 2022). Hexane-free extraction methods, including water extraction and supercritical fluid extraction (SFE), which employ carbon dioxide at high pressure to recover significant molecules without the use of hazardous solvents, have been improved lately. While addressing the growing demand for healthier oils in culinary uses and establishing *Brassica* oils as an indispensable component in both food and non-food sectors, these modern techniques improve oil quality by reducing saturated fatty acids and augmenting health benefits (Ahangari et al. 2021). Technological developments in oil extraction from *Brassica* crops cover both traditional methods and advanced ecologically friendly ones. Though judged ineffective, conventional methods, including mechanical pressing, are simple and reasonably affordable. Because of the utilized chemical solvents, chemical operations negatively affect the environment, oil quality, and human health even when they provide higher oil recovery (Beszterda and Nogala-Kałucka 2019). Innovative technologies that solve the constraints of conventional methods have evolved from enzyme-assisted extraction, microwave-assisted extraction, ultrasonic-assisted extraction, and supercritical fluid extraction. The impact of extraction parameters, including temperature, solvent-to-solid ratio, time, and particle size, on oil output has been investigated using response surface methodology (RSM) employing Box-Behnken design (Usman et al. 2023; Nde and Foncha 2020). Extraction of supercritical carbon dioxide is pertinent; for rapeseed oil extracted using n-hexane Soxhlet, no considerable changes in fatty acid profiles were observed. The aqueous method is suggested to keep both oil and protein from *Brassica napus* in very good condition. Environmental factors in the oil extraction processing of *Brassica* crops contribute to reducing the environmental impacts and supporting sustainable agricultural practices (Sun et al. 2021; Raboanatahiry et al. 2021). *Brassica* species, including *Brassica juncea* and *Brassica napus*, need to be cultivated through sustainable agricultural practices, such as crop rotation, organic fertilization, and natural pest management. These methods not only help enhance soil quality but also reduce reliance on harmful chemicals, therefore protecting ecological integrity and biodiversity (Zandberg et al. 2022). The method of extracting oil may have a significant impact on environmental outcomes; for instance, lower temperature cold-pressed methods that avoid the need for chemical solvents minimize pollution and maintain the nutritional integrity of the oil. Recyclable organic manures are also oil extraction residues or by-products, such as mustard cake, supporting a circular economy, which reduces waste and boosts soil fertility (Ancuţa and Sonia 2020; Cravotto et al. 2023). Furthermore, proper techniques could be combined with renewable energy sources to lower the carbon footprint of the *Brassica* oil production process. Studies revealed that improvement in these processes increases oil yield; it will also abide by the standard global sustainability criteria, making *Brassica* a better choice for biodiesel production toward environmental sustainability. Understanding the environmental dynamics related to *Brassica* oil extraction will build future food security and ecological well-being due to all climate change challenges to agricultural productivity (Saldukaitė et al. 2022; Borges et al. 2023).

5 Nutritional Benefits and Market Demand for *Brassica* Oils

In *Brassica* oils, along with other essential fatty acids, tocopherols, and phytosterols, omega-3 fatty acids are present in high amounts. Specific *Brassica* species provide different ratios of fatty acids, whether unsaturated or saturated. The oil extracted from mustard (*Brassica juncea*) is said to contain good amounts of omega-6 and omega-3 polyunsaturated fatty acids (PUFAs) along with a proper balance of saturated and unsaturated fatty acids (Rai et al. 2022). It contains specific monounsaturated fatty acids (MUFAs), such as erucic acid, oleic acid, gondoic acid, nervonic acid, and docosahexaenoic acid (DHA). Linoleic acid, linolenic acid, DHA, and EPA are polyunsaturated fatty acids. Besides these fatty acids, palmitic acid has its small part (Rahman et al. 2024). Additionally, rapeseed (*Brassica napus*) oil presents high MUFAs and fewer PUFAs and has a high level of tocopherols, phytosterols, and omega-3 fatty acids (Goyal et al. 2021). Oleic acid is a commonly found fat in rapeseed oil and can comprise a few types, whereby in some cases, it forms up to 75%. Next ranked are linoleic acid and γ-linolenic acid, both polyunsaturated. Rich in monounsaturated fatty acids (MUFA), rapeseed oil has been associated with lower incidences of various malignancies, such as skin, breast, and colon cancer, and coronary heart disease. Part of these polyunsaturated fatty acids, particularly gamma-linolenic acid and linoleic acid, is said to improve cardiovascular function. Involved in these functions are α-linolenic acid, which also lowers cholesterol, controls weight, reduces triglycerides, and protects the heart and brain. α-Linolenic acid, on the other hand, has medical effects, such as fighting cancer, preventing blood clots, protecting the heart and brain, and preventing diabetes (Shen et al. 2023). Canola oil contains, ultimately, 61.6% monounsaturated fatty acids (MUFA), 30.2% polyunsaturated fatty acids (PUFA), and 7.6% saturated fatty acids (SFA). Of all the fatty acids, oleic acid makes up 57.9%; α-linolenic acid makes up 12.7%; linoleic acid makes up 17.5%. Compared to n-6 (21.7%) and n-9 (59.1%), n-3 (9.5%) is lower (Dobrzyńska and Przysławski 2020). For the synthesis of omega-3 fatty acids—including ALA, SDA, EPA, DPA, and DHA—it provides a scalable, sustainable terrestrial solution. One hectare of DHA canola could provide omega-3 oil equal to that from 10,000 kg of fish. This would make it a cheap, safe, and reliable alternative to omega-3 oil that comes from fish (MacIntosh et al. 2021). Functional food is a more specific term that usually applies to food products for which nutritional value improves the health of consumers beyond its usual value (Alongi and Anese 2021). Consumption of *Brassica* species has shown considerable influence in the prevention, treatment, and management of various diseases, which include cancer, hypertension, diabetes, chronic inflammatory conditions, oxidative stress, etc. The specific advantages result from particular molecules, such as polyphenols or glucosinolates (GSL), bioactive agents in rapeseed oil, having an active effect on human life (Ramirez et al. 2020). Such *Brassica* crops are majorly known in the oilseed industry for their production of oil, along with highly nutritious, high oil-yielding crops, such as *Brassica napus*, *Brassica rapa*, and *Brassica juncea.* They are also known for omega-3 fatty acids like alpha-linolenic acid (ALA), which has been associated with cardiovascular benefits, and oils being high in monounsaturated fatty acids, especially oleic acid (Tian et al. 2023). Canola oil, extracted from *Brassica napus*, is one of the healthiest among all oils and contains only about 7% of saturated fat, making it one of the most preferred cooking oils in the food industry for heart-healthy uses (Rahim et al. 2023). *Brassica* oils are also good sources of vitamin E, which acts as an antioxidant improving skin conditions and lessening oxidative stress in the body (Kaur et al. 2019). The healthy fatty acid composition makes it well-suited as a stable cooking oil with a high smoke point so that harmful chemicals are not produced during heating. With more and more consumers searching for plant-based, functional health products, this cost-effective source has made a strong contribution toward its global demand for use in *Brassica* oil formulations. The increasing interest in plant-based diets has increased the demand market for *Brassica* oils, allowing easy and flexible incorporation into practically any cuisines (Afzal et al. 2022; Ivan et al. 2024). Highly processed foods, ranging from cooking oils to salad dressings, margarine, and baking products, are increasingly using *Brassica* oils. Such dietary nutrients in food not only add health benefits but also improve taste, texture, and flavor, which is why they are also added to many functional foods (Goyal et al. 2021). For instance, those foods tend to have an extra omega-3 for brain and heart benefits. More and more consumers are using *Brassica* oil because of the increasing understanding people have of the relationship that exists between nutrition and disease prevention as well as fear of the dangers posed by animal fats; this perception is mostly among those who care for their health, or are vegetarians or follow a vegan diet. Besides serving their main nutritional purpose, *Brassica* oils are growing in importance in biofuels, thereby enhancing the economic and ecological viability of these crops (Botella-Martínez et al. 2023). *Brassica* oils, particularly from canola, have been regarded as the best feedstock for biodiesel synthesis with the increasing demand for sustainable energy sources. Their chemical properties, especially high concentrations of unsaturated fatty acids within the oil, promote efficient transesterification into biodiesel with very little processing demands. Thus, *Brassica* oils are a good alternative to conventional fossil fuels, significantly reducing greenhouse gas emissions and supporting the transition to a sustainable energy model (He and Shrestha 2020; Guo et al. 2022). Growing biofuels from *Brassica* is also thought to benefit the ecological balance because they can grow in several places with little soil damage and normally need a much lower application of synthetic inputs, such as fertilizers and pesticides, compared with other oilseed crops like soy. Additionally, crops of *Brassica* fit very well in the crop rotation scheme, enhancing soil health and reducing the need for chemical inputs. The cultivation of *Brassica* oils for biodiesel feedstock therefore serves a dual purpose; it confronts energy consumption at a global level and encourages farming sustainability. Apart from their nutritional importance, *Brassica* crops have become an increasingly important commodity in oilseed production because of their contribution to the global challenges of sustainable development. All the mentioned applications, be it functional foods, diets based on plants, or the biofuel industry, show how versatile these crops are by their significant role in promoting healthier diets and sustainable energy futures. Currently, the rising demand for *Brassica* oils in multiple industries is proving that these crops are key in realizing environmentally sustainable and health-oriented practices worldwide (Larkin 2022; Brust 2019; Aslan 2023).

6 Breeding and Genetic Improvements in *Brassica* Crops

Genetic improvement and breeding strategies in *Brassica* crops have become important for enhancing disease resistance, pest management, and productivity in agriculture (Amas et al. 2023). Hybrid breeding has transformed production efficiency of *Brassica* crops through hybrid vigor, or heterosis. Hybrid cultivars have the advantages of better yield, resilience to environmental stresses, and disease resistance when compared with their parent lines. This application has greatly increased oilseed production worldwide through the development of hybrid canola cultivars (Mohd Saad et al. 2021). These hybrids can achieve a 20 to 30% higher yield than conventional varieties due to their faster growth and the ability to withstand a wide range of environmental conditions. Male sterility systems have also been developed, leading to controlled pollination as a result of the hybridization in *Brassica* crops. This method restricts self-pollination of hybrid lines while ensuring consistent seed production. Therefore, hybrid cultivars have become a critically modern aspect of *Brassica* breeding initiatives, allowing for the increased use of oil from vegetables and protein-rich animal feed globally (Gupta et al. 2024). Diseases, such as clubroot (*Plasmodiophora brassicae*) and blackleg (*Leptosphaeria maculans*), are most recurrently found in *Brassica* crops, and, as a result, *Brassica* breeding strategies are crucial in disease resistance. The field of genomics-assisted breeding has progressed significantly with the introduction of disease-resistant gene discovery for many *Brassica* species. High-quality genome references have been developed for several *Brassica* species, enabling researchers to map genes associated with resistance traits (Kopec et al. 2021; Thomas et al. 2022). Analyses of sequences from the *Brassica* genomes recently revealed the presence of various resistance gene analogs

(RGAs), indicative of large duplications found between species. Strategies can combine the genetic redundancy into the pool of potential resistance genes. An example would be breeding through marker-assisted selection (MAS) after having identified blackleg resistance genes like Rlm2 and Rlm9 (Amas et al. 2023). Moreover, genomic selection (GS) techniques allow breeders to make predictions for the performance of new hybrids based on the latter's genetic basis. Genomic and phenotypic data make it quicker for breeders to breed disease-resistant varieties along with high yield potential (Mohd Saad et al. 2021). Another important aspect of developing *Brassica* crops is the pest resistance along with a disease resistance capability. Aphids and diamondback moths are the insect pests that have a significant effect on the yield and quality. Traditional pest management techniques heavily involve the application of chemical pesticides, although there are tendencies of these methods to create pest resistance as well as environmental problems. Some obstacles have been removed in terms of direct transfer of genes which are offering pest resistance to *Brassica* crops, through genetic transformation technologies (Zhang et al. 2024). Transgenic lines producing insecticidal proteins from *Bacillus thuringiensis* (Bt) will help in pest control in a system more efficiently without threatening beneficial insects. This biotechnology method is a sustainable practice and less dependent on chemical pesticides (Kalendar et al. 2022). The tremendous growth of genetic engineering has vastly improved the efficiency of breeding programs to improve *Brassica* crops. Introducing the desired trait is mostly applied to a variety of *Brassica* species through Agrobacterium-mediated transformation (Poveda et al. 2020). This method permits accurate gene insertion without modification of the overall genomic makeup of the plant. The latest advances in CRISPR/Cas9 technology have additionally revolutionized genetic engineering by making possible targeted gene editing (Zhang et al. 2024). Using this method, scientists can insert favorable genes precisely or eliminate undesired characteristics. For example, through alteration of fatty acid profiles in canola, oil quality has been improved through this new biotechnology. This technology permits rapid progress in producing disease-resistant cultivars by focusing on specific genes associated with pathogen susceptibility. Simultaneously modifying multiple genes further increases the possibility of developing resilient cultivars that can thrive under highly varying environmental conditions (Mohd Saad et al. 2021; Chen et al. 2024). Integration of genomic resources has also eased the identification of agronomically important genes conferring abiotic stress tolerance (salinity, drought) which is vital for adapting *Brassica* crops to changing climatic conditions (Poveda et al. 2020). Breeders can also select traits that improve resilience to environmental stressors using the pangenome construction that offers a detailed account of genetic diversity within *Brassica* species. Continuous efforts in research on breeding approaches are important in ensuring that development systems of *Brassica* crops are sustainable in the future as the demand for vegetable oils continues to increase globally, notwithstanding opposing forces of insect pressures and global climate change (Thomas et al. 2024).

7 Economic and Environmental Impact

Species from the genus *Brassica* (*Brassica napus*, *Brassica rapa*, and *Brassica juncea*) comprise one of the most critical groups of oilseed crops in the global market, wherein the seeds have oil contents of 40–50%. In addition, they are one of the largest sources of multiple uses, such as biofuels and edible oils (Ayadi et al. 2022). They are initially high-yielding and increasingly oriented toward meeting global demand for vegetable oils; the adoption of farming methods and breeding strategies would profit farmers and increase their output (Murphy 2024; Bist et al. 2024). Moreover, such crops can be grown as income-generating crops with much lower investment in inputs compared to popular oilseeds like sunflowers or soybeans, and can also be grown in a huge variety of climates ranging from mild to semi-arid (Foolen-Torgerson et al. 2024; Sher et al. 2023). In general, *Brassica* improves agricultural sustainability; with their deep roots, they can improve soil structure and fertility by making nutrients more available while preventing soil erosion. This way, they play an important role in crop rotation systems. Moreover, they require less pesticide and chemical fertilizer in their production, thereby contributing to a lesser overall environmental impact from the types of farming methods (Jat et al. 2021; Singh et al., 2024; Cárceles Rodríguez et al. 2022). *Brassica* oils are feedstocks for biofuels, which in effect reduce dependence on fossil fuels, alongside providing solutions to climate change problems. The continued research on genetic improvements would ultimately lead to avenues for better environmental tolerance and further enhancement of oil generation. This puts an emphasis on the relevance of *Brassica* crops in attaining sustainable agriculture in the long run, one that marries environmental health with economic benefits (Schroeder et al. 2023; Irfan et al. 2023).

8 Challenges Facing *Brassica* Oilseed Production

The primary source of edible oil around the world is *Brassica* planting. *Brassica napus* and *Brassica juncea* constitute a major part of the oilseed industry. These plants are said to be good as they possess minimal toxins and an ample amount of antioxidants (Zandberg et al. 2022). Rising climate change effects and issues of the unstable market, pests, and disease pressure continue to influence the farming of these commodities. Studies indicate that in some climatic extremes, yields could drop due to irregularities associated with climate change, which could lead to extreme events and situations, such as drought or excessive rainfall, thus decimating crop establishment and production (Kumar et al. 2022; Clarke et al. 2022). The oilseed economy, particularly, has periodic downturns induced by high international market demand, trade regulations, and competition from other crops. Economic uncertainty then prevails on farmers, especially when price is their only saving grace (Vinnichek et al. 2019). Geopolitical wars and pandemics trigger supply chain disruptions that interfere with the availability of strategic agricultural inputs, such as seeds and fertilizers, thus aggravating their production woes. Throughout almost the entire developmental phase, pests and diseases are highly hazardous to *Brassica* crops. Big players are the pathogens like *Plasmodiophora brassicae* (clubroot) and *Leptosphaeria maculans* (blackleg), which can cause great reductions in crop yield if no control measures are put in place (Geoffrey 2024). Therefore, the integrated pest management (IPM) solution should be the way to resolve these issues, but this usually faces problems from pests developing resistance to chemical control (Deguine et al. 2021). Hence, there must be cooperation in research, policymaking, and implementation of sustainable agricultural practices to make a crop return to both food and biofuel markets in response to the challenges it confronts.

9 Future Directions and Innovations in the *Brassica* Oilseed Industry

The fact that *Brassica* crops produce a lot of edible oil, especially major species like *Brassica napus* and *Brassica juncea,* shows how important they are to the oilseed industry. These oilseeds are clearly important in agriculture as they hold the second place in total production and control edible oil output in India. The 72 million tons of oilseed *Brassica* produced worldwide over 35 million hectares demonstrate its economic importance. Precision agriculture and big data analytics are revolutionizing *Brassica* farming with efficiency in resource use and increase in yields; hence, decisions on irrigation and pest control can be made. The adaptability of *Brassica* oils is increasingly creating new avenues of utilization in cosmetics and medicines, thereby providing new market opportunities. High-yielding, disease-resistant cultivars are needed to meet the rising demand for edible oils and biofuels in expanding markets and help farmers adapt to climate change. Growth in the *Brassica* oilseed industry will simultaneously improve food security and financial stability.

10 Conclusion

In conclusion, the *Brassica* crop assumes an important position in the global oilseed industry since it is the one that is more in demand through the application of various vegetable oils and biofuels and their numerous applications. *Brassica napus, Brassica rapa, Brassica juncea,* and *Brassica carinata* species of *Brassica* offer totally different, unique nutritional and oil characteristics in the market. Currently, there is a transition in the industry due to a number of consumers' preferences toward healthier, more sustainable options, plant-based diets, differential geopolitical dynamics, and many others. The effects of all these factors, climate change, and trade policies call for flexible, resilient, and adaptable agricultural technologies. Although conventional oil extraction methods are still widely utilized, modern techniques are definitely very much needed today in order to satisfy the rising demand but also enhance the quality of oil production. Thus, while the oilseed market is growing, especially among the emerging economies, continuing innovation and understanding of these complex and interacting forces will be essential for stakeholders to ensure food security and capitalize on opportunities that will be there for long-term growth in the *Brassica* oilseed sector.

References

Abid R, Islam M, Saeed H, Ahmad A, Imtiaz F, Yasmeen A, Rathore HA (2022) Antihypertensive potential of *Brassica rapa* leaves: an *in vitro* and *in silico* approach. Front Pharmacol 13:996755

Afzal MF, Khalid W, Armghan Khalid M, Zubair M, Akram S, Kauser S, Al-Farga A (2022) Recent industrials extraction of plants seeds oil used in the development of functional food products: a review. Int J Food Prop 25(1):2530–2550

Ahangari H, King JW, Ehsani A, Yousefi M (2021) Supercritical fluid extraction of seed oils–a short review of current trends. Trends Food Sci Technol 111:249–260

Alongi M, Anese M (2021) Re-thinking functional food development through a holistic approach. J Funct Foods 81:104466

Amas JC, Thomas WJW, Zhang Y, Edwards D, Batley J (2023) Key advances in the new era of genomics-assisted disease resistance improvement of *Brassica* species. Phytopath 113(5):771–785

Ancuţa P, Sonia A (2020) Oil press-cakes and meals valorization through circular economy approaches: a review. App Sci 10(21):7432

Aslan V (2023) An overview of biodiesel produced from 2nd generation feedstock: mustard seed types. Bioenergy Res 16(3):1380–1400

Ayadi J, Debouba M, Rahmani R, Bouajila J (2022) Brassica genus seeds: a review on phytochemical screening and pharmacological properties. Molecules 27(18):6008

Bashyal M, Mulvaney MJ, Lee D, Wilson C, Iboyi JE, Leon RG, Boote KJ (2021) Brassica carinata biomass, yield, and seed chemical composition response to nitrogen rates and timing on southern coastal plain soils in the United States. GCB Bioenergy 13(8):1275–1289

Batool U, Nawaz R, Ahmad S, Irshad MA, Irfan A, Gaafar ARZ, Bourhia M (2024) Physico-and phytochemical properties of Brassica juncea as affected by agroclimatic conditions. Sci Rep 14(1):797

Beszterda M, Nogala-Kałucka M (2019) Current research developments on the processing and improvement of the nutritional quality of rapeseed (*Brassica napus* L.). Eur J Lipid Sci Technol 121(5):1800045

Bist RB, Bist K, Poudel S, Subedi D, Yang X, Paneru B, Chai L (2024) Sustainable poultry farming practices: a critical review of current strategies and future prospects. Poultry Sci:104295

Borges CE, dos Santos V, Veloso R, Da Conceição CA, Mendes DS, Ramirez-Cabral NY, Shabani F, Da Silva RS (2023) Forecasting Brassica napus production under climate change with a mechanistic species distribution model. Sci Rep 13(1):12656

Botella-Martínez C, Pérez-Álvarez JÁ, Sayas-Barberá E, Rodríguez N, de Vera C, Fernández-López J, Viuda-Martos M (2023) Healthier oils: a new scope in the development of functional meat and dairy products: a review. Biomol 13(5):778

Braun J (2018) Bioeconomy–the global trend and its implications for sustainability and food security. Global Food Sec 19:81–83

Brust GE (2019) Management strategies for organic vegetable fertility. In: *Safety and practice for organic food*. Academic Press, pp 193–212

Cárceles Rodríguez B, Durán-Zuazo VH, Soriano Rodríguez M, García-Tejero IF, Gálvez Ruiz B, Cuadros Tavira S (2022) Conservation agriculture as a sustainable system for soil health: a review. Soil Sys 6(4):87

Carré P, Citeau M, Robin G, Estorges M (2016) Hull content and chemical composition of whole seeds, hulls and germs in cultivars of rapeseed (*Brassica napus*). OCL 23(3):302

Chen F, Chen L, Yan Z, Xu J, Feng L, He N, Liu C (2024) Recent advances of CRISPR-based genome editing for enhancing staple crops. Front Plant Sci 15:1478398

Clarke B, Otto F, Stuart-Smith R, Harrington L (2022) Extreme weather impacts of climate change: an attribution perspective. Envir Res Clim 1(1):012001

Cravotto C, Fabiano-Tixier AS, Claux O, Abert-Vian M, Tabasso S, Cravotto G, Chemat F (2022) Towards substitution of hexane as extraction solvent of food products and ingredients with no regrets. Foods 11(21):3412

Cravotto C, Claux O, Bartier M, Fabiano-Tixier AS, Tabasso S (2023) Leading edge technologies and perspectives in industrial oilseed extraction. Molecules 28(16):5973

Czwartkowski K, Wierzbic A, Golimowski W (2022) Quality, key production factors, and consumption volume of niche edible oils marketed in the European Union. Sustainability 14(3):1846

Deguine JP, Aubertot JN, Flor RJ, Lescourret F, Wyckhuys KA, Ratnadass A (2021) Integrated pest management: good intentions, hard realities. A review. Agro Sustain Dev 41(3):38

Dobrzyńska MA, Przysławski J (2020) The effect of camelina oil (α-linolenic acid) and canola oil (oleic acid) on lipid profile, blood pressure, and anthropometric parameters in postmenopausal women. Archives Med Sci 17(6):1566

FAOSTAT (2025). https://www.fao.org/faostat/en/#data/QCL

Foolen-Torgerson KL, Meijering JV, van Voorn GAK (2024) Estimating farmers' net change in profit when using insect frass as an input for brassica crops. J Insects Food Feed 11(2):373–399

Gagour J, Ahmed MN, Bouzid HA, Oubannin S, Bijla L, Ibourki M, Hajib A, Koubachi J, Harhar H, Gharby S (2022, 2022) Proximate composition, physicochemical, and lipids profiling and elemental profiling of rapeseed (*Brassica napus* L.) and sunflower (*Helianthus annuus* L.) grown in Morocco. eCAM:3505943

Geoffrey RDV (2024) Pests and pathogens. Veg Brassicas Related Crucifers 40:295

Goyal A, Tanwar B, Sihag MK, Kumar V, Sharma V, Soni S (2021) Rapeseed/canola (*Brassica napus*) seed. Health Attributes and Food Applications, Oilseeds, pp 47–71

Guo C, Bai Z, Wang X, Zhang W, Chen X, Lakshmanan P, Chen X (2022) Spatio-temporal assessment of greenhouse gas emission from rapeseed production in China by coupling nutrient flows model with LCA approach. Food Energy Sec 11(3):398

Gupta S, Kumar A, Janeja HS, Prakash A, Anand R (2024) Genetic engineering in Indian mustard (Brassica juncea L.): current Progress and future directions for enhanced crop improvement. J Advan Bio Biotech 27(5):739–751

He BB, Shrestha D (2020) Production of biodiesel from oilseeds: canola/rapeseed. In: Achieving carbon negative bioenergy systems from plant materials. Burleigh Dodds Science Publishing, pp 217–242

Irfan M, Xianhua L, Shauket A, Jafir M, Ahmad A, Ahmad SJN, Ahmad JN (2023) Promoting energy crops to replace fossil fuel use. In: Strategizing agricultural Management for Climate Change Mitigation and Adaptation. Springer International Publishing, Cham, pp 69–87

Ivan I, Posan P, Marin M, Bahaciu GV, Bololoi IȘ, Nicolae CG (2024) Exploring unconventional plant-based ingredients and their influence on sustainable trout farming practices

Jat RS, Singh VV, Sharma P, Rai PK (2019) Oilseed brassica in India: demand, supply, policy perspective and future potential. OCL 26:8

Jat RS, Choudhary RL, Singh HV, Meena MK, Singh VV, Rai PK (2021) Sustainability, productivity, profitability and soil health with conservation agriculture based sustainable intensification of oilseed brassica production system. Sci Rep 11(1):13366

Kalendar R, Orbovic V, Egea-Cortines M, Song GQ (2022) Recent advances in plant genetic engineering and innovative applications. Front Plant Sci 13:1045417

Kaur R, Sharma AK, Rani R, Mawlong I, Rai PK (2019) Medicinal qualities of mustard oil and its role in human health against chronic diseases: a review. Asian J Dairy Food Res 38(2):98–104

Kaur M, Sharma HK, Kumar N (2022) Processing of oilseeds 12. *Agro-processing and food*. Engineering 483

Kephe PN, Ayisi KK, Petja BM, Mulaudzi AP, Mabitsela KE (2020) Factors influencing the production of oilseed crops among smallholder farmers in Limpopo Province. OCL 27:41

Kopec PM, Mikolajczyk K, Jajor E, Perek A, Nowakowska J, Obermeier C, Karlowski WM (2021) Local duplication of TIR-NBS-LRR gene marks clubroot resistance in Brassica napus cv. Tosca. Front Plant Sci 12:639631

Kumar L, Chhogyel N, Gopalakrishnan T, Hasan MK, Jayasinghe SL, Kariyawasam CS, Ratnayake S (2022) Climate change and future of Agri-food production. In: Future foods. Academic Press, pp 49–79

Kwon HY, Choi SI, Park HI, Choi SH, Sim WS, Yeo JH, Cho JH, Lee OH (2020) Comparative analysis of the nutritional components and antioxidant activities of different *Brassica juncea* cultivars. Foods 9(6):840
Larkin RP (2022) Long-term effects of compost amendments and brassica green manures in potato cropping systems on soil and crop health and productivity. Agronomy 12(11):2804
Licata M, Farruggia D, Di Miceli G, Salamone F, Iacuzzi N, Tuttolomondo T (2024) Productivity of two brassica oilseed crops in a Mediterranean environment and assessment of the qualitative characteristics of raw materials for bioenergy purposes. Heliyon 10(5):e26818
MacIntosh SC, Shaw M, Connelly M, Yao ZJ (2021) Food and feed safety of NS-B5ØØ27-4 omega-3 canola (*Brassica napus*): a new source of long-chain omega-3 fatty acids. Front Nut 8:716659
McVetty PB, Mietkiewska E, Omonov T, Curtis J, Taylor DC, Weselake RJ (2016) Brassica spp. oils. In: *Industrial oil crops*. AOCS Press, pp 113–156
Mohd Saad NS, Severn-Ellis AA, Pradhan A, Edwards D, Batley J (2021) Genomics armed with diversity leads the way in brassica improvement in a changing global environment. Front Gen 12:600789
Murphy DJ (2024) Sustainable vegetable oils. In: Functional dietary lipids. Woodhead Publishing, pp 31–47
Nde DB, Foncha AC (2020) Optimization methods for the extraction of vegetable oils: a review. PRO 8(2):209
Poudel P, Petropoulos SA, Di Gioia F (2023) Plant tocopherols and phytosterols and their bioactive properties. In: Natural secondary metabolites: from nature, through science, to industry. Springer, Cham, pp 285–319
Poveda J, Francisco M, Cartea ME, Velasco P (2020) Development of transgenic *brassica* crops against biotic stresses caused by pathogens and arthropod pests. Plants 9(12):1664
Raboanatahiry N, Li H, Yu L, Li M (2021) Rapeseed (Brassica napus): processing, utilization, and genetic improvement. Agronomy 11(9):1776
Rahim MA, Ayub H, Sehrish A, Ambreen S, Khan FA, Itrat N, Rocha JM (2023) Essential components from plant source oils: a review on extraction, detection, identification, and quantification. Molecules 28(19):6881
Rahman M, Khatun A, Liu L, Barkla BJ (2024) Brassicaceae mustards: phytochemical constituents, pharmacological effects, and mechanisms of action against human disease. Int J Mol Sci 25(16):9039
Rai PK, Yadav P, Kumar A, Sharma A, Kumar V, Rai P (2022) *Brassica juncea*: a crop for food and health. In: The Brassica juncea genome. Springer, Cham, pp 1–13
Ramirez D, Abellán-Victorio A, Beretta V, Camargo A, Moreno DA (2020) Functional ingredients from Brassicaceae species: overview and perspectives. Int J Mol Sci 21(6):1998
Rani H, Sharma S, Bala M (2021) Technologies for extraction of oil from oilseeds and other plant sources in retrospect and prospects: a review. J Food Pro Eng 44(11):13851
Redda ZT, Laß-Seyoum A, Yimam A, Barz M, Jabasingh SA (2024) Characterization of hexane-defatted Brassica carinata oilseed meals to explore their potential for valorization towards a sustainable circular bioeconomy. Waste Biomass Valor 15(2):1185–1197
Saldukaitė L, Šarauskis E, Zabrodskyi A, Adamavičienė A, Buragienė S, Kriaučiūnienė Z, Savickas D (2022) Assessment of energy saving and GHG reduction of winter oilseed rape production using sustainable strip tillage and direct sowing in three tillage technologies. Sustain Energy Tech Assess 51:101911
Schroeder K, Davis JB, Xiao F, Zhang Z (2023) Developing non-food grade brassica biofuel feedstock cultivars with high yield, oil content, and oil quality that are suitable for low input production dryland systems (No. DOE DE-SC0016366). University of Iowa, Iowa City
Serrano C, Oliveira MC, Lopes VR, Soares A, Molina AK, Paschoalinotto BH, Barata AM (2024) Chemical profile and biological activities of Brassica rapa and Brassica napus ex situ collection from Portugal. Foods 13(8):1164
Shekhawat PK, Goyal A, Akhatar J, Sharma S, Kaur N, Bharti B, Atri C (2023) Genetic analysis of the variation for mineral accumulation in the leaves and seeds of natural germplasm of Brassica

rapa L.(AA) and the its derived forms extracted from an allotetraploid B. juncea L.(AABB). Plant Physiol Biochem 203:108084
Shen J, Liu Y, Wang X, Bai J, Lin L, Luo F, Zhong H (2023) A comprehensive review of health-benefiting components in rapeseed oil. Nutrients 15(4):999
Sher A, Allah SU, Sattar A, Ijaz M, Qayyum A, Manaf A, Sarwar B (2023) Agronomic seed production and oil quality evaluation of various *Brassica* species grown under semi-arid climatic condition. Turkish J Field Crops 28(1):57–63
Singh NK, Sachan K, Manoj BP, Panotra N, Katiyar D (2024) Building soil health and fertility through organic amendments and practices: a review. Asian J Soil Sci Plant Nut 10(1):175–197
Sun Q, Shi J, Scanlon M, Xue SJ, Lu J (2021) Optimization of supercritical-CO2 process for extraction of tocopherol-rich oil from canola seeds. Lwt 145:111435
Tass MA, Bhat AA, Riyaz I, Hakim IA (2024) Russia-Ukraine war and its impact on global oil and food prices. J Advan Zoology 45:1145–1151
Thomas WJ, Amas JC, Wu T, Cantila AY, Ton LB, Batley J (2022) The impact of genomics in identifying blackleg disease resistance genes in canola. In: *Next-generation sequencing and agriculture*. CABI, GB, pp 24–38
Thomas WJ, Amas JC, Dolatabadian A, Huang S, Zhang F, Zandberg JD, Batley J (2024) Recent advances in the improvement of genetic resistance against disease in vegetable crops. Plant Physiol 196(1):32–46
Tian M, Bai Y, Tian H, Zhao X (2023) The chemical composition and health-promoting benefits of vegetable oils—a review. Molecules 28(17):6393
Tileuberdi N, Turgumbayeva A, Yeskaliyeva B, Sarsenova L, Issayeva R (2022) Extraction, isolation of bioactive compounds and therapeutic potential of rapeseed (*Brassica napus* L.). Molecules 27(24):8824
Usman M, Nakagawa M, Cheng S (2023) Emerging trends in green extraction techniques for bioactive natural products. PRO 11(12):3444
Vinnichek L, Pogorelova E, Dergunov A (2019) Oilseed market: Global trends. In: IOP conference series: earth and environmental science (Vol. 274, No. 1). IOP Publishing, p 012030
Wandhekar S, Pawar V, Shinde S, Gangakhedkar P (2023) Extraction of oil from oilseeds by cold pressing: a review. Indian Food Ind Mag 4:63–69
Wang J (2016) Analysis and comparison of the factors influencing worldwide four kinds of vegetable oil trade: based on gravity model. Mod Econ 7(02):173–182
Weselake RJ, Fell DA, Wang X, Scofield S, Chen G, Harwood JL (2024) Increasing oil content in brassica oilseed species. Progress Lipid Res:101306
Willer H, Travnicek J, Schlatter B (2020) Current status of organic oilseeds worldwide– statistical update. OCL 27:62
Yadav S, Teng PY, Choi J, Singh AK, Kim WK (2022) Nutrient profile and effects of carinata meal as alternative feed ingredient on broiler performance, tight junction gene expression and intestinal morphology. Poultry Sci 101(2):101411
Yaheliuk S, Fomych M, Rechun O (2024) Global market trends of grain and industrial crops. For Comm Bull 1(17):134–145
Zandberg JD, Fernandez CT, Danilevicz MF, Thomas WJW, Edwards D, Batley J (2022) The global assessment of oilseed brassica crop species yield, yield stability and the underlying genetics. Plants 11(20):2740
Zhang L, Meng S, Liu Y, Han F, Xu T, Zhao Z, Li Z (2024) Advances in and perspectives on transgenic technology and CRISPR-Cas9 gene editing in broccoli. Genes 15(6):668
Zhang J, Zhang M, Chen K, Deng D (2025) Improvement strategies for fats and oils used in future food processing based on health orientation and sustainability: research progress, challenges and solutions. Crit Rev Food Sci Nut 65(1):47–63

Origin and Importance of *Brassica* Species

Akshay Kumar Vats, Lokender Kashyap, Anchal Sharma, Anjana Kumari, and Aditi Sharma

Abstract

Brassica as Chinese *shijing* was mentioned in Ancient Indian texts, viz. Upanishads and Brahmanas around 1500 BCE. Around 3000 BCE, Chinese texts were evident in their documentation describing the diversity of *Brassica* species. It is also described by Linnaeus in 1735 in *Species Plantarum,* indicating its significant contribution to the Kingdom Plantae. He discussed various species of *Brassica* such as coles, rapes, cabbages, and mustards. *Brassica* offers a variety of its forms as food, even its leaves and fruits are eaten fresh, processed or cooked as food, as fodder, and feed. The journey of *Brassica* is fascinating, witnessing stable hybrids raised from crossing diverse genotypes and intermating populations. In its breeding journey, *Brassica* has undergone hybridization, selection, genetic gain and loss, speciation, genome duplication, as well as polyploidization many times. Most renowned scientists, namely Aristotle, Olumella, Pliny, Confucius, Linnaeus, Otto Eugen, Schultz, and Korean Botanist Woo-Jang-Choon

A. K. Vats
Department of Plant Breeding and Genetics, Dr. Kalam Agricultural College, Arrabari, Bihar Agricultural University, Bhagalpur, Bihar, Kishanganj, India

L. Kashyap (✉)
Faculty of Agriculture, Guru Kashi University, Talwandi Sabo, Bathinda, Punjab, India

A. Sharma
Department of Mathematics and Humanities, Maharishi Markandeshwar (Deemed to be) University, Mullana, Ambala, India

A. Kumari
Department of Applied Sciences and Humanities, Ambala College of Engineering and Applied Research, Haryana, India

A. Sharma
College of Horticulture & Forestry, Dr Y S Parmar University of Horticulture & Forestry, Nauni, Solan, Himachal Pradesh, India

A. Kumar, B. Koul (eds.), *Crop Improvement Strategies in Brassica species: Applied Science*, https://doi.org/10.1007/978-981-95-3861-4_3

(Popularly known in scientific texts as Nagaharu U), enlightened evolution and origin of *Brassica* species ahead of their times. This chapter summarizes major reports and evolutionary pathways of *Brassica* species and its known wild relatives in a figurative way to understand.

Keywords

Brassica · Origin and diversity · Evolution of *Brassica* · Phylogeny of *Brassica* · Hybridization in *Brassica*

1 Origin and History

Brassica genera belongs to the Family: Brassicaceae, Tribe: Brassiceae, which comprises around 372 genera and around 4060 species. This is commonly called as "The Cabbage family," "The Mustards," or "The Crucifers" (Gomez-Campo 1980). Every part of the plant has its own economic importance, edible roots, leaves, buds, stem, seeds, and flowers. The *Brassica* family is a potential source of condiments and oil products. The source of cytoplasmic male sterility genes for hybrid seed production in *Brassica* is mainly wild relatives (Branca et al. 2012). Wild relatives in the *Brassica* family are also a source of resistant genes against diseases and pests of economic importance (Brullo et al. 2013). Mucilaginous seeds (e.g., some species of *Sinapis*), hairy cotyledons (e.g., some species of *Sinapis*), and glabrous parts of the plant (e.g, in Moricandia) are some of the visible characters that maintains variability among different species of the *Brassica* family (Han et al. 2021). Some *Sinapis* species with having chromosome number 9 have no mucilaginous seeds in comparison to some *Sinapis* species with a chromosome number 12 (Sarilar et al. 2013). Glabrous *Moricandia* species are more prone to aphids when compared to hairy types. Similarly, there is lot of variability in color of mature seeds (yellow to black) and flower shape in different species (white to violet) (Al-Shehbaz 1985).

2 Distribution and Taxonomy of *Brassica*

2.1 *Brassica nigra* (L.) Koch ($n = 8$)

Brassica nigra is widely cultivated in the Mediterranean region. It has a chromosome number ($n = 8$) (Tsunoda 1980). It is considered to have evolved from *Sinapis arvensis* (closest progenitor). Mostly it is found in the roadside fields of Morocco, Rhodes, Ethiopia, Turkey, Sicily, and Crete cities. Generally, plants of *B. nigra* reach 2 m in height. It does not need any vernalization to induce flower initiation.

2.2 *Brassica oleracea* L. ($n = 9$)

This species has a chromosome number (n = 9). Diversity of *B. oleracea* is found around the southern part of Greece, South southwestern part of Turkey, Italy, Yugoslavia, and Kriti area (Snogerup 1980). It appears woody and is perennial in its life cycle. It has two subspecies: *B. villosa* ssp. *cretica* and *B. rupestris* ssp. *incana.* Some endemic local forms found in western Italy include *B. macrocarpa* Guss. *B. insularismoris,* earlier popularly categorized as *B. atlantica* in Tunisia, and have white petals which differ from *B. villosa* genetically. *B. Montana* Pourr. grouped under *B. oleracea* is perennial in growth habit and found in the northern parts of Spain, Italy, and Southern France coastal regions (Farnham et al. 2008). *B. olearacea* L. is found in South and Southwestern Britain and Western France coasts (Giles 1941). It is also perennial in nature. *Brassica hilarionis* Holm is popularly and acceptably known as *Coincya monensis,* a pink-flowered species of *Brassica* found in the Kyrenia mountains. However, *B. oleracea* is a highly diverse species and may be subdivided into many subgroups that maintain unique flavor, use, and characteristics, for example:

1. Cabbage group (*B. olearacea* var. *capitata*): It includes green cabbage, red cabbage, savoy cabbage, Napa cabbage, etc.
2. Broccoli group (*B. oleracea* var. *italica*): It includes calabrese broccoli, sprouting broccoli, purple broccoli, white broccoli, etc.
3. Cauliflower group (*B. olearacea* var. *botrytis*): It includes white cauliflower, purple cauliflower, green cauliflower, orangecauliflower, etc.
4. Kale group (*B. olearacea* var. *sabellica*): It includes curly kale, lacinato or dinosaur kale, red Russian kale, Siberian kale, etc.
5. Collard greens group (*B. olearacea* var. *viridis*): It includes Georgia collard greens, Texas collard greens, champion collard greens.
6. Kohlrabi group (*B. olearacea* var. gongylodes): It includes green kohlrabi, purple kohlrabi, white kohlrabi, etc.
7. Brussles sprouts group (*B. olearacea* var. *gemmifera*): It includes Long Island improved brussels sprouts, Jade cross brussels sprouts, Falstaff brussels sprouts, etc.
8. Gai Lan group (*B. olearacea* var. *alboglabra*): It includes Chinese broccoli, Gai Lan, and Chinese kale, etc.
9. Romanesco group (*B. olearacea* var. *romanesca*): It includes Romanesco broccoli, Romanesco cauliflower.

Different theories regarding wild ancestors/progenitors of *B. oleracea* have been proposed from time to time. According to one widely accepted theory, *B. oleracea* originated in *B. montana* (Snogerup 1980). However, key progenitors in the case of *B. oleracea* are *B. incana* (progenitor of some *Brassica* cultivars like broccoli and cauliflower), *B. rupestris* (Progenitor of kale and collard greens), and *B. villosa* (considered to be the progenitor of cabbage and savoy cabbage) (Snogerup 1980).

2.3 *Brassica rapa* L. (*n* = 10)

Highly diverse species and is considered native to Europe and Asia. It is cool-season crop and is often cultivated in fall or early spring. It may tolerate light frosts. It is synonymous with *B. campestris* which was the older name of *B. rapa*. It is believed that *B. rapa* moved from Mongolia to China to Japan. Its origin is considered to be the Mediterranean region (Tsunoda 1980). It grows well under low temperature conditions. It is cultivated as an oilseed crop in India, Sweden, Canada, and Finland. Being a highly diverse species, it can also be subgrouped under seven groups as detailed below:

1. Broccoli raab (*Brassica rapa* var. *rapa*): This group is very similar in its characteristics to the Broccoli group, rather it has smaller florets and is stronger in flavor.
2. Turnip (*Brassica rapa* var. *rapa*): It is grown as a root vegetable and is mostly used in salads, soups, and stews.
3. Rutabaga (*Brassica rapa* var. *napobrassica*): It is a fertile cross between turnip and cabbage and is used in stews, soups, and mash.
4. Bok choy (*Brassica rapa* var. *chinensis*): Its origin is China and is a type of Chinese cabbage. It is mostly used in salads and in fried form.
5. Napa cabbage (*Brassica rapa* var. *pekinensis*): Its origin is in China and is used in salads and to make kimchi.
6. Komatsuna (*Brassica rapa* var. *perviridis*): It originated in Japan. It is a type of Japanese mustard spinach and is mostly used in salads and is usually fried.
7. Tatsoi (*Brassica rapa* var. *narinosa*): It is a highly cold-tolerant species and is a type of Chinese cabbage. It is also used as a fried dish or mostly in the form of salads. It is also of Chinese origin.

2.4 Brassica carinata A. Braun (*n* = 17)

It is self-incompatible and amphidiploid species (n = 17), which is considered to be developed from the interspecific cross between *B. nigra* (n = 8) and *B. oleracea* (n = 9) (Geraci et al. 2001). It originated in Ethiopia and Eritrea and is also called Ethiopian mustard. It is lower in yields when compared to other *Brassica* species. However, it is high in oil content, hence used as a producer of vegetable oils and biofuels. It is susceptible to diseases like downy mildew and white rust (Gómez-Campo 1999). It is a drought and a frost-tolerant crop. It has high nutritional value, i.e., high fiber content, high protein content, high in vitamins, etc. It is mostly restricted to Ethiopia and Eritrea; therefore, it is not well adapted to other environments. It is consumed as food (leafy vegetables), feed (forage purpose), and for oil (seeds with high oil content) in Ethiopia and Eritrea.

2.5 *Brassica juncea* (L.) Czern & Coss (*n* = 18)

It is self-incompatible and amphidiploid species (*n* = 18), which is considered to have been developed from the interspecific cross between *B. nigra* (*n* = 8) and *B. rapa* (*n* = 10). It is believed to have originated in the Near Eastern region and Southern Iran, because a great diversity of their wild forms (i.e., *B. nigra* and *B. rapa*) are found in this region. It is cultivated as an oilseed crop in India (also known as Indian Brown Mustard) and Ukraine, and as a leafy vegetable in China (Gustafsson and Lannér-Herrera 1997). Its root type forms are also cultivated in China. Seeds of *B. juncea* are brown in color in the Indian environment and yellow in China. Indian brown mustard contains 3-butenyl glucosinolates in its seeds, while Chinese mustard has only 2-propenyl (allyl) glucosinolates (Zhu et al. 2023). Chinese mustard types have trace quantities of 3-butenyl glucosinolates. *B. juncea* is also cultivated for condiment purposes in Western countries like Western Canada. However, it is high in oil content, hence used as a producer of vegetable oils and biofuels (Gustafsson 1982). It is susceptible to diseases like downy mildew and white rust. It is a drought and a frost-tolerant crop (Gómez-Campo et al. 2005).

2.6 *Brassica napus* L. (*n* = 19)

It is self-incompatible and amphidiploid species (*n* = 19), which is considered to have developed from the interspecific cross between *B. oleracea* (*n* = 9) and *B. rapa* (*n* = 10). Most wild forms of *B. napus* L. (viz. subsp. *napus*, *rapifera* and *oleifiera*) are found in the Mediterranean region, Europe, and Asian region (Tsunoda 1980). Beaches of Gothland, coastal regions of Sweden, New Zealand, and the Netherlands dominate in diversity in wild forms of *B. napus* L. *B. napus* is high yielding and contains high oil content, that is why it is cultivated as an oilseed species in different parts of the world (Hodgkin 1995). Root type (viz. Rutabaga) and fodder type *B. napus* genotypes have also been reported.

3 Rapeseed, Mustard, and Raya

Rapeseed mustard and Raya are related to the *Brassica* genus, which includes various crops used for oilseed, food, vegetables, and other purposes. All three are cool-season crops.

Rapeseed: It is also known as oilseed rape (Liu et al. 2014). It is cultivated for oil purposes, as rapeseed is oil rich. The flowers of rapeseed are bright yellow in color, when compared to Mustard and Raya. It grows up to height of (90 to 150 cm) (Crisp 1982). Rapeseed is widely cultivated as an annual plant in different parts of Europe, Asia, and North America for its oil-rich seed used in making vegetable oils, animal feed, and biodiesel (Boukema and van Hintum 1999). In Canada (Saskatchewan and Alberta), rapeseed genotypes are canola type (i.e., having less than 2% erucic acid and low glucosinolates). High oleic and low linolenic acid is desirable in rapeseed genotypes called "HOLL" type genotypes (Kiefer et al. 2013). A HOLL type canola variety (highly nutritional) is a major breeding objective in breeding *Brassica napus* genotypes in most countries (Kaur et al. 2014).

Mustard: It includes species like brown mustard (*B. juncea*), black mustard (*B. nigra*), and white mustard (*Sinapis alba*) (Maggioni et al. 2013). Seeds and leafy parts of mustard are mainly used as spices and condiments. It grows up to the height of 60 to 180 cm. It is widely cultivated for seed or leafy vegetable purposes (Massie et al. 1996).

Raya: It includes *B. Juncea* species, cool-season crop grown in India and south southeastern parts of Asia (Vavilov 1926). It is usually grown for seed purposes, spicy condiments, in curries, marinades, sauces, chutneys, pickles, salads and dressings. It is also called Brown Indian Mustard (Zamani-Nour et al. 2013). It may grow up to 5 feet (90–150 cm). It may also be consumed as a leafy vegetable in daily use.

4 The "Triangle of U"

Korean Botanist Woo-Jang-Choon (popularly known as Nagaharu, U) beautifully described the hybridization that took place in three *Brassica* species, i.e., (*B. nigra, B. olearacea,* and *B. rapa*) to form self-incompatible amphidiploid species i.e., (*B. carinata, B. juncea,* and *B. napus*) (Fig. 1) (Table 1) (Nagaharu and Nagaharu 1935). He deduced that *B. nigra* (Black mustard) originated in the Mediterranean region. *B. oleracea* is known to be the progenitor of European cole vegetables. It is domesticated in Mediterranean coasts through the movement of people and cultures (Maggioni 2015). *B. rapa* is considered as progenitor of Asian *Brassica* vegetables (De Candolle 1824, 1885). It is believed to have originated in the high plateau regions of Iran, Iraq, and Turkey. *B. rapa* is known to withstand high temperatures, desert climates. Natural crossing among close wild relatives and wild species of *Brassica* has undergone polyploidization to develop amphidiploid species (Guo et al. 2014). It made base building blocks for today's cultivated species of *Brassica*.

5 Adaptability of *Brassica* species

Kale and carinata types of *Brassica* species are the types that thrive best in cool environments (Borgen et al. 1979; Bothmer et al. 1995). However, *B. carinata* (the Abyssinian or Ethiopian types) are mostly heat or drought-tolerant types. *Brassica juncea* can be subcategorized into subspecies (Guo et al. 2014)

(a) *Juncea*: i.e., seed mustard types, which are utilized in the form of condiments and oilseeds.
(b) *Integrifolia* type, i.e., leaf mustard types: which is utilized because of variation in their leaf morphology.

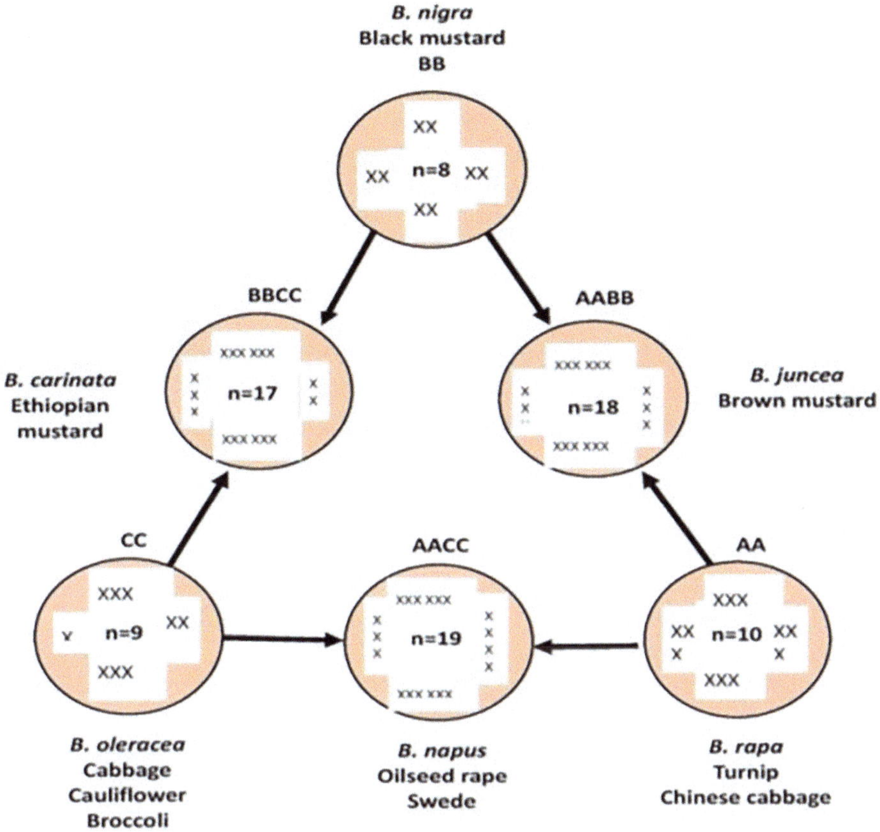

Fig. 1 Evolutionary relationships between diploid and amphidiploid *Brassica* species. (The 'Triangle of U': Nagaharu Nagaharu and Nagaharu 1935)

Table 1 Genome size of various *Brassica* species

Brassica species	Chromosome number	Genome size	Types
Brassica nigra (BB)	$n = 8$	515 Mb	Black mustard
Brassica oleracea (CC)	$n = 9$	490 Mb	Cabbage, cauliflower, broccoli
Brassica rapa (AA)	$n = 10$	350 Mb	Turnip, Chinese cabbage
Brassica carinata (BBCC)	$n = 17$	300 Mb	Ethiopian mustard
Brassica juncea (AABB)	$n = 18$	930 Mb	Brown mustard
Brassica napus (AACC)	$n = 19$	1130 Mb	Oilseed rape, Swede rape

Nagaharu and Nagaharu (1935)

(c) *Napiformis* type, i.e., root forms, which is utilized for its diverse root forms.
(d) *Tumida* type, i.e., stem types, which is used for its edible stem.

Similarly, *B. napus* has three subspecies

(a) *B. napus* ssp. *oleifera*: Oilseed rape.
(b) *B. napus* ssp. *rapifera*: Rutabaga or Swede rape.
(c) *B. napus* ssp. *pabularia*: Leaf rape or Siberian Kale type.

It is believed that "A" subgenome found in *B. napus* evolved from its ancestor common to the ancestor of European Turnip, and that of "C" subgenome evolved from the common ancestor of Kohlrabi, Chinese Kale, cauliflower, and broccoli (Song et al. 2024).

6 Wild Relatives of *Brassica* Species

Various species with their distribution throughout the world are listed as wild relatives of *Brassica* species (Table 2) (De candolle 1821; Linnaeus 1753; de Tournefort 2012). It is believed that these wild relatives were naturally hybridized to form amphidiploid stable species in *Brassica*, which are mostly domesticated in different parts of the world (Chalhoub et al. 2014). These species include *Brassica fruticulosa* ssp. *cossoniana* and ssp. *fruticulose* having haploid chromosomes as $n = 8$. These species are distributed among pine trees of Mediterranean coastal areas and also moved to the Middle Atlas of Morocco (Baillon 1871; Schulz 1919; Branca et al. 2012). Similarly, *B. maurorum* and *spinescens,* having a chromosome number ($n = 8$), are considered to have originated in the Northern parts of Africa. Species *maurorum* is also extended in Morocco and Algeria lands. Species *oxyrrhina* and *barrelieri* are also considered to have originated in Moroccan region, Iberian region, Southern Portugal, and Spain. Species *tournefortii* is also believed to have originated in the Mediterranean coasts, as well as India (de Tournefort 2012). *Brassica gravinae* originated in Algeria. *Brassica elongata* is well adapted to south-Eastern lands of Europe and Iran (Arias and Pires 2012). *Brassica amplexicaulis* is also believed to be found in the Algerian region (Arias et al. 2014).

7 Hypothesis About Evolution of *Brassica* Species

Based on few reports (Kim et al. 2023; Coves et al. 2023; Beretta et al. 2021, and Gepts 2023) and analysis of chloroplast, mitochondrial and nuclear DNA clones of artificially synthesized analogues of *B. napus, juncea,* and *carinata,* here are some inferences drawn about genetic alterations in amphidiploid species of *Brassica* (Allender et al. 2007; Al-Shehbaz et al. 2006) (Fig. 2).

He deduced as (Fig. 3):

1. *Brassica* species experienced some altered gene expressions (Lloyd et al. 2018).

Table 2 Wild relatives and allied genera of various *Brassica* species and their distribution

Sr. No.	*Brassica* species	Plant morphology	Distribution/center of diversity	Chromosome number
1.	*Erucastrum virgatum*		Northern Africa, Sicily, Southern Italy	$2n = 14$
2.	*Diplotaxis acris*		Sindian deserts and the Sahara deserts	$2n = 14$
3.	*Hirschfeldia incana*		Mediterranean region	$2n = 14$
4.	*Brassica fruticulosa ssp. cossoniana*		Saharan coasts, the Middle Atlas mountains in Morocco	$2n = 16$
5.	*Brassica spinescens*		Northern Africa	$2n = 16$

(continued)

Table 2 (continued)

Sr. No.	*Brassica* species	Plant morphology	Distribution/center of diversity	Chromosome number
6.	*Brassica maurorum*		Morocco, Algeria, North Africa	$2n = 16$
7.	*Brassica fruiticulosa ssp. fruticulose*		Mediterranean coastal areas	$2n = 16$
8.	*Erucastrum leucanthum*		Northern Africa	$2n = 16$
9.	*Brassica barrelieri*		Morocco, Algeria, Iberia	$2n = 18$
10.	*Sinapis arvensis*		Mediterranean coastal areas	$2n = 18$

(continued)

Table 2 (continued)

Sr. No.	*Brassica* species	Plant morphology	Distribution/center of diversity	Chromosome number
11.	*Sinapis turgida*		Mediterranean coastal areas	$2n = 18$
12.	*Diplotaxis tenuisiliqua*		Northern Africa	$2n = 18$
13.	*Erucastrum cardaminoides*		Canary Islands	$2n = 18$
14.	*Raphanus sativus*		Southern Asia	$2n = 18$
15.	*Raphanus raphanistrum*		Mediterranean coastal areas, Marmara coasts, Bosporus Strait	$2n = 18$

(continued)

Table 2 (continued)

Sr. No.	*Brassica* species	Plant morphology	Distribution/center of diversity	Chromosome number
16.	*Raphanus rapistrum rugosum*		Mediterranean coastal areas, Marmara coasts, Bosporus Strait	$2n = 18$
17.	*Raphanus sativus* var. *oleiferus*		Mediterranean coastal areas, Marmara coasts, Bosporus Strait	$2n = 18$
18.	*Brassica oxyrrhina*		Morocco, Spain, Southern Portugal	$2n = 18$
19.	*Brassica repanda*		Algeria	$2n = 20$
20.	*Brassica tournefortii*		Mediterranean coastal areas, India	$2n = 20$

(continued)

Table 2 (continued)

Sr. No.	*Brassica* species	Plant morphology	Distribution/center of diversity	Chromosome number
21.	*Brassica trichocarpa*	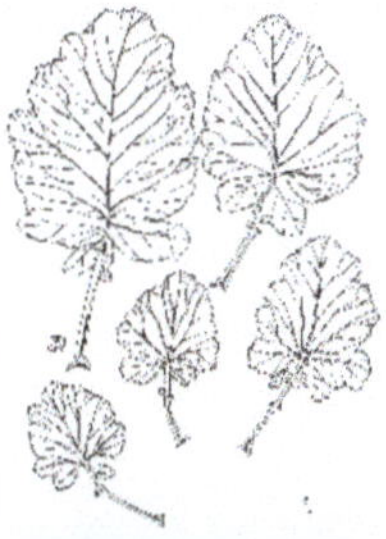	Mediterranean region, Sicily, western Asia, North America, and Australia	$2n = 20$
22.	*Sinapidendron* spp.		Madeira	$2n = 20$
23.	*Brassica amplexicaulis*		South East Algeria	$2n = 22$
24.	*Brassica elongata*		South Eastern Europe, Iran, Western Asia	$2n = 22$
25.	*Eruca sativa* syn. *Eruca vesicaria ssp. sativa*		Southern Europe, North Africa, India, and Western Asia	$2n = 22$

(continued)

Table 2 (continued)

Sr. No.	*Brassica* species	Plant morphology	Distribution/center of diversity	Chromosome number
26.	*Sinapis alba*		Europe and Mediterranean coastal areas	$2n = 24$
27.	*Hutera* spp.		Iberia	$2n = 24$
28.	*Diplotaxis harra*		Sindian deserts and the Sahara deserts	$2n = 26$
29.	*Erucastrum laevigatum*		Northern Africa, Sicily, Southern Italy	$2n = 56$

2. Some parental fragments were lost during evolution and novel fragments were observed in early generations of amphidiploids (Song et al. 1993).
3. Some transposable elements were also altered (Sarilar et al. 2013).
4. *B. nigra* has its evolutionary pathway common with *Sinapis arvensis* and it is likely to be the progenitor of *B. nigra* (Song et al. 1993).

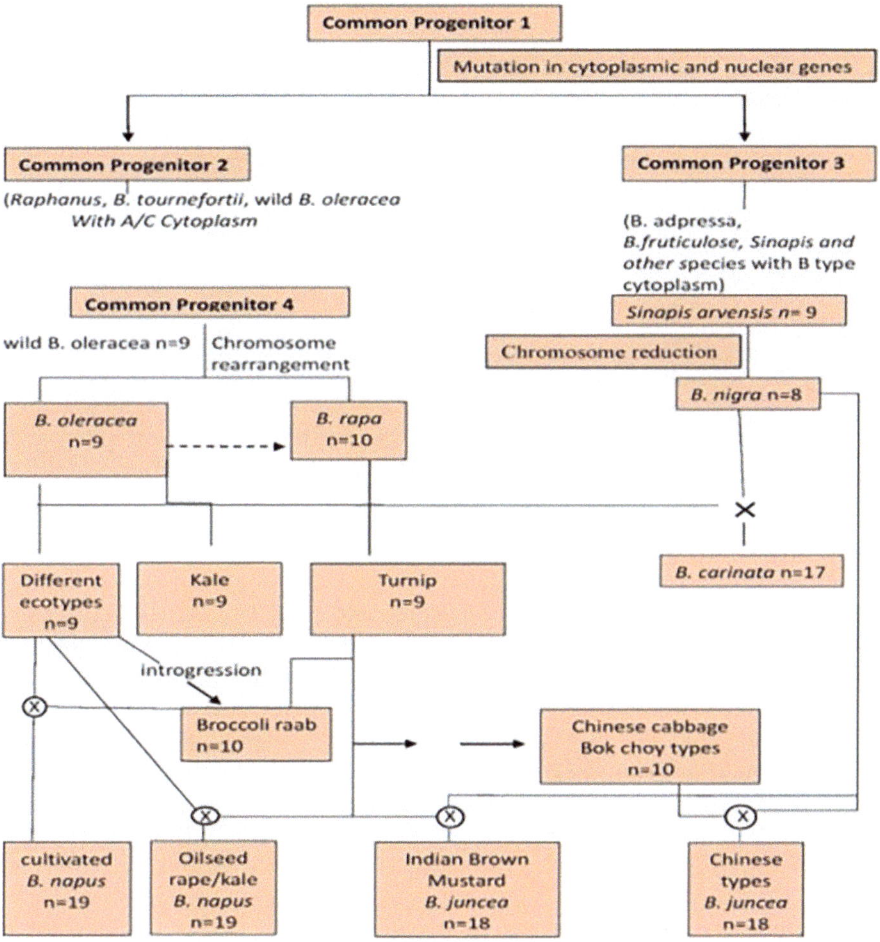

Fig. 2 Hypothesis of genomic evolution in *Brassica* species (Song et al. 1993)

8 Conclusions

Brassica and its wild relatives are much more diverse in their genotypic configuration throughout the world. Various amphidiploids in *Brassica* species are more likely cross breeds in nature, leading to more heterozygosity. Production of *Brassica* in India is lower in comparison to China and Canada, due to low seed yields. The average yield in Europe is around 3.0 tons per hectare, which is twice the yields achieved in Australia and Canada due to the favorable climate in Europe. Major breeding objectives for *Brassica* include breeding for zero erucic acid and low glucosinolates (for meal quality) (Cartea et al. 2019). High unsaturated fatty acids (i.e., linolenic acid and alpha linolenic acid) and low glucosinolate content (i.e., 2-propenyl or allyl and 3-butenyl glucosinolate) in the meal quality of *Brassica* is a

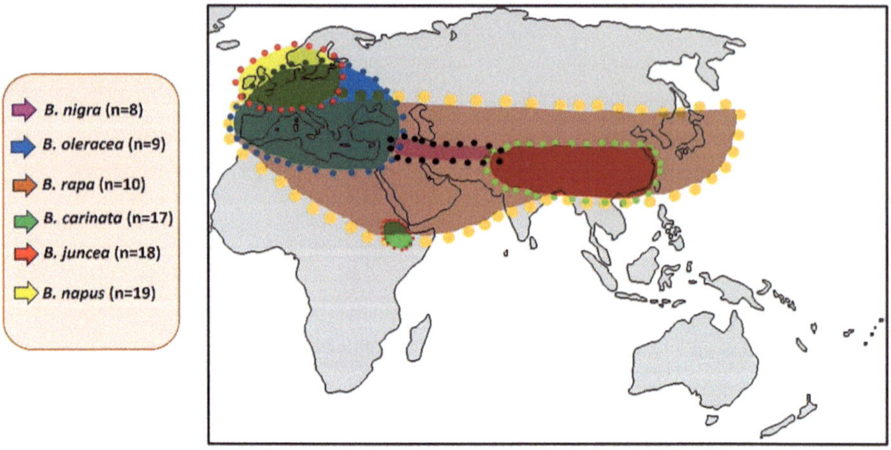

Fig. 3 Geographical distribution of major *Brassica* species

desirable trait for animal feed purposes (Zhu et al. 2023). Double low (i.e., low glucosinolate and low erucic acid) genotypes are being promoted in *Brassica* species to make it an export-quality feed ration (Zhu et al. 2023). HOLL genotypes (i.e., high oleic acid and low linolenic acid) are desirable and are achievable with reference to such a huge diversity present in the genus. In contrast to canola type genotypes (i.e., low erucic acid genotypes), HEAR (high erucic acid rapeseed) genotypes (having more than 55% erucic acid) are being preferred in the production of biodegradable diesel and lubricants in steam engines, which attracts a limelight in the fuel industry and is in great demand (Zhu et al. 2023). It will capture a huge market share in the near future. Diversity is key to the scope of selection in plant breeding. No milestone is too far to achieve in breeding of *Brassica* genotypes to achieve major breeding objectives due to the diversity present in the genus.

References

Allender CJ, Allainguillaume J, Linn J, King GJ (2007) Simple sequence repeats reveal uneven distribution of genetic diversity in chloroplast genomes of *Brassica* L. and n = 9 wild relatives. Theor Appl Genet 114:609–618

Al-Shehbaz IA (1985) The genera of Brassiceae (Cruciferae; Brassicaceae) in the southeastern United States. J Arnold Arbor 66:279–351

Al-Shehbaz IA, Beilstein MA, Kellogg EA (2006) Systematics and phylogeny of the Brassicaceae (Cruciferae): an overview. Plant Syst Evol 259(2–4):89–120

Arias T, Pires C (2012) A full resolved chloroplast phylogeny of the Brassica crops and wild relatives (Brassicaceae: Brassiceae): Novel clade and potential taxonomic implications. Taxon 61:980–988

Arias T, Beilstein MA, Tang M, McKain MR, Pires C (2014) Diversification times among Brassica (Brassicaceae) crops suggest hybrid formation after 20 million years of divergence. Am J Bot 101(1):86–91

Baillon H (1871) Historie des plantes. Monographie des Cruciferes; Librairie de Hachette et C.ie, Paris, France, pp 188–195

Beretta HV, Torres-Palazzolo C, Ramírez DA, Camargo AB (2021) Brassica Vegetables: Rich Sources of Neuroprotective Compounds. In: Psychiatry and neuroscience update: from epistemology to clinical psychiatry–Vol. IV. Springer, Cham, pp 327–341

Borgen L, Rustan ØH, Elven R (1979) *Brassica bourgeaui* (Cruciferae) in the Canary Islands. Nor J Bot:255–264

Bothmer R, von Gustafsson M, Snogerup S (1995) Brassica sect. Brassica (Brassicaceae) II. Inter- and intraspecific crosses with cultivars of *B. oleracea*. Genet Resour Crop Evol 42:165–178

Boukema IW, van Hintum TJL (1999) Genetic resources. In: Gómez-Campo C (ed) Biology of Brassica Coenospecies. Elsevier Science B.V, Amsterdam, pp 461–479

Branca F, Argento S, Tribulato A (2012) Assessing genetic reserves in Sicily (Italy): the Brassica wild relatives case study. In: Maxted N, Dulloo ME, Ford-Lloyd B, Frese L, Iriondo J, Carvalho MAA (eds) Agrobiodiversity conservation: securing the diversity of crop wild relatives and landraces. CAB International, Wallingford, pp 52–58

Brullo C, Brullo S, Giusso Del Galdo G, Ilardi V (2013) *Brassica trichocarpa* (Brassicaceae), a new species from Sicily. Phytotaxa 122(1):45–60

Cartea E, Haro-Bailón AD, Padilla G, Obregón-Cano S, del Rio-Celestino M, Ordás A (2019) Seed oil quality of *Brassica napus* and *Brassica rapa* germplasm from Northwestern Spain. Foods 8:292. https://doi.org/10.3390/foods8080292

Chalhoub B, Denoeud F, Liu S, Parkin IA, Tang H, Wang X, Chiquet J, Belcram H, Tong C, Wincker P (2014) Early allopolyploid evolution in the post-Neolithic Brassica napus oilseed genome. Science 345(6199):950–953

Coves S, Soengas P, Velasco P, Fernández JC, Cartea ME (2023) New vegetable varieties of *Brassica rapa* and *Brassica napus* with modified glucosinolate content obtained by mass selection approach. Front Nutr 10:1198121

Crisp P (1982) The use of an evolutionary scheme for cauliflowers in the screening of genetic resources. Euphytica 31:725–734

De Candolle AP (1821) Cruciferae Syst Nat 2:139–700

De Candolle MAP (1824) Memoirs on the different species, races and varieties of the genus Brassica (cabbage) and on the genera allied to it, which are cultivated in Europe. In: Transactions of the Horticultural Society of London, pp 1–43

De Candolle A (1885) Origin of cultivated plants. D. Appleton and Company, New York, p 468

De Tournefort JP (2012) Historia Plantarum: Species Hactenus Editas Aliasque Insuper Multas Noviter Inventas & Descriptas Complectens. London, Henricum Faithorne, pp 1686–1704

Farnham MW, Davis EH, Morgan JT, Smith JP (2008) Neglected landraces of collard (*Brassica oleracea* L. var. *viridis*) from the Carolinas (USA). Genet Resour Crop Evol 55:797–801

Gepts P (2023) Biocultural diversity and crop improvement. Emerg Top Life Sci 7:151–196

Geraci A, Divaret I, Raimondo FM, Chèvre AM (2001) Genetic relationships between Sicilian wild populations of Brassica analysed with RAPD markers. Plant Breed 120:193–196

Giles WF (1941) Cauliflower and broccoli. What they were and where they came from. J Roy Hort Soc 66:265–278

Gomez-Campo C (1980) Morphology and morpho-taxonomy of the tribe Brassiceae. In: Tsunoda S, Hinata K, Gomez-Campo C (eds) Brassica crops and wild allies, biology and breeding. Japan Scientific Societies Press, Tokyo, pp 3–31

Gómez-Campo C (1999) Taxonomy. In: Gómez-Campo C (ed) Biology of Brassica Coenospecies. Elsevier, Amsterdam, pp 3–32

Gómez-Campo C, Aguinagalde I, Ceresuela JL, Lázaro A, Martinez-Laborde JB, Parra-Quijano M, Simonetti E, Torres E, Tortosa ME (2005) An exploration of wild *Brassica oleracea* L. germplasm in Northern Spain. Genet Resour Crop Evol 52:7–13

Guo Y, Chen S, Li Z, Cowling WA (2014) Center of origin and centers of diversity in an ancient crop, *Brassica rapa* (Turnip rape). J Hered 105(4):555–565. https://doi.org/10.1093/jhered/esu021

Gustafsson M (1982) Germplasm conservation of wild (n=9) Mediterranean *Brassica* species. Sveriges Utsädesförenings Tidskrift 92:133–142

Gustafsson M, Lannér-Herrera C (1997) Overview of the *Brassica oleracea* complex: their distribution and ecological specificities. Bocconea 7:27–37

Han S, Sebastin R, Wang X, Lee KJ, Cho G-T, Hyun DY, Chung J-W (2021) Identification of Vicia species native to South Korea using molecular and morphological characteristics. Front Plant Sci 12:608559. https://doi.org/10.3389/fpls.2021.608559

Hodgkin T (1995) Cabbages, kales, etc. In: Smartt J, Simmonds NW (eds) Evolution of crop plants, 2nd edn. Longman, London, pp 76–82

Kaur P, Banga S, Kumar N, Gupta S, Akhatar J, Banga SS (2014) Polyphyletic origin of *Brassica juncea* with *B. rapa* and *B. nigra* (Brassicaceae) participating as cytoplasm donor parents in independent hybridization events. Am J Bot 101:1157–1166

Kiefer M, Schmickl R, German D, Lysak M, Al-Shehbaz IA, Franzke A, Mummenhoff K, Stamatakis A, Koch MA (2013) BrassiBase: Introduction to a novel knowledge database on Brassicaceae evolution. Plant Cell Physio 55(1):e3. https://doi.org/10.1093/pcp/pct158

Kim SH, Lee GA, Subramanian P, Hahn BS (2023) Quantification and Diversity Analyses of Major Glucosinolates in Conserved Chinese Cabbage (*Brassica rapa* L. ssp *pekinensis*) Germplasms. Foods 12:1243. https://doi.org/10.3390/foods12061243

Linnaeus C (1753) Species Plantarum. Impensis Laurentii Salvii, Stockholm, pp 667–668

Liu S et al (2014) The *Brassica oleracea* genome reveals the asymmetrical evolution of polyploid genomes. Nat Commun 5:3930. https://doi.org/10.1038/ncomms4930

Lloyd A, Blary A, Charif D, Charpentier C, Tran J, Balzergue S, Jenczewski E (2018) Homoeologous exchanges cause extensive dosage-dependent gene expression changes in an allopolyploid crop. New Phytol 217(1):367–377

Maggioni L (2015) Domestication of *Brassica oleracea* L. Doctoral Thesis, Swedish University of Agricultural Sciences, Alnarp (Thesis No. 2015: 74

Maggioni L, Bas N, Poulsen G, Branca F, Ralli P, von Bothmer R, Lipman E (2013) Availability of wild Brassica sect. In: Brassica accessions in genebanks. Poster presented at the EUCARPIA Genetic Resources section meeting, Alnarp, pp 11–13

Massie IH, Astley D, King G (1996) Patterns of genetic diversity and relationships between regional groups and populations of Italian landrace cauliflower and broccoli (*Brassica oleracea* L. var. *botrytis* L. and var. *italica* Plenck). In: Dias JS, Crute I, Monteiro AA (eds) Proceedings of the ISHS Brassica Symposium – IX Crucifer Genetics Workshop, Acta Hort, vol 407, pp 45–53

Nagaharu U, Nagaharu N (1935) Genome analysis in Brassica with special reference to the experimental formation of *B. napus* and peculiar mode of fertilization. J Japan Bot 7:389–452

Sarilar V, Palacios PM, Rousselet A, Ridel C, Falque M, Eber F, Alix K (2013) Allopolyploidy has a moderate impact on restructuring at three contrasting transposable element insertion sites in resynthesized *Brassica napus* allotetraploids. New Phytol 198(2):593–604

Schulz OE (1919) Cruciferae-Brassiceae. In: Pflanzenreich D, Engler HGA (eds) Pars prima. Wilhelm Englmann, Leipzig, pp 1–290

Snogerup S (1980) The wild forms of the *Brassica oleracea* group (2n = 18) and their possible relations to the cultivated ones. In: Tsunoda S, Hinata K, Gomez-Campo C (eds) Brassica crops and wild allies, biology and breeding. Japan Scientific Societies Press, Tokyo, pp 121–132

Song K, Tang K, Osborn TC (1993) Development of synthetic *Brassica amphidiploids* by reciprocal hybridization and comparison to natural amphidiploids. Theor Appl Genet 86:811–821

Song H, Hua Y, Zhou T, Yue C, Huang J, Feng Y (2024) Genomic identification of CYP450 enzymes and new insights into their response to diverse abiotic stresses in *Brassica napus*. Plant Mol Biol Report:1–19

Tsunoda S (1980) Eco-physiology of wild and cultivated forms in Brassica and allied genera. In: Tsunoda S, Hinata K, Gomez-Campo C (eds) Brassica crops and wild allies, biology and breeding. Japan Scientific Societies Press, Tokyo, pp 109–120

Vavilov NI (1926) Studies on the origin of cultivated plants. In: Bulletin of applied botany, genetics and plant breeding, pp 1–248

Zamani-Nour S, Clemens R, Möllers C (2013) Cytoplasmic diversity of *Brassica napus* L., *Brassica oleracea* L. and *Brassica rapa* L. as determined by chloroplast microsatellite markers. Genet Resour Crop Evol 60:953–965

Zhu B, Liang Z, Zang Y, Zhu Z, Yang J (2023) Diversity of glucosinolates among common brassicaceae vegetables in China. Hortic Plant J 9:365–380

Phytochemicals Composition and Biomedical Properties of *Brassica* spp

S. T. Gopukumar, A. Ganesh Kumar, K. Sunitha kumari, Sinouvassane Djearamane, and Ranjithkumar Rajamani

Abstract

Brassica species from the Brassicaceae family are important agricultural plants, including mustard, cabbage, broccoli, cauliflower, and kale, along with rapeseed, and they receive widespread cultivation throughout the world. The phytochemical content of *Brassica* species consists of glucosinolates together with polyphenols and flavonoids and carotenoids and alkaloids and tannins and phytosterols and vitamins C and E, and essential minerals, which together create their functional properties. Biomedical applications such as antimicrobial, anti-inflammatory, antioxidant, and anticancer properties of the *Brassica* species play a vital role in nutraceuticals and functional foods in daily life. Nowadays, researchers focus more on the increasing applications of natural resources like seaweed extracts and chitosan-based agricultural products to increase the quality

S. T. Gopukumar
Center for Global Health Research, Saveetha Medical College and Hospital, Saveetha Institute of Medical and Technical Sciences (SIMATS), Saveetha University, Chennai, Tamil Nadu, India

A. Ganesh Kumar
Department of Microbiology, Saveetha Medical College and Hospital, Saveetha Institute of Medical and Technical Sciences (SIMATS), Saveetha University, Chennai, Tamil Nadu, India

K. Sunitha kumari
Department of Botany, PSGR Krishnammal College for Women, Peelamedu, Coimbatore, Tamil Nadu, India

S. Djearamane
Department of Biomedical Science, Faculty of Science, Universiti Tunku Abdul Rahman (UTAR), Kampar, Perak, Malaysia

R. Rajamani (✉)
Department of Pharmacology, Saveetha Medical College and Hospital, Saveetha Institute of Medical and Technical Sciences (SIMATS), Saveetha University, Chennai, Tamil Nadu, India

A. Kumar, B. Koul (eds.), *Crop Improvement Strategies in Brassica species: Applied Science*, https://doi.org/10.1007/978-981-95-3861-4_4

parameters of phytochemicals in agricultural crops, which help to improve healthcare applications. In this chapter, we discussed the phytochemicals and biomedical applications of Brassica species. Furthermore, we highlighted the role of natural resources in enhancing the development of *Brassica* species for sustainable agriculture.

Keywords
Brassica species · Phytochemicals · Nutraceuticals · Sustainable agriculture · Disease prevention

1 Introduction

The Brassicaceae Burnett family, also known as the "mustard family" (syn. Cruciferae A.L. de Jussieu), encompasses approximately 340 genera and 4636 recognized species (Francis et al. 2021). Among these, the *Brassica* genus is the most prominent and economically significant (El-Esawi 2021). The family's former name, Cruciferae, is derived from the cross-shaped petal arrangement of mustard flowers. A key distinguishing feature of *Brassica* species is their fruit, which undergoes developmental modifications to enhance seed distribution. These species are characterized by conduplicate cotyledons—cotyledons folded lengthwise around the radicle—and segmented fruits with seeds in one or both segments, setting them apart from other dicotyledons (Gómez-Campo 1980).

The evolutionary origins of the *Brassica* genus are believed to lie in the Mediterranean and Saharan regions, areas marked by hot, dry summers and mild winters. Over time, species within this genus have adapted to thrive in colder climates, likely due to selective pressures from the Mediterranean basin's challenging environment, which favored the development of salt- and drought-tolerant taxa (Gómez-Campo 1980). The genus includes 37 species, ranging from invasive wild plants to cultivated crops, most of which are annuals or biennials (Rakow 2004). One defining feature of *Brassica* plants is their high concentration of sulfur-containing glucosinolates, which hydrolyze into mustard oils that contribute to the vegetables' distinct smells and flavors. Depending on the chemical structure of the glucosinolate, the resulting isothiocyanates (ITCs) can have aromatic, aliphatic, or indolic side chains. Research suggests that ITCs may help reduce the incidence of various cancers (Dinkova-Kostova and Kostov 2012).

In addition to glucosinolates, *Brassica* plants provide an array of phytochemicals critical for human health and well-being (Önder 2018). Phytochemicals, though non-nutritive, play a protective role in human physiology. They act as antioxidants by scavenging free radicals, which are byproducts of biological reactions, thus preventing oxidative damage (Uttara et al. 2009). These compounds shield essential biomolecules from oxidative stress and contribute to the prevention of neurological, cardiac, and other physiological disorders. Numerous *Brassica* species have been studied for their phytochemical profiles and antioxidant capacities (Fig. 1).

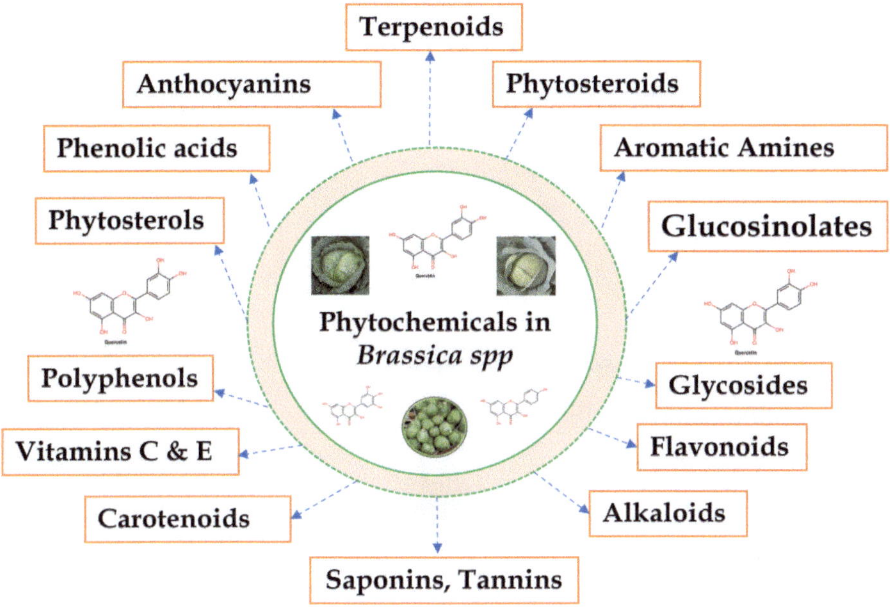

Fig. 1 Phytochemicals in *Brassica* spp

The bioactive compounds frequently found in *Brassica* species include polyphenols, phenolic acids, flavonoids, carotenoids (such as lutein, zeaxanthin, and β-carotene), alkaloids, tannins, saponins, anthocyanins, phytosterols, chlorophyll, glucosinolates, phytosteroids, terpenoids, glycosides, vitamins C and E, and aliphatic and aromatic amines (Danlami et al. 2016; Chauhan et al. 2016; Al Shahawany et al. 2016; Kim et al. 2016; Sharma et al. 2017; Singh and Malik 2017).

These compounds collectively enable *Brassica* plants to exert significant biological activities, offering potential therapeutic benefits for a wide range of diseases (Fig. 2). The edible parts of these plants have demonstrated antimicrobial, antibacterial, antidiabetic, antimalarial, anti-aging, anti-ulcer, anti-hyperglycemic, anti-hyperlipidemic, anti-proliferative, neuroprotective, anti-genotoxic, and antioxidant properties. However, the phytochemical content in *Brassica* plants is influenced by various factors, including plant variety, growth conditions, stage of maturity, harvesting time, postharvest storage, industrial processing, and culinary preparation methods (Aires 2015; Lafarga et al. 2018). Species such as *Brassica juncea* (Indian mustard), *Brassica napus* (rapeseed), *Brassica rapa* (turnip), *Brassica nigra* (black mustard), and *Brassica oleracea* (including cabbage, broccoli, and kale) are particularly rich in phytochemicals and antioxidants, making them valuable components of a healthy diet. Their broad spectrum of bioactivities highlights their potential for use in disease prevention and treatment.

From traditional uses in medicine to emerging roles in functional foods, *Brassica* plants continue to be a subject of interest in both nutritional and biomedical research. The current chapter offers an extensive overview of the research development on the

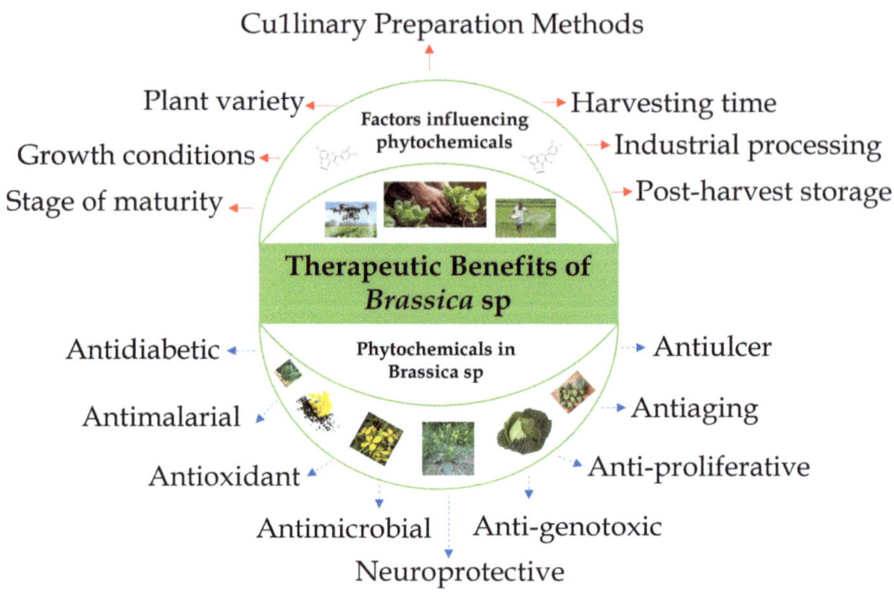

Fig. 2 Biological properties and factors influencing phytochemical content

Brassica species, phytochemical screening, pharmacological properties, effect of natural compounds on the development of *Brassica* species through an eco-friendly approach.

2 Cultivation of *Brassica* Species in India

India cultivates *Brassica* species, which are referred to as rapeseed-mustard among the most essential oilseed plants in the country. Rapeseed-mustard serves as a fundamental agricultural component that strengthens sustainability alongside providing security for food and advancing economic growth. The four *Brassica* species *Brassica juncea*, *Brassica napus*, *Brassica rapa*, together with *Brassica carinata,* make important contributions to India's oilseed production sector (Meena et al. 2020). *Brassica* species mainly grow throughout India's northern and central districts because they flourish in semi-arid and temperate climatic environments (Sharma and Singh, 2021). Crop rotation practices enable these important plants to increase soil fertility, especially because they contain high oil content and valuable nutrients (Gupta et al. 2022).

The growth of *Brassica* species reaches its peak performance zone when supplied with dry conditions that maintain temperatures between 10 °C and 25 °C. Soil types that work best comprise loamy or sandy loam textures where the soil pH ranges between 5.5 and 7.5. Germination, along with crop establishment, suffers negative consequences when waterlogging occurs and when soils have excessive moisture content. Improving yield depends strongly on the management of soil fertility through proper organic amendments and balanced fertilizer applications according to Yadav et al. (2023).

Veteran researchers in India have developed multiple upgraded *Brassica* agricultural types to boost farming output and protect against diseases while increasing oil production. Pusa Bold (*Brassica juncea*) shows high yields through its resilience to moisture stress conditions. The short maturation timeframe makes Pusa Bold suitable as a multi-crop system because of its quick development period. The oil content of Pusa Bold reaches between 38 and 40 percent, which makes it an ideal choice for commercial agricultural practices. NRCDR-2 shows disease resistance against white rust and Alternaria blight at a major level. This variety achieves high seed yields while showing tolerance to delayed planting which allows different cropping schemes suitable to farmers (Verma and Sharma, 2018). GSC-7 demonstrates excellent performance in irrigated areas because it excels in recovering maximum oil content. The medium chronological period of this variety, together with its pest-resistance capability, makes it an attractive choice. The variety has achieved fame among farmers who can maintain dependable harvests because they possess sufficient water resources.

Brassica crops need to be sown through line sowing or broadcasting techniques according to recommended rates between 4 and 6 kg/ha. The planting rate for each type of crop needs distinct specifications according to species requirements between 4 and 6 kg/ha. The proper establishment of crops depends on October–November seed planting as it maximizes yield optimization (Jain et al. 2023). *Brassica* productivity receives substantial benefits from proper fertilization with nitrogen (80–100 kg/ha) alongside phosphorus (40–50 kg/ha) and potassium (30–40 kg/ha) as per. Farmyard manure (FYM), along with biofertilizers, provides several advantages to soil health and available nutrients as shown in Kumar and Gupta (2022). The flowering period and pod formation stage require two or three irrigation events for *Brassica* plants. The control of weeds in *Brassica* crops can be achieved through combined methods, which consist of hand removal of weeds followed by Pendimethalin herbicide treatment at 1.0 kg/ha concentration and the application of mulch. The productivity of *Brassica* crops is adversely affected by the mustard aphid (*Lipaphis erysimi*) and diseases including white rust (*Albugo candida*) and Alternaria blight (*Alternaria brassicae*) (Rimmer et al. 2007). The combination of resistant plant varieties and appropriate usage of Imidacloprid insecticide with biological control agents proves as a good strategy for effective management of pest issues.

The production of *Brassica* is threatened by three major consequences of climate change which include rising temperatures and erratic rainfall alongside increased pest outbreaks. The success of risk reduction for climate-related problems depends on implementing climate-resilient varieties together with conservation tillage, along with water-efficient irrigation systems. Because of recent advances in genomic-assisted breeding combined with CRISPR-based genome editing and hybrid seed technology, *Brassica* farmers gain better outcomes in productivity, together with improved stress tolerance, according to Kumar and Gupta (2022). The combination of precision farming systems and drone crop surveillance, and digital advisory

Table 1 Major oilseed *Brassica* species cultivated in India

Species	Common name	Approximate cultivation (%)
Brassica juncea	Indian mustard	~90%
Brassica rapa	Brown/yellow sarson	Minor
Brassica napus	Gobhi sarson	Minor
Brassica carinata	Karan rai	Minor

systems will boost yield production (Joshi et al. 2023). *Brassica* crop production in India presents a vital opportunity for national edible oil fulfillment while maintaining ecological sustainability in agriculture. The advancement of *Brassica* cultivation in India depends upon the combination of technological interventions and policy support that addresses pest control problems and climate adaptation needs, and input management requirements. The Indian agriculture sector produces Indian mustard (*Brassica juncea*) as the main category of oilseed *Brassica* species, while it dominates nearly 90% of the rapeseed-mustard cultivating lands (Table 1).

Brassica rapa varieties like brown sarson and yellow sarson, together with *Brassica* napus types that include *Gobhi sarson* and *Brassica carinata* known as karan rai, exist, yet they are grown on much smaller areas. *Brassica juncea* stands as the dominant oilseed *Brassica* species in India because the other species receive only minor cultivation compared to juncea (Gupta et al. 2022). The production levels of oilseed *Brassica* have increased substantially throughout numerous decades in India. The production rose from 0.76 million tons in 1950–1951 to 7.98 million tons in 2016–2017. The production yield rose from 368 kg/ha to 1324 kg/ha throughout this time frame. An important expansion occurred simultaneously with productivity improvements. India holds the ranking as the third-largest producer of oilseed *Brassica* worldwide after Canada and China but must still import substantial edible oil amounts to satisfy its domestic consumption (Table 2). During 2016–2017 India became the 11.2% largest global importer of edible oils by procuring 15.3 million tons valued at 103.2 million USD. The present situation offers vast potential to boost domestic cultivation while decreasing foreign oil procurement.

Proper inputs combined with technological interventions and policy frameworks create extensive production and market opportunities in Indian rapeseed-mustard cultivation for traditional and non-traditional areas. The strategy would create higher returns for farmers while supporting the national economic growth. India shows great potential to expand both oilseed *Brassica* cultivation and processing, reflecting the global edible oil market growth, accompanied by rising national domestic demand. Postharvest technologies, together with modern agricultural practices and value addition investments, will develop the export market potential and improve the marketability of these crops. The member crops of *Brassica*, such as mustard, along with rapeseed cabbage and broccoli, experience multiple diseases that harm their production quality severely. *Alternaria brassicae* and *Alternaria brassicicola* cause Alternaria Leaf Spot through black circular lesions with rings that appear on leaves, stems, and pods, which results in defoliation and diminished

Table 2 *Brassica* species cultivation and marketing data

Category	Value
Total *Brassica* production (2016–17)	7.98 million tons
Yield per hectare (2016–17)	1324 kg/ha
India's rank in global production	3rd (after Canada and China)
Edible oil imports (2016–17)	15.3 million tons
Total world import share (%)	11.20%
Total import value (USD)	103.2 million USD
Potential marketing opportunity	High (demand for domestic production and export potential)
Key export markets	Middle East, Southeast Asia, European Union

seed yield. Blackleg (*Leptosphaeria maculans*) causes stem sunken lesions that damage plant structure and leads to lodging which severely reduces yield (West et al. 2001). White fungal growth on the underside of affected leaves with yellowing creates a downy mildew appearance (*Peronospora parasitica*), which negatively affects plant growth rate and seed quality (Meena et al. 2010).

The fungal pathogen *Albugo candida* causes White Rust by creating white pustules, which deform plant parts and make seeds non-viable. Clubroot (*Plasmodiophora brassicae*) represents one of the most destructive soil-borne diseases because it produces swollen roots that cause damage to nutrition uptake and lead to plant wilting and poor growth (Dixon 2009). The fungal pathogen *Sclerotinia sclerotiorum* creates soft white fungal growth that attacks stems and causes plant collapse as well as economic damage (Derbyshire and Denton-Giles 2016). Powdery Mildew *(Erysiphe cruciferarum)* produces white fungal growth on leaf surfaces that weakens plant photosynthesis and causes damage (Braun et al. 2019).

Plant decay and yield reduction occur because of bacterial Black Rot (*Xanthomonas campestris pv. campestris*), which creates V-shaped yellow leaf lesions (Vicente and Holub, 2013). *Xanthomonas campestris* leads to Bacterial Leaf Spot, which develops as necrotic darkened spots that damage the plant health substantially (Ignatov et al. 2020). Plant growth drastically decreases through the appearance of leaf mottling and distortion, and stunted growth, which results from the viral infection Turnip Mosaic Virus (TuMV) (Walsh and Jenner 2002). The management of diseases in *Brassica* crops includes selective variety use combined with crop rotation practices and field maintenance, as well as fungicide application and irrigation improvements to stop moisture-related pathogen spread. The establishment of preventive and control measures enables *Brassica* farmers to protect their crops from harm and achieve better plant health (Kumar et al. 2014). Clubroot (*Plasmodiophora brassicae*) stands as the most damaging and unmanageable disease affecting *Brassica* species. This pathogen infects the roots of mustard, cabbage, rapeseed, and cauliflower, causing clubroot disease that develops in the soil. The inability to control clubroot disease is mainly caused by resting spores in the soil that can survive for over 10–20 years, thus crop rotation becomes useless (Dixon 2009). The quick spread of this disease happens by means of contaminated soil, water, farm tools, and wind-carried dust particles and falls into a category of diseases that prove extremely hard to restrain (Donald and Porter 2014).

Clubroot pathogenic advantage is amplified through favorable environmental conditions that include acidic, along with moist soils having a pH level below 7.0 (Hwang et al. 2012). *Brassica* varieties with resistance traits experience rapid evolutionary changes in *Plasmodiophora brassicae,* which render persistent cross-breeding activities progressively difficult (Rahman et al. 2014). *Plasmodiophora brassicae* cannot be eradicated from infected fields because researchers have not successfully developed chemical fungicides against this pathogen (Kumar et al. 2014). The severe negative effects of clubroot damage *Brassica* crops through root swelling and galling, and root deformation that blocks water and nutrient absorption (Dixon 2009). The infection of plants results in wilting along with yellow leaves and growth stunting until crop failure occurs in heavily affected fields (Donald and Porter, 2014). The entire *Brassica* cultivation area becomes unusable for production unless proper control measures are used to combat the disease (Hwang et al. 2012).

The complete elimination of clubroot seems impossible, but specific disease management methods combine to suppress the impact of this disease. Soil disease severity remains suppressed when farmers apply agricultural lime to reach a pH value above 7.2 (Rahman et al. 2014). The spread of clubroot pathogens decreases when farmers control soil drainage and moisture because the pathogen needs wet conditions to thrive (Donald and Porter 2014). Resistant *Brassica* plant varieties provide moderate protection against clubroot disease development (Kumar et al. 2014). Proper farm hygiene requires complete equipment sterilization while stopping the dispersal of infected soil because these preventive steps help control pathogen transmission (Hwang et al. 2012). Clubroot continues to be among the hardest *Brassica* diseases to manage since it survives for extended periods in soil while evolving quickly against resistance genes and has limited chemical control options available. For the effective management of clubroot while sustaining *Brassica* cultivation, farmers need to implement multiple cultural, genetic, and management techniques integrating soil practices.

3 Role of Sea Weeds in Agriculture

The macroscopic marine algae that grow in coastal areas, known as seaweeds, have become important in agricultural practices because of their multiple beneficial compounds, which serve as organic fertilizers, help condition the soil, and stimulate plant growth. The variety of nutrients with polysaccharides and phytohormones, and bio-stimulants in marine resources provides better soil health while increasing crop productivity (Khan et al. 2009). Modern sustainable farming embraces these seaweed assets because they provide both organic nutrient inputs and improve the health of the soil while strengthening crops against environmental challenges. The illustration (Fig. 3) shows how seaweed impacts agricultural progress, starting with the harvesting phase and leading to better crop production. The process starts with harvesting seaweed to create usable forms of extraction products that include

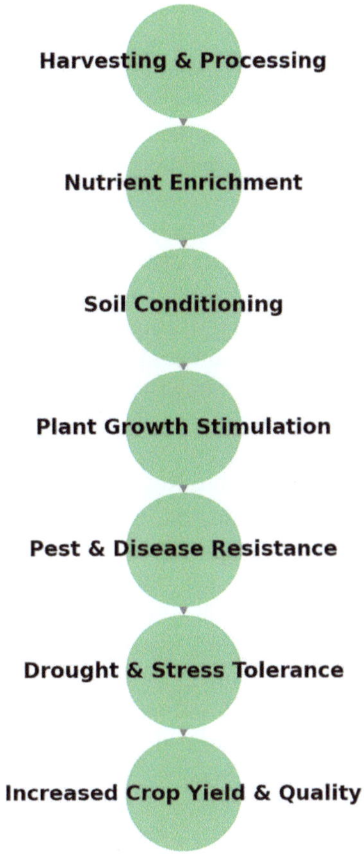

Fig. 3 Steps involved in the role of harvesting to improve crop yield

bio-fertilizers, liquid extracts, and powders. The manufactured products from seaweed support nutrient enrichment through multiple component supply, including essential mineral substances such as nitrogen, potassium, and phosphorus for improving soil fertility. Soil conditioning happens next because seaweed enhances both structure and water retention and microbial activity, which results in an ideal plant growth environment.

The plant growth stimulation process demonstrates how seaweed provides auxins and cytokinins, which lead to superior seed germination and healthier root development. The application of seaweed provides plants with pest and disease resistance by feeding them compounds that fortify their defense mechanisms, which decreases the need for chemical pesticides. Through its incorporation, seaweed allows crops to gain resistance against environmental stressors that include both drought conditions and variable stresses from temperature drifts and salt intrusion. The positive effects of seaweed application in agriculture increase both crop quantity and product quality while enabling sustainable environmental agricultural methods (Battacharyya et al. 2015). A diagram efficiently illustrates the complete process where seaweed functions as a driver of healthy plant development and improved production levels.

Soil amendment represents a significant agricultural contribution that seaweeds provide to farming operations. Seaweed-based fertilizers applied to soil enhance soil structure as well as retain water effectively and cultivate diverse microbial ecosystems that maintain soil fertility (Craigie 2011). The polysaccharides found in seaweeds feature natural chelating properties that link essential nutrients into a form that plants can easily absorb through binding mechanisms. The organic components in seaweed extracts improve soil cation exchange capacity while simultaneously increasing the ability of crops to retain nutrients (Battacharyya et al. 2015). The use of seaweed applications leads to better soil air circulation, which avoids compaction issues while creating an optimal environment for beneficial microorganisms to thrive, thus improving nutrient absorption and rooting health of plants.

Laboratory tests show that plant growth stimulation occurs from phytohormone compounds present in seaweed extracts, which contain auxins along with cytokinins and gibberellins. Seaweed compounds function as controllers for seed germination and root formation, and flower development procedures (Zodape 2001). Scientific research demonstrates that seaweed-based biostimulants extend root lengths, which improves plant stabilization and nutrient intake to produce better crop development and health (Khan et al. 2009). The use of seaweed formulations produces documented evidence that proves their ability to create systemic resistance in plants, enabling them to defend against pathogens as well as various abiotic stressors (Sharma et al. 2014). Seaweeds act as natural agents that help plants survive under saline conditions, thus enabling crop growth in areas affected by saline soil. This advantage works best in areas with salt-affected soils adjacent to coastal arid zones.

Seaweed-based fertilizers serve the environment by promoting sustainability through the use of alternative synthetic agrochemicals. A distinguishing trait of seaweed formulations over chemical fertilizers is their biodegradability because they enhance the natural microbial balance of soils (Craigie, 2011). The replacement of traditional farming approaches with sustainable methods preserves soil productivity along with reducing harmful discharge from agricultural areas into fresh and saltwater ecosystems (Battacharyya et al. 2015). Due to their production method, seaweed-based fertilizers help lower greenhouse gas emissions from synthetic fertilizers, which make them suitable sustainably beneficial fertilizer option for contemporary farming needs. The use of seaweed as a biofertilizer stands essential in supporting organic farming practices. Athletic organic farming incorporates natural amendments while widely adopting seaweed-based products because they deliver effective crop health improvements in addition to avoiding synthetic agricultural inputs (Sharma et al. 2014). Current studies have demonstrated that seaweed-derived preparations elevate crop nutritional benefits by adding vitamins and minerals, which leads to enhanced human health results (Zodape 2001). Laboratory studies demonstrate the excellent potential of seaweed extracts to extend food shelf life because they decrease microbial growth while strengthening fruits and vegetables to prevent spoilage and decrease food waste.

The demand for resilient agricultural systems under current climate change has made seaweed implementations stand out as potential adaptive solutions. The drought resistance properties, as well as soil moisture retention and stress tolerance abilities of crops are critical elements that make seaweed extracts essential for sustainable food production, according to Khan et al. (2009). Research efforts to discover fresh agricultural applications of seaweed indicate an expanding agricultural impact that will transform conventional farming worldwide into sustainable agro-ecological and regenerative systems. Experts study genetically modified seaweeds to achieve higher nutritional density as well as better management capabilities for soils and crops.

Biofertilizers and biostimulants, joined by organic pesticides, are turning into commercial products made from seaweed-based materials. Today, many agricultural operations include seaweed extracts in their farming systems, and scientific research demonstrates their effectiveness in increasing the yields of several important plants, including grains, vegetables, and fruit-bearing crops (Craigie 2011). Agricultural field trials demonstrate that farmers throughout the world see value in seaweed product implementation because it leads to better germination rates and stronger stress tolerance, and higher yields at market value. Government entities, together with environmental agencies, are establishing seaweed cultivation as an environmentally friendly approach by promoting coastal communities to establish sustainable seaweed farms that deal with agricultural and environmental problems. The agricultural sector is likely to expand its use of seaweed technologies through research and technological improvements to offer farmers eco-friendly production-enhancing solutions. Future investigations about seaweed utilization in agriculture will concentrate on developing enhanced extraction techniques for higher bioactive compound retention and developing standardized application procedures for diverse crop needs, and evaluating extended agricultural effects on soil biology and crop health. Modern agricultural systems will achieve a sustainable future of global food production by implementing seaweed-based solutions. The agricultural implementation of seaweed has progressed notably during the period from 2022 through early 2025, especially through its use as an additive in livestock feed to lower methane production levels. Scientific studies prove that drinking Red seaweed *Asparagopsis taxiformis* extract by cattle drastically minimizes their methane emissions. A research team at the University of California, Davis revealed that incorporating modest quantities of seaweed into beef cattle nutrients could minimize methane gas release by 82% (UC Davis 2021). The PLOS ONE publication showed that *A. taxiformis* supplementation within beef steer feeding resulted in greater than 80% methane emission reduction from the digestive tract. Various organizations remain focused on recognizing such scientific breakthroughs for commercial implementation purposes (Roque et al., 2021). Sea Stock commercialized a patented technique to extract bromoform from Asparagopsis to make a feed additive that lowers methane emissions while producing natural substances for food additives and cosmetics used to enhance revenue for farmers. The seaweed cultivation market demonstrates quick expansion because it will grow from $22.45 billion in 2024 to $25.38 billion in 2025 due to rising environmental awareness about climate change and new use cases emerging Australian (2023). Sustainable livestock farming practice advancement and climate change mitigation efforts occur at both the regional and global levels. Seaweed-based feed additives show great potential for reducing agricultural greenhouse gas emissions and providing farmers with supplementary financial gains (The Business Research Company, 2024).

The agricultural sector adopted new business opportunities for seaweed production during the years from 2022 through 2025, enabled by its proven potential as organic fertilizers and livestock additives, along with biostimulants. The bioactive compound-rich seaweed fertilizers improve both soil quality and plant growth and minimize the need for chemical fertilizers (Craigie 2022). Scientists have demonstrated that *Ascophyllum nodosum* liquid extracts raise crop productivity and minimize stress reactions, which establish them as vital sustainable agriculture components (Pise & Sabale, 2023). The use of *Asparagopsis taxiformis* seaweed supplements in livestock feed has been scientifically validated to decrease methane emissions by 90% thereby creating profitable opportunities for sustainable animal feeding systems (Kinley et al., 2023). A Business Research Company analysis indicates that this market sector will grow from $22.45 billion in 2024 to reach $25.38 billion in 2025 (The Business Research Company, 2024). The expanding seaweed market provides substantial business prospects to companies that manage seaweed cultivation and perform processing operations alongside developing innovative products. Governments, along with startup entities, continue to finance seaweed cultivation within climate-smart agricultural programs while expanding its business landscape.

4 Chitosan's Effects on Plant Growth

Chitosan, as a natural pesticide, lowers the quantity of eggs and improves the appearance of females (Sahab et al. 2015). According to Ningsih et al. (2020), chitosan can prevent *Spodoptera litura* from eating. It is a powerful fungicide that may prevent the growth and reproduction of mycelium in *Colletotrichum*, *Gloeosporioides*, Alternaria spp., *Fusarium oxysporum*, and *F. graminearum*, in addition to its insecticidal properties (Barrera-Necha et al. 2018). The capacity of chitosan to promote plant growth is one of its main advantages. In *Brassica* crops, chitosan has been shown to enhance a number of growth metrics, including as root and shoot development, leaf area, and total biomass. Applying chitosan improves physiological functions like nutrition uptake, photosynthesis, and cell division while also activating plant defense mechanisms. Numerous investigations have demonstrated that chitosan stimulates the production of compounds that promote plant growth, such as phytohormones including auxins, cytokinins, and gibberellins. Under both typical and stressful circumstances, these hormones aid in controlling the growth and development of plants, resulting in better growth. As an illustration, Rabea et al. (2003) found that adding chitosan to broccoli and cabbage increased the plants' fresh weight, root length, and height. In a similar vein,

El-Tanbouly et al. (2008) found that *Brassica oleracea* grew better after receiving chitosan treatments, showing notable increases in leaf area and shoot biomass in comparison to untreated controls.

5 Chitosan in *Brassica*

Utilizing materials that are readily available, readily degradable, environmentally safe, and abundant in nature is one way to boost agricultural output while reducing environmental harm. Chitosan is one of the substances that may be useful in boosting agricultural output (Maluin and Hussein, 2020). Natural sources of chitosan, a biopolymer made from chitin, include the cell walls of fungi and the exoskeletons of arthropods like crustaceans. Chitosan has drawn interest as a plant growth enhancer and biocontrol agent due to its many uses, especially in agriculture (Fig. 4).

Chitin, the second most prevalent natural polysaccharide biopolymer after cellulose, undergoes deacetylation to produce chitosan (Meyers and Bligh 1981). Chitosan's natural, biodegradable, biocompatible, and non-toxic qualities make it useful in a wide range of industrial settings. In practically every aspect of life, chitosan has been explored extensively, particularly in agriculture. Numerous studies have examined how *Brassica* species, which include commercially significant crops like cauliflower, broccoli, and cabbage, react to chitosan treatment. Chitosan has the ability to stimulate plant growth. For instance, it can raise the whole fresh mass of roots and leaves, as well as the activity of plant peroxidase and leaf catalase in *Capsicum annuum* (Asgari-Targhi et al. 2018). In a different study, Zayed et al. (2017) showed that chitosan might boost *Phaseolus vulgaris* seed flowering after 72 hours in salt, as well as Zea mays plant height and leaf width. Furthermore,

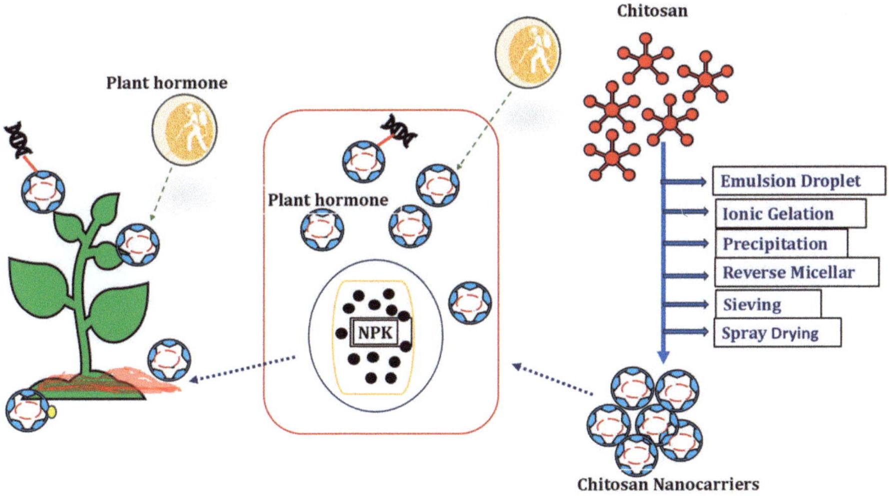

Fig. 4 Chitosan and nano-chitosan in agricultural applications

Sunarka et al. (2015) found that chitosan could enhance the *Salacca zalacca* plant's percentage of fruit set, relative leaf water content, and chlorophyll content. It could also increase chlorophyll, leaf breadth, and dry leaf weight, as well as maximize photochemical efficiency and photochemical results.

6 Enhancing Disease Resistance in *Brassica*

As a biocontrol agent, chitosan has shown great promise, especially against plant diseases. Plant defense responses, such as the synthesis of antimicrobial substances like phytoalexins, enzymes (such as chitinases and glucanases), and reactive oxygen species (ROS), are known to be triggered by it. Together, these compounds increase the plant's resistance to pathogens, which lowers the frequency and severity of disease. Chitosan has been demonstrated to increase resistance to bacterial and fungal infections in *Brassica* species. For example, a study by Raghavendra et al. (2011) found that treating *Brassica* plants with chitosan effectively protected them from *Alternaria brassicae*, a fungal pathogen that causes Alternaria leaf spot disease. This disease can significantly reduce productivity and is a major concern for *Brassica* crops. When compared to untreated controls, the chitosan-treated plants showed fewer lesions and better overall plant health. Furthermore, it has been discovered that chitosan increases the activity of enzymes involved in plant defense, such as polyphenol oxidases and peroxidases, which help the plant fight against pathogen attacks. These enzymes' enhanced activity helps to decrease disease-related damage and stop infections from spreading throughout plant tissues.

7 Impact on Crop Yield

Chitosan treatment in *Brassica* crops has been associated with increased crop output, particularly in biotic and abiotic stress situations. Chitosan increases plant growth and disease resistance, which raises *Brassica* species' overall output. Applying chitosan has been shown in numerous studies to increase the yields of *Brassica* crops, such as broccoli, cauliflower, and cabbage. In many studies, it has been proven that chitosan treatment enhanced the size, weight, and overall production of cabbage plants in a field experiment by Abdel-Mawgoud et al. (2010). In a similar vein, Omer et al. (2016) found that applying chitosan to broccoli greatly improved both its output and quality, with treated plants showing higher weight per plant and more marketable heads. Chitosan's multifarious action is responsible for its beneficial impacts on crop productivity. Chitosan not only promotes disease resistance and plant growth, but it also makes plants more resilient to environmental stressors like salinity and drought. Because of this, chitosan is a useful tool for raising *Brassica* crop productivity, especially in areas with less than optimum environmental conditions. When large quantities of chitosan (2.5–4.5%) are used, it has been found that the leaves of *Brassica juncea* plants that are two- or three-weeks old exhibit phytotoxicity. According to a study, using high concentrations of chitosan

(2.5–4.5%) causes phytotoxicity in plants of all ages. Therefore, the optimal concentration of chitosan to increase the chlorophyll content is 0.5%, and it has no harmful effects on *Brassica juncea*. By triggering a range of plant defense mechanisms, chitosan application changes the characteristics of plants. The buildup of phytoalexin, pathogenesis-related proteinase inhibitors, and other enzymes, such as the peroxidase enzyme, are examples of the plant's defense mechanisms (El Hadrami et al., 2010). The application of chitosan concentrations to plants is to minimize negative side effects and give plants a beneficial effect. A promising biostimulant for improving *Brassica* crop growth, productivity, and disease resistance is chitosan. It is a desirable alternative for sustainable farming methods because of its capacity to promote plant growth, activate defense mechanisms, and enhance stress tolerance. Chitosan may be crucial in raising *Brassica* species' productivity while lowering dependency on artificial fertilizers and pesticides as the need for efficient and ecologically friendly agricultural inputs grows. Future studies should concentrate on improving chitosan dosages and administration techniques, as well as investigating any possible synergistic effects with other biocontrol agents and agronomic techniques.

8 Conclusion

The Brassicaceae family includes *Brassica* species that demonstrate great importance in farm production and health care applications because of their multiple phytochemical content and various biological characteristics. There is global interest in cultivating mustard together with cabbage and broccoli and kale and cauliflower, and rapeseed because these plants enhance food resources and sustainable agricultural methods as well as human health benefits. *Brassica* species contain bioactive compounds, which include glucosinolates and their hydrolysis products. These compounds show antioxidant and antimicrobial and anticancer, and anti-inflammatory capabilities that help prevent diseases while developing functional foods. Various phytochemical features in *Brassica* species depend on both genetic inheritance together with environmental elements, and specific agricultural management strategies that also influence the amount and accessibility of important chemical compounds. Researching these differences enables better exploitation of these compounds for pharmaceuticals and nutraceuticals as well as biotechnological applications. The importance of *Brassica* species in sustainable agriculture grows stronger because they improve soil quality while permitting crop product variety and help fuel biofuel industries. The combination of biofertilizers with organic amendments and biostimulants that utilize chitosan and seaweed extracts serves to enhance their position in environmentally friendly agricultural systems. The upcoming research must expand as it aims to unlock *Brassica* species' complete agricultural, nutritional, and medicinal value. The advancement of *Brassica* cultivation relies on research that optimizes growing methods and enhances bioengineering applications, and investigates new medical applications from *Brassica* compounds. *Brassica* species can become a vital instrument for global health services through coordinated approaches between traditional practices and modern scientific innovations in order to achieve sustainable food production and climate-adapted agriculture. *Brassica* species present an essential strategic solution to address current food security, along with health and environmental problems, and they should remain an important research subject for developing novel policies and commercial ventures.

References

Abdel-Mawgoud AMR, Ghoneim KM, El-Azab HM (2010) Impact of chitosan on growth, yield, and disease resistance in cabbage. Acta Hortic 849:327–332

Aires A (2015) Brassica composition and food processing. In: Preedy V (ed) Processing and impact on active components in food. Academic Press, San Diego, pp 17–25

Al Shahawany AW, Al Hattab ZN, Al Tahhan SF (2016) Qualitative and quantitative analysis of Sinigrin in different parts *in vitro* and *in vivo* of *Brassica nigra* plants. Biomed Biotechnol 4:19–24

Asgari-Targhi G, Iranbakhsh A, Ardebili ZO (2018) Potential benefits and Phytotoxicity of bulk and Nano-chitosan on the growth, morphogenesis, physiology, and micropropagation of capsicum Annuum. Plant Physiol Biochem 127:393–402. https://doi.org/10.1016/j.plaphy.2018.04.013

Barrera-Necha LL, Correa-Pacheco ZN, Bautista-Baños S, Hernández-López M, Jiménez JEM, Mejía AFM (2018) Synthesis and characterization of chitosan nanoparticles loaded botanical extracts with antifungal activity on Colletotrichum Gloeosporioides and Alternaria species. Adv Microbiol 8(04):286–296. https://doi.org/10.4236/AIM.2018.84019

Battacharyya D, Babgohari MZ, Rathor P, Prithiviraj B (2015) Seaweed extracts as biostimulants in horticulture. Sci Hortic 197:39–48

Braun U, Cook RTA, Inman AJ (2019) The powdery mildews (Erysiphales) of Europe. CBS Biodiversity Series

Chauhan ES, Tiwari A, Singh A (2016) Phytochemical screening of red cabbage (*Brassica oleracea*) powder and juice-a comparative study. J Med Plant Stud 4:196–199

Craigie JS (2011) Seaweed extract stimuli in plant science and agriculture. J Appl Phycol 23(3):371–393

Craigie JS (2022) Seaweed extract biostimulants: agricultural applications and future directions. Phycologia 61(2):231–248

Danlami U, Orishadipe Abayomi T, Lawal DR (2016) Phytochemical, nutritional and antimicrobial evaluations of the aqueous extract of brassica Nigra (Brassicaceae) seeds. American. J Appl Chem 4:161–163. https://doi.org/10.11648/j.ajac.20160404.17

Derbyshire MC, Denton-Giles M (2016) The control of Sclerotinia stem rot on oilseed rape: current practices and future opportunities. Plant Pathol 65(6):859–877

Dinkova-Kostova AT, Kostov RV (2012) Glucosinolates and isothiocyanates in health and disease. Trends Mol Med 18(6):337–347. https://doi.org/10.1016/j.molmed.2012.04.003

Dixon GR (2009) The occurrence and economic impact of clubroot (*Plasmodiophora brassicae*) and prospects for control. Crop Prot 28(9):757–762

Donald EC, Porter IJ (2014) Integrated management of clubroot in brassicas. Plant Pathol 63(1):1–10

El Hadrami A, Adam LR, El Hadrami I, Daayf F (2010) Chitosan in plant protection. Mar Drugs 8:968–987. https://doi.org/10.3390/MD8040968

El-Esawi MA (2021) Taxonomic relationships and biochemical genetic characterization of brassica resources: towards a recent platform for germplasm improvement and utilization. Annual Res Rev Biol 8:1–11. https://doi.org/10.9734/ARRB/2015/20645

El-Tanbouly NS, Kassem AH, Ismail EA (2008) Effect of chitosan on growth and yield of broccoli (*Brassica oleracea* var. italica). Int J Agri Biol 10(3):266–270

Francis A, Lujan-Toro B, Warwick S, Macklin J, Martin S (2021) Update on the Brassicaceae species checklist. Biodivers Data J 9:e58773. https://doi.org/10.3897/BDJ.9.e58773

Gómez-Campo C (1980) Morphology and morphotaxonomy of the tribe Brassiceae. In: Tsunoda S, Hinata K, Gómez-Campo C (eds) *Brassica* crops and wild allies. Japan Scientific Societies Press, Tokyo, pp 3–31

Gupta R, Sharma V, Singh K (2022) Enhancing soil fertility through crop rotation: the case of *Brassica* species. Agri Sci J 56(4):245–260

Hwang SF, Strelkov SE, Feng J, Gossen BD, Howard RJ (2012) *Plasmodiophora brassicae*: a review of an emerging pathogen and clubroot disease in Canada. Can J Plant Pathol 34(3):321–338

Ignatov AN et al (2020) Xanthomonas campestris pv. campestris: insights into the bacterial leaf spot of brassicas. J Plant Pathol 102(1):93–104

Jain P, Kumar R, Verma S (2023) Agronomic interventions for optimizing mustard productivity in India. Field Crop Res 178(2):195–208

Joshi A, Rao P, Patel N (2023) Precision farming in *Brassica* species: role of digital technology. J Agri Technol 45(1):120–138

Khan W, Rayirath UP, Subramanian S, Jithesh MN, Rayorath P, Hodges DM, Prithiviraj B (2009) Seaweed extracts as biostimulants of plant growth and development. J Plant Growth Regul 28(4):386–399

Kim HW, Ko HC, Baek HJ, Cho SM, Jang HH, Lee YM, Kim JB (2016) Identification and quantification of glucosinolates in Korean leaf mustard germplasm (*Brassica juncea var. integrifolia*) by liquid chromatography-electrospray ionization/tandem mass spectrometry. Eur Food Res Technol 242:1479–1484. https://doi.org/10.1007/s00217-016-2648-6

Kinley RD, de Nys R, Vucko MJ (2023) Asparagopsis taxiformis as a methane-reducing feed supplement for livestock. Anim Feed Sci Technol 290:115555. https://doi.org/10.1016/j.anifeedsci.2023.115555

Kumar S, Gupta P (2022) Advances in breeding for climate resilience in mustard. Plant Biotechnol J 40(3):89—8102

Kumar P et al (2014) White rust disease of rapeseed-mustard and its integrated management strategies. Indian J Agric Sci 84(4):478–485

Lafarga T, Bobo G, Viñas I, Zudaire L, Simó J, Aguiló-Aguayo I (2018) Steaming and sous-vide: effects on antioxidant activity, vitamin C, and total phenolic content of brassica vegetables. International journal of gastronomy and food. Science 13:134–139. https://doi.org/10.1016/j.ijgfs.2018.05.007

Maluin FN, Hussein MZ (2020) Chitosan-based Agronanochemicals as a sustainable alternative in crop protection. In: Molecules. MDPI AG. https://doi.org/10.3390/molecules25071611

Meena PD et al (2010) Alternaria blight: a chronic disease in rapeseed-mustard. Journal of oilseed. Brassica 1(1):1–11

Meena H, Sharma P, Yadav B (2020) Status and trends in oilseed production: focus on *Brassica* species. Indian J Agric Sci 90(5):321—32338

Meyers SP, Bligh D (1981) Characterization of Astaxanthin pigments from heat-processed crawfish waste. J Agric Food Chem 29(3):505–508. https://doi.org/10.1021/JF00105A017

Ningsih S, Putra NS, Trisyono YA (2020) Feeding inhibition by chitosan on larvae of Spodoptera litura (Lepidoptera: Noctuidae). J Perlindungan Tanaman Indonesia 24(2):182. https://doi.org/10.22146/jpti.44032

Omer MH, El-Kholy MA, Badr SA (2016) The role of chitosan in enhancing the yield and quality of broccoli under field conditions. J Appl Hortic 18(1):28–34

Önder A (2018) Coriander and its phytoconstituents for the beneficial effects. In: El-Shemy HA (ed) Potential of essential oils. IntechOpen, Croatia, pp 165–185. https://doi.org/10.5772/intechopen.78656

Pise NM, Sabale AB (2023) Effect of Ascophyllum nodosum extract on crop productivity. J Food Sci Technol 60(4):456–467. https://doi.org/10.1007/s13197-023-05788-x

Rabea EI, Badawy MEI, Abo El-Khair SA (2003) Chitosan as antimicrobial agent: applications and potential. Res J Microbiol 2(4):260–271

Raghavendra AS, Channarayappa R, Nanjappa H (2011) Effect of chitosan on disease resistance in *Brassica* species. Plant Pathol J 10(1):21–29

Rahman H, Peng G, Yu F, Falk KC, Kulkarni M, Selvaraj G (2014) Genetic resistance to Plasmodiophora brassicae in Brassica napus and breeding strategies to improve clubroot resistance. Can J Plant Pathol 36(1):17–28

Rakow G (2004) Species origin and economic importance of *Brassica*. In: Pua EC, Douglas CJ (eds) Brassica, Biotechnology in agriculture and forestry, 54. Springer, Berlin/Heidelberg. https://doi.org/10.1007/978-3-662-06164-0_1

Rimmer SR, Shattuck VI, Buchwaldt L (2007) Diseases of *Brassica* species. The American Phytopathological Society Press

Roque BM, Venegas M, Kinley RD, de Nys R, Duarte TL, Yang X, Kebreab E (2021) Red seaweed Asparagopsis taxiformis reduces enteric methane in beef steers. PLoS One 16(3):e0247820. https://doi.org/10.1371/journal.pone.0247820

Sahab A, Waly A, Sabbour M, Nawar LS (2015) Synthesis, antifungal and insecticidal potential of chitosan (CS)-G-poly (acrylic acid) (PAA) nanoparticles against some seed borne fungi and insects of soybean. Int J Chem Tech Res 8(2):589–598

Sharma K, Singh R (2021) Climatic suitability and soil requirements for mustard cultivation in India. Climatic Change Agri 12(3):175–190

Sharma HSS, Fleming C, Selby C, Rao JR, Martin T (2014) Plant biostimulants: a review on the processing of macroalgae and use for crop management to reduce abiotic and biotic stresses. J Appl Phycol 26(1):465–490

Sharma A, Kumar V, Kanwar MK, Thukral AK, Bhardwaj R (2017) Phytochemical profiling of the leaves of *Brassica juncea* L. using GC-MS. Int Food Res J 24(2):547–551

Singh Y, Malik CP (2017) Phenols and their antioxidant activity in Brassica juncea seedlings growing under $HgCl_2$ stress. J Microbiol Biotechnol Res 1:124–130

Sunarka IK, Rai IN, Kartini DNL (2015) Pengaruh Konsentrasi Antitranspiran Chitosan Terhadap Pembuahan dan Produksi Salak Gula Pasir di Luar Musim. Agrotrop 5(1):30–36

The Australian (2023) Sea Stock patents unique approach to Asparagopsis seaweed production. Retrieved from https://www.theaustralian.com.au/nation/seastock-patents-unique-approach-to-asparagopsis-seaweed-production/news-story/c8a10a66c0a957e181ff7f-2cb7c201b0

The Business Research Company (2024) Seaweed cultivation global market report 2024. Retrieved from https://www.thebusinessresearchcompany.com/report/seaweed-cultivation-global-market-report

University of California, Davis (2021) Feeding cattle seaweed reduces their greenhouse gas emissions by 82 percent. UC Davis News. Retrieved from https://www.ucdavis.edu/climate/news/feeding-cattle-seaweed-reduces-their-greenhouse-gas-emissions-82-percent

Uttara B, Singh AV, Zamboni P, Mahajan RT (2009) Oxidative stress and neurodegenerative diseases: a review of upstream and downstream antioxidant therapeutic options. Curr Neuropharmacol 7:65–74. https://doi.org/10.2174/157015909787602823

Verma D, Sharma K (2018) Genetic improvement of *Brassica* species: recent developments and challenges. J Crop Sci 35(2):10–118

Vicente JG, Holub EB (2013) Xanthomonas campestris pv. campestris: the cause of black rot disease of brassicas. Mol Plant Pathol 14(1):1–16

Walsh JA, Jenner CE (2002) Turnip mosaic virus and the quest for durable resistance. Mol Plant Pathol 3(6):289–300

West JS, Kharbanda PD, Fitt BBDL (2001) Epidemiology and management of *Leptosphaeria maculans* and *L. biglobosa* in oilseed rape. Eur J Plant Pathol 107(4):377–397

Yadav L, Mishra S, Rathore H (2023) Nutrient management strategies for higher yield in mustard. Soil Sci Plant Nutr 29(4):410–426

Zayed MM, Elkafafi SH, Zedan AMG, Dawoud SFM (2017) Effect of Nano chitosan on growth, physiological and biochemical parameters of Phaseolus vulgar sunder salt stress. J Plant Prod 8(5):577–585. https://doi.org/10.21608/JPP.2017.40468

Zodape ST (2001) Seaweed as a biofertilizer. J Sci Ind Res 60(5):378–382

Phytochemical and Pharmaceutical Applications of *Brassica* Plant

Lakhwinder Singh, Kanik Kumar Bansal, and Roushan Kumar

Abstract

Brassica species are gaining global attention for their health benefits, particularly due to bioactive compounds like glucosinolates and sulforaphane, which contribute to cancer prevention, neuroprotection, and metabolic health. However, commercializing *Brassica*-derived nutraceuticals and pharmaceuticals presents challenges in regulatory approval, sustainable cultivation, and technological advancements. Strict safety and efficacy assessments by regulatory bodies such as the FDA and EFSA are essential to ensure scientific validation of health claims and determine safe consumption levels. Sustainable farming methods, including hydroponics, aeroponics, and vertical farming, are improving bioactive compound accumulation while minimizing environmental impact. These approaches optimize growth conditions and enhance phytochemical composition, making *Brassica* crops more suitable for functional food applications. Additionally, artificial intelligence (AI) and machine learning (ML) are transforming *Brassica* research by accelerating genomic selection for high-yielding, nutrient-rich varieties. AI-driven precision agriculture further refines cultivation techniques, maximizing phytochemical content and improving efficiency in pharmaceutical development. This chapter explores the commercialization of *Brassica*-based nutraceuticals and pharmaceuticals, emphasizing the importance of regulatory frameworks, sustainable farming practices, and innovative technologies in unlocking their full potential.

L. Singh · K. K. Bansal
Faculty of Agriculture, Guru Kashi University, Talwandi Sabo, Bathinda, Punjab, India

R. Kumar (✉)
Department of Pharmaceutical Sciences and Technology, Maharaja Ranjit Singh Punjab Technical University, Bathinda, Punjab, India

A. Kumar, B. Koul (eds.), *Crop Improvement Strategies in Brassica species: Applied Science*, https://doi.org/10.1007/978-981-95-3861-4_5

Keywords

Brassica plants · Phytochemicals · Glucosinolates · Sulforaphane · Pharmaceuticals · Sustainable farming · Aeroponics · Artificial intelligence · Machine learning · Nutraceuticals · Precision agriculture.

1 Introduction

1.1 Genetic Diversity and Evolution

Brassica, a genus within the Brassicaceae family, consists of several economically and nutritionally valuable species cultivated globally. Key members of this group include *Brassica oleracea* (cabbage, broccoli, cauliflower, kale), *Brassica rapa* (turnip, pak choi, Chinese cabbage), *Brassica nigra* (black mustard), *Brassica juncea* (Indian mustard), *Brassica carinata* (Ethiopian mustard), and *Brassica napus* (rapeseed, canola). These species exhibit remarkable genetic diversity, displaying variations in morphology, biochemical properties, and environmental adaptability. Such diversity has facilitated their successful cultivation across diverse agro-climatic regions (Chalhoub et al. 2014).

1.2 The Triangle of U: Evolutionary Relationships

A pivotal model explaining the evolutionary pathways of *Brassica* species is the "Triangle of U," introduced by Nagaharu U in 1935. This concept elucidates the genetic relationships among six major species, demonstrating how three diploid species—*B. rapa* (AA), *B. nigra* (BB), and *B. oleracea* (CC)—underwent natural hybridization to give rise to three allotetraploid species—*B. juncea* (AABB), *B. carinata* (BBCC), and *B. napus* (AACC) (Parkin et al. 2010). Extensive research using molecular markers, cytogenetic techniques, and whole-genome sequencing has confirmed these evolutionary connections, reinforcing the significance of polyploidy in shaping *Brassica* genomes (Nagaharu 1935).

1.3 Genetic Variability and Domestication

The domestication of *Brassica* crops has led to substantial morphological and biochemical diversification. Selective breeding over centuries has resulted in subspecies with distinct characteristics, including compact heads (cabbage), enlarged floral structures (broccoli, cauliflower), and leafy forms (kale, collard greens) (Maggioni 2015). However, the domestication process has also contributed to a narrowing genetic base, necessitating the incorporation of wild relatives in modern breeding programs to enhance resilience and adaptability (Warschefsky and Riesberg 2021).

Wild *Brassica* species serve as reservoirs of beneficial genes conferring resistance to environmental stresses. Notably, *B. villosa* and *B. incana* harbor resistance traits against downy mildew and black rot, offering potential solutions for sustainable crop improvement (Quezada-Martinez et al. 2021). Recent advancements in genetic technologies, such as genome-wide association studies (GWAS) and CRISPR/Cas9-mediated gene editing, are accelerating the development of superior *Brassica* varieties with improved resistance, nutritional content, and environmental adaptability (Von Wettberg et al. 2018).

1.4 Phylogenetic Insights and Molecular Evolution

The application of molecular phylogenetics has provided deeper insights into the evolutionary history of *Brassica* species (Nei & Kumar 2000). Whole-genome sequencing studies have revealed that polyploidy events and genome rearrangements have significantly contributed to the genetic variability observed within the genus (Li et al. 2017). Comparative genomics between *Brassica* species and Arabidopsis thaliana, a model plant, has facilitated the identification of key genes involved in secondary metabolite biosynthesis, stress tolerance, and developmental regulation, thereby expanding the understanding of functional genomics in these crops (Chalhoub et al. 2014).

1.5 Implications for Medicinal and Nutritional Value

The rich genetic diversity of *Brassica* species translates into significant variation in their phytochemical composition, particularly in bioactive compounds such as glucosinolates, flavonoids, and phenolic acids. These secondary metabolites are recognized for their anticancer, antioxidant, and anti-inflammatory properties, contributing to the health-promoting benefits of *Brassica*-derived foods (El-Esawi 2017). The variation in glucosinolate profiles across different species influences their chemoprotective potential, underscoring the need for precise genetic characterization to optimize their nutritional and medicinal applications (Cartea and Velasco 2008). Figure 1 clearly shows the benefits of plant parts along with their medicinal uses.

2 Ethnobotanical Use of *Brassica* in Traditional Medicine

2.1 Ayurvedic Medicine

The significance of *Brassica* species in Ayurveda lies in their role in maintaining the balance of the three doshas—Vata, Pitta, and Kapha. *Brassica juncea* (Indian mustard) has traditionally been utilized for alleviating respiratory ailments, joint discomfort, and skin infections. Mustard seed oil is commonly applied to enhance blood circulation and ease muscular pain (Bosi et al. 2022).

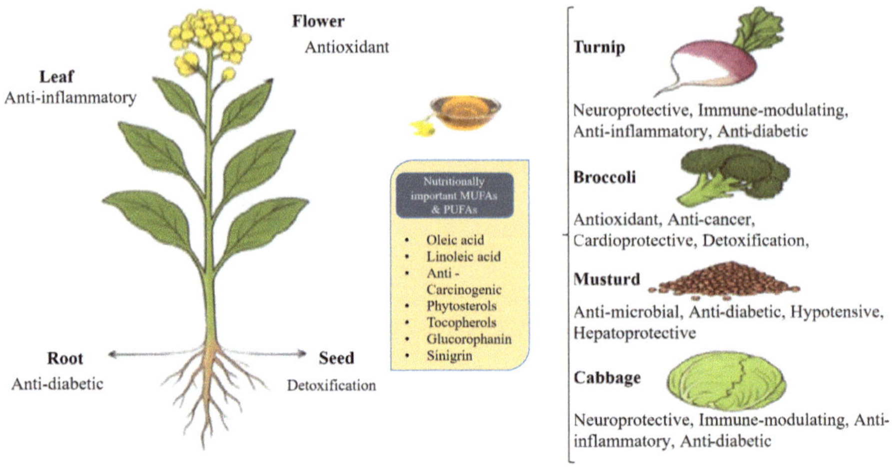

Fig. 1 Phytochemical composition and medicinal applications of *Brassica*

2.2 Traditional Chinese Medicine (TCM)

In Traditional Chinese Medicine, *Brassica* species are known for their warming and detoxifying properties. Chinese cabbage (*B. rapa*) is consumed to expel heat and phlegm, promoting better digestion and respiratory health. Mustard seeds are often incorporated into herbal preparations for addressing asthma, bronchitis, and colds (Au et al. 2008).

2.3 Unani Medicine

Within Unani medicine, mustard seeds (*B. nigra*) are valued as stimulants and expectorants. They are included in remedies aimed at boosting metabolism and managing arthritis. Extracts derived from cabbage are also recommended for digestive wellness and liver function support (Rehman et al. 2015). The comprehensive review of different parts of *Brassica* plant parts with their examples in traditional medicines is given in Fig. 2.

3 Phytochemistry of *Brassica* Crops

3.1 Sulforaphane Bioavailability and Stability

Sulforaphane (SFN), an isothiocyanate produced from glucoraphanin, is predominantly found in cruciferous vegetables such as broccoli, Brussels sprouts, and cabbage. It is widely recognized for its antioxidative, anti-inflammatory, and chemoprotective benefits (Zhang et al. 2024). However, SFN bioavailability is

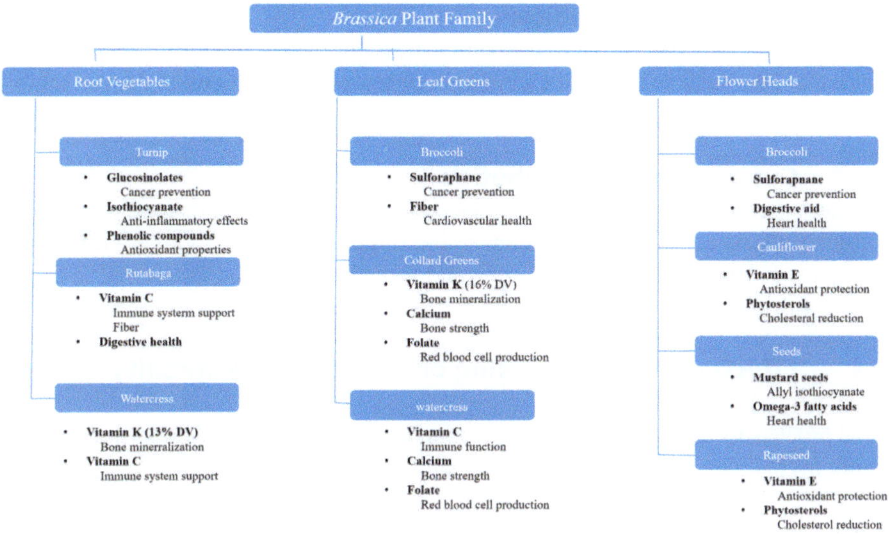

Fig. 2 Comprehensive flowchart of ethnobotanical use of *Brassica* plant parts in traditional medicine

influenced by several factors, including enzymatic hydrolysis, food processing techniques, gastrointestinal conditions, and individual metabolism. This section discusses the primary factors affecting SFN release and stability, focusing on the role of myrosinase enzyme activity and food processing in regulating SFN bioavailability (Ullah 2015).

3.2 Myrosinase Enzyme Activity and Sulforaphane Release

The enzyme myrosinase (EC 3.2.1.147), classified as a β-thioglucosidase, is responsible for breaking down glucoraphanin into SFN when plant tissues are disrupted (Mahn et al. 2020). This enzyme is housed in distinct cellular compartments within *Brassica* plants and becomes active when mechanical damage (such as chopping or chewing) or microbial fermentation in the gut occurs (Bones and Rossiter 1996).

3.3 Factors Affecting Myrosinase Activity

3.3.1 pH and Temperature Sensitivity

Myrosinase functions optimally within a pH range of 5.5 to 6.5, with its activity significantly reduced at extreme pH levels. Exposure to temperatures above 70°C causes the enzyme to denature, leading to a decrease in SFN formation (Van Eylen 2008).

3.3.2 Cofactors and Inhibitors

Ascorbic acid plays a crucial role in enhancing myrosinase activity, whereas divalent metal ions such as Fe^{2+} and Cu^{2+} and high-temperature processing act as inhibitors (Gardner 2015).

3.3.3 Gut Microbiota Contribution

In cases where dietary myrosinase is inactivated due to cooking or food processing, colonic bacteria can facilitate the conversion of glucoraphanin, though with limited efficiency (~10-20%) (Sikorska-Zimny and Beneduce 2021).

3.4 Effects of Food Processing on Sulforaphane Stability

Food processing techniques have a profound impact on SFN bioavailability by modifying myrosinase activity and glucoraphanin stability.

3.4.1 Boiling and Microwaving

Extended boiling of cruciferous vegetables results in substantial SFN losses (30-60%) due to thermal breakdown and leaching into the cooking water. However, short-duration microwaving (<2 min) preserves SFN content better than boiling (Houghton 2018).

3.4.2 Steaming and Blanching

Brief steaming (1-3 min) is beneficial in retaining myrosinase activity and enhancing glucoraphanin availability. Conversely, blanching at temperatures exceeding 90°C inactivates myrosinase while preserving glucoraphanin, which can later be converted by gut microbiota (Sun et al. 2021).

3.4.3 Fermentation and Sprouting

Lactic acid fermentation of *Brassica* vegetables (e.g., kimchi, sauerkraut) boosts SFN production by safeguarding enzymatic activity and altering glucosinolate composition (Luang-In et al. 2014). Sprouting enhances SFN bioavailability, as broccoli sprouts contain up to 100 times higher SFN precursors than mature broccoli due to increased myrosinase activity and glucoraphanin levels (Fahey et al. 1997).

3.5 Enhancing Sulforaphane Bioavailability

3.5.1 Various Approaches Can Optimize SFN Absorption and Stability

Blending Raw and Cooked *Brassica*

Mixing raw mustard seed powder (rich in myrosinase) with cooked broccoli enhances SFN production (Cramer 2012) (Table 1).

Table 1 Phytochemical constituents and pharmaceutical benefits of *Brassica* species

S. No.	*Brassica* species	Common name	Major phytochemicals	Pharmaceutical applications	References
1.	*Brassica oleracea*	Cabbage, broccoli, cauliflower, kale, Brussels sprouts	Glucosinolates (sulforaphane, indole-3-carbinol), flavonoids, vitamin C, carotenoids, polyphenols	Antioxidant, anticancer, anti-inflammatory, cardioprotective, detoxification, gut health improvement	Syed et al. (2023), Ağagündüz et al. (2022) and Baldelli et al. (2025)
2.	*Brassica rapa*	Turnip, pak choi, Chinese cabbage, tatsoi	Phenolic acids, carotenoids, alkaloids, anthocyanins, glucosinolates	Neuroprotective, immune-modulating, anti-inflammatory, antidiabetic	Ayadi et al. (2022) and Ramirez et al. (2020)
3.	*Brassica juncea*	Indian mustard, mustard greens	Sinigrin, isothiocyanates, omega-3 fatty acids, essential oils	Anti-microbial, anti-diabetic, hypotensive, hepatoprotective	Ayadi et al. (2022) and Rahman et al. (2024)
4.	*Brassica napus*	Rapeseed, canola, swede turnip	Tocopherols, polyphenols, omega-6 fatty acids, phytosterols	Cardiovascular health, lipid metabolism regulation, cholesterol-lowering, anti-inflammatory	Kumar et al. (2023) and Siddiqui et al. (2023)
5.	*Brassica nigra*	Black mustard	Allyl isothiocyanate, tannins, saponins, flavonoids	Anti-fungal, anti-inflammatory, pain relief, digestive stimulant	Rahman et al. (2024) and Kelechi et al. (2009)
6.	*Brassica carinata*	Ethiopian mustard	Erucic acid, flavonoids, lutein, sulfur-containing compounds	Anti-cancer, cardioprotective, anti-aging, skin health enhancer	Ayadi et al. (2022) and Zhang et al. (2025)
7.	*Brassica tournefortii*	(Asian mustard, Sahara mustard)	Sulforaphane, polyphenols, alkaloids, essential oils	Neuroprotection, anti-inflammatory, antimicrobial	Kumar et al. (2023) and Ayadi et al. (2022)

- **Encapsulation Methods:** Techniques such as liposomal and nano-emulsion encapsulation shield SFN from degradation and enhance its intestinal absorption (Yuan et al. 2017).
- **Dietary Enhancers:** Pairing SFN with dietary fats improves solubility and absorption, while probiotics aid gut microbial conversion of glucoraphanin (Atwell et al. 2015).

4 Comparative Phytochemical Analysis of *Brassica oleracea* and *Brassica rapa*

Brassica, a genus within the Brassicaceae family, comprises nutritionally and pharmacologically valuable vegetables. Among them, *Brassica oleracea* (e.g., broccoli, cabbage, kale) and *Brassica rapa* (e.g., turnip, bok choy) are widely cultivated for their edible and medicinal properties (Li et al. 2020). These species exhibit diverse phytochemical compositions, including glucosinolates, flavonoids, phenolic acids, and essential vitamins that contribute to their numerous health benefits (Cartea et al. 2010; Dominguez-Perles et al. 2018). The comparison and the analysis of phytoconstituents of *Brassica oleracea* and *Brassica rapa* are depicted in Fig. 3.

4.1 Glucosinolate Profiles and Breakdown Products

Glucosinolates, sulfur-rich secondary metabolites, are characteristic of *Brassica* species and undergo enzymatic hydrolysis to produce biologically active compounds such as isothiocyanates, thiocyanates, and nitriles (Verkerk et al. 2009).

4.2 Glucosinolate Content in *Brassica oleracea*

Species within *B. oleracea*, particularly broccoli and kale, are rich in aliphatic and indolic glucosinolates. Sulforaphane, derived from glucoraphanin, is a predominant isothiocyanate in broccoli, known for its detoxifying and anticancer potential (Traka

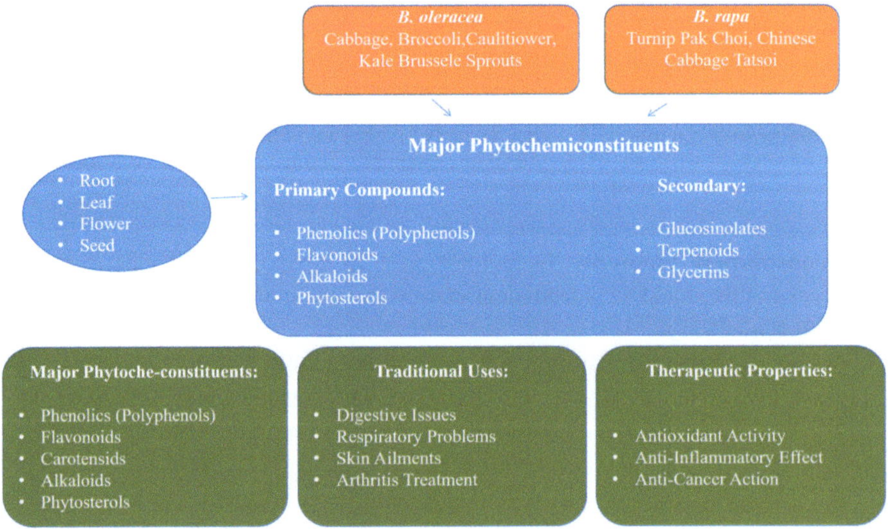

Fig. 3 Analysis and comparison of phytoconstituents of *Brassica oleracea* and *Brassica rapa*

and Mithen 2009). Other notable glucosinolates in this group include sinigrin, glucobrassicin, and neoglucobrassicin (Hanschen and Schreiner 2017).

4.3 Glucosinolate Content in *Brassica rapa*

B. rapa species, such as turnip and bok choy, generally contain lower glucosinolate levels than *B. oleracea*. However, they are particularly abundant in gluconasturtiin, a precursor of phenethyl isothiocyanate (PEITC), known for its anti-inflammatory and chemopreventive properties. Turnip is distinguished by its high glucotropaeolin content (Clarke et al. 2020).

4.4 Phenolic Compounds in *Brassica oleracea*

Broccoli and kale contain a wealth of flavonoids such as quercetin, kaempferol, and myricetin. They also feature phenolic acids, including ferulic acid and sinapic acid, which contribute to their notable antioxidant potential (Hounsome et al. 2009). Kale, in particular, has a diverse flavonoid composition, making it one of the most nutrient-dense *Brassica* vegetables (Khalid et al. 2023).

5 Vitamin and Mineral Content

Both *B. oleracea* and *B. rapa* species serve as excellent sources of essential vitamins and minerals, contributing significantly to human nutrition.

5.1 Vitamin Composition

B. oleracea vegetables, notably broccoli and kale, provide abundant vitamin C, vitamin K, and provitamin A (β-carotene) (Jeffery and Araya 2009). Among green leafy vegetables, kale possesses one of the highest concentrations of vitamin K. *B. rapa* species, while containing moderate amounts of vitamin C, are particularly rich in folate, which is crucial for DNA synthesis and fetal development (Cartea and Velasco 2008).

5.2 Mineral Composition

B. oleracea species exhibit high levels of calcium, magnesium, and potassium, which play vital roles in bone health and cardiovascular functions. Conversely, *B. rapa* species are characterized by elevated phosphorus and selenium levels, which are essential for antioxidant enzyme activity (Verkerk et al. 2009).

6 Antioxidant Capacity and Health Benefits

The antioxidant properties of *Brassica* vegetables are primarily linked to their polyphenolic content, vitamins, and glucosinolate-derived bioactive compounds.

6.1 Antioxidant Activity in *Brassica oleracea*

The significant presence of flavonoids and glucosinolates in *B. oleracea* enhances its antioxidative potential. Compounds such as sulforaphane and kaempferol exhibit potent free radical scavenging abilities, thereby protecting cells from oxidative stress-related damage (Traka and Mithen 2009).

6.2 Antioxidant Activity in *Brassica rapa*

Despite containing lower glucosinolate levels than *B. oleracea*, *B. rapa* species display strong antioxidant properties due to their rich anthocyanin and hydroxycinnamic acid content, particularly in turnip greens (Podsędek 2007).

7 Pharmaceutical and Therapeutic Applications of *Brassica* Active Phytoconstituents

Brassica species contain a wide range of bioactive compounds, including glucosinolates, flavonoids, and polyphenols, which exhibit significant pharmaceutical applications. These compounds have been extensively investigated for their impact on epigenetic mechanisms, neurodegenerative diseases, and metabolic disorders. This section highlights three primary pharmaceutical applications of *Brassica*-derived compounds in disease prevention and treatment. The pharmaceutical and medicinal applications of active phytochemicals are depicted in Fig. 4.

7.1 Sulforaphane and Epigenetic Modulation

Sulforaphane (SFN), an isothiocyanate derived from glucoraphanin hydrolysis, has been identified as a potent epigenetic regulator. It plays a pivotal role in modifying DNA methylation and histone modifications, which contribute to cancer prevention by repressing oncogenes and activating tumor-suppressor genes (Clarke et al. 2020).

7.1.1 DNA Methylation Changes

DNA methylation, which is facilitated by DNA methyltransferases (DNMTs), serves as a critical gene expression regulatory process. Research indicates that SFN inhibits DNMT1 and DNMT3A, leading to the reactivation of hypermethylated tumor suppressor genes, such as p16INK4a, BRCA1, and DAPK1, particularly in

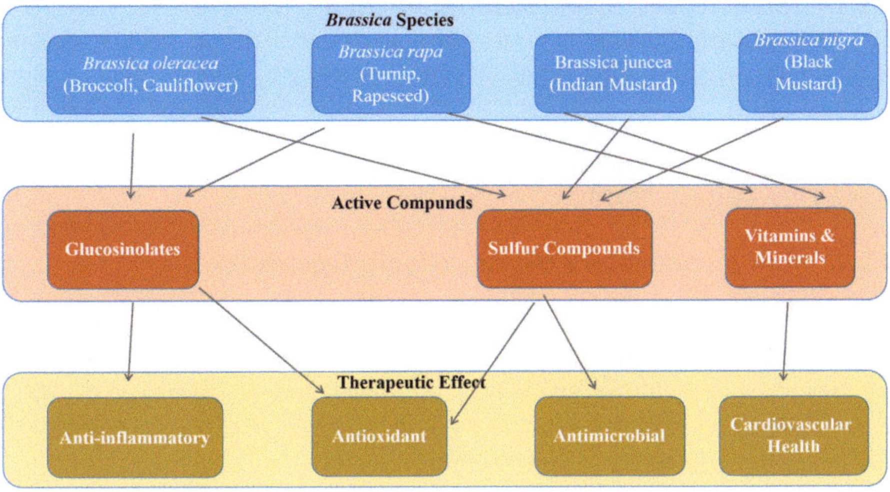

Fig. 4 Active phytochemical constituents and therapeutic benefits of different *Brassica* plant species

breast and prostate cancer models (Park et al. 2021). These findings suggest that dietary SFN consumption may help reverse abnormal methylation patterns, restoring normal cellular function and reducing cancer susceptibility.

7.1.2 Histone Modifications

Post-translational histone modifications, such as acetylation and methylation, play a crucial role in regulating chromatin structure and gene expression. SFN acts as an inhibitor of histone deacetylases (HDACs), particularly HDAC1, HDAC3, and HDAC6, promoting increased acetylation of histones H3 and H4. This chromatin relaxation facilitates the transcriptional activation of genes associated with apoptosis and cell cycle arrest, reinforcing SFN's role in cancer chemoprevention (Kumar et al. 2022).

7.2 Glucosinolate-Derived Compounds in Neurodegeneration

Neurodegenerative disorders, such as Parkinson's disease (PD) and Alzheimer's disease (AD), involve oxidative stress, neuroinflammation, and protein aggregation. Glucosinolate-derived isothiocyanates, including SFN and allyl isothiocyanate (AITC), have demonstrated neuroprotective effects through activation of the nuclear factor erythroid 2-related factor 2 (NRF2) pathway, thereby reducing oxidative damage and neurotoxicity (Zhao et al. 2022).

7.2.1 NRF2 Activation and Neuroprotection

The NRF2-Keap1 signaling pathway plays an essential role in cellular defense against oxidative stress. SFN has been found to enhance NRF2 expression, leading to the activation of antioxidant response element (ARE)-dependent genes, including NQO1, HO-1, and GCLC. Studies in PD models suggest that SFN mitigates dopaminergic neuronal loss and inhibits α-synuclein aggregation, a pathological hallmark of PD (Gao et al. 2019).

7.2.2 Anti-inflammatory Effects in Neurodegeneration

Neuroinflammation, driven by microglial activation and cytokine release, contributes to neuronal damage in AD. SFN has been shown to suppress the production of pro-inflammatory cytokines (IL-6, TNF-α, and IL-1β) by inhibiting NF-κB signaling (Spagnuolo et al. 2018). Additionally, SFN facilitates the autophagic clearance of amyloid-beta (Aβ) plaques, thereby reducing AD pathology and improving cognitive function in preclinical models (Kim et al. 2021).

7.3 Metabolic Syndrome Management

Metabolic syndrome (MetS) encompasses a cluster of disorders, including obesity, insulin resistance, hypertension, and dyslipidemia, all of which elevate the risk of cardiovascular diseases and type 2 diabetes. Bioactive compounds present in *Brassica* vegetables, particularly SFN and indole-3-carbinol (I3C), have shown potential in regulating lipid metabolism and inflammatory responses associated with MetS (Castelão-Baptista et al. 2021).

7.3.1 Regulation of Lipid Metabolism

Brassica-derived SFN has been shown to modulate lipid metabolism by activating peroxisome proliferator-activated receptor alpha (PPARα), a key regulator of fatty acid oxidation (He et al. 2021). PPARα activation enhances hepatic β-oxidation, reducing triglyceride accumulation and lowering circulating LDL cholesterol levels. Additionally, SFN upregulates AMP-activated protein kinase (AMPK), promoting glucose uptake and improving insulin sensitivity (Wang et al. 2022).

7.3.2 Anti-Inflammatory Role in Obesity-Linked Inflammation

Obesity-induced chronic inflammation is a major driver of MetS. SFN and I3C suppress adipose tissue inflammation by inhibiting pro-inflammatory mediators such as **TNF-α**, **IL-6**, and **MCP-1** (Kushwaha et al. 2022). Furthermore, SFN attenuates NF-κB activation, preventing macrophage infiltration into adipose tissue and reducing systemic inflammation (Baptista et al. 2021).

8 Nutraceutical and Functional Food Applications

Brassica species have gained significant recognition for their dual role in pharmaceutical and nutraceutical applications. These vegetables are rich in bioactive compounds such as glucosinolates, sulforaphane, flavonoids, and polyphenols, which contribute to various health benefits. Recent developments in food science have focused on improving the bioavailability and stability of these bioactive compounds to maximize their therapeutic effects. This section explores key strategies, including encapsulation technologies, fermented *Brassica* products, and clinical trials evaluating their functional food applications (Tundis et al. 2017).

8.1 Encapsulation Techniques for Enhancing Bioavailability of *Brassica* Bioactives

Encapsulation techniques have been extensively researched to improve the stability and absorption of *Brassica*-derived phytochemicals. Given their vulnerability to degradation caused by heat, pH variations, and enzymatic activity in the digestive tract, nanoencapsulation and microencapsulation techniques have been developed as effective solutions (Zabot et al. 2022).

8.1.1 Nanoencapsulation Strategies

Nanoencapsulation has been extensively utilized to safeguard and regulate the controlled release of *Brassica* bioactives. Nanocarriers such as liposomes, polymeric nanoparticles, and lipid-based delivery systems have significantly enhanced the solubility and bioavailability of sulforaphane and other glucosinolate derivatives. Studies have shown that nanoparticles loaded with sulforaphane exhibit superior intestinal absorption and prolonged circulation in the body, thereby amplifying its chemopreventive properties (Zhang et al. 2021).

8.1.2 Microencapsulation and Spray Drying Methods

Microencapsulation techniques such as spray drying, freeze drying, and emulsification have been applied to preserve *Brassica* bioactives (Mithen et al. 2011). These methods contribute to stabilizing glucosinolates and isothiocyanates while facilitating their incorporation into functional food formulations. Research findings suggest that microencapsulated *Brassica* extracts maintain their antioxidant properties even after food processing (Kim et al. 2021).

8.2 Fermented *Brassica* Products: Enhancing Probiotic and Anti-inflammatory Properties

Fermentation is a natural bioprocessing technique that enhances the bioactivity and availability of phytochemicals in *Brassica* vegetables. Fermented *Brassica* products, such as kimchi and sauerkraut, have been extensively studied for their role in promoting gut health and mitigating inflammation (Shahbazi et al. 2021).

8.2.1 Kimchi: A Functional Fermented *Brassica* Product

Kimchi, a traditional Korean fermented dish, is produced using *Brassica* species like Napa cabbage and radish, fermented with lactic acid bacteria (LAB). This fermentation process enhances the bioavailability of glucosinolates and their hydrolyzed derivatives, including sulforaphane and indole-3-carbinol. Research indicates that regular consumption of kimchi improves gut microbiota composition, decreases inflammatory markers, and supports immune function (Park et al. 2021).

8.2.2 Sauerkraut and its Probiotic Effects

Sauerkraut, a fermented cabbage preparation, is well-known for its probiotic benefits and positive impact on digestive health. Fermentation significantly boosts the levels of bioactive components, including flavonoids, phenolic acids, and vitamin C. Clinical research has indicated that consuming sauerkraut may promote digestive well-being and reduce the risk of chronic inflammatory conditions (Fijan et al. 2024).

8.3 Clinical Trials on *Brassica*-Based Supplements

Numerous clinical studies have investigated the health benefits and disease-preventive properties of *Brassica*-derived supplements. These trials provide critical insights into the bioavailability, efficacy, and safety of *Brassica* bioactives (Samec and Salopek-Sondi 2019).

8.3.1 Sulforaphane-Based Supplements in Chronic Disease Prevention

Clinical research has demonstrated that sulforaphane-rich broccoli sprout extracts exert protective effects against oxidative stress-related diseases, including cardiovascular disorders, neurodegenerative conditions, and metabolic syndrome. A randomized controlled study reported that daily supplementation with sulforaphane resulted in improved inflammatory and oxidative stress markers in individuals with type 2 diabetes (Nandini et al. 2020).

8.3.2 Glucosinolate-Derived Compounds in Cancer Prevention

Various intervention studies have underscored the role of glucosinolate-derived compounds in cancer prevention. A clinical study involving prostate cancer patients found that sulforaphane supplementation significantly decreased prostate-specific antigen (PSA) levels, suggesting its chemopreventive potential (Traka and Mithen 2009). Additionally, dietary intake of *Brassica* vegetables has been associated with a reduced risk of colorectal and breast cancers (Hayes et al. 2008).

8.3.3 Functional Foods Containing *Brassica* Bioactives

The incorporation of *Brassica* bioactives into functional food formulations has gained increasing commercial attention for their role in managing metabolic disorders. Clinical investigations have shown that dietary interventions with *Brassica*-enriched foods contribute to improvements in lipid metabolism, insulin sensitivity, and inflammatory markers in obese individuals (Artés-Hernández F et al. 2023).

9 Advanced Analytical and Biotechnological Approaches

Brassica plants are well known for their diverse phytochemical composition, particularly glucosinolates, flavonoids, and phenolic compounds, which contribute significantly to their health benefits. To optimize the extraction, quantification, and enhancement of these bioactives for therapeutic applications, advanced analytical and biotechnological methodologies are crucial. This section highlights three primary areas: high-throughput metabolomic profiling, CRISPR-based genetic modifications for glucosinolate enhancement, and biotransformation strategies aimed at utilizing *Brassica* agricultural waste (Ravanfar et al. 2017).

9.1 Metabolomics in *Brassica*

Research Metabolomics, a comprehensive systems biology approach, enables in-depth profiling of phytochemicals in *Brassica* species. Cutting-edge screening techniques such as gas chromatography-mass spectrometry (GC–MS), liquid chromatography-mass spectrometry (LC–MS), and nuclear magnetic resonance (NMR) spectroscopy have revolutionized the detection and quantification of bioactive compounds (Raza et al. 2020).

- **Untargeted vs. Targeted Metabolomics:** Untargeted metabolomics facilitates broad-spectrum identification of both known and novel metabolites within *Brassica* species, whereas targeted metabolomics provides precise quantification of specific bioactives like sulforaphane and kaempferol (Can 2023).
- **Metabolic Fingerprinting for Chemotype Classification:** The variation in glucosinolate profiles across different *Brassica* species has been mapped using principal component analysis (PCA) and hierarchical clustering, allowing for effective classification based on genetic and geographic influences.
- **Stress-Induced Phytochemical Variability:** Environmental stressors, such as drought, UV radiation, and pathogen attacks, play a pivotal role in influencing glucosinolate and flavonoid accumulation. Advanced metabolomic studies have identified correlations between abiotic stress conditions and increased bioactive production, offering valuable insights for breeding resilient *Brassica* varieties (Muthusamy and Lee 2024). The application of metabolomics significantly contributes to refining the phytochemical content of *Brassica* plants, making them more viable for nutraceutical and pharmaceutical advancements.

9.2 CRISPR for Glucosinolate Enhancement

CRISPR (Clustered Regularly Interspaced Short Palindromic Repeats) has emerged as a precise genome-editing tool for modifying biosynthetic pathways in *Brassica* plants to elevate glucosinolate concentrations.

- **Gene Editing for Higher Glucosinolate Yield:** Utilizing CRISPR-Cas9 to suppress negative regulators of glucosinolate biosynthesis, such as MYB28, has led to an increased concentration of beneficial bioactive compounds (Zhang et al. 2021).
- **Targeting Transcription Factors:** Transcription factors like MYB34 and MYB51 are central to the regulation of glucosinolate biosynthesis. CRISPR-induced overexpression of these genes has been associated with elevated levels of health-promoting indole and aliphatic glucosinolates (Augustine et al. 2014).
- **Biofortification Strategies:** Genetic modifications using CRISPR have also been employed to enhance the bioavailability of critical *Brassica* phytochemicals, such as sulforaphane, through targeted modifications of myrosinase enzyme activity (Wang et al. 2023). This innovative genome-editing approach offers significant potential for developing *Brassica* crops with superior nutritional and therapeutic properties.

9.3 Biotransformation of *Brassica* Waste

Brassica agricultural residues, including stems, leaves, and seed meal, are rich in bioactive compounds and can be efficiently utilized through biotransformation techniques.

- **Enzyme-Assisted Hydrolysis for Bioactive Extraction:** The use of enzymatic hydrolysis involving cellulases and pectinases has been shown to enhance the efficiency of glucosinolate extraction from *Brassica* waste, making them viable pharmaceutical precursors (Ungureanu et al. 2022).
- **Microbial Fermentation for Functional Ingredient Production:** The fermentation of *Brassica* residues using lactic acid bacteria and Saccharomyces cerevisiae has been demonstrated to enhance the release of bioactive isothiocyanates, increasing their bioavailability and antioxidant activity (Shinali et al. 2024).
- **Green Chemistry Approaches for Sustainable Utilization:** Advanced green extraction techniques, such as supercritical fluid extraction and ultrasound-assisted extraction, have been successfully employed to recover glucosinolates and polyphenols from *Brassica* waste, promoting sustainable functional food production (Matteo 2017).

10 Future Perspectives and Challenges

The increasing recognition of *Brassica* plants for their rich phytochemical and pharmaceutical potential calls for a strategic approach to their cultivation, processing, and commercial utilization. Although significant strides have been made in comprehending their bioactive constituents, hurdles persist in regulatory frameworks, sustainable production methods, and the data-driven enhancement of phytochemical yields. This section delves into the regulatory framework for *Brassica*-based nutraceuticals and pharmaceuticals, novel sustainable cultivation methods, and the integration of artificial intelligence (AI) for optimizing phytochemical biosynthesis.

10.1 Regulatory Landscape of *Brassica*-Based Products

The commercialization of *Brassica*-derived functional foods and pharmaceutical products requires thorough safety and efficacy evaluations mandated by global regulatory bodies.

- **Food and Drug Administration (FDA) Approval Pathways:** In the United States, *Brassica*-based products categorized as dietary supplements must comply with the Dietary Supplement Health and Education Act (DSHEA), while pharmaceutical formulations must undergo Investigational New Drug (IND) and New Drug Application (NDA) approvals (Naghshineh et al. 2014).
- **European Food Safety Authority (EFSA) Guidelines:** In the European Union, any health-related claims for *Brassica*-derived bioactives, such as glucosinolates and sulforaphane, require scientific validation under EFSA's regulations, specifically Regulation (EC) No. 1924/2006 (EFSA 2023).
- **Codex Alimentarius and Global Harmonization:** The Codex Alimentarius establishes international benchmarks for ensuring the safety and quality of *Brassica*-based food products, particularly in regulating glucosinolate concentrations (EFSA 2023).
- **Toxicological Assessments and Risk Analysis:** While *Brassica* phytochemicals offer health benefits, certain compounds like indole glucosinolates may pose cytotoxic risks at elevated levels. Regulatory agencies mandate toxicological evaluations through in vitro and in vivo models to determine safe consumption thresholds (Traka and Mithen 2009). Comprehensive regulatory policies are crucial to guaranteeing the safety, effectiveness, and commercial success of *Brassica*-derived bioactives in international markets.

10.2 Sustainable Urban Farming of *Brassica*

With rising global interest in nutrient-rich crops, sustainable urban farming innovations have emerged as effective solutions for large-scale *Brassica* cultivation while mitigating environmental challenges.

- **Hydroponic Systems for Controlled Cultivation:** Soil-less hydroponic cultivation with nutrient-enriched solutions has been shown to enhance bioactive compound accumulation in *Brassica* plants. Research indicates that hydroponically grown *Brassica* varieties exhibit increased glucosinolate and flavonoid concentrations due to optimized nutrient absorption (Rouphael and Colla 2020).
- **Aeroponics for Enhanced Phytochemical Synthesis:** In aeroponic cultivation, where plant roots are suspended in an oxygenated mist, antioxidant capacity and secondary metabolite synthesis in *Brassica* species improve significantly due to enhanced nutrient uptake and reduced oxidative stress (Borges et al. 2018).
- **Vertical Farming and LED Light Spectra Optimization:** The integration of vertical farming with hydroponics and LED lighting maximizes space efficiency. Studies suggest that blue and red light spectra modulate phytochemical synthesis in *Brassica* plants, leading to increased sulforaphane and kaempferol levels (KOHDA et al. 2022).
- **Carbon-Neutral and Circular Economy Models:** Incorporating *Brassica* cultivation into carbon-neutral farming approaches, such as wastewater recycling and biochar soil amendments, minimizes environmental impact while ensuring sustainable production (Rathore et al. 2022). These advanced agricultural methods facilitate year-round *Brassica* production with elevated bioactive compound synthesis, reducing reliance on conventional farming techniques.

10.3 AI and Machine Learning in *Brassica* Research

The advent of artificial intelligence (AI) and machine learning (ML) in agricultural and pharmaceutical research has transformed predictive modeling for optimizing phytochemical biosynthesis in *Brassica* plants.

- **AI-Based Genomic Selection for High-Yielding Varieties:** Deep learning algorithms analyze *Brassica* genomic data to identify high-glucosinolate-producing cultivars, streamlining breeding efforts (Li et al. 2024).
- **Machine Learning for Environmental Stress Prediction:** ML-driven models utilizing climate and soil datasets predict abiotic stress factors (e.g., drought, salinity) affecting phytochemical accumulation in *Brassica* plants, allowing for adaptive cultivation strategies (Li et al. 2024).
- **Big Data Analytics for Precision Agriculture:** AI-powered analysis of multispectral imaging and sensor data facilitates precision agriculture techniques, ensuring optimal light, water, and nutrient conditions to maximize sulforaphane and flavonoid production (Rai et al. 2024).
- **AI in Drug Discovery and Phytochemical Synergy Analysis:** Machine learning techniques predict the bioavailability, pharmaceutical potential, and synergistic interactions of *Brassica*-derived compounds, expediting nutraceutical and drug development (Ayadi et al. 2022). The integration of AI and ML in research provides opportunities to refine *Brassica* phytochemical content, improve precision farming practices, and fast-track the discovery of bioactive compounds for therapeutic applications.

11 Conclusion

Brassica plants possess remarkable potential in nutrition, medicine, and sustainable agriculture due to their diverse phytochemical profile. Compounds such as glucosinolates, sulforaphane, and flavonoids offer notable health advantages, including antioxidant, anticancer, and neuroprotective effects. However, maximizing their pharmaceutical applications necessitates a comprehensive approach that integrates innovative cultivation techniques, precise food processing strategies, and stringent regulatory measures. Sustainable farming practices, including aeroponics, hydroponics, and precision agriculture, have been shown to enhance the accumulation of bioactive compounds while reducing environmental impact. Additionally, advancements in technology, such as AI-driven breeding programs and big data analytics, contribute to improved productivity and optimized phytochemical synthesis in *Brassica* crops. Despite these technological advancements, key challenges persist in enhancing the bioavailability of critical compounds like sulforaphane, which is affected by enzymatic activity, food processing methods, and interactions with gut microbiota. The successful commercialization of nutraceuticals derived from *Brassica* requires extensive scientific validation, compliance with regulatory standards, and increased consumer awareness regarding their health benefits and safety. Future research should emphasize genetic enhancement strategies, personalized nutrition models, and the development of novel pharmaceutical formulations to unlock the full therapeutic potential of these functional plants. Through the integration of biotechnology, sustainable farming techniques, and pharmacological innovations, *Brassica* species can play a crucial role in promoting global health and enhancing food security.

References

Ağagündüz D, Şahin TÖ, Yılmaz B, Ekenci KD, Duyar Özer Ş, Capasso R (2022) Cruciferous vegetables and their bioactive metabolites. Evid Based Complement Alternat Med 2022:1534083. https://doi.org/10.1155/2022/1534083

Artés-Hernández F, Martínez-Zamora L, Cano-Lamadrid M, Hashemi S, Castillejo N (2023) Genus *Brassica* by-products revalorization with green technologies to fortify innovative foods: a scoping review. Foods 12(3):561. https://doi.org/10.3390/foods12030561

Atwell LL, Zhang Z, Mori M, Farris PE, Vetto JT, Naik AM, Shannon J (2015) Sulforaphane bioavailability and chemopreventive activity in women scheduled for breast biopsy. Cancer Prev Res 8(12):1184–1191. https://doi.org/10.1158/1940-6207.CAPR-15-0137

Au DT, Wu J, Jiang Z, Chen H, Lu G, Zhao Z (2008) Ethnobotanical study of medicinal plants used by Hakka in Guangdong, China. J Ethnopharmacol 117(1):41–50. https://doi.org/10.1016/j.jep.2008.01.006

Augustine R, Arya GC, Nambiar DM, Kumar R, Bisht NC (2014) Translational genomics in *Brassica* crops: challenges, progress, and future prospects. Plant Biotechnol Rep 8:65–81. https://doi.org/10.1007/s11816-013-0306-4

Ayadi J, Debouba M, Rahmani R, Bouajila J (2022) *Brassica* genus seeds: a review on phytochemical screening and pharmacological properties. Molecules 27(18):6008. https://doi.org/10.3390/molecules27186008

Baldelli S, Lombardo M, D'Amato A, Karav S, Tripodi G, Aiello G (2025) Glucosinolates in human health. Foods 14:912. https://doi.org/10.3390/foods14060912

Baptista S, Teixeira A, Castro L, Cunha M, Serrão C, Rodrigues A, Duarte I (2021) Physician burnout in primary care during the COVID-19 pandemic: a cross-sectional study in Portugal. J prim care community health 12:21501327211008437. https://doi.org/10.1177/21501327211008437

Bones AM, Rossiter JT (1996) The myrosinase-glucosinolate system, its organization and biochemistry. Physiol Plant 97(1):194–208. https://doi.org/10.1111/j.1399-3054.1996.tb00492.x

Borges CV, Junior SS, Ponce FS, Lima GPP (2018) Agronomic factors influencing *brassica* productivity and phytochemical quality. In: El-Esawi MA (ed) *brassica* germplasm—characterization, breeding and utilization. IntechOpen, pp 57–74. https://doi.org/10.5772/intechopen.76901

Bosi G, De Felice S, Wilkinson MJ, Allainguillaume J, Arru L, Nascimbene J, Buldrini F (2022) *Brassica* and Sinapis seeds in medieval archaeological sites: an example of multiproxy analysis for their identification and ethnobotanical interpretation. Plants 11(16):2100. https://doi.org/10.3390/plants11162100

Can H (2023) Melatonin application at different doses changes the physiological responses in favor of cabbage seedlings (*Brassica oleracea* var. capitata) against flooding stress. Gesunde Pflanz 75(6):2733–2745. https://doi.org/10.1007/s10343-023-00693-7

Cartea ME, Velasco P (2008) Glucosinolates in brassica foods: bioavailability in food and significance for human health. Phytochem Rev 7(2):213–229. https://doi.org/10.1007/s11101-007-9072-2

Cartea ME, Francisco M, Soengas P, Velasco P (2010) Phenolic compounds in *Brassica* vegetables. Molecules 16(1):251–280. https://doi.org/10.3390/molecules16010251

Castelão-Baptista JP, Barros A, Martins T, Rosa E, Sardão VA (2021) Three in one: the potential of *brassica* by-products against economic waste, environmental hazard, and metabolic disruption in obesity. Nutrients 13(12):4194. https://doi.org/10.3390/nu13124194

Chalhoub B, Denoeud F, Liu S, Parkin IA, Tang H, Wang X et al (2014) Early allopolyploid evolution in the post-Neolithic Brassica napus oilseed genome. Science 345(6199):950–953. https://doi.org/10.1126/science.1253435

Clarke WE, Chen L, Higgins EE, Kagale S, Koh CS, Parkin IA (2020) Narrow genetic base shapes population structure and linkage disequilibrium in an industrial oilseed crop, *Brassica* carinata A Braun. Sci Rep 10(1):12629. https://doi.org/10.1038/s41598-020-69499-y

Cramer J (2012) The bioavailability of sulforaphane from broccoli products in men and its epigenetic activity. Doctoral dissertation, University of Illinois at Urbana-Champaign

Domínguez-Perles R, Abellán Á, León D, Ferreres F, Guy A, Oger C, Gil-Izquierdo Á (2018) Sorting out the phytoprostane and phytofuran profile in vegetable oils. Food Res Int 107:619–628. https://doi.org/10.1016/j.foodres.2018.02.064

EFSA E (2023) European food safety authority

El-Esawi MA (2017) Genetic diversity and evolution of *brassica* genetic resources: from morphology to novel genomic technologies–a review. Plant Genet Resour 15(5):388–399. https://doi.org/10.1017/S1479262117000266

Fahey JW, Zhang Y, Talalay P (1997) Broccoli sprouts: an exceptionally rich source of inducers of enzymes that protect against chemical carcinogens. Proc Natl Acad Sci U S A 94(19):10367–10372. https://doi.org/10.1073/pnas.94.19.10367

Fijan S, Fijan P, Wei L, Marco ML (2024) Health benefits of kimchi, sauerkraut, and other fermented foods of the genus *brassica*. Appl Microbiol 4(3):1165–1176. https://doi.org/10.3390/applmicro4031165

Gao P, Xiang D, Quilichini TD, Venglat P, Pandey PK, Wang E, Gillmor CS, Datla R 2019 Gene expression atlas of embryo development in Arabidopsis. Plant Reproduction 32: 93–104. https://doi.org/10.1007/s00497-019-00364-x

Gardner AM (2015) Variation and quantitative trait loci analysis of myrosinase activity and phenolic compound accumulation in *Brassica oleracea* var. *italica*. Doctoral dissertation, University of Illinois at Urbana-Champaign

Hanschen FS, Schreiner M (2017) Isothiocyanates, nitriles, and epithionitriles from glucosinolates are affected by genotype and developmental stage in *Brassica oleracea* varieties. Front Plant Sci 8:1095. https://doi.org/10.3389/fpls.2017.01095

Hayes JD, Kelleher MO, Eggleston IM (2008) The cancer chemopreventive actions of phytochemicals derived from glucosinolates. Eur J Nutr 47:73–88. https://doi.org/10.1007/s00394-008-2009-8

He H, Wang HH, Liu J, Liu X, Li W, Wang Y (2021) Research progress and application of single-atom catalysts: a review. Molecules 26(21):6501. https://doi.org/10.3390/molecules26216501

Houghton C (2018) The role of phytochemicals in modulating intrinsic human cellular defence processes. Unpublished manuscript

Hounsome N, Hounsome B, Tomos D, Edwards-Jones G (2009) Changes in antioxidant compounds in white cabbage during winter storage. Postharvest Biol Technol 52(2):173–179. https://doi.org/10.1016/j.postharvbio.2008.11.003

Jang JE, Kim J, Purevdorj E, Son YJ, Nho CW, Yoo G (2021) Effects of long light exposure and drought stress on plant growth and glucosinolate production in pak choi (*Brassica rapa* subsp. *chinensis*). Food Chem 340:128167. doi:https://doi.org/10.1016/j.foodchem.2020.128167

Jeffery EH, Araya M (2009) Physiological effects of broccoli consumption. Phytochem Rev 8:283–298. https://doi.org/10.1007/s11101-008-9106-4

Kelechi O, Nwanebu F, Ndubuisi U, Orji N (2009) Antibacterial qualities of *Cucurbita pepo* and *Brassica nigra* oils. J Med Plants Res 3(5):351–354

Khalid W, Ikram A, Nadeem MT, Arshad MS, Rodrigues SDO, Pagnossa JP et al. (2023). Effects of traditional and novel cooking processes on the nutritional and bioactive profile of Brassica oleracea (Kale). J Food Process Preserv 2023(1):2827547. https://doi.org/10.1155/2023/2827547

Kohda J, Sasaki R, Suzuki R, Kugimiya A, Nakano Y, Takano Y (2022) Effects of light-emitting diode irradiation pattern on the anthocyanin content of broccoli sprouts in a plant factory for the production of functional foods. Jpn J Food Eng 23(3):71–83. https://doi.org/10.11301/jsfe.23.71

Kumar R, Saini DK, Kumar M, Priyanka V, Akhatar J, Kaushik D, Kaushik P (2022) Revealing the genetic architecture of yield-related and quality traits in Indian mustard [*Brassica juncea* (L.) Czern. and Coss.] using meta-QTL analysis. Agronomy 12(10):2442. https://doi.org/10.3390/agronomy12102442

Kumar A, P N, Kumar M, Jose A, Tomer V, Oz E, Proestos C, Zeng M, Elobeid T, Sneha K, Oz F (2023) Major phytochemicals: advances in health benefits. Molecules 28(2):887. https://doi.org/10.3390/molecules28020887

Kushwaha A, Goswami L, Lee J, Sonne C, Brown RJ, Kim KH (2022) Selenium in soil-microbe-plant systems: Sources, distribution, toxicity, tolerance, and detoxification. Crit Rev in Environ Sci Technol 52(13):2383–2420. https://doi.org/10.1080/10643389.2021.1883187

Li P, Zhang S, Li F, Zhang S, Zhang H, Wang X, Borm TJ (2017) A phylogenetic analysis of chloroplast genomes elucidates the relationships of the six economically important *Brassica* species comprising the triangle of U. Front Plant Sci 8:111. https://doi.org/10.3389/fpls.2017.00111

Li L, Gao Y, Zhao K, Kong L, Yu S, Li R, Yu X (2020) Comparative analysis of basic helix-loop-helix gene family among *Brassica oleracea*, *Brassica rapa*, and *Brassica napus*. BMC Genomics 21:1–18. https://doi.org/10.1186/s12864-020-07016-2

Li Z, Shah F, Xiong L, Zhang J, Wu W (2024) Unmanned aerial vehicles (UAVs)-based crop lodging susceptibility and seed yield assessment during different growth stages of rapeseed (*Brassica napus*). Comput Electron Agric 221:108980. https://doi.org/10.1016/j.compag.2024.108980

Luang-In V, Narbad A, Nueno-Palop C, Mithen R, Bennett M, Rossiter JT (2014) The metabolism of methylsulfinylalkyl-and methylthioalkyl-glucosinolates by a selection of human gut bacteria. Mol Nutr Food Res 58(4):875–883. https://doi.org/10.1002/mnfr.201300377

Maggioni L (2015) Domestication of *Brassica oleracea* L. NordGen, Report No 2015:74

Mahn A, Quintero J, Castillo N, Comett R (2020) Effect of ultrasound-assisted blanching on myrosinase activity and sulforaphane content in broccoli florets. Catalysts 10(6):616. https://doi.org/10.3390/catal10060616

Matteo R (2017) Non-food *brassicas* for green chemistry purposes through a biorefinery approach. Unpublished manuscript
Mithen R, Armah C, Traka M (2011) Cruciferous vegetables and biological activity of isothiocyanates and indoles. In: Vegetables, whole grains, and their derivatives in cancer prevention. Springer, pp 1–30. https://doi.org/10.1007/978-94-007-0230-5_1
Muthusamy M, Lee SI (2024) Abiotic stress-induced secondary metabolite production in *brassica*: opportunities and challenges. Front Plant Sci 14:1323085. https://doi.org/10.3389/fpls.2024.1323085
Nagaharu U (1935) Genome analysis in brassica with special reference to the experimental formation of B. napus and peculiar mode of fertilization. Jpn J Bot 7:389–452
Naghshineh NM, Brown S, Cederna PS, Levi B, Lisiecki J, D'Amico RA, Hume KM, Seward W, Rubin JP (2014) Demystifying the U.S. Food and Drug Administration: understanding regulatory pathways. Plast Reconstr Surg 134(3):559–569. https://doi.org/10.1097/PRS.0000000000000477
Nandini DB, Rao RS, Deepak BS, Reddy PB (2020) Sulforaphane in broccoli: the green chemoprevention!! Role in cancer prevention and therapy. J Oral Maxillofac Pathol 24(2):405. https://doi.org/10.4103/jomfp.JOMFP_320_20
Nei M, Kumar S (2000) Molecular evolution and phylogenetics. Oxford University Press
Park JS, Shin YH, Park YD (2021) DNA methylation level changes in transgenic Chinese cabbage (*Brassica rapa* ssp. pekinensis) plants and their effects on corresponding gene expression patterns. Genes 12(10):1563. https://doi.org/10.3390/genes12101563
Parkin IA, Clarke WE, Sidebottom C, Zhang W, Robinson SJ, Links MG, Sharpe AG (2010) Towards unambiguous transcript mapping in the allotetraploid *Brassica* napus. Genome 53(11):929–938. https://doi.org/10.1139/g10-074
Podsędek A (2007) Natural antioxidants and antioxidant capacity of *brassica* vegetables: a review. LWT-Food Sci Technol 40(1):1–11. https://doi.org/10.1016/j.lwt.2005.07.023
Quezada-Martinez D, Addo Nyarko CP, Schiessl SV, Mason AS (2021) Using wild relatives and related species to build climate resilience in *Brassica* crops. Theor Appl Genet 134(6):1711–1728. https://doi.org/10.1007/s00122-021-03808-x
Rahman M, Khatun A, Liu L, Barkla BJ (2024) *Brassica*ceae mustards: phytochemicals and pharmacological effects. Int J Mol Sci 25(16):9039. https://doi.org/10.3390/ijms25169039
Rai N, Pathak H, Mahecha MV, Buckmaster DR, Huang Y, Overby P, Sun X (2024) A case study on canola (*Brassica* napus L.) potential yield prediction using remote sensing imagery and advanced data analytics. Smart Agric Technol 9:100698. https://doi.org/10.1016/j.sat.2024.100698
Ramirez D, Abellán-Victorio A, Beretta V, Camargo A, Moreno DA (2020) Functional ingredients from *Brassica*ceae species. Int J Mol Sci 21(6):1998. https://doi.org/10.3390/ijms21061998
Rathore SS, Babu S, Shekhawat K, Singh VK, Upadhyay PK, Singh RK, Zaki FM (2022) Oilseed *Brassica* species diversification and crop geometry influence the productivity, economics, and environmental footprints under semi-arid regions. Sustainability 14(4):2230. https://doi.org/10.3390/su14042230
Ravanfar SA, Orbovic V, Moradpour M, Abdul Aziz M, Karan R, Wallace S, Parajuli S (2017) Improvement of tissue culture, genetic transformation, and applications of biotechnology to *Brassica*. Biotechnol Genet Eng Rev 33(1):1–25. https://doi.org/10.1080/02648725.2017.1367695
Raza A, Hafeez MB, Zahra N, Shaukat K, Umbreen S, Tabassum J, Hasanuzzaman M (2020) The plant family *Brassica*ceae: introduction, biology, and importance. In: The plant family *Brassica*ceae: biology and physiological responses to environmental stresses. Springer, pp 1–43. https://doi.org/10.1007/978-3-030-46859-8_1
Rehman K, Khan MA, Ullah Z, Chaudhary HJ (2015) An ethnobotanical perspective of traditional medicinal plants from the Khattak tribe of Chonthra Karak, Pakistan. J Ethnopharmacol 165:251–259. https://doi.org/10.1016/j.jep.2015.02.032
Rouphael Y, Colla G (2020) Biostimulants in agriculture. Front Plant Sci 11:40. https://doi.org/10.3389/fpls.2020.00040

Šamec D, Salopek-Sondi B (2019) Cruciferous (*Brassica*ceae) vegetables. In: Nonvitamin and nonmineral nutritional supplements. Academic Press, pp 195–202. https://doi.org/10.1016/B978-0-12-812491-8.00022-7

Shahbazi R, Sharifzad F, Bagheri R, Alsadi N, Yasavoli-Sharahi H, Matar C (2021) Anti-inflammatory and immunomodulatory properties of fermented plant foods. Nutrients 13(5):1516. https://doi.org/10.3390/nu13051516

Shinali TS, Zhang Y, Altaf M, Nsabiyeze A, Han Z, Shi S, Shang N (2024) The valorization of wastes and byproducts from cruciferous vegetables: a review on the potential utilization of cabbage, cauliflower, and broccoli byproducts. Foods 13(8):1163. https://doi.org/10.3390/foods13081163

Siddiqui SA, Khan S, Mehdizadeh M, Bahmid NA, Adli DN, Walker TR, Perestrelo R, Câmara JS (2023) Phytochem Food Packag Heliyon 9(11):e21196. https://doi.org/10.1016/j.heliyon.2023.e21196

Sikorska-Zimny K, Beneduce L (2021) The metabolism of glucosinolates by gut microbiota. Nutrients 13(8):2750. https://doi.org/10.3390/nu13082750

Spagnuolo C, Moccia S, Russo G L (2018) Anti-inflammatory effects of flavonoids in neurodegenerative disorders. Eur J med chem 153:105–115. https://doi.org/10.1016/j.ejmech.2017.09.001

Sun J, Wang Y, Pang X, Tian S, Hu Q, Li X, Lu Y (2021) The effect of processing and cooking on glucoraphanin and sulforaphane in *Brassica* vegetables. Food Chem 360:130007. https://doi.org/10.1016/j.foodchem.2021.130007

Syed RU, Moni SS, Break MK, Khojali WM, Jafar M, Alshammari MD, Abdelsalam K, Taymour S, Alreshidi KSM, Elhassan Taha MM, Mohan S (2023) Broccoli: a multi-faceted vegetable for health. Antibiotics 12(7):1157. https://doi.org/10.3390/antibiotics12071157

Traka M, Mithen R (2009) Glucosinolates, isothiocyanates and human health. Phytochem Rev 8:269–282. https://doi.org/10.1007/s11101-008-9103-7

Tundis R, Tenuta MC, Loizzo MR, Bonesi M, Menichini F, Duthie G (2017) Natural compounds and vegetable powders improve the stability and antioxidant properties of *Brassica* napus L. var. oleifera (rapeseed) oil. Eur J Lipid Sci Technol 119(4):1600228. https://doi.org/10.1002/ejlt.201600228

Ullah MF (2015) Sulforaphane (SFN): an isothiocyanate in a cancer chemoprevention paradigm. Medicine 2(3):141–156. https://doi.org/10.3390/medicines2030141

Ungureanu C, Fierascu I, Fierascu RC (2022) Sustainable use of cruciferous wastes in nanotechnological applications. Coatings 12(6):769. https://doi.org/10.3390/coatings12060769

Van Eylen D (2008) Effect of high pressure/temperature processing on the myrosinase-glucosinolate system in broccoli. Doctoral dissertation, Katholieke Universiteit Leuven

Verkerk R, Schreiner M, Krumbein A, Ciska E, Holst B, Rowland I, Dekker M (2009) Glucosinolates in *brassica* vegetables: the influence of the food supply chain on intake, bioavailability, and human health. Mol Nutr Food Res 53(S2):S219–S219. https://doi.org/10.1002/mnfr.200800164

Von Wettberg EJ, Chang PL, Başdemir F, Carrasquila-Garcia N, Korbu LB, Moenga SM et al (2018) Ecology and genomics of an important crop wild relative as a prelude to agricultural innovation. Nat Commun 9(1):649. https://doi.org/10.1038/s41467-018-03067-4

Wang W, Pang J, Zhang F, Sun L, Yang L, Siddique KH (2022) Transcriptomic and metabolomics-based analysis of key biological pathways reveals the role of lipid metabolism in response to salt stress in the root system of *Brassica* napus. Plant Growth Regul 97(1):127–141. https://doi.org/10.1007/s10725-022-00836-1

Wang Y, Song S, Hao Y, Chen C, Ou X, He B, Chen R (2023) Role of BraRGL1 in regulation of *Brassica rapa* bolting and flowering. Hortic Res 10(8):uhad119. https://doi.org/10.1093/hr/uhad119

Warschefsky EJ, Rieseberg LH (2021) Laying the groundwork for crop wild relative conservation in the United States. Proc Natl Acad Sci U S A 118(4):e2024375118. https://doi.org/10.1073/pnas.2024375118

Yuan D, Li W, Hua Y, King GJ, Xu F, Shi L (2017) Genome-wide identification and characterization of the aquaporin gene family and transcriptional responses to boron deficiency in *Brassica* napus. Front Plant Sci 8:1336. https://doi.org/10.3389/fpls.2017.01336

Zabot GL, Schaefer Rodrigues F, Polano Ody L, Vinícius Tres M, Herrera E, Palacin H, Olivera-Montenegro L (2022) Encapsulation of bioactive compounds for food and agricultural applications. Polymers 14(19):4194. https://doi.org/10.3390/polym14194194

Zhang T, Shang C, Du T, Zhuo J, Wang C, Li B et al. (2021) Cytoprotection of probiotics by nanoencapsulation for advanced functions. Trends in Food Science & Technology 142:104227 https://doi.org/10.1016/j.tifs.2023.104227

Zhang Y, Zhang W, Zhao Y, Peng R, Zhang Z, Xu Z, Deng J (2024) Bioactive sulforaphane from cruciferous vegetables: advances in biosynthesis, metabolism, bioavailability, delivery, health benefits, and applications. Crit Rev Food Sci Nutr 1(21). https://doi.org/10.1080/10408398.2024.1008000

Zhang X, Jia Q, Jia X, Li J, Sun X, Min L, Liu Z, Ma W, Zhao J (2025) *Brassica* vegetables: a nutritional goldmine. Hortic Res 12(2):uhae302. https://doi.org/10.1093/hr/uhae302

Zhao Y, Huang S, Wang N, Zhang Y, Ren J, Zhao Y, Feng H (2022) Identification of a biomass-unaffected pale green mutant gene in Chinese cabbage (*Brassica rapa* L. ssp pekinensis). Sci Rep 12(1):7731. https://doi.org/10.1038/s41598-022-11780-3

Advancements in Cultivation Practices of *Brassica* Crops: Ancient to Current Trends

Monika Chaudhary and Dilfuza Jabborova

Abstract

The production of *Brassica* crops, which include economically important species like broccoli, cabbage, and rapeseed, has transformed tremendously from ancient times to contemporary biotechnology innovations. Through history, these crops were cultivated for their nutritional and medicinal properties, with the earliest traces found in ancient Egyptian, Greek, and Roman cultures. The conventional agricultural practices, such as rain-fed cultivation, broadcasting by hand, and organic manuring, formed the basis of environmentally friendly practices. The Green Revolution (1960s–1980s) turned the tide with the introduction of HYVs, chemical inputs, and mechanization, dramatically increasing productivity. Yet, these developments also created environmental issues, such as soil degradation and misuse of pesticides.

The latter half of the twentieth and early twenty-first centuries were marked by revolutionary biotechnology advances in the form of hybridization, genetic engineering, and tissue culture, which led to increased disease resistance and stress tolerance in Brassica crops. Recent emphasis is on precision agriculture, using CRISPR-Cas9 genome editing for genes such as drought resistance and AI-based tools for maximized resource utilization. Climate-resilient measures such as no-till agriculture and biofertilizers, coupled with sustainable water management practices, are mitigating the impact of climate change. The future points to the use of nanotechnology for targeted application of agrochemicals, robotics for efficient farm management, and genomic selection for enhanced crops. They

M. Chaudhary
School of Biotechnology Engineering & Life Sciences, Shobhit Institute of Engineering & Technology (NAAC Accredited Grade 'A', Deemed to-be-University) NH-58, Meerut, Uttar Pradesh, India

D. Jabborova (✉)
Institute of Genetics and Plant Experimental Biology, Kibray, Uzbekistan

A. Kumar, B. Koul (eds.), *Crop Improvement Strategies in Brassica species: Applied Science*, https://doi.org/10.1007/978-981-95-3861-4_6

are meant to align production with ecological balance, stabilizing food amidst global uncertainty.

Keywords

Brassica crops · Domestication · Green revolution · Biotechnology · CRISPR-Cas9 · Precision agriculture · Sustainability · Climate-smart farming

1 Introduction

The genus *Brassica* is part of the *Brassicaceae* family, which includes over 3700 species and 338 genera, and comprises several economically significant crops, including broccoli, cabbage, cauliflower, kale, turnip, and rapeseed (canola). Historically, these plants have been utilized as food and medicinal sources by ancient civilizations such as the Greeks, Romans, Indians, and Chinese. Today, *Brassica* species are grown in more than 150 countries as both vegetable and oilseed crops, offering edible roots, leaves, stems, buds, flowers, and seeds (Salehi et al. 2021).

These species are valued for their rich nutritional profiles and are extensively cultivated and consumed globally, playing a vital role in agriculture and nutrition. Certain *Brassica* species are cultivated as cover crops or forage, with their leaves and stems serving as feed for livestock. These plants are distinguished by their four-petaled flowers and generally feature elongated, green leaves. Adapted to cooler climates, they exhibit notable frost tolerance, making them suitable for cultivation in such environments.

1.1 History

The cultivation of *Brassica* species dates back thousands of years, playing a significant role in the historical and cultural development of various societies and civilizations. Early evidence of *Brassica* cultivation has been documented in ancient Egyptian, Greek, and Roman cultures, where it was valued both as a food source and for its medicinal properties. In ancient Rome, *Brassica* was regarded as a symbol of prosperity and abundance, frequently featured in elaborate feasts and banquets. During the Middle Ages, it became a staple crop in Europe, widely used in culinary preparations such as soups, stews, and braised dishes. Additionally, *Brassica* is a polyploid genus, characterized by multiple sets of chromosomes. This genetic trait has facilitated its evolutionary adaptability, enabling the genus to thrive in diverse environments and contributing to its extensive diversification.

In contemporary agriculture, *Brassica* is cultivated globally across Europe, Asia, Africa, and the Americas. Its remarkable adaptability allows it to grow in a wide range of climatic conditions, from temperate to tropical regions, underscoring its importance as a versatile and resilient crop (Abbott 2025).

1.2 Genetics

Brassica is a polyploid genus, meaning its species possess multiple sets of chromosomes. The chromosome number varies across species, with some being diploid (2n), tetraploid (4n), or hexaploid (6n) (Wu et al. 2022; Li et al. 2023). The genus is characterized by a relatively large genome size, with certain species exceeding 1 billion base pairs. The *Brassica* genome is complex and diverse, containing numerous genes involved in the synthesis and metabolism of key plant compounds, such as glucosinolates and flavonoids. Additionally, the mustard family (*Brassicaceae*), to which *Brassica* belongs, is believed to have evolved and diversified during the early Cenozoic era and is closely related to the *Amaranthaceae* and *Chenopodiaceae* families.

The *Brassica* genome has been the subject of extensive research, leading to the development of various genetic resources and tools, including genetic maps, gene expression data, and DNA sequences. The complete sequencing of the *Brassica* genome has provided significant insights into the genetic mechanisms underlying important agronomic traits, such as disease resistance and stress tolerance, further enhancing its value in agricultural research and breeding programs (Mohd Saad et al. 2021).

1.3 Botanical Classification of *Brassica*

Six species of the genus *Brassica*, namely *Brassica oleracea L., Brassica rapa L., Brassica napus L., Brassica carinata A.* Braun, *Brassica nigra* (L.) K. Koch, and *Brassica juncea* (L.) Czern., are widely used as oilseed, condiments, fodder, or vegetable crops throughout the world. *Brassica oleracea*, the major vegetable species, includes varieties such as broccoli and cauliflower (*var. botrytis*), cabbage (*var. capitata*), kale (*var. acephala*), kohlrabi (*var. gongylodes*), and *Brussels sprouts* (*var. gemmifera*) (Šamec et al. 2017). *Brassica rapa* contains vegetable forms, such as turnip, Chinese cabbage, and pak choi, along with forage and oilseed types.

1.4 Cultivation Practices of *Brassica* Crops

Brassica vegetables, including the *Brassica oleracea* (i.e. cabbage, broccoli, cauliflower), *Brassica rapa* (e.g. turnip, bok choy), and *Brassica juncea* (mustard greens) crops, are essential in world agriculture because of their nutrient content and flexibility. Production of these crops successfully requires thorough knowledge of their agronomic needs, such as adaptability to climate, soil preferences, planting techniques, nutrient regulation, control of pests and diseases, and harvesting methods (Sobus 2022).

1.4.1 Climate and Soil Requirements

Brassicas prefer cooler climates, and their best growth is achieved in temperatures between 18 °C and 25 °C. They need a reasonable amount of soil moisture during growth and a dry, clear environment at the time of ripening. The crops are ideally suited to locations with annual precipitation ranging from 350 to 450 mm. Soil brassicas do well in well-drained, friable light loam soils with high organic matter content, having a pH of 6.5–7.0. They cannot tolerate waterlogging and have moderate tolerance to salinity and alkalinity (Agriculture in India).

1.4.2 Seedbed Preparation and Sowing

An ideal seedbed is essential for the growth of brassicas. The field must be deeply ploughed using a soil-turning plough followed by cross harrowing to attain a fine tilth. Planking should follow every ploughing so that the ground is well-pulverized and levelled, allowing for the best seed germination. The sowing dates differ based on the particular crop and local climate; nevertheless, sowing after the final spring frost or much earlier than the initial autumnal frost is most advisable.

1.4.3 Nutrient Management

Brassica crops are nutrient-hungry, especially needing sufficient nitrogen (N), phosphorus (P), and potassium (K) for their best growth. Soil test levels of phosphorus and potassium must be within the optimal range before planting. If phosphorus is less than 55 pounds per acre, application during *Brassica* growth may be justified since it can enhance yield when nitrogen is not growth-limiting. Phosphorus applications, though, can lower the crude protein level of the forage. For two or more grazings, a further 70 pounds of nitrogen must be applied between 60 and 80 days after planting to maximize yield and crude protein content (Penn State Extension 1993).

1.4.4 Pest and Disease Management

Brassicas are attacked by a range of pests and diseases, such as aphids, flea beetles, cabbage worms, bacterial soft rot, and leaf spot. Strategies for integrated pest management include regular surveillance, rotation of crops, and, when these become necessary, the selective use of insecticides and fungicides. Brassicas also release bio-toxic compounds when decomposing, which have wide activity against bacteria, fungi, insects, nematodes, and weeds, and could act as a natural fumigant when mixed with the soil (eorganic 2010).

1.4.5 Harvesting and Post-harvest Handling

Harvesting times for brassica crops vary by the particular species and use. For leafy types such as mustard greens, detopping early in the young crop stage (40–50 days after planting) can increase seed yield through increased branching. Such a practice can add supplemental income without reducing seed production. For turnips and other root crops, the time of harvesting may be when roots are of desired size, usually 60–70 days after planting. Sensible post-harvest handling, such as cooling and storage, is vital in order to preserve quality and prolong shelf life.

1.4.6 Cover Cropping and Soil Health

In addition to their function as food crops, some *Brassica* species are used as cover crops to build soil health. Their quick growth makes them a great soil cover, inhibiting erosion and weeds. Brassicas will root to depths of six feet or more, salvaging nutrients below the rooting depth of most crops. In order to optimize biomass production and scavenging of nutrients in the fall, brassicas need to be planted before winter cereal cover crops in most climates.

2 Evolution of *Brassica* Cultivation Practices

The history of *Brassica* cultivation practices has made remarkable progress from primitive to recent technological developments. The initial domestication in South and East Asia entailed cultural incorporation and rain-fed agriculture dependency, which formulated the basis of *Brassica* crop diversity. The Green Revolution (1960s–1980s) was a revolutionary period when high-yielding varieties (HYVs), dwarf cultivars, and chemical inputs were introduced and greatly improving the yield. Later biotechnological developments brought genetic alteration, tissue culture, and integrated pest management (IPM) into practice to achieve better disease resistance and crop yield. In contemporary agriculture (2010 to date), new technologies like CRISPR gene editing, AI-based farming, climate-resilient innovations, and precision irrigation are leading to sustainable cultivation patterns. These technologies help to enhance crop endurance, maximize resource efficiency, and reduce environment-related problems in *Brassica* cultivation (Fig. 1).

3 Ancient and Traditional Cultivation Practices

3.1 Early Domestication of *Brassica* Species

Domestication of *Brassica* species is a significant event in agricultural history as such plants are now a basic fare to a wide variety of cooking cultures and agricultural systems around the globe. From wild species to domesticated cultivars, the process takes several millennia and different locations.

3.2 Origins and Centres of Domestication

The *Brassica* species include a number of species such as *Brassica rapa, Brassica oleracea*, and *Brassica nigra* with varying domestication histories. *Brassica rapa*, such as turnips and Asian leafy greens, has been domesticated multiple times over a wide region extending from the Mediterranean to India, with indications of the earliest cultivation being approximately 2000 BCE. Prehistoric Chinese almanacs from around 3000 BCE and Indian literature from around 1500 BCE refer to these crops, and it is proposed that they must have been of significance in their societies from the

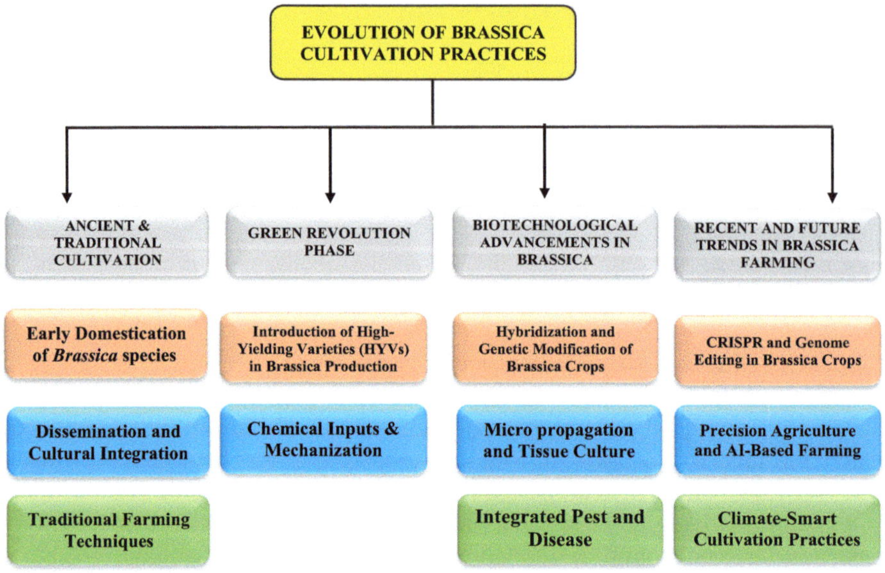

Fig. 1 Historical and modern advances in *Brassica* cultivation

earliest times (Wells 2024). *Brassica oleracea*, including species such as cabbage, broccoli, and kale, has a more disputed origin. Some have speculated domestication on European Atlantic coasts, with other suggestions being a Mediterranean origin, spreading to other regions later (Maggioni 2015).

3.3 Selection and Cultivation Practices

Early cultivators employed selective breeding to establish desirable traits in *Brassica* crops. Indian and Chinese farmers valued their crops for productivity and oil yield to meet food and medicinal demands. Selection remained largely empirical and based on the observable characteristics such as seed weight, oil yield, and vigour of the plants. This inductive form of genetic selection facilitated the diversity of cultivars currently available.

3.4 Dissemination and Cultural Integration

Trade networks and cultural contacts encouraged the dissemination of *Brassica* crops. Turnips (*Brassica rapa subsp. rapa),* for instance, were cultivated in northern Europe and became a staple in ancient Rome. The tolerance of *Brassica* species for diverse climates and soils made them versatile and thus extensively adopted and integrated into indigenous food systems and farming systems.

3.5 Genetic Insights and Evolutionary Pathways

Latest genetic studies have provided a greater understanding of *Brassica* crop domestication and evolution. Experiments have indicated that *Brassica rapa* most likely originated 4000 to 6000 years ago in the Hindu Kush of Central Asia with a triploid genome that was credited with its morphological variability. Diploid forms emerged as a consequence of domestication processes, giving rise to the varying vegetables and oilseed crops cultivated today (McAlvay et al. 2021).

3.6 Traditional Farming Techniques for *Brassica* Crops

Traditional farming methods for *Brassica* crops have evolved over centuries, driven by environmental conditions, cultural practices, and the need for sustainable agriculture. These methods, based on indigenous knowledge, emphasize harmony with nature and resource-use efficiency.

3.6.1 Rain-Fed Agriculture

In the majority of regions, *Brassica* production in the past relied on rain-fed agriculture, depending on seasonal rainfall to meet the water requirements of the crops. Such reliance on seasonal rainfall was common where irrigation systems were not in place or were rather underdeveloped. The rains instigated cultivation in accordance with when the crops needed an amount of water that enabled sowing. As helpful as it maintained man-made irrigation needs at a minimum, crop production was accompanied by high susceptibilities to rainfall changes. Such circumstances exposed agriculture to dry season stress, particularly droughts.

3.6.2 Manual Broadcasting and Sowing

Seeds were most often sown by a labour-intensive method, with broadcasting being one of the most widespread techniques. Farmers broadcast seed across levelled land evenly, a process that required skill to achieve uniform plant density. Labour intensive as it was, broadcasting allowed quick coverage of vast areas by seed without the requirement for specialized equipment. Seeds in certain cases were covered with soil superficially by using crude tools or by scratching twigs over the ground to provide improved contact between seeds and soil and enable increased germination.

3.6.3 Application of Farmyard Manure and Natural Compost

Fertility of the soil was maintained by applying organic inputs such as farmyard manure (FYM) and compost. They added nutrients to the soil as and when needed, assisted in structuring the soil, and provided moisture retention. FYM was used after it was collected as animal dung, bedding material, and crop residue, and left for decomposition as compost. Thus, the obtained compost was utilized on the farm during the preparation of land, which promoted nutrient recycling in a sustainable manner and reduced dependence on chemical fertilizers (Fao 2015).

3.6.4 Mixed Cropping and Intercropping

For maximum land use and minimum chances of total crop failure, farmers used mixed cropping and intercropping systems. *Brassica* crops were commonly grown together with cereals like wheat or legumes like peas and lentils. The crops offered a variety of benefits:

- *Resource Use Efficiency*: Different crops consume varying amounts of nutrients and water, thereby allowing complementary use of available resources.
- *Disease and Pest Management*: Plant diversities disrupt pest and disease life cycles, consequently reducing the cases of infestation compared to monoculture systems.
- *Improving Soil Fertility*: Legume intercrops that fix nitrogen from the atmosphere improve soil fertility, while bordering *Brassica* plants benefit from such fertilizer influences (Singh et al. 2010).

4 Green Revolution Phase (1960s–1980s): Increased Yield and Productivity in *Brassica* Cultivation

The Green Revolution of the 1960s and 1980s was a revolution in world agriculture, with the introduction of high-yielding varieties (HYVs), the increased use of chemical inputs, and mechanization (Fig. 2). While the primary emphasis has been laid on staple cereals such as wheat and rice, the *Brassica* crops from mustard to rapeseed also underwent substantial developments during this period (Swaminathan 2006).

4.1 Introduction of High-Yielding Varieties (HYVs) in *Brassica* Production

The Green Revolution, 1960s–1980s, was a period of revolutionary change in world agriculture by the innovation and spread of high-yielding varieties (HYVs) of food staple crops. Though the greatest emphasis has always been on cereals such as rice and wheat, much also happened in the improvement of *Brassica* crops such as rapeseed and mustard. This was to meet the objective of increasing oilseed production to meet the increasing demand of rapidly growing populations.

4.1.1 Selective Breeding Programs for *Brassica* Cultivars with High Oil Content

Selective breeding in the Green Revolution was aimed at creating *Brassica* cultivars of high oil content and better agronomic characteristics. The conventional varieties were discovered to have yield and oil concentration limitations, which challenged scientists to cross and select individuals with desirable characters. This was achieved by a process of selecting parent lines with desirable traits like disease resistance, drought tolerance, and high seed oil content, and crossing them to breed these characteristics together in new varieties. The HYVs so bred not only produced more oil

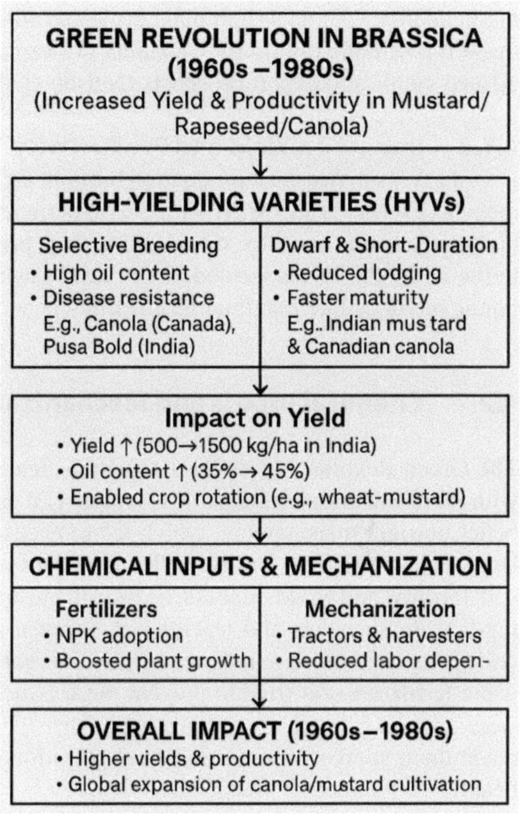

Fig. 2 Impact of Green Revolution on *Brassica* Cultivation

but were also more tolerant of varying climatic conditions, thereby increasing the geographical range of cultivation of *Brassica* (Khush 2001).

4.1.2 Evolution of Dwarf and Short-Duration Types in India and Canada

The impact of the Green Revolution in India on *Brassica* crops was characterized by the evolution of dwarf and short-duration mustard types. The new crop types were designed to accommodate multiple cropping systems such that farmers could produce mustard as one crop in a rotation without putting much into the farm calendar. The short growth period enabled planting and harvesting at the right time, thereby minimizing the potential for losses in the crop from inclement weather. Dwarf varieties were also less susceptible to lodging (falling over) since lodging is a prevalent problem in tall varieties of plants, leading to yield losses.

Canada's own contribution is the creation of canola, coined from the "Canadian oil, low acid." Canola was created through selective breeding by Canadian researchers that led to the traditional rapeseed being altered, eliminating unwanted compounds like erucic acid and glucosinolates, toxic to humans and animals. Such a breakthrough produced a crop with better nutritional values and better marketability. Canola's short plant cycle and adaptation to cooler climate conditions made it a suitable option for Canadian growers, considerably boosting the nation's oilseed yields and export prospects (Nelson et al. 2019; Pingali 2012).

4.1.3 Impact on Yield and Productivity

Use of HYVs in *Brassica* production brought significant yield and oil yield enhancements. Growers using such varieties experienced better per-acre seed yields and higher extraction rates for oil, meaning they became more profitable and secure in terms of food. HYVs yielded better results when cross-applied with better agronomic methods like maximal fertilization, pest control, and watering.

4.2 Chemical Inputs and Mechanization

The Green Revolution (1960s–1980s) drastically transformed *Brassica* cultivation with mechanization and chemical inputs. HYVs of mustard and rapeseed required better nutrient management, which led to excessive use of synthetic fertilizers at a large scale, particularly nitrogen (N), phosphorus (P), and potassium (K), to enhance soil fertility and yield. Excessive use of nitrogen led to acidification of the soil, nutrient deficiencies, and leaching of nitrates, an environmentally degrading process. Integrated nutrient management with organic manure supplemented with inorganic fertilizers was tried to prevent these issues. Pesticides, apart from fertilizers, were applied extensively for pest and disease control, but their indiscriminate use made them sources of environmental pollution, pest resistance, and even potential health risk (Ouda and Mahadeen 2008).

Mechanization changed *Brassica* cultivation with seed drills offering steady sowing, enhanced germination, better spacing of plants, and maximum yield at least cost, increasing the production. Mechanization movement overall contributed to the requirements for tractors, harvesters, and threshers, and offered intensive cultivation. Parallel development of irrigation systems was an instrumental step toward increasing *Brassica* production, particularly in non-conventional mustard lands. In India, canal and dam multipurpose constructions enabled frequent supply of water and production of *Brassica* vegetables in regions with undesirable rainfall conditions. This expanded production increased the quantity and yield of food, as well as contributed to water deficit as well as salinization issues; the conservation of water became necessary. In short, while chemical inputs and mechanization drastically raised *Brassica* productivity in the Green Revolution, they did not come without their environmental and resource management issues. Any eventual plan will have to marry sustainable agriculture practices with providing long-term sustainability, combining productivity with ecosystem conservation (Ozer 2003; Mahto et al. 2024).

5 Biotechnological Advancements in *Brassica* Cultivation (1990s–2000s)

The late twentieth and early twenty-first centuries marked significant biotechnological progress in *Brassica* crop cultivation, encompassing hybridization, genetic modification, tissue culture, micro propagation, and integrated pest and disease management (IPM).

5.1 Hybridization and Genetic Modification of *Brassica* Crops

Hybridization has been a foundation for improvement in *Brassica* crops, allowing the integration of preferred features from diverse varieties. Interspecific hybridization, which involves the crossing of different species in the *Brassica* genus, has been instrumental in the incorporation of features like disease resistance and enhanced productivity. For example, hybrids between *Brassica napus* (canola) and *Brassica rapa* have been researched for their promise to integrate beneficial features of the two parent species.

The development of genetic modification (GM) also continued to transform *Brassica* production. Transgenic varieties of *Brassica* that expressed such traits as herbicide resistance were, by the 1990s, developed to this end. For instance, glyphosate-resistant canola was introduced, which created the possibility for enhanced control of weeds and decreased use of mechanical cultivation. Nevertheless, there have been fears over the gene flow from GM crops to the wild relatives, with research showing possible hybridization involving transgenic *B. napus* and other allied species like *B. rapa* (Pandian et al. 2024).

5.2 Micropropagation and Tissue Culture

Methods of tissue culture have greatly facilitated the propagation and genetic enhancement of *Brassica* crops. Micropropagation, which comprises in vitro growth of plant tissues or cells, allows for efficient multiplication of disease-free and genetically homogeneous plantlets. The process has been effectively used on a number of *Brassica* crops, such as cabbage (*B. oleracea*), that allows for large-scale production and preservation of high-quality genotypes (Butt et al. 2015). Somatic embryogenesis, a tissue culture technique, has been used to regenerate plants from somatic cells, providing a route for genetic transformation and breeding. In *Brassica juncea* (Indian mustard), procedures have been developed for the induction of somatic embryos that mature into whole plantlets, providing a platform for the introduction of new traits by genetic engineering (Faisal et al. 2021).

5.3 Integrated Pest and Disease Management (IPM)

Integrated Pest Management (IPM) programs have been designed to manage the problem of pests and diseases in *Brassica* production. IPM focuses on the application of a combination of tactics, such as cultural management, biological control, and selective application of chemical insecticides, to control pest populations in a sustainable manner. The incorporation of biotechnological interventions, such as the creation of pest-resistant *Brassica* genotypes using genetic engineering, has made IPM programs more effective. For instance, transgenic *B. napus* lines with insecticidal proteins have been engineered to make them resistant to certain pests and minimize the application of chemical insecticides, and reduce environmental effects (Munshi et al. 2007).

6 Recent and Future Trends in *Brassica* Farming (2010–Present): Precision and Sustainable Farming

The past decade witnessed incredible advancements in *Brassica* farming, which has all been about precision and sustainable farming. Some of the major breakthroughs have been the application of CRISPR and genome editing technologies, precision agriculture with artificial intelligence (AI), climate-smart farming practices, and sustainable water management (Fig. 3).

6.1 CRISPR and Genome Editing in *Brassica* Crops

Gene-editing technologies like CRISPR-Cas9 have revolutionized *Brassica* crop breeding with the power to make targeted changes in plant genomes. Researchers used CRISPR-Cas9 to design heat- and drought-resistant mustard crops by editing specific genes for stress responses. A case in point is the editing of *Brassica rapa* BrERF4 gene for enhancing salt and drought tolerance (Zhu et al. 2016). Apart from this, CRISPR-Cas9 was also used to reduce the uric acid content in *Brassica napus* (canola) and thus produce varieties with improved-quality oil. A mutation evolved in the BnaFAE1 gene significantly reduced the uric acid content, enhancing canola oil quality (Liu et al. 2022; Ton et al. 2020).

Besides this, genome editing has enabled the ability to create disease-resistant *Brassica* lines. Genome targeting of disease susceptibility genes, such as the genes for the fatty acid desaturase, has improved resistance to bacterial and fungal pathogens and reduced chemical pesticides use.

6.2 Precision Agriculture and AI-Based Farming

With the integration of AI and precision agricultural methods, *Brassica* cultivation steps have been redefined. The use of multispectral imaging sensor-drones in crop

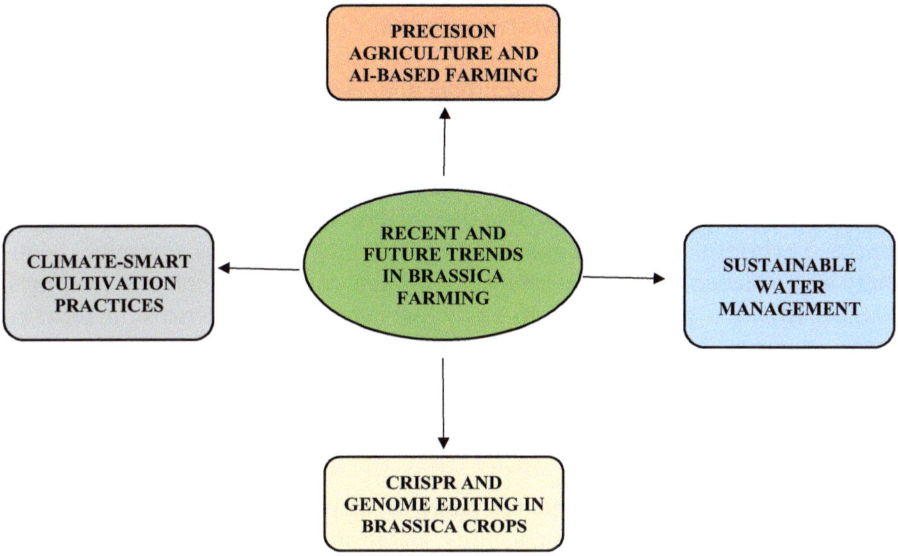

Fig. 3 Recent and future trends in *Brassica* farming

health monitoring is applied to detect stress indications such as nutrient deficiency, pests, and disease in the preliminary stage. This aids in intervening specifically, allowing for maximum usage efficiency of the resources and environmental footprint minimization.

Big data analysis and artificial intelligence are fundamental drivers of disease outbreak forecasting and fertilizer application optimization. Machine learning algorithms analyse large data sets, including weather patterns, soil composition, and historical crop yields, to provide meaningful outputs. For example, disease occurrence probability predictions are possible with AI-based models, which enable preventive management interventions.

6.3 Climate-Smart Cultivation Practices

To counter the effects of climate change, *Brassica* farmers are embracing climate-resilient agricultural practices. No-till and minimum-till crop management systems are utilized to improve soil health, carbon sequestration, and reduce greenhouse gas emissions. The practices also increase soil water storage, which is of utmost importance for crop stability in case of drought. Application of bio-fertilizers and organic supplements, such as mycorrhiza and Azotobacter, has gained importance in minimizing the reliance on chemical fertilizers. Such biological fertilizers enhance soil fertility, promote activity of beneficial microorganisms, and enhance plant nutrient absorption, all contributing toward sustainable *Brassica* production systems (Banerjee et al. 2023).

6.4 Sustainable Water Management

Water must be efficient in *Brassica* production, especially where there is climate change. Micro-irrigation technologies like drip and sprinkler irrigation are employed to conserve water and keep the soil at maximum water content. These provide water directly into the rooting area of the plant, minimizing evaporation loss and enhancing the efficiency of water use. Parallel to this, drought tolerance breeding in *Brassica* hybrids through traditional and biotechnological means guarantees continuity of production under water stress. For instance, overexpression of some genes, like BrERF4, has been reported to enforce greater drought tolerance in *Brassica* crops (Zhu et al. 2016).

7 Future Perspective

The future of *Brassica* production is ready to adopt state-of-the-art technologies that will increase efficiency, sustainability, and resilience. Nanotechnology is leading the way with the potential for targeted delivery systems of nutrients and pesticides. Nanoparticles can be used to encapsulate agrochemicals and provide controlled release, limiting environmental pollution. For example, biopolymer-based nanoparticles have been engineered to provide localized delivery of fertilizers and pesticides, reducing soil and air pollution. Besides, iron oxide nanoparticle application as a carrier of fertilizers makes it possible for controlled-release to improve the efficiency of nutrient uptake and crop yields (Sampathkumar et al. 2020; Rana et al. 2024). Agriculture is being revolutionized by robotics and automation, and it has vast implications for the production of *Brassica*. Using robots and intelligent sensors with functionalities of AI, such as planting, weeding, and harvesting, is now possible in an accurate way. For example, weeding robots identify the presence of weeds with the use of image recognition technology and kill them, and hence less herbicides are employed. Crop-harvesting robots, using cameras and robotic arms, can determine when to harvest at the optimal time for quality and reduce labour dependency. The technologies not only address the labour deficit but also enable more environmentally friendly farming by reducing wastage of resources (Bernier 2023). Very sophisticated genomic technology will accelerate the breeding of quality *Brassica* varieties using genomic selection. With the examination of genetic markers associated with desired characteristics such as resistance to disease, drought tolerance, and enhanced nutritional content, breeders can now make specific selections, thereby reducing the time it takes to develop new varieties. The process enhances the precision of breeding programs, which leads to *Brassica* crops that are well adapted to varying environmental conditions and consumer preferences.

8 Conclusion

The history of *Brassica* farming has been characterized by continuous innovations, from the classical to contemporary biotechnology. Each phase, from ancient domestication to the Green Revolution and precision agriculture today, has improved yield, quality, and sustainability. Genome editing, precision agriculture with artificial intelligence, and climate-smart agriculture are now defining *Brassica* farming of the future. In the years to come, the combination of nanotechnology, robotics, and genomic selection will play a dominant role in revolutionizing agriculture by maximizing resource utilization, minimizing environmental footprint, and creating crop varieties that are resilient. All these technologies will be crucial in addressing climate change and promoting global food security.

References

Abbott L (2025) Brassica https://cultivationag.com/brassica/

Agriculture in India (2022) How to cultivate and grow mustard greens (condiment crops). Retrieved from https://www.agricultureinindia.net/cultivation/mustard-greens/how-to-cultivate-and-grow-mustard-greens-condiment-crops-agriculture/18589

Banerjee S, Mukherjee A, Kundu A (2023) The current scenario and future perspectives of transgenic oilseed mustard by CRISPR-Cas9. Mol Biol Rep 50. https://doi.org/10.1007/s11033-023-08660-6

Bernier C (2023, July 17) Weeding robots: redefining sustainability in agriculture. *HowToRobot*. https://howtorobot.com/expert-insight/weeding-robots-redefining-sustainability-agriculture

Butt SJ, Varis S, Nasir IA, Sheraz S, Shahid A (2015) Micro propagation in advanced vegetable production: a review. Adv Life Sci 2(2):48–57

eOrganic (2010) Brassica cover crops for biofumigation and soil improvement. Retrieved from https://eorganic.org/node/2554

Faisal M, Abdel-Salam EM, Alatar AA, Qahtan AA (2021) Induction of somatic embryogenesis in *Brassica juncea* L. and analysis of regenerants using ISSR-PCR and flow cytometer. Saudi J Biol Sci 28(1):1147–1153

FAO (2015). https://www.fao.org/fileadmin/templates/nr/sustainability_pathways/docs/Compilation_techniques_organic_agriculture_rev.pdf

Khush GS (2001) Green revolution: the way forward. Nat Rev Genet 2(10):815–822

Li F, Gong Y, Mason AS, Liu Q, Huang J, Ma M, Fu D (2023) Research progress and applications of colorful brassica crops. Planta 258(2):45

Liu Y, Du Z, Lin S, Li H, Lu S, Guo L, Tang S (2022) CRISPR/Cas9-targeted mutagenesis of BnaFAE1 genes confers low-erucic acid in Brassica napus. Front Plant Sci 13:848723

Maggioni L (2015) Domestication of *Brassica oleracea L* (No. 2015: 74)

Mahto R, Singh RK, Singh JP, Tiwari RK, Vishwakarma DK, Obaidullah AJ, Yadav AK (2024) Evaluation of *Brassica* species for growth, yield and heat use efficiency under nitrogen nutrition and iron sulphide nanoparticles application. Sci Hortic 333:113278

McAlvay AC, Ragsdale AP, Mabry ME, Qi X, Bird KA, Velasco P, An H, Pires JC, Emshwiller E (2021) *Brassica rapa* domestication: untangling wild and feral forms and convergence of crop Morphotypes. Mol Biol Evol 38(8):3358–3372

Mohd Saad NS, Severn-Ellis AA, Pradhan A, Edwards D, Batley J (2021) Genomics armed with diversity leads the way in brassica improvement in a changing global environment. Front Genet 12:600789

Munshi M, Roy P, Kabir M, Ahmed G (2007) In vitro regeneration of cabbage (*Brassica oleracea L. var. Capitata*) through hypocotyl and cotyledon culture. In: Plant tissue culture and biotechnology, p 17
Nelson ARLE, Ravichandran K, Antony U (2019) The impact of the green revolution on indigenous crops of India. J Ethnic Foods 6(1):8
Ouda B, Mahadeen A (2008) Effect of fertilizers on growth, yield, yield components, quality and certain nutrient contents in broccoli (*Brassica oleracea*). Int J Agric Biol 10
Ozer H (2003) Sowing date and nitrogen rate effects on growth, yield and yield components of two summer rapeseed cultivars. Eur J Agron 19(3):453–463
Pandian S, Ban YS, Shin EK, Thamilarasan SK, Muthusamy M, Oh YJ, An HK, Sohn SI (2024) Introgression of herbicide-resistant gene from genetically modified *Brassica napus* L. to *Brassica rapa* through backcrossing. Plants 13(20):2863
Penn State Extension (1993) Use of brassica crops to extend the grazing season. Retrieved from https://extension.psu.edu/use-of-brassica-crops-to-extend-the-grazing-season
Pingali PL (2012) Green revolution: impacts, limits, and the path ahead. Proc Natl Acad Sci 109(31):12302–12308
Rana L, Kumar M, Rajput J, Kumar N, Sow S, Kumar S, Raigar BL (2024) Nexus between nanotechnology and agricultural production systems: challenges and future prospects. Discover Appl Sci 6(11):1–19
Salehi B, Quispe C, Butnariu M, Sarac I, Marmouzi I, Kamle M, Martorell M (2021) Phytotherapy and food applications from brassica genus. Phytother Res 35(7):3590–3609
Šamec D, Pavlovic I, Salopek-Sondi B (2017) White cabbage (*Brassica oleracea var. capitata F. alba*): botanical, phytochemical and pharmacological overview. Phytochem Rev 16(1):117–135
Sampathkumar K, Tan KX, Loo SCJ (2020) Developing nano-delivery systems for agriculture and food applications with nature-derived polymers. Iscience 23(5):101055
Singh R, Kumar M, Singh A (2010) Brassica based intercropping systems- a review. Agric Rev 31:253–266
Sobus K (2022) Garden Savy https://gardensavvy.com/gardening-blog/plants/growing-brassica-vegetables/
Swaminathan MS (2006) An evergreen revolution. Crop Sci 46(5):2293–2303
Ton LB, Neik TX, Batley J (2020) The use of genetic and gene technologies in shaping modern rapeseed cultivars (*Brassica napus* L.). Genes 11(10):1161
Wells R (2024) Origins and diversity of brassica and its relatives. In: Vegetable brassicas and related crucifers. CABI, GB, pp 1–41
Wu J, Liang J, Lin R, Cai X, Zhang L, Guo X, Wang X (2022) Investigation of brassica and its relative genomes in the post-genomics era. Hortic Res 9:uhac182
Zhu M, Monroe JG, Suhail Y, Villiers F, Mullen J, Pater D, Assmann SM (2016) Molecular and systems approaches towards drought-tolerant canola crops. New Phytol 210(4):1169–1189

Brassica Crop Improvement Through Seed Technology

Satya Prakash, Adesh Kumar, and Khalid Ali Khan

Abstract

Brassica crops are essential for global agriculture, contributing to food, oil production, and industrial applications. This chapter explores the evolving landscape of seed technology and genetic improvements within *Brassica* crops. Focusing on both qualitative and quantitative traits, the discussion highlights innovative breeding strategies aimed at enhancing seed health, oil quality, and overall crop yield. Through advanced genetic techniques, such as marker-assisted selection and CRISPR/Cas9, significant strides have been made in improving disease resistance, stress tolerance, and nutrient content. However, challenges persist within the seed technology and oil industries, including issues of seed quality maintenance, genetic diversity, and environmental adaptability. By addressing these challenges and incorporating emerging innovations, the future of Brassica crop improvement holds promising prospects for sustainable agriculture and global food security.

Keywords

Brassica · Seed technology · CRISPR/Cas9 · Oil quality · Crop yield

S. Prakash
Department of Genetics and Plant Breeding, Lovely Professional University, Phagwara, Punjab, India

A. Kumar (✉)
Department of Plant Pathology, Lovely Professional University, Phagwara, Punjab, India
e-mail: adesh.19078@lpu.co.in

K. A. Khan
Center of Bee Research and its Products (CBRP), King Khalid University, Abha, Saudi Arabia

A. Kumar, B. Koul (eds.), *Crop Improvement Strategies in Brassica species: Applied Science*, https://doi.org/10.1007/978-981-95-3861-4_7

1 Introduction

The *Brassica* genus, part of the Brassicaceae family, comprises a variety of economically essential crops, including canola (*Brassica napus*), mustard (*B. juncea*), cabbage (*B. oleracea*), and turnip (*B. rapa*), which hold significant value in global food systems (Sharma et al. 2022a, b; Yirmibeş and Ulger 2024). These crops are indispensable in the realms of food security, agricultural sustainability, and industrial applications, serving multifaceted roles as oilseeds, vegetables, animal feed, and biofuel sources (Ibrahim et al. 2024; Gray 2024). With the rapid global expansion of oilseed demand, *Brassica* species have become a primary supplier of edible oils, characterized by high nutritional value and bioactive components such as omega-3 fatty acids, antioxidants, and phytochemicals, which are beneficial for human health (Shen et al. 2023; Konopka et al. 2023; Axentii and Codina 2024). The past five decades have seen an exponential increase in the production of B. napus (rapeseed/canola), largely driven by a growing demand for stable vegetable oil sources, alongside innovations in agronomic practices, breeding technologies, and integrated pest and disease management strategies. In the 2020/2021 period, global production of *Brassica napus* (rapeseed/canola) reached approximately 72 million tons, driven by the demand for a stable source of cooking oil and advancements in farming practices, breeding programs, and improved disease and pest management strategies. Projections indicate that the global supply is expected to witness a remarkable increase to around 89 million tons in 2022/2023, reflecting a 17.2% increase compared to the previous year. This growth is attributed to factors such as increased consumption and the adoption of efficient farming techniques (Borges et al. 2023), further consolidating its position as a key oilseed crop. However, the productivity of other *Brassica* mustard species, including *B. juncea*, *B. nigra*, and *B. carinata*, remains variable, highlighting the need for continued research into genetic enhancement and crop management to overcome yield inconsistencies and improve overall crop performance.

Brassica species exhibit exceptional adaptability to a wide range of agro-climatic conditions, thriving in both temperate and subtropical regions. Many varieties demonstrate resilience to abiotic stresses such as drought, cold, and soil salinity, reinforcing their role in sustainable agriculture (Panjabi et al. 2019; Meena et al. 2025). Additionally, *Brassica* crops contribute to soil health through biofumigation and green manure applications, reducing reliance on chemical inputs (Tagele et al. 2021; Walker et al. 2022). Recent technological advancements, including CRISPR-Cas9 gene editing, marker-assisted selection, and omics technologies, are transforming *Brassica* breeding, enabling the development of cultivars with enhanced yield potential, improved oil composition, increased disease resistance, and greater environmental resilience (Ton et al. 2020; Sachan et al. 2024). Given their economic significance and wide-ranging applications, *Brassica* crops remain critical for addressing global agricultural challenges, ensuring food security, and promoting sustainable farming practices. Continuous research and breeding innovations are essential to further enhance their productivity and adaptability in response to changing climatic conditions.

Genetic enhancement for increased yield in *Brassica* crops has been a central focus of breeding initiatives. Techniques such as hybridization and marker-assisted selection (MAS) have facilitated the development of high-yielding varieties with superior oil content, enhanced disease resistance, and improved tolerance to abiotic stresses (Sunagar and Pandey 2024; Kumar et al. 2022). For instance, hybrid varieties of Indian mustard (*Brassica juncea*) and canola-quality rapeseed (*Brassica napus*) have demonstrated significant yield advantages due to heterosis breeding (Singh and Prakash 2024; Gupta 2016). Biotechnological advancements have further contributed to trait improvement, incorporating resistance to major diseases such as blackleg (*Leptosphaeria maculans*) and clubroot (*Plasmodiophora brassicae*) (Neik et al. 2017). Additionally, quantitative trait loci (QTL) mapping has facilitated the precise transfer of favorable alleles for yield enhancement and stress tolerance while maintaining agronomic performance. For example, a study identified QTLs associated with heat stress tolerance in *Brassica* rapa, which can be utilized to develop heat-tolerant cultivars (Chen et al. 2022).

Climate resilience and adaptation are critical for Brassica crops, given their widespread cultivation under diverse agro-climatic conditions. Drought and heat stress significantly impact seed germination, flowering, and seed development, thereby reducing yield potential. Efforts to develop drought- and heat-tolerant *Brassica* varieties have led to the introduction of early-maturing and water-efficient mustard cultivars suitable for rainfed conditions (Singh et al. 2022; Sharma 2020). Similarly, cold-tolerant rapeseed-mustard varieties have been developed to extend the growing season in temperate regions (Ilyas et al. 2022; Raza et al. 2024). Seed coating technologies and priming treatments further improve germination and seedling establishment under extreme environmental conditions. For instance, seed priming has been shown to enhance drought tolerance in *Brassica napus* by improving seed germination and early seedling growth (Gad et al. 2021). The integration of advanced breeding techniques, biotechnological innovations, and climate-resilient strategies has significantly enhanced the productivity and adaptability of *Brassica* crops. Ongoing research and breeding efforts are essential to further improve these crops' performance in the face of evolving environmental challenges.

Advancements in seed improvement have significantly enhanced the nutritional profiles of *Brassica* oilseeds and vegetables. Biofortification efforts have led to the development of mustard (*Brassica juncea*) varieties with reduced erucic acid and glucosinolate levels, rendering mustard oil safer for human consumption (Thakur et al. 2020). Additionally, high-oleic acid canola (*Brassica napus*) varieties have been developed to improve oil stability and associated health benefits. In vegetable *Brassica* crops, breeding programs targeting increased glucosinolate content in broccoli (*Brassica oleracea* var. *italica*) and cabbage (*Brassica oleracea* var. *capitata*) have enhanced their antioxidant and anticancer properties (Ross 2024). Yellow-seeded *B. juncea* varieties with improved seed meal quality have also been introduced, making them more suitable for animal feed due to lower fiber content (Jayaraman et al. 2016; Rad-Spice et al. 2018; Sharma et al. 2022a, b). These developments highlight the critical role of biofortification in addressing micronutrient deficiencies and improving human health through the consumption of *Brassica* crops. Ongoing research and breeding innovations are essential to further enhance the nutritional quality and health benefits of these crops.

Seed quality and viability are critical determinants of successful crop establishment and high yields in *Brassica* species. Optimal seed performance is characterized by uniform germination, robust seedling vigor, and enhanced disease resistance, all of which are essential for maximizing productivity. Technologies such as seed priming, pelleting, and coating have been developed to improve seed performance by enhancing nutrient uptake and providing protection against pathogens (Ma 2019; Javed et al. 2022). For example, nano-organic coatings have been shown to improve emergence percentage and seedling traits in *Brassica* napus by increasing antioxidant enzyme activities and reducing electrolyte leakage (Abrahimi et al. 2023). Nanotechnology-based seed treatments are being explored to further enhance seed vigor, improve resistance to seed-borne diseases, and increase germination rates under suboptimal conditions. Inorganic and organic nanoparticles are utilized as promising agents for seed priming, coating, and pelleting, offering potential benefits such as improved seedling growth and stress tolerance (Ali et al. 2023; Kloepper et al. 2004). Proper seed storage, certification, and testing protocols are essential for maintaining seed viability and genetic integrity over time, ensuring consistent field performance. Implementing these practices is crucial for sustaining agricultural productivity and meeting the growing global demand for *Brassica* crops.

Seeds are fundamental to the advancement of *Brassica* crop improvement, serving as the cornerstone for genetic enhancement, climate resilience, nutritional quality, and sustainable agricultural practices. Investing in high-quality, stress-tolerant, and biofortified *Brassica* seeds is imperative for enhancing productivity and ensuring food security, all while minimizing environmental impact. The integration of modern breeding technologies, such as genome editing and marker-assisted selection, alongside biodiversity conservation efforts and advanced seed treatments, plays a pivotal role in augmenting the adaptability and sustainability of *Brassica* crops globally. The conservation of *Brassica* genetic resources is crucial for maintaining genetic diversity, which is essential for future breeding endeavors. At present, at least 81,752 accessions of *Brassica* are recorded to be conserved in 148 holding institutes across 81 countries, underscoring the global commitment to preserving this valuable genetic pool. The strategic investment in seed quality and the adoption of innovative breeding and conservation practices are vital for the continued improvement and sustainability of *Brassica* crops, thereby contributing to global food security and environmental stewardship.

2 Seed Technological Advancements in *Brassica* Species Seed Production and Protection

2.1 Advancements in *Brassica* Seed Technology

Genomic innovations have significantly advanced *Brassica* seed improvement, driven by cutting-edge techniques such as CRISPR/Cas9 gene editing and pan-genome analysis. CRISPR/Cas9 has enabled precise alterations in key genes, such as BnFAD2, which plays a crucial role in fatty acid metabolism, leading to enhanced oleic acid content in *Brassica* oil (Ali and Zhang 2023). This improvement results in oils with greater oxidative stability, contributing to higher-quality products with extended shelf life. Comprehensive pangenome studies have uncovered structural variations and loci associated with stress resistance, offering valuable targets for the development of *Brassica* cultivars with superior resilience to environmental stressors (Pant et al. 2022). The polyploid nature of *Brassica napus*, which carries multiple sets of chromosomes, poses unique challenges in genetic manipulation. This complexity necessitates the use of multiplex genome editing strategies, where tandemly linked single-guide RNAs (sgRNAs) enable simultaneous modifications of homologous genes across multiple genomes (Mohd Saad et al. 2021). These genomic advancements have significantly enhanced seed vigor, oil quality, and stress resilience, optimizing crop performance under diverse environmental conditions. Seed vigor and viability are paramount for ensuring successful germination and early seedling establishment. Techniques like hydropriming, osmopriming, and biopriming have been shown to improve stress tolerance and extend seed longevity, promoting better emergence rates and robust early growth (Waqas et al. 2019). These treatments are particularly beneficial in mitigating the adverse effects of abiotic stresses, ensuring uniform seedling establishment even in fluctuating environmental conditions.

Hybrid breeding has also played a critical role in maximizing the yield potential of *Brassica* crops. The Barnase-Barstar gene system, employed in hybrids like *Brassica juncea* cultivar DMH-11, ensures genetic purity and facilitates the achievement of heterotic benefits. This system allows for the controlled production of hybrid seeds with superior agronomic traits, leveraging the power of hybrid vigor (Meena et al. 2025). In parallel, molecular marker-assisted selection (MAS) has revolutionized the process of seed quality improvement, enabling precise identification of parental lines and accurate verification of hybrid genotypes. This molecular approach ensures higher seed quality and consistency, laying the groundwork for the development of elite *Brassica* hybrids with optimized traits (Kumar et al. 2024).

Genomic-assisted breeding has revolutionized the improvement of disease resistance in *Brassica* species, utilizing advanced techniques like genomic selection (GS) and marker-assisted selection (MAS) to introgress critical resistance loci from wild relatives and other sources. In *Brassica napus*, loci for resistance to *Leptosphaeria maculans* (blackleg) such as Rlm and LepR have been successfully incorporated, providing enhanced defense against this devastating pathogen (Rouxel et al. 2024). Likewise, the development of clubroot resistance, caused by *Plasmodiophora brassicae*, has been significantly advanced through the stacking of multiple genes for resistance to downy mildew (*Hyaloperonospora brassicae*), which has provided new targets for breeding efforts aimed at controlling this pathogen (Zhang et al. 2022). CRISPR/Cas9 technology has been utilized to selectively resist genes, including CRa, CRb, Crr1, and Crr2, which work synergistically to improve disease tolerance (Saharan et al. 2022). In *B. oleracea*, the identification of candidate genes deactivates susceptibility genes, enabling precise resistance mechanisms (Boubakri 2023). For instance, the knockout of the BnWRKY70 gene in *Brassica* napus has resulted in enhanced resistance to blackleg disease, while the mutation of the BnaA9.WRKY28 gene in the same species has fortified its defense against clubroot (Kaur 2023).

Breeding efficiency in *Brassica* crops has also been drastically improved with the development of rapid-cycling *Brassica* rapa and B. napus lines. These fast-growing cultivars, such as the Wisconsin Fast Plants, complete their life cycle in just 35–45 days, accelerating the breeding process. These innovations enable breeders to rapidly cycle generations, thereby shortening breeding timelines and facilitating the development of new, high-performing cultivars in a fraction of the time (Saharan et al. 2023). Maintaining high seed vigor is essential for ensuring the success of these rapid-cycling lines. To optimize seed longevity and germination rates, precision priming techniques are employed. Methods like hydropriming and osmopriming, combined with controlled storage conditions (including cryopreservation and moisture reduction to ~6%), ensure that seeds retain their viability and can germinate quickly under optimal conditions (Arun et al. 2022). Additionally, the use of mutagenesis techniques such as ethyl methane sulfonate (EMS) and RNA interference (RNAi), along with polyploid manipulation, has further refined the selection of rapid-cycling genotypes, enhancing their suitability for commercial breeding applications (Kalia and Singh 2020). These genetic refinement strategies enable the creation of highly productive, disease-resistant *Brassica* varieties that can thrive in diverse agricultural settings.

2.2 Seed Coating and Seed Treatment Technologies

2.2.1 Seed Coating Technologies

Figure 1 gives an overview of seed coating, an important method in *Brassica* crops that consists of adding a thin protective covering on seeds to enhance germination, early plant growth, and general plant vigor. Seed coating has drawn immense interest based on its ability to increase seed performance and crop establishment under different environmental conditions. Evidence has shown the validity of seed coating in *Brassica* species. Dadlani (2025) indicated that coating seeds with fungicides and insecticides enhanced emergence and yield in seedlings. Likewise, Kumar et al. (2025) discovered that seed coating using plant growth-promoting rhizobacteria (PGPR) increased germination, stress survival, and biomass production. Seed coating using biofertilizers and drought-coping bacteria at hydro-primed significantly enhanced the performance of lentils under unsatisfactory growing conditions, something that could also be used with stress-tolerant *Brassica* growing. In general, seed coating is still a promising method for enhancing *Brassica* crop productivity and sustainability (Kumar et al. 2025). It remains an area of interest for research on how to optimize seed treatments based on specific environmental issues and crop needs.

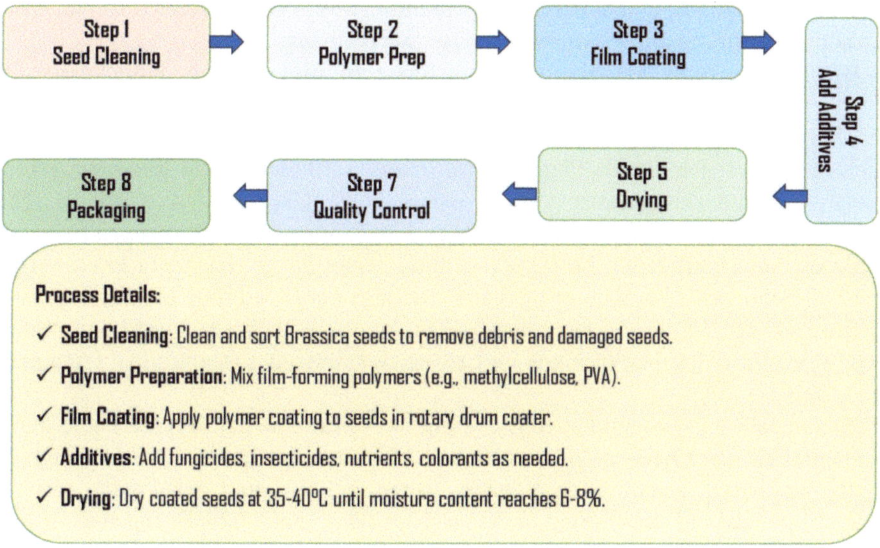

Fig. 1 *Brassica* seed film coating procedure

Film Coating

Film seed coating emerges as a dynamic adaptation strategy for optimizing *Brassica* crop performance under stress-prone environments. Film coating represents a precise and advanced seed enhancement technique, which involves the deposition of a microscopic, uniform layer of bioactive compounds onto the seed surface of *Brassica* species. Typically, the film constitutes up to 5% of the seed's total weight, ensuring that the active ingredients are applied in an efficient and controlled manner (Kavusi et al. 2023). The process relies on a liquid adhesive matrix that acts as the carrier, facilitating the dissolution and even distribution of bioactive substances onto the seed. To optimize the coating process, sophisticated delivery systems such as fluidized-bed treaters or pharmaceutical-grade coating drums are employed (Bahmid et al. 2024). These systems ensure consistent adhesion of the film while minimizing the loss of materials during application, contributing to an impressive recovery rate of over 90%, underscoring the method's high efficiency in seed treatment.

Upon sowing, *Brassica* seeds face a range of environmental stressors, both biotic and abiotic, which can hinder seedling establishment. Film coating technology has demonstrated a remarkable ability to enhance seed resilience against these stressors, providing protection from the imbibition phase through to the establishment of the seedling (Zaim et al. 2023). Among the numerous abiotic stress factors, waterlogging poses a significant threat to *Brassica* crops. When uniconazole (at a concentration of 0.0075%) was applied as a film coating to *Brassica* napus seeds, substantial protection against waterlogging stress was observed (Javed et al. 2022). This treatment not only improved the seeds' physiological resilience but also enhanced their morphological vigor. Specifically, antioxidant enzyme activity, including peroxidase, catalase, and superoxide dismutase, was notably increased, indicating the activation of the plant's stress-response mechanisms.

Biological agents are gaining prominence in film coating applications for *Brassica* species. Biological coatings, such as those derived from *Bacillus subtilis* and *Trichoderma harzianum*, offer protection by outcompeting or inhibiting pathogenic microbes. Paraburkholderia phytofirmans, formulated as a film coat, effectively mitigated drought stress in *Brassica* crops (Cardarelli et al. 2022). The functional synergy of Trichoderma fungi and *Pseudomonas fluorescens* in *Brassica* seed coatings significantly bolstered seedling performance (Rajendra Prasad et al. 2016). Trichoderma harzianum-mediated coatings enhanced root and shoot growth in *Brassica* species, while synergistic formulations with imidacloprid and metalaxyl-M conferred superior vigor and reduced pathogen susceptibility (Accinelli et al. 2016).

Nanotechnology-integrated film coatings have demonstrated remarkable agronomic benefits in *Brassica* species. Zinc and iron nanoparticles (750 ppm) significantly improved seedling emergence, vigor indices, and enzymatic activity (dehydrogenase and amylase) in *Brassica* crops (Dey et al. 2024; Shang et al. 2019). A polymer-mediated insecticide delivery extended the shelf life and biotic stress resilience of *Brassica* seeds (Singh et al. 2024). Film coatings utilizing chitosan and polyethylene glycol (PEG) fostered improved germination and vigor in *Brassica* species (Faizan et al. 2023).

The stability of agrochemicals in coated *Brassica* seeds is a critical consideration. Coating with polymers reduced the degradation rates of systemic insecticides such as imidacloprid and thiacloprid. However, the introduction of plant growth-promoting rhizobacteria (*Bacillus* spp.) within polymer matrices accelerated pesticide biodegradation, highlighting the dual role of coatings in stress mitigation and environmental sustainability.

Seed Encrusting

Encrusting involves the application of a seed coating that increases seed weight by 8–500% while maintaining the original shape of the seed (Javed et al. 2022). For successful encrusting in *Brassica* species, two critical phases, the before coating phase and the coating phase, play a significant role. The process begins with the BC phase, where the initial seed weight, powder, and binder are recorded. During the coating phase, small quantities of powder and binder are applied to the seeds, ensuring proper adhesion of the coating material. Studies indicate that encrusted *Brassica* seeds exhibit improved radicle emergence, maintaining vigor and ensuring uniform seedling establishment. Encrusting also helps mitigate the adverse effects of herbicides on germination. For instance, encrusting was effective in protecting *Brassica* seeds from chemical stressors, thereby enhancing seedling emergence and growth. Higher germination and seed vigor were observed in encrusted *Brassica* seeds before storage compared to seeds encrusted post-storage and non-treated seeds (Shilova et al. 1978).

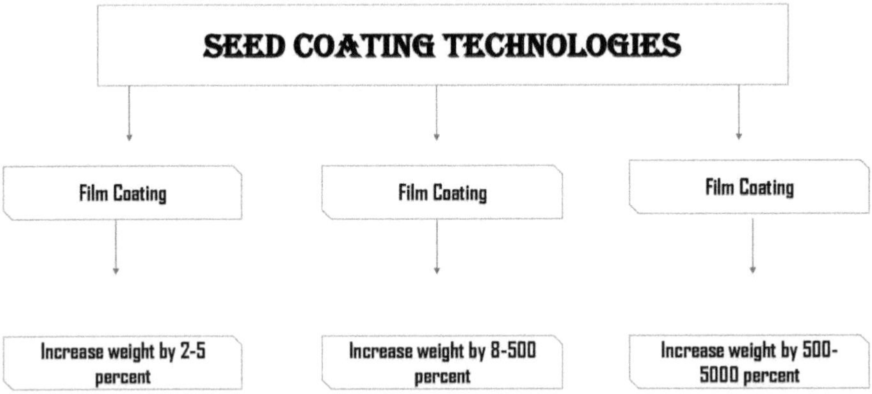

Fig. 2 Modern seed coating technologies and % seed weight increase

Seed Pelleting

Seed pelleting is the process of enlarging seed size by applying external natural or chemical materials, making the original seed shape unrecognizable. This technique has been successfully employed to enhance germination, seed vigor, and storage potential in various crops and holds great promise for *Brassica* species. Seed pelleting has been shown to improve early growth and yield parameters. In *Brassica* crops, nutrient-enriched pelleting using ammonium molybdate, ferrous sulfate, and beneficial microbial inoculants can promote root development, nutrient uptake, and overall plant establishment (Gopalsamy et al. 2025).

Disease resistance is another significant advantage of seed pelleting. Research on sesame has demonstrated that pelleting with plant growth-promoting rhizobacteria *Paenibacillus polymyxa* ZF129 significantly reduced club root disease and enhanced biological control efficiency (Abdukerim et al. 2023). Similar benefits can be expected in *Brassica* species, which are susceptible to damping-off caused by *Rhizoctonia solani* and *Pythium* spp. (McGehee et al. 2019) (Fig. 2).

3 Mechanism of Seed Coating

3.1 Coating Equipment

The efficiency of seed coating machinery for *Brassica* crops relies on the accurate addition of a specified amount of coating material to each seed with minimal physical injury during the process (Afzal et al. 2016). The width of the seed coating machinery, rotor speed, spinner settings, fineness of the coating agent, porosity of the seeds, water-holding capacity, and cementing properties have a significant impact on the coating process and the quality of the final seeds. Successful coating involves the equipment ensuring consistent seed rotation within the coating pan and enabling precise application of binders and powders onto the seeds (Sohail et al.

2022). Pharmaceutical coating methods and equipment, initially used in tablet coating, have been modified for seed treatment (Salawi 2022).

The use of protective agents like pesticides and fungicides in *Brassica* seed coating, as well as their equipment, has also been addressed by Lamichhane et al. (2020). Seed health improved by coating mechanisms and special machines has been reviewed by Greer et al. (2023). Commercial *Brassica* seed coating can be done on a large scale using equipment that has an open and wide surface area to give uniform application. This can be done with industrial-scale coating units like those utilized in the cement business, with wide, sloping drums that provide sufficient mixing time with a regulated rotational rate.

3.2 Adhesives

Seed coating in *Brassica* crops is successful if the coating material is ensured to be intact with the seed but dissolves easily when subjected to water. This can be attained by using adhesives like binders, glues, and sticking agents, which ensure the coating's integrity and do not lead to cracking or dusting upon handling and transport (Javed et al. 2022). Typically utilized seed coating integrity preservatives in *Brassica* are polyvinyl alcohol (Javed et al. 2022), soy flour (Qiu et al. 2020), and gum arabic (Tiamiyu et al. 2023).

Imbibition and radicle protrusion in *Brassica* seeds are greatly determined by the binding ability of the used binder, which determines the adhesion of coating agents. High water-holding capacity and porous materials allow for easy absorption of moisture and are regarded as more effective (Upretee et al. 2024; Woodstock 1988). Tested different binders and concluded that gum arabic gave the greatest physical strength with the least resistance to moisture uptake and radicle protrusion (Prasad et al. 2022). The efficacy of binders not only depends on their capacity to hold coating agents but also on their effect on the viability of plant growth-promoting rhizobacteria (Chakraborty 2020).

Owing to its accessibility, low toxicity, high water solubility, and low application rate, methyl cellulose has been recognized as a more efficient binder compared to gum arabic for the coating of lime on *Brassica* seeds (Shiam et al. 2025). Moreover, seed coatings with biochar and other activated carbon sources have been stabilized efficiently with polyvinyl alcohol (Zhang et al. 2022) to make them ready for enhancing *Brassica* seed performance.

3.3 Coating equipment

Past studies have investigated numerous seed coating materials that have a beneficial effect on seed quality and health, as well as creating a rhizobia habitat. In the case of *Brassica* crops, protective materials that are most commonly used are lime, gypsum, montmorillonite, kaolinite, rock phosphate, and bentonite, which improve rhizobia survival (Javed et al. 2022). Organic seed coating materials like Moringa leaf powder have been extensively utilized to enhance seed quality since they are rich in calcium, potassium, antioxidants, proteins, ascorbates, phenols, and cytokinin. The use of Moringa leaf extract has been found to improve shoot and root dry weight and the number of fruits per plant in *Brassica* crops such as rape and cabbage (Ariani et al. 2023; Liang et al. 2025). Other organic substances employed in coatings are peat, poultry manure, farmyard manure, mucilage, and bonemeal (Subbaiah 2019).

Despite the above, limited information exists about the properties, purity, and mechanism of action of bioactive compounds in such organic materials. The development of coating layers is a function of mean particle size and distribution and thus requires the knowledge of coating agent properties prior to application (Javed et al. 2022).

3.4 Coating Types

Seed coating in *Brassica* species enhances seed performance by improving germination, seedling vigor, and protection against environmental stresses. Common coating types, as presented in Table 1.

4 Seed Quality and Germination Enhancement Technologies

Figure 3 gives an overview of Seed Improvement Techniques, dividing them into Physical Enhancement and Chemical Enhancement techniques. Physical enhancement techniques enhance seed quality without the use of chemicals (Thakur and Singh 2024). The application of a magnetic field accelerates germination, guarantees uniform crop establishment, increases yield, decreases pathogen susceptibility, and favors poor-quality seeds (Sarraf et al. 2020). Plasma treatment is an eco-friendly and chemical-free method that enhances seed quality, changes enzymatic activity, and sterilizes seed surfaces without damaging seeds (Neha and Vasudevan 2021; Thakur and Singh 2024). Radiation exposure enhances early maturity, improves saline tolerance, increases grain yield, enhances product quality, sterilizes products of agriculture, and activates antioxidant mechanisms, which helps in improving stress resistance (Dai et al. 2025).

Chemical enhancement methods, however, include treatments that change seed biochemistry and physiology to increase performance. Hydropriming and osmo priming enhance seed imbibition and activate metabolic functions prior to sowing, while hormonal priming applies plant growth regulators to increase germination and early seedling growth. Nutrient priming provides key elements to promote seedling vigor (Hameed et al. 2025). Furthermore, coating covers seeds in covering layers, which are mostly with nutrients, pesticides, or biostimulants, while pelleting makes seeds larger for ease of handling and for more even planting (Javed et al. 2022). While physical improvement techniques provide chemical-free and

Table 1 Seed coating types, purpose, and main characteristics

S. no.	Coating type	Purpose	Key features
1	Inoculant coating	Increases beneficial microbial colonization	Uses carbon and adhesive sources to sustain microbial survival Peat, biochar, or polymers desiccation protect between coating and sowing, and in the soil before establishment Microbial strain selection and carrier material determine effectiveness Increases nitrogen fixation in *brassica* and general plant health
2	Lime coating	Enhances seedling establishment in acidic soils	Balances soil acidity for improved root growth Improves rhizobia survival and symbiotic nitrogen fixation Enhances seed imbibition and hastens radicle protrusion Enhances early seedling vigor and crop yield
3	Protective coating	Protects seeds against diseases and pests	Inhibits fungal infections of seed- and soil-borne fungi May incorporate fungicides, insecticides, and biocontrol products Other formulations have a bird, rodent, and insect-repellent effect New coatings yield both preventive and curative results
4	Nutrient coating	Provides necessary nutrients for initial development	Reduces nutrient leaching and loss Offers specific nutrients according to seed demand Increases seedling vigor and uniform emergence Will need an inert pre-coating layer to avoid nutrient toxicity
5	Hydrophilic coating	Improves water absorption and germination	Enhances seed hydration and oxygen exchange Accelerates the germination at dry conditions Excessive moisture may decrease emergence under some conditions Calcium peroxide or superabsorbent polymers can be employed
6	Hydrophobic coating	Enhances seed cover and water retention	Lowers excess water intake during waterlogging Improves seed longevity and storability Calcium carbonate and wax coatings enhance water retention Applied in dryland agriculture for efficient utilization of water

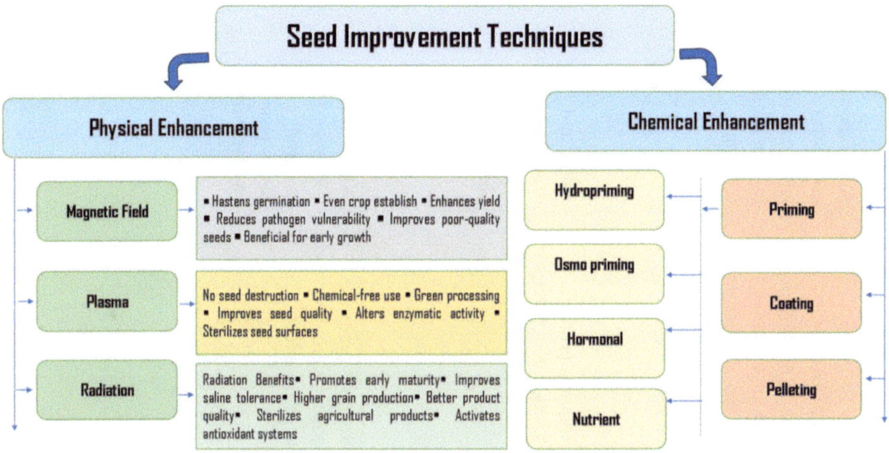

Fig. 3 Seed improvement techniques through physical and chemical enhancements

environmentally friendly solutions, chemical improvements offer specific enhancements in germination, seedling development, and protection. A combination of these methods can be used to maximize seed performance under various agricultural conditions.

5 Physiological Seed Enhancements

Seed priming is an essential physiological enhancement technique in *Brassica* species, promoting improved germination, seedling vigor, and stress tolerance. The priming methods outlined in Table 2 optimize seed performance by regulating water uptake, breaking dormancy, and enhancing nutrient availability, ensuring better crop establishment and resilience.

6 Biological Seed Enhancements

6.1 Bacterial Seed Agents

Plant growth-promoting rhizobacteria are those soil-living microorganisms that, when inoculated into seeds, soil, or roots, improve plant growth or decrease the incidence of soil-borne pathogens (Dutta et al. 2022). Plant growth-promoting rhizobacteria promote plant growth either directly or indirectly by nitrogen fixation, solubilization of phosphorus and zinc, and siderophore production, which enhances the uptake of iron (Wahab et al. 2024; Iqbal et al. 2024). They also synthesize phytohormones like auxins, cytokinins, and gibberellins for the promotion of plant

Table 2 Comprehensive seed priming techniques for *Brassica* species

S. no.	Priming type	Method	Materials used	Key benefits	Limitations/notes	Application conditions
1	Hydropriming	Soaking seeds in water and drying back to the initial moisture level	Water only	Low cost and eco-friendly Uniform germination and early vigor Improves drought tolerance Ensures faster crop establishment	Risk of non-uniform hydration leading to uneven emergence	Arid and semi-arid regions, large-scale farming
2	Osmo priming	Soaking seeds in osmotic solution under aerated conditions	Polyethylene glycol Potassium chloride Potassium nitrate Calcium chloride Magnesium sulfate Potassium phosphate	Improves germination rate and seedling uniformity Enhances stress tolerance Supplies essential nutrients during early growth	Requires precise solution concentration to avoid osmotic stress	Saline and drought-prone soils Early-season planting
3	Hormonal priming	Treating seeds with plant growth regulators	Gibberellic acid Indole-3-butyric acid Kinetin Moringa leaf extract Seaweed extract Chitosan	Breaks dormancy and promotes faster germination Enhances seedling vigor, hypocotyl elongation, and root growth Improves tolerance to abiotic stress	Requires precise concentration to avoid negative effects	When breaking dormancy is required Stress-prone environments

4	Nutrient priming	Pre-sowing soaking of seeds in micronutrient-enriched solution	Zinc Boron Molybdenum Manganese Copper	Enhances early seedling growth and root development Improves nutrient use efficiency Boosts disease resistance and seedling survival	Overdosing can lead to toxicity and reduced germination	Nutrient-deficient soils Low-fertility regions
5	Solid matrix priming	Incubating seeds in an inert medium with controlled hydration	Vermiculite Bentonite clay Peat moss	Provides slow and uniform hydration Enhances seedling emergence and establishment Improves resistance to environmental fluctuations	Requires specific matrix materials and controlled hydration conditions	When precise control of hydration is needed High-value seed production

growth and produce ACC-deaminase for controlling ethylene under stress (Kejela 2024).

Biopriming is a contemporary seed enhancement strategy that combines biological and physiological methodologies to enhance the growth, development, and defense of plants. It is used as an alternative to traditional measures for the management of seed- and soil-borne diseases. The inoculation of useful microorganisms, such as bacteria and fungi, by seed priming tends to hasten plant growth, increase vigor, and be useful under adverse soil conditions. Apart from disease suppression, biopriming with PGPR is a promising biofertilization method that decreases the use of synthetic fertilizers (Sooda et al. 2020). Plant growth-promoting bacteria commonly applied as co-inoculants with rhizobia comprise *Azotobacter*, *Azospirillum*, *Bacillus*, *Pseudomonas*, *Serratia*, and *Streptomyces* (Jalal et al. 2022).

6.1.1 Role of Bacterial Biopriming in Plant Growth Promotion

Plant growth-promoting bacteria (PGPB) are a heterogeneous collection of beneficial microbes that live in the rhizosphere, root surfaces, or plant tissues. They promote plant growth and impact protection against biotic and abiotic stress agents (Valente et al. 2025; Ongaga et al. 2025).

6.1.2 Role of Bacterial Biopriming in Disease Management

Seed improvement through biopriming means treating seeds by coating or immersing them in beneficial microorganisms or consortia of microorganisms and subjecting them to controlled temperatures and humidity levels for incubation. Different studies established that rhizobacteria-mediated biopriming significantly prevents disease through activation of a mechanism called ‘induced systemic resistance’ (ISR) in crop plants (Kaleh et al. 2025). Out of the widely researched bacterial families, *Bacillus* and *Pseudomonas* spp. are well-documented rhizosphere microorganisms with high disease suppression potential (Das et al. 2025). Biopriming seeds with BCAs like *Bacillus subtilis* and *Pseudomonas fluorescens* has been found to be an effective method of controlling seed and root rot pathogens (Sharma and Gautam 2025), offering a safe alternative to chemical fungicides without being toxic to humans, animals, or the environment.

6.1.3 Enhancing Abiotic Stress Tolerance Through Biopriming

Seed priming with beneficial microbes can enhance plant resilience to abiotic stress, particularly in arid and semi-arid environments (Hameed et al. 2025). Plant growth-promoting bacteria are well adapted to challenging conditions and help protect plants from environmental stressors, ultimately improving crop yields (Valente et al. 2025).

6.1.4 Fungal Seed Agents for Biopriming

Biopriming with useful fungi and bacteria is an eco-friendly method of seed improvement, stimulating plant development, enhancing yield, and reducing biotic and abiotic stress elements. It is a sustainable method of seed improvement that is being accepted as an alternative to chemical treatments, and it helps in general plant and soil health. Studies have proven that biopriming wheat seeds with Trichoderma sp. dramatically enhanced drought tolerance by improving photosynthetic parameters and redox states under stress (Singh et al. 2025).

6.2 Seed Coating and Pelleting

Seed treatments like film coating, pelleting, priming, and inoculation are used globally (Dadlani 2025) for enhancing seed handling, uniform distribution, germination, and storage. These treatments allow for coating the seeds with adhesive films, pesticides, plant growth stimulants, and biological agents. Seed coating is used as a carrier for chemical materials that enhance early seedling growth (Singh 2025). Various additives, such as growth regulators, microbial inoculants, micronutrients, fungicides, insecticides, and other protective agents, are added to the pellet to improve seed performance (Dadlani 2025).

Uniform film formation of the inert material onto the seed is important in providing a good coat. This has the added role of providing protective cover against plant- and seed-borne pests and diseases (Reff). In pharmaceutical processes, polymer films are applied to get uniform deposition of several substances over seeds. For commercial use, plasticizers, polymers, and pigments in liquid formulations are typically formulated as thin-film coatings (Jelani et al. 2025). Yet, the exact formulation of these coatings is usually proprietary knowledge of the manufacturing firms. Coating materials usually consist of binding agents, fillers (e.g., polyvinyl alcohol, gypsum, and clay), and an intermediate layer (e.g., clay, polyvinyl acetate, or vermiculite). Another method, seed agglomeration, allows for the sowing of several seeds from the same batch or various batches, varieties, or species in one unit (Martynková et al. 2024; Fu et al. 2024).

7 Seed Enhancements and Plant Development

7.1 Modulating Seedling Growth

Seedling vigor is the key to effective crop establishment. Seed treatments prior to sowing overcome germination limitations under regular and stressful conditions. Different methods, such as chemical priming, nutrient supply, plant compounds, and plant growth-promoting bacteria, have been found to have positive impacts in both greenhouse and field environments (Timofeeva et al. 2024).

General priming techniques, such as osmo priming and on-farm priming, are standardized for crops depending on the duration of soaking and concentration. Chemical priming compounds are polyamines (spermine, spermidine, putrescine), calcium chloride ($CaCl_2$), potassium chloride (KCl), sodium chloride (NaCl), potassium phosphate (KH_2PO_4), potassium nitrate (KNO_3), polyethylene glycol (PEG), and hydro-absorbers (e.g., humic acid, biplantol). Natural molecules, including nitric oxide (NO), hydrogen sulfide (H2S), hydrogen peroxide (H2O2), ascorbate, salicylic acid, melatonin (Mel), and moringa leaf extracts with high cytokinin levels, have also been used for seed priming. The treatments tend to increase levels of endogenous molecules to initiate growth in seeds (Thapa et al. 2022; Singh et al. 2015).

Seed priming advantages *Brassica* crops by enhancing germination rate, decreasing germination time, and increasing uniform emergence. Variations in germination rates based on seed coating thickness and composition may affect seed reserve mobilization efficiency. Hydro-absorbers are suggested for coated seeds to improve nutrient availability. Nutrient priming enhances seed nutrient content to facilitate early seedling growth. Manganese (Mn) priming enhances stand establishment, growth, yield, and grain content over soil application (Singh et al. 2015).

7.2 Effects on Crop Phenology

Primed seeds generally germinate quickly and attain developmental phases like tillering, flowering, and maturity earlier than control seeds. This adaptability is advantageous in delayed planting situations due to unfavorable weather conditions. Early and dense crop stands capture more resources, resulting in better root growth, greater leaf area, and increased photo-assimilation, ultimately resulting in greater yields. Combining seed priming with crop management methods such as planting geometry, irrigation, and fertilization can even maximize the use of resources more efficiently (Mondal and Bose 2021; Sarkar et al. 2021).

8 Seed Storage and Preservation Technologies

8.1 Importance of Seed Storage in Agriculture

Seed banks are critical for plant genetic material conservation, providing the necessary genetic diversity for breeding programs. Such diversity increases the adaptability of plants to changing environments and the ability to resist pests and diseases (Salgotra and Chauhan 2023). Seed banks, through their preservation of seeds of different species, ensure that biodiversity and ecological resilience are maintained (Ray and Bordolui 2021). Seed banks provide important tools for researchers and breeders in developing high-yielding, stress-resistant, and disease-resistant crops (Meena et al. 2025). Additionally, seed banks serve as insurance against natural disasters and climate change, maintaining food security through crop regeneration (Wambugu et al. 2023). They also ensure the conservation of endangered species to avoid extinction. In general, seed banks aid climate-resilient agriculture, global food security, and scientific research by serving as material for studying plant genetics and evolution (Mba and Ogbonnaya 2022) (Fig. 4).

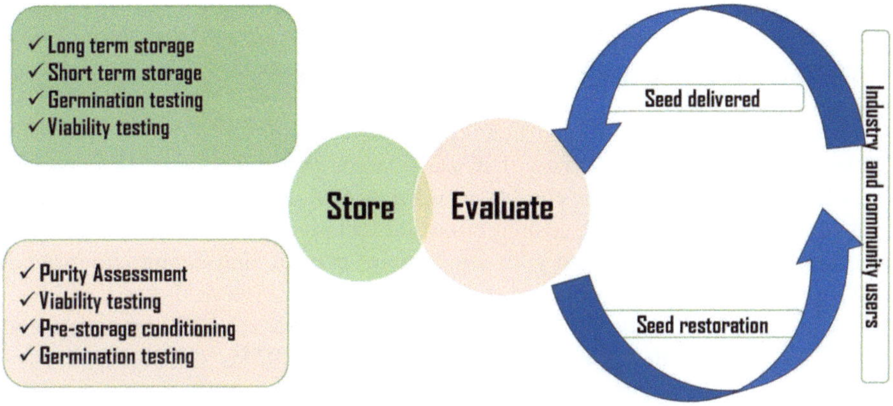

Fig. 4 Various components and supply chain of seed storage technologies

8.1.1 Environmental Impact of Traditional Seed Storage

Conventional seed storage techniques have both positive and negative environmental effects. Positively, they conserve biodiversity by keeping locally adapted seeds and have a lower carbon footprint compared to contemporary refrigerated storage. Negatively, they do not have sophisticated pest control and climate management, which can decrease seed viability and cause crop loss. Exposure to unstable temperature and humidity levels reduces seed shelf life, and climate change worsens these conditions. Also, inefficient utilization can cause seed spoilage and necessitate that farmers sow more seeds. Integration of conventional systems with new technologies might improve storage efficacy without losing their advantages (Sharma et al. 2021).

8.1.2 Traditional Seed Storage Methods

Farmers used centuries-old, low-cost methods of seed preservation to maintain viability and future seedling success. These simple, inexpensive methods provide support for preserving crop diversity. Seed drying, in many cases, beneath the sun or in well-ventilated places, removes moisture to limit microbial development. Different cultures employed specialized structures, African raised-floor granaries, Asian bamboo granaries, or clay pots and earthen jars, to control airflow and temperatures. Cloth or jute bags minimize the fungal threats. Layering seeds in sand or ash, ventilation using handwoven baskets, and smoke treatment for its antifungal properties, particularly for tubers, are other methods. The process of traditional seed saving also encompasses choosing seeds from the best plants. Though successful in most rural communities, these are supported by modern controlled-environment seed banks. Both conventional and cutting-edge approaches collectively play a central role in preserving agricultural biodiversity as well as securing food (Rahman et al. 2021).

Cold Storage

Cold storage is a technique of seed viability and vigor preservation through low temperatures. Seed storage experiments (SSE) assist in identifying the best temperature and moisture levels for seed longevity (Corbineau 2024). Cold storage effectiveness varies with storage conditions, seed origin, and type of dormancy (Maleki et al. 2021). Studies show that seeds can be preserved longer when stored in cool, low-oxygen, or anoxic conditions (Ciacka et al. 2022). Various cold storage methods have been described, such as controlled relative humidity and temperature conditions, humid cold storage with temperature control, and dehumidifier-aided cold storage. Although conventional storage methods preserve seed-associated bacterial microbiomes (Mohammed et al. 2021), cold storage is still expensive and optimal for low-volume-high-value seeds (Kumar et al. 2024).

Drying and Desiccation

Desiccation and drying are fundamental seed storage methods for ensuring viability and longevity. Accurate drying aids seeds to resist freezing, enhances storage life, inhibits attacks by pests and pathogens, and slows germination (Dadlani et al. 2023). For long-term seed storage, seeds need to be dried down to around 15% eRH. Desiccants of good effect include silica gel, charcoal, and seeds such as rice or maize. Whereas silica gel is often employed for desiccating seeds at room temperature, calcium chloride is utilized when storage is done under refrigeration or freezing. Employing salts in a desiccator gives a more controlled drying process than silica gel (Zhang and Tao 1989; Kadirvel et al. 2025). In removing seeds from cold storage, they should be brought to room temperature prior to opening the container to avoid condensation of moisture (Beckles et al. 2025).

Chemical and Physical Treatments for Seed Preservation

Chemical treatment, including fungicides, aids in seed viability and storage fungi control. Liquid seed treatment is widely applied because it has a uniform application, is inexpensive, and lacks further drying requirements. Seeds are stored at low moisture content and temperature in hermetically sealed containers for long-term storage. The longest storage is in liquid nitrogen canisters (Wan et al. 2025). Physical procedures, such as electromagnetic waves, magnetic fields, ultrasounds, and ionizing radiation, provide a substitute for chemical treatments. These processes can improve seed vitality while lowering fertilizer consumption and reducing pollution of domestically produced raw materials (Yadav et al. 2025).

8.2 Emerging Sustainable Seed Storage Technologies

Cryopreservation is a trustworthy technique for the long-term preservation of plant genetic material, including those with recalcitrant seeds (Patra 2025; Malik et al. 2025). It is an efficient ex-situ conservation process that supersedes the limitations of traditional seed banks and clonal propagation (Malakar et al. 2025). The process requires storing biological specimens in liquid nitrogen, hence one of the most viable means of conserving diverse plant germplasm. Through sample viability post-thaw, cryopreservation supports indefinite storage without incurring additional costs for in vitro culture, space, risk of contamination, and human operator mistakes (Aabling et al. 2025). An efficient cryogenic protocol development still poses a problem, however, demanding stable and consistent preservation of samples. Making cost-effective protocols simpler and uniform, coupled with knowledge dissemination, can drive rapid progress in the area. The Millennium Seed Bank Partnership is investigating cryopreservation as a potential alternative to standard seed banking methods for storing endangered species that cannot be kept using conventional seed banking methods. Rapid drying of seeds or seed components followed by freezing in liquid nitrogen is the process that reduces the impact of seed collection on populations in the wild (Ballesteros et al. 2025).

8.3 Natural Seed Coatings

Natural seed coatings composed of biodegradable and eco-friendly materials are coming forward as one of the principal technologies for green seed storage. These coatings enhance germination of seeds, improve nutrient absorption, and enhance plant growth overall by providing a reservoir of useful microbes. Some of the common materials employed for natural seed coatings are chitosan, alginate, and gelatin. Chitosan, a natural polymer obtained from crustacean exoskeletons, has been found to enhance germination, promote plant growth, and develop resistance against biotic and abiotic stresses. Alginate, derived from brown seaweed, facilitates seed germination and nutrient uptake. Gelatin, a product of animal bone and skin collagen, has also been used to enhance seed germination and plant growth. Natural seed coating provides several advantages in seed conservation in a sustainable manner. These coatings shield the seeds from extrinsic stress conditions like drought, salinity, and high-temperature stress (Dadlani 2025). They can be used as vehicles for bioactive substances like plant growth regulators, fungicides, and insecticides, which guarantee enhanced seed vigor and crop performance. Through seed encapsulation in natural materials, these coatings represent a promising, environmentally friendly way of enhancing seed viability, insulating against environmental stresses, and promoting crop yield. Their uses in sustainable agriculture underscore their importance as a fundamental innovation in new seed storage technology.

8.4 Biotechnology and Sustainable Seed Storage

Biotechnology is crucial in sustainable seed storage through the improvement of seed resistance to environmental stresses, viability, and storage life. Efficient seed storage is important for preserving plant genetic diversity, ensuring economically important crops, and the provision of global food security. Seeds are grouped according to their storage behavior into two broad categories: orthodox and recalcitrant seeds. Orthodox seeds can withstand drying and low-temperature storage, where they can stay viable for many years without the loss of germination capacity. Recalcitrant seeds, on the other hand, are very sensitive to drying and temperature changes, where long-term storage is difficult and special preservation methods are needed (Patra 2025; Cai et al. 2025).

Genetic engineering can also improve seed survivability by introducing desiccation tolerance, oxidative stress tolerance, and microbial resistance traits. Moreover, molecular markers and omics technologies enable improved seed selection and viability testing to ensure the conservation of high-quality germplasm. Through the combination of biotechnology with conventional and advanced storage methods, sustainable seed conservation can be attained to complement global efforts in maintaining biodiversity, improving crop resilience, and ensuring future food production.

8.4.1 Molecular Approaches to Seed Longevity

Seed longevity relies on the capability to repair DNA damage, involving DNA repair pathway genes (Tan et al. 2025). Enzymes like superoxide dismutase and catalase serve to reduce oxidative damage, a major cause of seed deterioration (Barrero et al. 2025). In addition, epigenetic changes like DNA methylation and histone modifications control gene expression during aging seeds (Li et al. 2025). The involvement of hormonal pathways, especially abscisic acid (ABA) and gibberellins, also affects seed longevity (Tan et al. 2025). Molecular biology, genomics, transcriptomics, and bioinformatics advancements are aiding scientists in the determination of critical genes and pathways participating in seed deterioration and developing strategies for better storage (Pagnotta 2025).

Nanotechnology coatings increase stress protection and resistance to pathogens and enhance seed storage quality (Dadlani 2025). DNA barcoding and electronic storage preserve genetic information and make seed identification and retrieval easy for future application (Kimta et al. 2025). These treatments are important for the preservation of seed quality, viability, and genetic purity and hence contribute to sustainable agriculture and conservation of biodiversity.

9 Conclusion

The manuscript highlights significant progress in *Brassica* crop improvement through advanced seed technology and genetic strategies. Key developments, including marker-assisted selection (MAS), hybrid breeding, and CRISPR gene editing, have enhanced traits such as yield, disease resistance, and abiotic stress tolerance. These advancements have led to the creation of high-yielding, nutritionally enriched cultivars, addressing food security and health concerns. Despite these advances, challenges in seed production remain, including environmental stresses and seed viability issues. Technologies like seed priming, nanotechnology treatments, and advanced seed coating techniques have been integrated into breeding programs to improve seed performance under challenging conditions. However, overcoming these constraints still requires continued innovation. Looking ahead, sustainable and ethical approaches to seed improvement are essential. Focusing on biodiversity conservation, ecological farming practices, and equitable access to improved seeds is crucial for addressing global food security challenges. By combining technological progress with environmental responsibility, the *Brassica* seed industry can support sustainable agriculture. Continued investment in scientific research and farmer-oriented innovations will be key to ensuring that *Brassica* crops can thrive amid climate change and growing global populations.

References

Aabling RR, Rusan M, Møller AMJ, Munk-Pedersen N, Holm C, Elmengaard B, Møller BK (2025) A narrative review on manufacturing methods employed in the production of mesenchymal stromal cells for knee osteoarthritis therapy. J Biomed 13(2):509. https://doi.org/10.3390/biomedicines13020509

Abdukerim R, Xiang S, Shi Y, Xie X, Li L, Chai A, Fan T (2023) Seed pelleting with gum arabic-encapsulated biocontrol bacteria for effective control of clubroot disease in pak choi. Plants 12(21):3702. https://doi.org/10.3390/plants12213702

Abrahimi F, Taghvaei M, Mastinu A (2023) Nano-organic coatings improve early vigor of *Brassica napus* L. seeds in water deficit. Agronomy 13(2):390. https://doi.org/10.3390/agronomy13020390

Accinelli C, Abbas HK, Little NS, Kotowicz JK, Mencarelli M, Shier WT (2016) A liquid bioplastic formulation for film coating of agronomic seeds. J Crop Prot 89:123–128. https://doi.org/10.1016/j.cropro.2016.07.010

Afzal I, Rehman HU, Naveed M, Basra SMA (2016) Recent advances in seed enhancements. In: New challenges in seed biology-basic and translational research driving seed technology, pp 47–74. https://doi.org/10.5772/64791

Ali E, Zhang K (2023) CRISPR-mediated technology for seed oil improvement in rapeseed: challenges and future perspectives. Front in Plant Sci 14:1086847. https://doi.org/10.3389/fpls.2023.1086847

Ali K, Mubasher HM, Sher A, Sattar A, Manaf A (2023) seed priming for abiotic stress tolerance. In: Climate-resilient agriculture, vol 2: agro-biotechnological advancement for crop production, pp 641–665. https://doi.org/10.1007/978-3-031-37428-9_27

Ariani LN, Estiasih T, Sunarharum WB, Khatib A (2023) Potential of moringa (Moringa oleifera) leaf powder for functional food ingredients: a review. Czech J Food Sci 41(1). https://doi.org/10.17221/221/2022

Arun MN, Hebbar SS, Senthivel T, Nair AK, Padmavathi G, Pandey P, Singh A (2022) Seed priming: the way forward to mitigate abiotic stress in crops, vol 11. IntechOpen, London, p 173

Axentii M, Codina GG (2024) Exploring the nutritional potential and functionality of hemp and rapeseed proteins: a review on unveiling anti-nutritional factors, bioactive compounds, and functional attributes. Plants 13(9):1195. https://doi.org/10.3390/plants13091195

Bahmid NA, Suloi ANF, Anwar M, Prasetyo DJ, Wahono SK (2024) Incorporation of ground mustard seeds in an active packaging film: structural properties and controlled release of allyl isothiocyanate. IJFST 59(7):4979–4991. https://doi.org/10.1111/ijfs.17232

Ballesteros D, Davies RM, Peach JL, Pence VC (2025) Longevity, cryopreservation, and propagation of carnivorous plants seeds: insights from 13 species in long-term ex situ collections. Ann Bot:mcaf058. https://doi.org/10.1093/aob/mcaf058

Barrero JM, Buitink J, Chen SC, Da Silva EA, Hay FR, Kawakami N, Leprince O, Ojiewo CO, Pawłowski TA, Pérez HE, Vijay D (2025) Celebrating the 25th Anniversary of the International Society for Seed Science. Seed Science Research 35(1):1–8. https://doi.org/10.1017/S0960258525000029

Beckles DM, Kader A, Reid MS, Jiang CZ, Cantwell MI (2025) Postharvest technology of horticultural crops. In: Vegetable crops, herbs, and flowers, vol 21661, 4th edn. UCANR Publications

Borges CE, dos Santos V, Veloso R, Da Conceicao CA, Mendes DS, Ramirez-Cabral NY, Shabani F, Da Silva RS (2023) Forecasting Brassica napus production under climate change with a mechanistic species distribution model. Sci Rep 13(1):12656. https://doi.org/10.1038/s41598-023-38910-3

Boubakri H (2023) Recent progress in CRISPR/Cas9-based genome editing for enhancing plant disease resistance. Gene 866:147334. https://doi.org/10.1016/j.gene.2023.147334

Cai H, Shao J, Shen Y (2025) Seed storability in forest trees: research progress and future perspectives. Forests 16(3):467. https://doi.org/10.3390/f16030467

Cardarelli M, Woo SL, Rouphael Y, Colla G (2022) Seed treatments with microorganisms can have a biostimulant effect by influencing germination and seedling growth of crops. Plants 11(3):259. https://doi.org/10.3390/plants11030259

Chakraborty AP (2020) Carrier based bioformulations of PGPR-characteristics, shelf life and application in improving health status of crop plants-a mini review. Int J Res Rev 7:88–98

Chen S, Hayward A, Dey SS, Choudhary M, Witt Hmon KP, Inturrisi FC, Cowling WA (2022) Quantitative trait loci for heat stress tolerance in *Brassica rapa* L. are distributed across the genome and occur in diverse genetic groups, flowering phenologies and morphotypes. Genes 13(2):296. https://doi.org/10.3390/genes13020296

Ciacka K, Staszek P, Sobczynska K, Krasuska U, Gniazdowska A (2022) Nitric oxide in seed biology. Int J Mol Sci 23(23):14951. https://doi.org/10.3390/ijms232314951

Corbineau F (2024) The effects of storage conditions on seed deterioration and ageing: how to improve seed longevity. Seeds 3(1):56–75. https://doi.org/10.3390/seeds3010005

Dadlani M (2025) Seed enhancement technologies. In: Indian seed sector: evolution, technology, trade and impact, pp 207–235. https://doi.org/10.1007/978-981-96-0714-3_8

Dadlani M, Gupta A, Sinha SN, Kavali R (2023) Seed storage and packaging. In: Seed science and technology. Springer, pp 239–266. https://doi.org/10.1007/978-981-19-5888-5

Dai Q, Zhang L, Jiang S, Su B, Li Z, Shuai Y, Wang J (2025) Improved salt tolerance in *Brassica napus* L. overexpressing a synthetic Deinocuccus stress-resistant module DICW. Int J Mole Sci 26(6):2500. https://doi.org/10.3390/ijms26062500

Das P, Kashyap S, Sharma I, Ray SK, Agarwala N (2025) Exploration of endophytic and rhizospheric bacteria of invasive plant *Xanthium strumarium* L. reveals their potential in plant growth promotion and bacterial wilt suppression. Braz J Microbiol:1–23. https://doi.org/10.1007/s42770-024-01599-1

Dey S, Ghosh N, Nath S, Gopal G, Paul S, Mukherjee A, Kundu R (2024) Application of multimetallic nanoparticles in agriculture: the more, the better? Biocatal Agric Biotechnol:103238. https://doi.org/10.1016/j.bcab.2024.103238

Dutta P, Muthukrishnan G et al (2022) Plant growth-promoting rhizobacteria (PGPR) and its mechanisms against plant diseases for sustainable agriculture and better productivity. Biocell 46(8):1843–1859. https://doi.org/10.32604/biocell.2022.019291

Faizan M, Karabulut F, Alam P, Yusuf M, Tonny SH, Adil MF, Hayat S (2023) Nanobionics: a sustainable agricultural approach towards understanding plant response to heavy metals, drought, and salt stress. Nano 13(6):974. https://doi.org/10.3390/nano13060974

Fu T, Wang W, Fang G (2024) Thermal properties and applications of form-stable phase change materials for thermal energy storage and thermal management: a review. Energy Storage 6(1):e533. https://doi.org/10.1002/est2.533

Gad M, Chao H, Li H, Zhao W, Lu G, Li M (2021) QTL mapping for seed germination response to drought stress in *Brassica napus*. Front Plant Sci 11:629970. https://doi.org/10.3389/fpls.2020.629970

Gopalsamy A, Tamilmani E, Shanmugam K, Koilpitchai NN, Durairaj V, Mylsamy P, Ranganathan U (2025) Seeds of excellence: review on impact of seed quality enhancement on babygreens biomass production. J Agric Food Res 19:101597. https://doi.org/10.1016/j.jafr.2024.101597

Gray S (2024) The role of business in the plant-based oils sector–a socioenvironmental analysis. https://hdl.handle.net/10315/42617

Greer SF, Surendran A, Grant M, Lillywhite R (2023) The current status, challenges, and future perspectives for managing diseases of *brassicas*. Front Microbiol 14:1209258. https://doi.org/10.3389/fmicb.2023.1209258

Gupta SK (2016) Brassicas. In: Breeding oilseed crops for sustainable production, pp 33–53. https://doi.org/10.1016/B978-0-12-801309-0.00003-3

Hameed A, Hussain S, Nisar F, Rasheed A, Shah SZ (2025) Seed priming as an effective technique for enhancing salinity tolerance in plants: mechanistic insights and prospects for saline agriculture with a special emphasis on halophytes. Seeds 4(1):14. https://doi.org/10.3390/seeds4010014

Ibrahim AA, Abd-Ellatif S, Abdel Razik ESS, Salem KF (2024) Biodiversity of cereal crops and utilization in food and nutritional security. In: Sustainable utilization and conservation of plant genetic diversity, pp 31–61. https://doi.org/10.1007/978-981-99-5245-8_2

Ilyas M, Khan WA, Ali T, Ahmad N, Khan Z, Fazal H, Rizwan M (2022) Cold stress-induced seed germination and biosynthesis of polyphenolics content in medicinally important *Brassica* rapa. Phytomed Plus 2(1):100185. https://doi.org/10.1016/j.phyplu.2021.100185

Iqbal MZ, Singh K, Chandra R (2024) Recent advances of plant growth promoting Rhizobacteria (PGPR) for eco-restoration of polluted soil. Cleaner Eng Technol:100845. https://doi.org/10.1016/j.clet.2024.100845

Jalal A, Filho MCMT, da Silva EC, da Silva Oliveira CE, Freitas LA, do Nascimento V (2022) Plant growth-promoting bacteria and nitrogen fixing bacteria: sustainability of non-legume crops. In: Nitrogen fixing bacteria: sustainable growth of non-legumes, pp 233–275. https://doi.org/10.1007/978-981-19-4906-7_11

Javed T, Afzal I, Shabbir R, Ikram K, Zaheer MS, Faheem M, Iqbal J (2022) Seed coating technology: an innovative and sustainable approach for improving seed quality and crop performance. J Saudi Soc Agric Sci 21(8):536–545. https://doi.org/10.1016/j.jssas.2022.03.003

Jayaraman B, MacIsaac J, Anderson D (2016) Effects of derived meals from juncea (*Brassica juncea*), yellow and black seeded canola (*Brassica napus*) and Mult carbohydrase enzymes supplementation on apparent metabolizable energy in broiler chickens. Anim Nutr 2(3):154–159. https://doi.org/10.1016/j.aninu.2016.05.001

Jelani K, Rawat S, Singh PK, Yahya MZA, Yusuf SNF, Diantoro M, Tomar R (2025) Low viscous Imidazolium ionic liquid infused polyvinyl alcohol polymer electrolyte for light-emitting electrochemical device. Energy Storage 7(3):e70162. https://doi.org/10.1002/est2.70162

Kadirvel V, Palanisamy Y, Ganesan ND (2025) Active packaging system – an overview of recent advances for enhanced food quality and safety. Packag Technol Sci 38(2):145–162. https://doi.org/10.1002/pts.2863

Kaleh AM, Singh P, Ooi Chua K, Harikrishna JA (2025) Modulation of plant transcription factors and priming of stress tolerance by plant growth-promoting bacteria: a systematic review. Ann Bot 135(3):387–402. https://doi.org/10.1093/aob/mcae166

Kalia P, Singh S (2020) Accelerated improvement of cole vegetable crops. In: Accelerated plant breeding, volume 2: vegetable crops. Springer International Publishing, Cham, pp 101–135

Kaur K (2023) A multi-omics approach for genetic and molecular analysis of clubroot resistance in *Brassica napus* canola. Int J Mol Sci. 23(16):9280

Kavusi E, Shahi Khalaf Ansar B, Dehghanian Z, Asgari Lajayer B, Nobaharan K, Ma Y, Glick BR (2023) Delivery of beneficial microbes via seed coating for medicinal and aromatic plant production: a critical review. J Plant Growth Regul 42(2):575–597. https://doi.org/10.1007/s00344-022-10597-2

Kejela T (2024) Phytohormone-producing rhizobacteria and their role in plant growth. In: New insights into phytohormones. IntechOpen

Kloepper JW, Ryu CM, Zhang S (2004) Induced systemic resistance and promotion of plant growth by *Bacillus spp.* Phytopathology 94(11):1259–1266. https://doi.org/10.1094/PHYTO.2004.94.11.1259

Konopka I, Tańska M, Dąbrowski G, Ogrodowska D, Czaplicki S (2023) Edible oils from selected unconventional sources a comprehensive review of fatty acid composition and phytochemicals content. Appl Sci 13(23):12829. https://doi.org/10.3390/app132312829

Kumar R, Saini DK, Kumar M, Priyanka V, Akhatar J, Kaushik D, Kaushik P (2022) Revealing the genetic architecture of yield-related and quality traits in Indian mustard [*Brassica juncea* (L.) Czern. and Coss.] using meta-QTL analysis. Agronomy 12(10):2442. https://doi.org/10.3390/agronomy12102442

Kumar R, Das SP, Choudhury BU, Kumar A, Prakash NR, Verma R, Mishra VK (2024) Advances in genomic tools for plant breeding: harnessing DNA molecular markers, genomic selection, and genome editing. Biol Res 57(1):80. https://doi.org/10.1186/s40659-024-00562-6

Kumar P, Vashistha H, Kumar P, Chandra H, Dubey RC (2025) Plant growth promoting Rhizobacteria (PGPR)-based nanoparticles in agriculture. In: Nanofertilizers in agriculture: synthesis, mechanisms, and effect on plants, pp 117–140

Lamichhane JR, You MP, Laudinot V, Barbetti MJ, Aubertot JN (2020) Revisiting sustainability of fungicide seed treatments for field crops. Plant Dis 104(3):610–623. https://doi.org/10.1094/PDIS-06-19-1157-FE

Li X, Liu Q, Liu J (2023) Long Non-Coding RNAs: Discoveries, Mechanisms, and Research Strategies in Seeds. Genes 14(12):2214. https://doi.org/10.3390/genes14122214

Liang K, Cao J, Wang J, Cao F, Su E (2025) A comprehensive review of Moringa oleifera leaves and seeds with a focus on their food-related aspects. J Food Compos Anal:107247. https://doi.org/10.1016/j.jfca.2025.107247

Ma Y (2019) Seed coating with beneficial microorganisms for precision agriculture. Biotechnol Adv 37(7):107423. https://doi.org/10.1016/j.biotechadv.2019.107423

Malakar M, Jayasavitha K, Nithya R (2025) State-of-the-art breeding of peace lily (*Spathiphyllum spp*.). In: Breeding of ornamental crops: potted plants and shrubs, pp 329–395. https://doi.org/10.1007/978-3-031-80060-3_11

Maleki K, Soltani E, Arabhosseini A, Aghili Lakeh M (2021) A quantitative analysis of primary dormancy and dormancy changes during burial in seeds of *Brassica napus*. Nord J Bot 39(8). https://doi.org/10.1111/njb.03281

Malik SK, Bansal S, Malhotra EV, Singh AM, Singh GP (2025) Indian cryo gene bank conserving diverse plant genetic resources for the last three decades: achievements and way forward. Cryobiology 118:105205. https://doi.org/10.1016/j.cryobiol.2025.105205

Martynková GS, Barabaszová KČ, Hundáková M, Pazourková L, Holešová S, Kupková J, Kratošová G (2024) Decorated clays for polymer nanocomposites. In: Chemical physics of polymer nanocomposites: processing, morphology, structure, thermodynamics, rheology, vol 2, pp 679–710. https://doi.org/10.1002/9783527837021.ch20

Mba C, Ogbonnaya FC (2022) Utilizing plant genetic resources to develop climate resilient crops. In: Agricultural biotechnology, biodiversity and bioresources conservation and utilization. CRC Press, pp 373–404

McGehee CS, Raudales RE, Elmer WH, McAvoy RJ (2019) Efficacy of bio fungicides against root rot and damping-off of microgreens caused by *Pythium spp*. Crop Protection 121:96–102. https://doi.org/10.1016/j.cropro.2018.12.007

Meena BL, Meena HS, Singh VV, Meena MD, Sharma HK, Rai PK, Kumar P (2025) Breeding climate resilient rapeseed–mustard in climate change era: current breeding strategies and prospects. In: Breeding climate resilient and future ready oilseed crops, pp 303–332. https://doi.org/10.1007/978-981-97-7744-0_11

Mohammed M, Alqahtani N, El-Shafie H (2021) Development and evaluation of an ultrasonic humidifier to control humidity in a cold storage room for postharvest quality management of dates. Foods 10(5):949. https://doi.org/10.3390/foods10050949

Mohd Saad NS, Severn-Ellis AA, Pradhan A, Edwards D, Batley J (2021) Genomics armed with diversity leads the way in *Brassica* improvement in a changing global environment. Front Genet 12:600789. https://doi.org/10.3389/fgene.2021.600789

Mondal S, Bose B (2021) Seed priming: an interlinking technology between seeds, seed germination and seedling establishment. In: Plant reproductive ecology-recent advances. https://doi.org/10.5772/intechopen.100804

Neha T, Vasudevan SN (2021) Plasma treatment and seed quality advancement: a review. Agric Rev 42(2):197–202. https://doi.org/10.18805/ag.R-2024

Neik TX, Barbetti MJ, Batley J (2017) Current status and challenges in identifying disease resistance genes in *Brassica napus*. Front Plant Sci 8:1788. https://doi.org/10.3389/fpls.2017.01788

Ongaga EG, Muli JK, Kamau PK, Budambula NL (2025) Endophytic microflora of crotalaria: their diversity and role in plant growth promotion. Curr Microbiol 82(5):1–18. https://doi.org/10.1007/s00284-025-04181-6

Panjabi P, Yadava SK, Kumar N, Bangkim R, Ramchiary N (2019) Breeding *Brassica juncea* and B. rapa for sustainable oilseed production in the changing climate: progress and prospects. In: Genomic designing of climate-smart oilseed crops, pp 275–369. https://doi.org/10.1007/978-3-319-93536-2_6

Pant P, Hamsa S, Kaur J (2022) Advances in breeding strategies for improving stress tolerance in *Brassicas*. In: The Brassica juncea genome. Springer International Publishing, Cham, pp 439–469. https://doi.org/10.1007/978-3-030-91507-0_24

Pagnotta MA (2025) Molecular Breeding for Abiotic Stress Tolerance in Crops: Recent Developments and Future Prospectives. Int J Mol Sci 26(18):9164.https://doi.org/10.3390/ijms26189164

Patra S (2025) Recalcitrant seeds: a review of research on the key factors affecting and some important management strategies for extending longevity during storage. J Stress Phys Biochem 21(1):190–205

Prasad N, Thombare N, Sharma SC, Kumar S (2022) Gum arabic–a versatile natural gum: a review on production, processing, properties and applications. Ind Crop Prod 187:115304. https://doi.org/10.1016/j.indcrop.2022.115304

Qiu Y, Amirkhani M, Mayton H, Chen Z, Taylor AG (2020) Biostimulant seed coating treatments to improve cover crop germination and seedling growth. Agron 10(2):154. https://doi.org/10.3390/agronomy10020154

Rad-Spice M, Rogiewicz A, Jankowski J, Slominski BA (2018) Yellow-seeded *B. napus* and *B. juncea* canola. Part 1. Nutritive value of the meal for broiler chickens. Anim Feed Sci Technol 240:66–77. https://doi.org/10.1016/j.anifeedsci.2018.03.017

Rahman M, Liu L, Barkla BJ (2021) A single seed protein extraction protocol for characterizing *Brassica* seed storage proteins. Agron 11(1):107. https://doi.org/10.3390/agronomy11010107

Rajendra Prasad S, Kamble UR, Sripathy KV, Udaya Bhaskar K, Singh DP (2016) Seed biopriming for biotic and abiotic stress management. In: Microbial inoculants in sustainable agricultural productivity: vol. 1: research perspectives. Springer, New Delhi, pp 211–228

Ray J, Bordolui SK (2021) Role of seed banks in the conservation of plant diversity and ecological restoration. Res Rev J Environ Sci 3(2):1–16

Raza MA, Sohail H, Hassan MA, Sajad S, Xing Y, Song J (2024) Cold stress in *brassica* vegetables: Morpho-physiological and molecular responses underlying adaptive mechanism. Sci Hortic 329:113002. https://doi.org/10.1016/j.scienta.2024.113002

Ross IA (2024) The anticarcinogenic properties of brassica vegetables. In: Plant-based therapeutics, volume 2: the brassicaceae family. Springer Nature Switzerland, Cham, pp 175–259. https://doi.org/10.1007/978-3-031-63681-3_5

Rouxel T, Peng G, Van de Wouw A, Larkan NJ, Borhan H, Fernando WD (2024) Strategic genetic insights and integrated approaches for successful management of blackleg in canola/rapeseed farming. Plant Pathol 73(9):2260–2280. https://doi.org/10.1111/ppa.14018

Sachan DS, Naimuddin SK, Patra D, Subha L, Senthilkumar T, Chittibomma K, Prasad SV (2024) Advancements in enhancing oil quality in rapeseed and mustard: a comprehensive review. J Exp Agric Int 46(5):181–193. https://doi.org/10.9734/jeai/2024/v46i52369

Saharan GS, Mehta NK, Meena PD, Saharan GS, Mehta NK, Meena PD (2022) Inheritance of disease resistance. In: Genomics of crucifer's host-resistance. Springer Nature Singapore, Singapore, pp 195–264. https://doi.org/10.1007/978-981-16-0862-9_3

Saharan GS, Mehta DN, Meena PD (2023) Genomics of crucifer's host-pathosystem. Springer. https://doi.org/10.1007/978-981-19-3812-2

Salawi A (2022) Pharmaceutical coating and its different approaches, a review. Polymers 14(16):3318. https://doi.org/10.3390/polym14163318

Salgotra RK, Chauhan BS (2023) Genetic diversity, conservation, and utilization of plant genetic resources. Genes 14(1):174. https://doi.org/10.3390/genes14010174
Sarkar D, Rakshit A, Al-Turki AI, Sayyed RZ, Datta R (2021) Connecting bio-priming approach with integrated nutrient management for improved nutrient use efficiency in crop species. Agriculture 11(4):372. https://doi.org/10.3390/agriculture11040372
Sarraf M, Kataria S, Taimourya H, Santos LO, Menegatti RD, Jain M, Ihtisham M, Liu S (2020) Magnetic field (MF) applications in plants: an overview. Plants 9(9):1139. https://doi.org/10.3390/plants9091139
Shang Y, Hasan MK, Ahammed GJ, Li M, Yin H, Zhou J (2019) Applications of nanotechnology in plant growth and crop protection: a review. Molecules 24(14):2558. https://doi.org/10.3390/molecules24142558
Sharma P (2020) Responses and tolerance of *Brassicas* to high temperature. In: The plant family Brassicaceae: biology and physiological responses to environmental stresses, pp 277–310. https://doi.org/10.1007/978-981-15-6345-4_9
Sharma D, Gautam HR (2025) Bio-priming seed treatment for biological control of damping-off in solanaceous vegetable crops. Eur J Plant Pathol:1–11. https://doi.org/10.1007/s10658-025-03033-0
Sharma P, Roy M, Roy B (2021) Determining the effect of storage, moisture and temperature on vigour and viability of rapeseed and mustard. Curr Topic Agric Sci 5:39–51. https://doi.org/10.9734/bpi/ctas/v5/15153D
Sharma HK, Kumar A, Singh VV, Meena HS, Priya Medha, Meena BL, Rai PK (2022a) Genetic resources of *brassicas*. In: Cash crops: genetic diversity, erosion, conservation and utilization, pp 285–337. https://doi.org/10.1007/978-3-030-74926-2_9
Sharma S, Bala M, Kaur G, Tayyab S, Feroz SR (2022b) Chemical composition of oil and cake of *Brassica juncea*: implications on human and animal health. In: The *Brassica juncea* genome, pp 29–55. https://doi.org/10.1007/978-3-030-91507-0_3
Shen J, Liu Y, Wang X, Bai J, Lin L, Luo F, Zhong H (2023) A comprehensive review of health-benefiting components in rapeseed oil. Nutrients 15(4):999. https://doi.org/10.3390/nu15040999
Shiam MAH, Islam MS, Ahmad I, Haque SS (2025) A review of plant-derived gums and mucilages: structural chemistry, film forming properties and application. J Plast Film Sheeting. https://doi.org/10.1177/87560879251316553
Shilova OA, Khamova TV, Panova GG, Udalova OR, Artemyeva AM, Kornyukhin DL, Kopitsa GP (2023) sol-gel-derived functional coatings for pre-sowing seed treatment. Coatings 13(12):1978
Singh M (2025) Seed treatments and coatings for enhanced germination and vigor Dr. Monika Singh. In: Recent trends in dry land agriculture, p 429
Singh PK, Prakash S (2024) Advances in breeding for oil quality enhancement in Indian mustard (*Brassica spp.* L.): achievements, challenges, and research opportunities. Plant Cell Biotechnol Mol Biol 25(5–6):52–63. https://doi.org/10.56557/pcbmb/2024/v25i5-68689
Singh H, Jassal RK, Kang JS, Sandhu SS, Kang H, Grewal K (2015) Seed priming techniques in field crops-a review. Agric Rev 36(4):251–264. https://doi.org/10.18805/ag.v36i4.6662
Singh M, Singh VV, Singh N (2022) Drought tolerance in rapeseed-mustard: conventional and molecular approaches. In: Genomic designing for abiotic stress resistant oilseed crops, pp 199–218. https://doi.org/10.1007/978-3-030-90044-1_5
Singh A, Rajput VD, Ghazaryan K, Rajput P (2024) Nanotechnology-based genome editing: empowering sustainable agriculture for. In: Sustainable agriculture: nanotechnology, biotechnology, management and food security, p 173
Singh R, Lakra N, Sheoran AR, Luhach A (2025) Enhancing resistance to Sclerotinia stem rot in *Brassica juncea* through seed priming with bio-control agents. Austin J Plant Bio 11(1):1057
Sohail M, Pirzada T, Opperman CH, Khan SA (2022) Recent advances in seed coating technologies: transitioning toward sustainable agriculture. Green Chem 24(16):6052–6085
Sooda M, Kumarb V, Rawalc R (2020) Seed biopriming a novel method to control seed borne diseases of crops. In: Biocontrol agents and secondary metabolites, pp 181–223. https://doi.org/10.1016/B978-0-12-822919-4.00008-9

Subbaiah PV (2019) Review on vermicompost, poultry manure, farmyard manure, biogas digest, biochar, urban compost and biofertilizers as potential alternate nutrient sources for sustainable agriculture. Int J Chem Stud 7(4):255–258

Sunagar R, Pandey MK (2024) Genomic approaches for enhancing yield and quality traits in mustard (*Brassica spp.*): a review of breeding strategies. J Adv Biol Biotechnol 27(6):174–185. https://doi.org/10.9734/jabb/2024/v27i6877

Tan S, Cao J, Li S, Li Z (2025) Unraveling the Mechanistic Basis for Control of Seed Longevity. Plants 14(5):805. https://doi.org/10.3390/plants14050805

Tagele SB, Kim RH, Shin JH (2021) Interactions between *Brassica* biofumigants and soil microbiota: causes and impacts. J Agric Food Chem 69(39):11538–11553. https://doi.org/10.1021/acs.jafc.1c03776

Thakur R, Singh PK (2024) Assessment of thermo physical seed treatments in controlling seed borne diseases and enhancing seed quality parameters in vegetable crops: a review. Plant Arch 24(2). https://doi.org/10.51470/PLANTARCHIVES.2024

Thakur AK, Parmar N, Singh KH, Nanjundan J (2020) Current achievements and future prospects of genetic engineering in Indian mustard (*Brassica juncea* L. Czern & Coss.). Planta 252:1–20. https://doi.org/10.1007/s00425-020-03461-8

Thapa S, Baral B, Shrestha M, Dahal KC (2022) Effect of different priming methods on germination behaviour of broadleaf mustard cv. Marpha Chauda Paate. Trop Agrobiodiver 3(2):52–59. https://doi.org/10.26480/trab.02.2022.52.59

Tiamiyu QO, Adebayo SE, Yusuf AA (2023) Gum Arabic edible coating and its application in preservation of fresh fruits and vegetables: a review. Food Chem Adv 2:100251. https://doi.org/10.1016/j.focha.2023.100251

Timofeeva AM, Galyamova MR, Sedykh SE (2024) How do plant growth-promoting bacteria use plant hormones to regulate stress reactions? Plants 13(17):2371. https://doi.org/10.3390/plants13172371

Ton LB, Neik TX, Batley J (2020) The use of genetic and gene technologies in shaping modern rapeseed cultivars (*Brassica napus* L.). Genes 11(10):1161. https://doi.org/10.3390/genes11101161

Upretee P, Bandara MS, Tanino KK (2024) The role of seed characteristics on water uptake preceding germination. Seeds 3(4):559–574. https://doi.org/10.3390/seeds3040038

Valente IDL, Wancura JH, Zabot GL, Mazutti MA (2025) Endophytic and Rhizospheric microorganisms: an alternative for sustainable, organic, and regenerative bioinput formulations for modern agriculture. Microorganisms 13(4):813. https://doi.org/10.3390/microorganisms13040813

Wahab A, Bibi H, Batool F, Muhammad M, Ullah S, Zaman W, Abdi G (2024) Plant growth-promoting rhizobacteria biochemical pathways and their environmental impact: a review of sustainable farming practices. J Plant Growth Regul 104(2):637–662. https://doi.org/10.1007/s10725-024-01218-x

Walker BA, Powell SM, Tegg RS, Doyle RB, Hunt IG, Wilson CR (2022) Soil microbial community dynamics during ryegrass green manuring and *brassica* biofumigation. Appl Soil Ecol 179:104600. https://doi.org/10.1016/j.apsoil.2022.104600

Wambugu PW, Nyamongo DO, Kirwa EC (2023) Role of seed banks in supporting ecosystem and biodiversity conservation and restoration. Diversity 15(8):896. https://doi.org/10.3390/d15080896

Wan C, Shi C, Zhu S, Fang S, Qiu L, Shi G, Wang K (2025) Comprehensive design and preliminary experiments of liquid hydrogen storage tank for trucks. IJR 169:279–293. https://doi.org/10.1016/j.ijrefrig.2024.10.037

Waqas M, Korres NE, Khan MD, Nizami AS, Deeba F, Ali I, Hussain H (2019) Advances in the concept and methods of seed priming. In: Priming and pretreatment of seeds and seedlings: implication in plant stress tolerance and enhancing productivity in crop plants. Springer Singapore, Singapore, pp 11–41

Woodstock LW (1988) Seed imbibition: a critical period for successful germination. J Seed Tech:1–15

Yadav S, Giri R, Ledwani L, Sharma A (2025) Eco-friendly physical seed treatments for sustainable agriculture: Enhancing plant health and performance. In AIP confer proceed (3191, 1). https://doi.org/10.1063/5.0247541

Yirmibeş B, Ulger N (2024) Botanical characteristics, nutritional value, and economic importance of the *brassica* genus. J Agric Food Ecol 2(1):9–14. https://doi.org/10.5281/zenodo.13627029

Zaim NSHBH, Tan HL, Rahman SMA, Abu Bakar NF, Osman MS, Thakur VK, Radacsi N (2023) Recent advances in seed coating treatment using nanoparticles and nanofibers for enhanced seed germination and protection. J Plant Growth Regul 42(12):7374–7402. https://doi.org/10.1007/s00344-023-11038-4

Zhang XY, Tao KL (1989) Silica gel seed drying for germplasm conservation-practical guidelines. Plant Genet Res News 75(76):1–5

Zhang K, Khan Z, Yu Q, Qu Z, Liu J, Luo T, Luo L (2022) Biochar coating is a sustainable and economical approach to promote seed coating technology, seed germination, plant performance, and soil health. Plants 11(21):2864. https://doi.org/10.3390/plants11212864

Innovative Strategies in *Brassica* Crop Improvement: Harnessing Genetics, Biotechnology, and Robotics

Etalesh Goutam, Bharti, Satya Prakash, and Akshay Kumar

Abstract

The chapter presents a multidisciplinary approach to addressing the challenges of global food security and climate change through advanced agricultural technologies. It emphasizes the integration of genetic tools, biotechnological innovations, and robotics to enhance the productivity, resilience, and sustainability of *Brassica* crops. CRISPR/Cas genome editing is highlighted for its role in introducing targeted genetic modifications to improve crop performance under biotic and abiotic stresses. Biotechnological innovations are explored for their ability to accelerate breeding cycles by optimizing environmental conditions, allowing for the rapid development of superior crop varieties. Furthermore, phenotyping technologies using high-throughput imaging and artificial intelligence (AI) are described as critical tools for precise trait evaluation and selection. The transformative role of robotics in modern agriculture is another focal point. Automated systems for planting, monitoring, and harvesting are shown to enhance efficiency while reducing labour costs. Robotics integrated with AI-driven data analytics supports informed decision-making in crop management, ensuring optimal resource use. By combining these strategies, the chapter provides a roadmap for

E. Goutam
Department of Horticulture, Guru Kashi University, Bathinda, Punjab, India

Bharti
School of Agriculture Sciences & Engineering, Maharaja Ranjit Singh Punjab Technical University, Bathinda, Punjab, India

S. Prakash (✉)
Department of Genetics and Plant Breeding, Lovely Professional University, Phagwara, Punjab, India

A. Kumar
Agriculture and Environment Research Unit, School of Life and Medical Sciences, University of Hertfordshire, Hatfield, England, UK

A. Kumar, B. Koul (eds.), *Crop Improvement Strategies in Brassica species: Applied Science*, https://doi.org/10.1007/978-981-95-3861-4_8

developing climate-resilient *Brassica* varieties that meet future food demands sustainably while mitigating the impacts of environmental changes on agriculture.

Keywords
Artificial intelligence · *Brassica* crop improvement · CRISPR/Cas · Food security · Robotics

1 Introduction

The genus *Brassica* consists of around 40 species of flowering plants in the mustard family (*Brassicaceae*), many of which are important agricultural crops. *Brassicas* are native to Europe and temperate Asia and are especially common in the Mediterranean region; some are considered invasive species in areas outside their native range (Jabeen 2020; Raza et al. 2020; Petruzzello 2022; Ross 2024). Economically important members include broccoli, brown mustard, Brussels sprouts, cabbage, cauliflower, kale, kohlrabi, rape, rutabaga, and turnip. *Brassica* plants are pungent herbs with a diversity of growth habits. The plants generally have lobed or entire leaves, arranged alternately, and are often waxy and bluish green in colour. Many species are biennials. The four-petaled flowers are usually borne in clusters and bear dry fruits known as siliques or silicles, depending on their length (Branca and Cartea 2010). The *Brassica* genus encompasses a diverse group of agriculturally significant crops, including vegetables, oilseeds, and fodder plants. Among the most widely cultivated species are *Brassica oleracea* (e.g. cabbage, broccoli, cauliflower, kale), *B. rapa* (e.g. turnips, Chinese cabbage), *B. napus* (rapeseed/canola), and *B. juncea* (mustard). These crops are grown globally for their nutritional, economic, and industrial value (Greer et al. 2023). Table 1 states various *Brassica* members, their importance, and uses along with major producing countries.

Two *Brassica* species are commonly grown as rapeseed, *B. napus* and *B. rapa*. Rapeseed that has been bred to have low concentrations of both erucic acid and glucosinolates in the seed is called canola. Rapeseed is used as industrial oil, while canola is used for a wider range of products, including cooking oils and biodiesel. Annual or spring-type rapeseed belongs to the species *B. napus*, whereas winter-type or biennial rapeseed cultivars belong to the species *B. rapa*. Besides their use as an oil crop, these species are also used for forage. If pest suppression is an objective, rapeseed should be used rather than canola since the breakdown products of glucosinolates are thought to be a principal mechanism for pest control with these cover crops. Rapeseed has been shown to have biological activity against plant parasitic nematodes as well as weeds.

Mustard is a name that is applied to many different botanical species, including white (*Sinapis alba* syn. *B. alba* or *B. hirta*) or yellow mustard, brown or Indian mustard (*B. juncea*), sometimes erroneously referred to as canola and black mustard (*B. nigra* L.). The glucosinolate content of most mustards is very high compared to the true *Brassicas*. Mustards have also been shown to suppress the growth of weeds. The true radish or forage radish (*Raphanus sativus*) has only been known as a

Table 1 Description of various *Brassica* species, importance and uses, and major production countries

Common name	Botanical name	Importance and uses	Major production countries	Reference(s)
Cabbage	*Brassica oleracea* var. *capitata*	High in vitamins like C & K, and globally consumed either fresh or cooked	China, India, Russia, South Korea, Ukraine, etc.	Smith (2007), FAO (2021) and Kumar et al. (2023)
Cauliflower	*Brassica oleracea* var. *botrytis*	Export-oriented, rich in antioxidants, and edible curd (flower head) is consumed after cooking	China, India, the United States, Italy, France, etc.	Rubatzky and Yamaguchi (2012), FAO (2021) and Swain et al. (2024)
Broccoli	*Brassica oleracea* var. *italica*	Healthy food, anti-cancerous, and can be used as salads and vegetables	China, India, the United States, Mexico, Spain, Italy, etc.	Nagraj et al. (2020) and Kumar et al. (2023)
Kale	*Brassica oleracea* var. *acephala*	Nutrient and antioxidant rich superfood, and used as a green leafy vegetable	Spain, USA, UK, Netherlands, Germany, Poland, China, and Mexico	Cartea et al. (2011) and Kumar et al. (2023)
Turnip	*Brassica rapa* subsp. *rapa*	Root vegetables, humans, and livestock use, tolerant to cold and consumed as salad and vegetables	United States, India, Germany, China, France, Russia, Japan, and South Korea	Variyar et al. (2015) and Paul et al. (2019)
Radish	*Raphanus sativus*	Root vegetables, salad, and culinary use	India, China, Japan, Germany, South Korea, and France	FAO (2021) and Lahari et al. (2022)
Mustard	*Brassica juncea*	Major oilseed in South Asia and used in pickles and as green manure	Nepal, Russia, India, Bangladesh, and Canada	Rakow (2004) and Rathore et al. (2015)
Rutabaga (Swede)	*Brassica napobrassica*	Dual-purpose crop can be used as a root crop and fodder	Sweden, Canada, Great Britain, northern Europe, USA, and UK	Shattuck and Proudfoot (1990) and Variyar et al. (2015)
Canola (Rapeseed)	*Brassica napus*	Biodiesel potential and edible oil production	The European Union (EU), Australia China, and India	Snowdon et al. (2007) and FAO (2021)

(continued)

Table 1 (continued)

Common name	Botanical name	Importance and uses	Major production countries	Reference(s)
Pak choi (Bok choy)	*Brassica rapa* subsp. *chinensis*	Staple in East Asian diets, quick growth and Asian leafy vegetable	South Korea, China, Japan, Philippines, Thailand, Vietnam, and Malaysia	Fahey (2003) and Kim et al. (2021)

cultivated species since ancient times. Cultivars developed for high forage biomass or high oilseed yield are also useful for cover crop purposes. Their rapid fall growth has the potential to capture nitrogen in large amounts and from deep in the soil profile. Radishes have been shown to alleviate soil compaction and suppress weeds. Turnips (*B. rapa* L. var. *rapa* (L.) Thell) are used for human and animal food because of their edible root. Turnip has been shown to alleviate soil compaction. While they usually do not produce as much biomass as other *Brassicas*, they provide many macrochannels that facilitate water infiltration.

Brassicas play a vital role in global agriculture due to their versatile uses. Vegetable *Brassicas* such as cabbage, cauliflower, and broccoli are staples in diets worldwide. In 2020 alone, 96.4 million tonnes of vegetable *Brassicas* were produced globally, with significant contributions from China, India, and Russia (Greer et al. 2023). Rapeseed (*B. napus*) is the second most used source for vegetable oil production, serving both the food and biofuel industries. *Brassicas* like turnips and forage radish are valuable for livestock feed due to their high biomass production. *Brassicas* are known for their rapid fall growth, great biomass production and nutrient scavenging ability. However, they are also attracting interest because of their possible pest management characteristics. Most *Brassica* species release chemical compounds that may be toxic to soil-borne pathogens and pests, such as nematodes, fungi, and some weeds. The mustards, unusually, have higher concentrations of these chemicals. *Brassicas* are increasingly used as winter or rotational cover crops in vegetable and specialty crop production. Some *Brassicas* have a large taproot that can break through hard soil pans better than the fibrous roots of cereal cover crops. *Brassicas* provide excellent soil coverage and up to 8000 lb. biomass/acre. Because of their fast fall growth, *Brassicas* are well-suited to capture soil nitrogen (N) remaining after crop harvest. The amount of nitrogen captured is mainly related to biomass accumulation and the amount of N available in the soil profile. Because they immobilize less nitrogen than some cereal cover crops, much of the N taken up can become available for uptake by main crops in early to late spring. *Brassicas* can root to depths of 6 ft or more, scavenging nutrients from below the rooting depth of most crops. To maximize biomass production and nutrient scavenging in the fall, *Brassicas* must be planted earlier than winter cereal cover crops in most regions, making them more difficult to fit into grain production rotations.

Mustards and rapeseed yield oils used in pharmaceuticals, cosmetics, lubricants, and other industrial products. Their deep root systems break up soil compaction and enhance water infiltration. For example, radishes and turnips alleviate both surface and deep soil compaction. *Brassicas* capture residual soil nitrogen effectively, reducing nutrient leaching and improving fertility for subsequent crops. Many species produce glucosinolates, biotoxin compounds that suppress soil-borne pathogens, nematodes, fungi, and weeds. This natural fumigation potential makes *Brassicas* valuable in integrated pest management systems. All *Brassicas* have been shown to release biotoxins that exhibit broad activity against bacteria, fungi, insects, nematodes, and weeds. *Brassica* cover crops are often mowed and incorporated to maximize their natural fumigant potential. This is because the fumigant chemicals are produced only when individual plant cells are ruptured. The timing and method of incorporation seem to be a key factor in how effective they are at suppressing pests. University studies have shown *Brassica* green manures to suppress rhizoctonia, scab, and verticillium in potatoes and phytophthora blight in cucurbits (Larkin and Griffin 2007). It has also shown efficacy in reducing nematode and weed populations in various cropping systems. The use of *Brassica* cover crops for disease and insect control is still very much a work in progress (Ji et al. 2012). The economic significance of *Brassicas* is profound viz., the gross value of vegetable brassicas like cabbage and broccoli tripled between 2000 and 2020 to $32.3 billion USD globally (Greer et al. 2023). Rapeseed oil is a major commodity in both food security and renewable energy markets.

1.1 Challenges in Improving *Brassica* Crops

The *Brassica* genus represents one of the most economically significant and nutritionally valuable plant families in global agriculture. Despite their importance, *Brassica* crops face challenges such as susceptibility to diseases like black rot, clubroot, and downy mildew. Integrated pest management strategies tailored to specific geographic regions are essential for minimizing yield losses (Greer et al. 2023). In 2020, global production of vegetable *Brassica* reached 96.4 million tonnes, representing a 10.6% increase from the previous decade, yet significant yield gaps persist. Under ideal conditions, cabbages could yield as much as 77 tonnes/ha, but realized yields average only 46.5 tonnes/ha, while rapeseed reaches merely one-fifth of its predicted 9.2 tonnes/ha maximum yield potential (Greer et al. 2023). This substantial gap between theoretical and actual yields underscores the multifaceted challenges in *Brassica* improvement, ranging from disease susceptibility to labour constraints and technological limitations.

1.1.1 Biological Diversity and Susceptibility to Diseases

The remarkable biological diversity of *Brassica* crops presents unique challenges for crop improvement efforts. *Brassicas* are distinguished by their varied edible parts, including roots, stems, leaves, and flowers, making them versatile crops for multiple culinary and industrial applications (Jabeen 2020; Raza et al. 2020). However, this diversity also creates vulnerabilities, as each plant part is susceptible to different diseases and pests, complicating breeding efforts for comprehensive resistance (Raza et al. 2020). Biotic stresses represent one of the most significant obstacles to *Brassica* improvement. These crops are susceptible to a wide array of pathogens, including fungi, oomycetes, viruses, bacteria, and nematodes (Laha et al. 2024). Major diseases such as black rot (caused by *Xanthomonas campestris* pv. *campestris*), clubroot (*Plasmodiophora brassicae*), downy mildew (*Hyaloperonospora parasitica*), and turnip yellows virus have been identified as particularly damaging to production (Greer et al. 2023).

The economic impact of these diseases is substantial, with pathogens causing yield losses up to 60% in many cases, and reaching as high as 90% in severe outbreaks, depending on geographic location (Laha et al. 2024). The complexity of host–pathogen interactions in *Brassica* species presents significant challenges for developing resistant varieties. Plants employ diverse strategies and mechanisms to combat biological stress, involving intricate networks of transcription factors that interact with specific cis-elements to regulate gene expression (Laha et al. 2024). Understanding these complex defence mechanisms and successfully translating this knowledge into improved crop varieties remains challenging. Additionally, distinguishing between primary and secondary infections when developing disease management strategies adds another layer of complexity to *Brassica* improvement efforts (Jabeen 2020; Raza et al. 2020).

1.1.2 Pathogen Evolution and Management Challenges

The rapid evolution of pathogens presents a moving target for breeders and researchers working on *Brassica* improvement. Chemical controls for diseases are becoming increasingly problematic due to environmental concerns, regulatory restrictions, and the evolution of resistance in pathogen populations. Many existing chemical controls have broad-spectrum activity, potentially harming beneficial organisms and disrupting ecosystem services that support sustainable production systems (Greer et al. 2023). Furthermore, cultural practices, while important for disease prevention, generally have limited efficacy in reducing or eliminating established diseases. Implementing effective crop rotations is challenging due to the wide host range of many *Brassica* pathogens and economic pressures that favour continuous or short-rotation cropping systems. These challenges necessitate integrated approaches to disease management that combine genetic resistance with cultural and biological control methods.

1.1.3 Genetic and Breeding Limitations

Genetic improvement of *Brassica* crops faces several significant challenges, including the complexity of the *Brassica* genome, the polygenic nature of many important traits, and limited genetic diversity in elite breeding materials. The *Brassica* genus has a complex evolutionary history involving genome duplication events, making genetic analysis and trait manipulation more challenging than in simpler crop systems. While traditional breeding approaches have been effective for many traits, they are time-consuming and resource-intensive when addressing complex challenges like disease resistance. Incorporating quantitative traits such as yield stability, drought tolerance, and durable disease resistance through conventional breeding remains challenging due to their complex genetic architecture and environmental interactions (Laha et al. 2024).

1.1.4 Molecular Approaches and Their Limitations

Modern molecular breeding approaches offer potential solutions but come with their own challenges. Exploiting genotyping traits and utilizing technologies such as RNAi, miRNA, siRNA, and lncRNA show promise for safeguarding Brassica crops from diseases and pest damage (Laha et al. 2024). However, effectively implementing these approaches requires a sophisticated understanding of gene regulatory networks and their interaction with environmental factors. The research infrastructure needed to support advanced breeding approaches is substantial. Jan Hoogland's work at Bejo's research centre highlights the importance of specialized departments in supporting breeding efforts, including phytopathology and content matter, tissue culture and cell biology, marker technology and genomics, seed pathology, and seed physiology (Jabeen 2020; Raza et al. 2020). Establishing and maintaining such comprehensive research capabilities represents a significant investment that may be beyond the reach of many breeding programmes, particularly in developing regions where *Brassica* crops hold significant importance for food security.

1.1.5 Labour and Harvesting Challenges

Labour availability and costs represent a growing challenge for *Brassica* crop production globally. Manual harvesting can account for up to 50% of total production costs, placing significant economic pressure on growers. Labour shortages in agricultural regions worldwide are further exacerbating this challenge, necessitating alternative approaches to crop management and harvesting. The diverse growth habits and harvesting requirements of different *Brassica* crops add complexity to developing universal mechanical or automated harvesting solutions. Differences in plant architecture, maturity timing, and the delicate nature of many *Brassica* vegetables require specialized approaches to harvesting technology that can adapt to these variations without compromising product quality.

1.1.6 Robotics and Automation

Robotics and automation offer potential solutions to the labour challenges facing *Brassica* production. The Automated *Brassica* harvest in Cornwall (ABC) project exemplifies efforts to develop cutting-edge technology for cauliflower harvesting and potentially other fieldwork operations (Klein et al. 2019). Dr. Martin Stoelen's vision of creating small, mobile machines"little helpers"that could perform harvesting tasks represents an innovative approach to addressing labour shortages. However, developing effective robotic systems for *Brassica* harvesting presents significant technical challenges. The "GummiArm" robot with its concept of 'variable stiffness'robotic arms with joints that can mimic human-like movementsillustrates the sophisticated engineering required for these applications (Stoelen et al. 2022). Implementing these technologies at scale will require not only technical advances but also viable business models that make the technology accessible to producers of varying scales.

1.1.7 Abiotic Stresses and Environmental Adaptation

Abiotic stresses pose additional challenges for *Brassica* crop improvement. Factors such as drought, temperature extremes, soil salinity, and nutrient imbalances significantly impact crop productivity and quality. The effects of climate change are further exacerbating these challenges, with increasing frequency and severity of extreme weather events affecting *Brassica* cultivation worldwide. Nutrient management presents particular challenges for *Brassica* crops. Both nutrient deficiencies and excesses (particularly nitrogen) can adversely affect plant health and productivity (Jabeen 2020; Raza et al. 2020). Developing varieties with improved nutrient use efficiency while maintaining high yields and quality is a significant breeding challenge that requires integrated approaches that consider both genetic factors and management practices.

1.1.8 Integrated Management Approaches

Developing effective integrated pest management (IPM) strategies tailored to specific geographic locations and production systems represents a significant challenge in *Brassica* improvement. Such strategies must balance efficacy, economic viability, environmental sustainability, and practical implementation by growers. The complexity of *Brassica* diseases extends beyond the diversity of pathogens to the different plant parts they affect, requiring management approaches that address multiple vulnerability points (Jabeen 2020; Raza et al. 2020). Understanding both plant developmental processes and pathogenic disease resistance mechanisms is essential for developing stress-tolerant and resistant crops (Laha et al. 2024). Bridging the gap between laboratory research and field application represents a significant challenge in translating scientific advances into practical improvements for *Brassica* producers.

1.1.9 Technological Implementation and Adoption

The adoption of advanced technologies for *Brassica* improvement presents both opportunities and challenges. While biotechnological approaches such as CRISPR/Cas9 gene editing offer powerful tools for enhancing crop quality and resistance, their application is constrained by technical limitations, regulatory frameworks, and public acceptance issues in many regions (Laha et al. 2024). Similarly, the implementation of robotic harvesting systems faces challenges related to cost, adaptability to diverse field conditions, and integration with existing farming systems. Business models involving contractor ownership of specialized equipment with farmers purchasing the service when required offer potential solutions but require careful economic analysis and adaptation to local contexts.

1.2 Scope of Genetic and Biotechnological Approaches in *Brassica* Improvement

Genetic and biotechnological innovations are revolutionizing *Brassica* crop improvement, addressing challenges such as polyploidy, disease susceptibility, and yield limitations. These approaches combine traditional breeding with cutting-edge technologies to enhance trait precision, accelerate development cycles, and overcome biological constraints.

1.2.1 Interspecific Hybridization and Genomic Insights

Interspecific hybridization has emerged as a powerful tool for expanding genetic diversity in *Brassica*. A novel *B. napus* population, developed from hundreds of crosses between *B. rapa* and *B. carinata*, demonstrated rich genetic diversity through introgressions, allelic recombination, and structural variations like nonrandom deletions/duplications in the C subgenome. Selective sweep analysis identified 15 genomic regions linked to agronomic traits and genome stability, illustrating how hybridization can rapidly introduce adaptive traits while restoring fertility (Zou et al. 2018). Such approaches enable breeders to bypass the genetic bottlenecks of traditional elite varieties.

1.2.2 CRISPR/Cas9 Gene Editing System

These are increasingly applied to *Brassica* species, despite challenges posed by their polyploid genomes. Key advancements include knockout of *SHATTERPROOF1/2* (*BnSHP1/BnSHP2* homeologs) in *B. napus* reduced shattering by tenfold (Yang et al. 2018; Zandberg et al. 2022). Editing of BnD14 (a strigolactone receptor) (*DWARF14 (D14)* homeologs) produced dwarfed plants with 40% more flowers, improving harvestability (Yang et al. 2018; Stanic et al. 2021; Zandberg et al. 2022). Mutations in CLAVATA3 (CLV3) pathway genes increased seed weight per silique by 74% through multilocular structure formation (Yang et al. 2018; Zandberg et al. 2022). In addition to the above, Table 2 addresses diverse applications of CRISPR/Cas9 in different *Brassica* members. However, the redundancy of homologous genes in allotetraploids like *B. napus* complicates editing efficacy, necessitating multiplex targeting. Base and prime editing technologies remain underutilized but hold promise for introducing single-nucleotide changes critical for complex traits like drought tolerance (Li et al. 2022) (Fig. 1).

Breakthroughs in genetic transformation are overcoming genotype dependence in *B. oleracea*. A curd peduncle-based *Agrobacterium*-mediated method achieved 11.87% transformation efficiency across 12 cauliflower and broccoli genotypes. This system leverages cambial cells' regenerative capacity, enabling high-throughput editing and functional genomics studies previously hindered by inefficient protocols (Sheng et al. 2022). Protoplast fusion facilitates the development of cytoplasmic male sterile (CMS) lines, critical for hybrid seed production. Recent advances integrate somatic hybridization with CRISPR to introduce traits like disease resistance and climate resilience. For example, CMS lines derived from protoplast fusion reduce hybrid production costs while enabling the stacking of beneficial mitochondrial and nuclear genomes (Romero-Muñoz and Pérez-Jiménez 2024).

1.2.3 Integrated Strategies and Future Directions

The convergence of these technologies is reshaping *Brassica* improvement, such as combining SNP array-based genomic selection with CRISPR-edited alleles to

Table 2 Multifaceted use of CRISPR/Cas9 technology for trait-specific improvement in *Brassica*

S. no.	Targeted trait	Targeted genes	Effect on phenotype	Homolog of *Arabidopsis*	References
1	Number of seeds per silique	*BnaEOD3*	42% increase in number of seeds/siliques	Enhancer of *DA1 (EOD3)*	Zandberg et al. (2022) and Khan et al. (2024)
2	Plant architecture and yield	*BnaMAX1*	Nearly 35% reduction in plant height, 3× more primary branches and approximately 65% increase in total silique number	Vigorous axillary growth	Zheng et al. (2024)
3	Early flowering	*BnaSVP*; *BnSDG8*; BnaSDG8.C	Nearly 50% reduction in arrival of flowering	Short vegetative phase (SVP)	Jiang et al. (2018) and Ahmar et al. (2022)
4	Pod shattering resistance	*BnJAG.C02, BnJAG.A07, BnJAG.A02, BnJAG.A08 BnJAG.C06*	Provide double resistance than earlier	*JAGGED (JAG)*	Zaman et al. (2019)
5	Yield related	*BnCLV3*	Multilocular silique increased	–	Yang et al. (2018) and Rauf et al. (2023)
6	Pod shattering resistance	*BnALC*	Enhanced resistance against pod shattering	–	Braatz et al. (2017) and Ahmad et al. (2023)
7	Plant architecture and branch angle	*BnaBP*; *BnaA03.BP*	Approximately 20% reduction in plant height and branch angle reduced from 84° to 14°	BREVIPEDICELLUS (BP)	Fan et al. (2021)
8	Pod shattering resistance	*BnIND*	Increased pod shatter resistance	*BnTT8*	Zhai et al. (2020)

(continued)

Table 2 (continued)

S. no.	Targeted trait	Targeted genes	Effect on phenotype	Homolog of *Arabidopsis*	References
9	Resistance against herbicides	*BnALS*; *BnRGA*; *BoALS*; *BoCENH3*; *BnIAA7*	Developed herbicide resistance with semi dwarf architecture of the plant	–	Wu et al. (2021), Cheng et al. (2021) and Li et al. (2022)
10	Induction of self-incompatibility	*BnS6-SMI2*	Creation of self-incompatible lines	–	Dou et al. (2021)
11	Creation of orange-coloured flowers	*BnaA09.ZEP*; *BnaC09.ZEP*	Production of orange-coloured flowers in rapeseed	–	Liu et al. (2020)
12	Induction of male sterility	*BnMS5*	Development of male sterility	–	Xin et al. (2020)
13	Sensitivity to boron	*BnA9.WRKY47*	Enhanced sensitivity to low boron stress	*BnaA3.NIP5;1*	Feng et al. (2020)
14	Fatty acid and oil composition	*BnLPAT5*; *BnTTG1*; *BnFAE*; *BnLPAT2*	Low erucic acid and increased oil content	–	Zhang et al. (2019) and Cheng et al. (2023)
15	Glucosinolate	*BnGTR2*; *BnGTR1*; *BoMYB28*	Smaller seeds with higher seed oil content and reduced levels of glucosinolates	–	Li et al. (2021), Neequaye et al. (2021) and Tan et al. (2022)
16	Phytic acid	*BnITPK*	Obtained low phytic acid mutants with enhanced free phosphorus	–	Sashidhar et al. (2020)
17	Self-incompatibility	*BnA5.ZML1*	Reduction in self-incompatibility	–	Duan et al. (2020)
18	Drought stress	*BnaRGA*	Resistance to drought stress	–	Yang et al. (2017)
19	Wax content and green leaf	*BoCER1*	Improved green colour in leaf with reduced wax content	–	Cao et al. (2021)

(continued)

Table 2 (continued)

S. no.	Targeted trait	Targeted genes	Effect on phenotype	Homolog of *Arabidopsis*	References
20	Methylation of pectin	*Bra007665*; *Bra014410*; *Bra003491*	Four homozygous mutants were identified	–	Xiong et al. (2019)

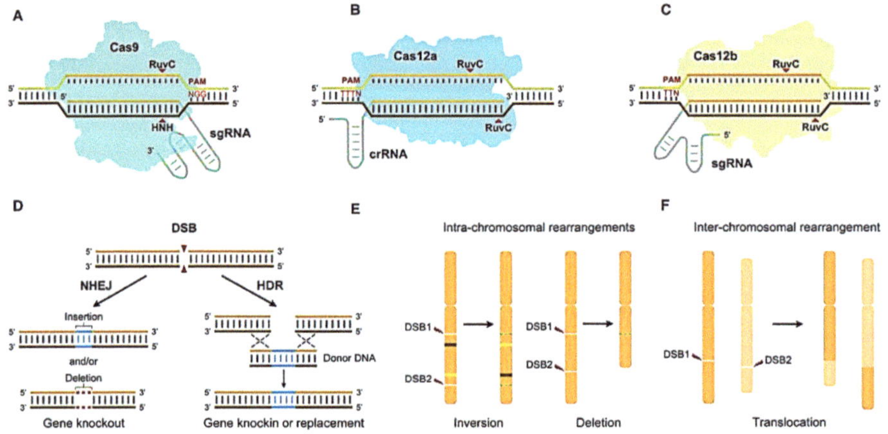

Fig. 1 Overview of the CRISPR/Cas genome editing systems and repair of double-strand breaks (DSBs). (Source: Li et al. 2022)This figure depicted, (**a**) is the CRISPR/Cas9, guided by the sequences on the sgRNA and PAM downstream of the target, Cas9 cleaves the double-stranded DNA, creating a blunt-ended DSB. (**b**) is the CRISPR/Cas12a, guided by the sequences on the CRISPR RNA (crRNA) and protospacer adjacent motif (PAM) upstream of the target. Cas12a cleaves the targeted DNA, generating a staggered DSB. (**c**) is the CRISPR/Cas12b, guided by the sequences on the single guide RNA (sgRNA) and PAM upstream of the target. Cas12b cleaves the targeted DNA, generating a staggered DSB. (**d**) is a single DSB that is repaired via the non-homologous end joining (NHEJ) or homology-directed repair (HDR) pathway, in the NHEJ pathway, there are variable insertions (blue line) and/or deletions (red dotted line) at the break, generating a knockout mutant. When there is a donor template around the DSB, it can be repaired via the HDR pathway, generating predefined, precisely repaired mutants. (**e**) is Two DSBs in the same chromosome can induce intra-chromosomal rearrangements, leaving indels (blue dashes) at the break and (**f**) is two DSBs in different chromosomes can induce inter-chromosomal rearrangements

pyramid yield-related traits (Zou et al. 2018). Deploying CMS lines enhanced through protoplast fusion and gene editing for seed production efficiency (Romero-Muñoz and Pérez-Jiménez 2024). Leveraging high-throughput phenotyping to validate edited traits under field conditions (Zandberg et al. 2022).

2 Exploring the Gene Pool of *Brassica* Species

2.1 Genetic Diversity and Variability in *Brassica* Crops

The genus *Brassica* encompasses a suite of agriculturally vital species ranging from oilseed rape (*B. napus*) and mustard (*B. juncea*) to cole crops (*B. oleracea*) and turnip (*B. rapa*). Underpinning the resilience and improvement potential of these crops is their inherent genetic diversity, which arises from both natural evolutionary processes and human-mediated selection. Molecular surveys employing SSRs, SNP arrays, and whole-genome resequencing have revealed that within-species variation often exceeds 60% of total observed diversity, indicating rich intra-specific heterogeneity that breeders can exploit (Zhang et al. 2023; Sameeullah et al. 2025). For instance, analyses of *B. juncea* germplasm uncovered that 65.9% of variation resides within populations, underscoring opportunities for heterosis breeding and trait introgression (Zhang et al. 2023).

At the inter-specific level, *Brassica* crops exhibit a classic "U-triangle" relationship, wherein three diploids (*B. rapa*, *B. nigra*, *B. oleracea*) have given rise to three allotetraploids (*B. juncea*, *B. napus*, *B. carinata*). This genomic architecture confers extensive allelic richness: pangenome studies of *B. napus* alone have identified thousands of presence–absence variants and novel subgenomic sequences introduced via synthetic resynthesis and wide crosses (Hu et al. 2021; MacNish et al. 2025). Such structural variation contributes not only to phenotypic diversity but also affects oil content, disease resistance, and abiotic stress tolerance, leading to the discovery of novel alleles for climate adaptation. Nevertheless, modern breeding bottlenecks have narrowed effective diversity in elite gene pools. Pedigree-based selection for yield and quality traits, particularly in *B. napus*, has led to a reduction in haplotype richness over successive breeding cycles. Genome-wide scans have detected selective sweeps around key agronomic loci evidence that despite high standing variation, usable diversity may be limited in advanced cultivars (Mohd Saad et al. 2021). To counteract this, pre-breeding efforts now emphasize the incorporation of underutilized landraces and exotic accessions, restoring allelic variants lost during domestication and modern improvement.

2.2 Importance of Wild Relatives and Germplasm Collections

Crop wild relatives (CWRs) of *Brassica* represent an indispensable reservoir of adaptive traits shaped by natural selection under diverse ecological pressures. Wild species such as *B. cretica*, *B. insularis*, and *B. montana* harbour alleles conferring resistance to pathogens (e.g. *Leptosphaeria maculans*), tolerance to drought and salinity, and novel metabolite profiles (Kioukis et al. 2020; Quezada-Martinez et al. 2021). Conservation of these taxa, both in situ and ex situ, is thus pivotal. Ex situ germplasm banks worldwide maintain over 70,000 *Brassica* accessions, yet inconsistencies in passport data and nomenclature complicate their utilization (Subramanian et al. 2023). Recent efforts in high-throughput genotyping and digital accession curation aim to rectify misidentifications and to stratify collections by genetic clusters, streamlining targeted germplasm mining.

The integration of CWR diversity into breeding pipelines follows two principal routes: direct introgression and pre-breeding via synthetic amphidiploids. Direct crosses between crops and CWRs are often impeded by pre- and post-zygotic barriers; however, embryo rescue and bridge species (e.g. using *B. rapa* to introgress *B. oleracea* alleles into *B. napus*) have facilitated successful gene transfer (Katche et al. 2019). Synthetic amphidiploids, generated by doubling the chromosome complement of inter-specific hybrids, restore fertility and enable recombination, effectively broadening the gene pool of target crops. For example, resynthesized *B. juncea* lines exhibit novel allelic combinations and enhanced resistance traits, albeit often with initial yield penalties that can be ameliorated through backcrossing and genomic selection (Garg et al. 2025). Beyond wild species, landraces and traditional varieties represent another critical tier of germplasm diversity. These locally adapted cultivars frequently carry region-specific adaptations such as cold tolerance in European turnip landraces or heat resilience in South Asian mustard, which may be absent in modern, broadly adapted cultivars (Mohd Saad et al. 2021). Characterization of these resources through genome-wide association studies (GWAS) and phenotype-environment modelling has begun to uncover genotype–environment associations that can guide breeding for future climates.

2.3 Gene Flow and Hybridization Strategies for Expanding the *Brassica* Gene Pool

Gene flow, both natural and managed, plays a dual role in shaping *Brassica* diversity. Spontaneous hybridization among sympatric species can lead to transgene escape, but also to the natural acquisition of adaptive traits. Studies have documented gene flow from *B. napus* into feral *B. rapa* populations at rates up to 1% under field conditions, mediated by pollen and seed dispersal vectors (Johannessen et al. 2006; Liu et al. 2013). While biosafety concerns motivate strategies to minimize unwanted introgression (e.g. chloroplast transformation, genetic use restriction technologies), controlled hybridization remains a cornerstone of pre-breeding.

Strategic inter- and intra-specific crosses exploit both conventional and biotechnological tools. Marker-assisted backcrossing (MABC) accelerates the recovery of elite backgrounds while retaining target alleles from wild donors. Meanwhile, genomic selection (GS) models trained on wide-cross populations can predict breeding values for complex traits such as drought tolerance, thus enabling the selection of superior recombinants even in early generations (Mohd Saad et al. 2021). Moreover, gene editing technologies (e.g. CRISPR/Cas) offer precision in introducing or modifying alleles identified in CWRs, bypassing linkage drag and reducing the number of backcross generations required.

Hybridization schemes increasingly incorporate bridge species and synthetic polyploids. For example, creating *B. napus* × *B. nigra* allohexaploids has unlocked B-genome diversity for oilseed rape improvement, while backcrossing these hexaploids to *B. napus* yields introgression lines with minimal linkage drag (Saban et al. 2023). Similarly, "nested association mapping" (NAM) populations derived from multiple wild donors provide high-resolution mapping of adaptive QTLs, facilitating their deployment across diverse breeding programmes.

3 Genetic Improvements in *Brassica* Crops

Brassica crops have undergone significant genetic advancements through conventional and modern biotechnological approaches. These improvements target enhanced yield, disease resistance, and nutritional quality, addressing global agricultural challenges.

3.1 Conventional Breeding Methods in *Brassica* Improvement

Conventional breeding in *Brassica* relies on Mendelian genetics and phenotypic selection to exploit natural genetic diversity. Hybridization with wild relatives (e.g. *B. oleracea* wild populations) has enriched gene pools, introducing traits like stress tolerance and improved morphology (Saban et al. 2023). For qualitative traits controlled by single genes (e.g. flower colour), traditional methods efficiently isolate desirable alleles. However, quantitative traits (e.g. yield, disease resistance) require iterative selection due to polygenic inheritance (Begna 2021). Recent studies highlight the role of crop-wild hybridization in *Brassica* domestication, revealing admixed origins of some cultivated varieties (Saban et al. 2023). Despite limitations in precision, conventional breeding remains foundational for trait introgression and maintaining genetic diversity.

3.2 Advances in Molecular Marker-Assisted Selection (MAS)

MAS has revolutionized *Brassica* breeding by enabling precise tracking of target loci. High-density molecular maps using SSR and SNP markers have identified quantitative trait loci (QTLs) for disease resistance, seed oil content, and inflorescence size (Snowdon and Friedt 2004; Zhao et al. 2022; Treccarichi et al. 2023), for example, QTL *qCaBR1* on chromosome C06 in *B. oleracea* confers resistance to black rot (caused by *Xanthomonas campestris*) (Lu et al. 2023). Markers linked to glucosinolates biosynthesis pathways facilitate selection for cancer-preventive compounds in broccoli (Han et al. 2021). GWAS in *B. napus* pinpointed 58 loci associated with seed oil content, enabling oilseed improvement (Zhao et al. 2022). Integration of MAS with genomic selection and high-throughput screening accelerates breeding cycles, particularly for complex polygenic traits (Amas et al. 2023; Mapari and Mehandi 2024).

3.3 CRISPR-Cas9 and Genome Editing in *Brassica* Species

CRISPR-Cas9 has emerged as a transformative tool for targeted mutagenesis in *Brassica*, like in *B. napus*, CRISPR achieved 65.3% mutation efficiency in T_0 generation plants, with 48.2% stably inherited in the next generation (T_1) (Yang et al. 2017; Dai et al. 2021). Simultaneous targeting of paralogous genes (e.g. fatty acid desaturases) enables trait stacking in oilseed crops (Yang et al. 2017). DNA-free editing via ribonucleoprotein (RNP) delivery to protoplasts yielded 62% mutated shoots in *B. napus*, bypassing regulatory hurdles (Sahab et al. 2024). Challenges persist in genotype-dependent regeneration, particularly in cabbage (*B. oleracea* var. *capitata*), where transformation efficiency varies widely (Gerszberg 2018).

3.4 Trait-Based Genetic Selection: Disease Resistance, Yield, and Nutritional Quality

Black rot resistance in *B. oleracea* is governed by QTLs on chromosomes C03 and C06, with candidate genes (e.g. receptor-like kinases) showing pathogen-responsive expression (Lu et al. 2023). Genomic selection models incorporating pathogen effector data improve prediction accuracy (Amas et al. 2023). In *B. napus*, QTL mapping identified 12 stable loci for 1000-seed weight, with *BnAP2* and *BnGRF* genes regulating seed size (Zhao et al. 2022). Hybrid breeding leveraging heterosis remains critical for yield gains in oilseed rape. Biofortification strategies target glucosinolates, vitamins, and antioxidants; for example, broccoli lines engineered for elevated levels show enhanced chemopreventive properties (Han et al. 2021). Agronomic biofortification with selenium and zinc increases mineral content in *B. oleracea* varieties (Zhang et al. 2025). These innovations underscore the synergy between traditional methods and cutting-edge technologies in *Brassica* improvement. Future directions include pangenome-assisted breeding, AI-driven trait prediction, and robotics-enabled phenotyping to address climate resilience and nutritional security.

4 Improvement in the Physiochemistry of *Brassica* Ingredients

4.1 Enhancing Nutritional Quality and Health Benefits of *Brassicas*

Brassica vegetables, including broccoli, cabbage, and mustard, are renowned for their rich nutritional profiles and associated health benefits. These crops are abundant in vitamins, minerals, dietary fibre, and bioactive compounds such as glucosinolates and phenolics. The consumption of *Brassica* vegetables has been linked to reduced risks of chronic diseases, including cardiovascular ailments and certain cancers (Li et al. 2025). To further enhance these health benefits, agronomic practices play a pivotal role. For instance, adequate sulphur fertilization has been shown to increase the concentration of health-promoting compounds like glucosinolates and phenolics in *Brassica* crops. Similarly, manganese supplementation can elevate vitamin C content, while proper potassium and calcium nutrition contribute to overall crop quality and nutritional value.

4.2 Modifying Anti-nutritional Compounds in *Brassica* Crops

While *Brassicas* offer numerous health benefits, they also contain anti-nutritional factors such as glucosinolates and erucic acid, which can adversely affect human and animal health. High levels of certain glucosinolates can interfere with thyroid function, and elevated erucic acid content has been associated with cardiac issues (Bisht and Augustine 2019; Sujith Kumar et al. 2020). Breeding programmes have successfully developed "double-zero" rapeseed varieties characterized by low erucic acid and reduced glucosinolate content. These advancements have been achieved through conventional breeding techniques and marker-assisted selection, leading to cultivars that are safer for consumption and possess improved nutritional profiles (Manikandan et al. 2024).

4.3 Biochemical Pathways and Their Role in Improving *Brassica* Traits

Understanding the biochemical pathways involved in the synthesis of key compounds is essential for the targeted improvement of *Brassica* crops. The glucosinolate-myrosinase system, for example, plays a crucial role in plant defence and human health. Upon tissue disruption, myrosinase enzymes hydrolyse glucosinolates into various products, including isothiocyanates, which have been recognized for their anticancer properties. Advances in genomics and molecular biology have facilitated the identification and manipulation of genes involved in these pathways. For instance, the FAE1 gene is pivotal in erucic acid biosynthesis. Mutations or downregulation of this gene have led to the development of low-erucic acid rapeseed varieties, enhancing the safety and quality of the oil produced (Knutzon et al. 1992; Bisht and Augustine 2019; Sujith Kumar et al. 2020; Duraiswamy et al. 2023; Manikandan et al. 2024).

4.4 Advances in *Brassica* Seed and Oil Composition

Recent biotechnological interventions have enabled the modification of *Brassica* seed oil composition to meet specific dietary and industrial requirements. Antisense expression of the stearoyl-acyl carrier protein desaturase gene, for example, has been employed to alter fatty acid profiles in *Brassica* seeds, resulting in increased stearic acid content. Such modifications enhance the stability and usability of the oil for various applications (Knutzon et al. 1992). Furthermore, biofortification strategies have been implemented to enrich *Brassica* seeds with essential nutrients and reduce anti-nutritional factors. These approaches encompass agronomic practices, conventional breeding, and genetic engineering techniques aimed at producing crops with improved oil quality, higher protein content, and enhanced mineral profiles (Sujith Kumar et al. 2020).

5 Biotechnological Approaches to *Brassica* Improvement

Advances in biotechnology have revolutionized the improvement of *Brassica* crops, enabling precise manipulation of genetic and biochemical pathways to enhance yield, stress tolerance, and nutritional quality.

5.1 Plant Tissue Culture and Somaclonal Variation in *Brassica*

In vitro regeneration systems form the backbone of many biotechnological interventions in *Brassica*. Efficient protocols for callus induction, organogenesis, and somatic embryogenesis have been established for major species such as *B. napus*, *B. oleracea*, and *B. rapa*, using optimized combinations of auxins (e.g. 2,4-D) and cytokinins (e.g. BAP) to achieve high-frequency shoot regeneration (Ravanfar et al. 2017).

Somaclonal variation is the genetic and epigenetic variability arising during tissue culture, serving as both a challenge and an opportunity. On one hand, unwanted variants can compromise clonal fidelity; on the other, they generate novel diversity for selection. Somaclonal variants in *Brassica* have exhibited altered oil content, improved disease resistance, and modified flowering time in R_1 progeny, with some lines showing significantly higher seed yield compared to controls (Jain et al. 1989). Mechanisms include chromosomal rearrangements (polyploidy, aneuploidy), point mutations, transposon activation, and DNA methylation changes (Park et al. 2021).

Screening and stabilization of somaclones involve molecular marker analyses (SSR, AFLP) to detect genetic changes, coupled with cytogenetic assays to monitor karyotype stability (Duta-Cornescu et al. 2023). Reducing subculture cycles and minimizing exposure to 2,4-D can help limit undesirable variation, while deliberate induction of variation under controlled conditions allows breeders to isolate beneficial mutants for field evaluation.

5.2 Genetic Transformation and Transgenic *Brassica* Varieties

Agrobacterium-mediated transformation is the predominant method for introducing foreign genes into *Brassica*. Protocols typically employ hypocotyl or cotyledon explants co-cultivated with *Agrobacterium tumefaciens* harbouring binary vectors. Key selectable markers include nptII (kanamycin resistance) and bar (glufosinate resistance), enabling efficient selection of transformants (Bhalla and Singh 2008; Sundar and Sakthivel 2008; Ravanfar et al. 2017).

Transgenic *Brassica* lines have been engineered for various specifications, such as herbicide tolerance (e.g. EPSPS gene for glyphosate resistance), insect resistance (Bt toxins such as Cry1Ac) (Koul 2022), disease resistance (overexpression of PR proteins, chitinases) and nutritional enhancement (altered fatty acid desaturases to increase omega-3 content) (Ravanfar et al. 2017). Gene editing technologies, particularly CRISPR/Cas9, have recently been applied to *Brassica* for targeted knockouts (e.g. FAD2 genes to modulate seed oil composition) and precise gene insertions, offering transgene-free varieties with enhanced consumer acceptance (Ravanfar et al. 2017).

5.3 Use of Metabolomics and Proteomics in *Brassica* Improvement

High-throughput metabolomic profiling (GC-MS, LC-MS) and proteomic analyses (2D-PAGE, iTRAQ, SWATH-MS) facilitate the identification of biomarkers associated with desirable traits such as stress tolerance and oil quality (Hu et al. 2015; Satrio et al. 2024). Metabolomics has revealed distinct glucosinolates and phenolic signatures correlated with pest resistance and human health benefits; for instance, elevated levels of sinigrin and glucoraphanin in certain *B. oleracea* lines confer enhanced insect deterrence and anticancer properties (Mitra and Gantait 2020). Proteomics studies have uncovered stress-responsive proteins-heat-shock proteins, antioxidative enzymes (SOD, CAT) upregulated in drought-tolerant *B. napus* cultivars, guiding marker-assisted selection (Kehr and Buhtz 2011; Mitra and Gantait 2020; Shu et al. 2022). Integration of "omics" data with quantitative trait loci (QTL) mapping and genome-wide association studies (GWAS) accelerates the discovery of candidate genes and pathways, enabling systems-level breeding strategies (Khan et al. 2021; Fan et al. 2025).

5.4 Synthetic Biology: Engineering *Brassica* for Enhanced Traits

Synthetic biology applies modular design principles to rewire metabolic pathways and regulatory networks in *Brassica* (Ravanfar et al. 2017). Strategies include reconstructing fatty acid biosynthesis pathways with heterologous desaturases to produce novel lipid profiles (e.g. long-chain polyunsaturated fatty acids) in seeds (Zhou et al. 2019). Engineering synthetic regulatory circuits to achieve tissue-specific or inducible expression of target genes, such as stress-inducible promoters driving antioxidant enzyme production (Khan et al. 2023). Spatial organization of enzymes via scaffold proteins to enhance flux toward valuable secondary metabolites like flavonoids and glucosinolates (Sharma et al. 2024). Designing Boolean logic gates to combine multiple traits (e.g. drought tolerance and pest resistance) within a single genetic construct (Singla 2020). Emerging tools like minimal synthetic chromosomes, RNA-based switches, and genome-scale metabolic models promise unprecedented control over *Brassica* phenotypes, moving beyond single-gene edits toward holistic, designer crops (Ton et al. 2020).

6 Integration of Artificial Intelligence in *Brassica* Crop Improvement

The integration of artificial intelligence (AI) into *Brassica* crop improvement has revolutionized traditional breeding and genomic research by enabling rapid analysis of complex datasets, predictive modelling of traits, and optimization of breeding cycles.

6.1 Artificial Intelligence (ML) Applications in *Brassica* Genomic Research

AI-driven tools are critical for managing and interpreting the vast genomic data generated in *Brassica* species. For instance, machine learning (ML) algorithms have mapped *Brassica* BAC end sequences onto the *Arabidopsis thaliana* genome, leveraging evolutionary conservation to identify orthologous genes and structural variants. Tools like ASTRA integrate annotation pipelines with gene ontology (GO) terms, enabling efficient discovery of molecular markers such as SSRs and SNPs (Love et al. 2004). Advanced ML models process whole-genome re-sequencing data to construct pan-genomes, capturing genetic diversity across *Brassica* accessions (Mohd Saad et al. 2021). This approach has identified loci linked to agronomic traits like drought tolerance and oil content. Convolutional neural networks (CNNs) achieve >90% accuracy in classifying *B. rapa* varieties using lateral-view phenotypic images, outperforming traditional top-view analyses (Jung et al. 2021). These models help correlate morphological traits with genomic signatures.

6.2 Machine Learning (ML) for Predicting Trait Inheritance and Crop Performance

Machine learning algorithms excel at decoding genotype-phenotype relationships in *Brassica* crops. For example, by rescaling transcriptomic data across developmental stages, ML identified divergent gene regulatory networks in rapid-cycling *B. rapa* varieties compared to *Arabidopsis*. Key differences in hormone signalling and mRNA stability pathways explain accelerated flowering in Bangladeshi cultivars (Morris et al. 2021). Random Forest and Support Vector Machine models predict 12 rapeseed traits, including yield and disease resistance, from RNA-seq data with mean squared errors as low as 0.016 (Khan et al. 2022). Pairwise trait correlations further refine breeding priorities. UAV swarms equipped with hyperspectral cameras and ML pipelines detect *Brassica* diseases like blackleg and Alternaria blight with 89% accuracy, enabling targeted interventions (Ali et al. 2024).

6.3 AI-Driven Approaches for Accelerating Breeding Programmes

AI synergizes with speed breeding to compress breeding cycles. ML models preselect genotypes with optimal marker scores for yield and stress tolerance, reducing field trial reliance (Mohd Saad et al. 2021). For example, predictive breeding values for *B. napus* oil content improved genetic gains by 23% in half the time. Deep learning assimilates phenomics, transcriptomics, and metabolomics data to simulate gene–environment interactions. This approach identified *B. rapa* accessions with dual resistance to heat and flea beetles (Rai 2022). CNNs classify root architecture and leaf morphology from 3D scans, enabling high-throughput selection of drought-tolerant lines (Jung et al. 2021). When combined with CRISPR-editing pipelines, these tools slash varietal development timelines from decades to years (Rai 2022).

7 Robotics and Automation in *Brassica* Genomics

Recent advancements in robotics and automation have begun to reshape the landscape of *Brassica* genomics, offering unprecedented precision in data collection, phenotypic analysis, and genomic breeding. By integrating robotic systems with high-throughput phenotyping and computational genomics, researchers are overcoming traditional bottlenecks in crop improvement, enabling the rapid identification of agronomically valuable traits and accelerating the development of resilient Brassica varieties (Li et al. 2020; Tan et al. 2024).

7.1 Robotic Systems in *Brassica* Cultivation and Phenotypic Data Acquisition

7.1.1 Automated Harvesting and Field Robotics

The integration of robotics into *Brassica* cultivation has addressed critical challenges such as labour shortages and the need for precision in harvesting. The Automated *Brassica* Harvesting in Cornwall (ABC) project exemplifies this shift, utilizing robots equipped with “variable stiffness” arms that mimic human flexibility to harvest cauliflower, broccoli, and kale. These systems employ 3D cameras and tactile sensors to distinguish ripe produce, achieving real-time modelling of crop geometry while minimizing damage to delicate tissues (Williams 2018). Such innovations not only streamline harvesting but also generate spatially resolved data on crop maturity and morphology, which can be cross-referenced with genomic databases to identify yield-related traits (Li et al. 2020; Anonymous 2024).

7.1.2 Sensor Networks and Precision Agriculture

Modern robotic platforms are increasingly embedded with multispectral sensors, LiDAR, and hyperspectral imaging systems to monitor *Brassica* fields at submetre resolutions. For instance, Syngenta's drone-based systems capture high-definition imagery to assess plant health, soil moisture, and pest infestations, creating datasets that inform irrigation and pesticide application schedules. These autonomous systems reduce agrochemical use by 15–30% while generating terabytes of phenotypic data linked to specific genomic markers, such as those governing drought tolerance or pathogen resistance (Li et al. 2020; Anonymous 2024).

7.2 High-Throughput Phenotyping (HTP) and Genomic Integration

7.2.1 Dynamic Trait Quantification via Robotic Platforms

High-throughput phenotyping (HTP) systems have revolutionized the dissection of *Brassica*'s genetic architecture. A landmark study using automated imaging pipelines quantified 43 dynamic traits (i-traits) across 12 developmental stages in *B. napus*, revealing temporal patterns of shoot growth and biomass accumulation (Li et al. 2020). By correlating these i-traits with yield outcomes, researchers identified 937 quantitative trait loci (QTLs), 36% of which exhibited consistency across growing seasons, a feat unattainable through manual phenotyping. Crucially, robotic platforms enabled the discovery of *BnFY1*, a QTL colocalizing with final yield parameters, providing a mechanistic link between early-stage vegetative vigour and reproductive success (Li et al. 2020).

7.2.2 Robotics-Enabled Genomic Selection

The fusion of robotic data acquisition with machine learning algorithms has refined genomic selection models in *Brassica*. For example, the *ASTRA* database integrates robotic field data with *B. rapa* and *A. thaliana* orthologs, allowing breeders to prioritize candidate genes for traits like branching angle or flowering time (Love et al. 2004; Yang et al. 2005). In the Korea *Brassica* Genome Project, robotic phenotyping of substitution lines validated the triplication history of the *Brassica* genome, pinpointing regions enriched for gene losses that influence agronomic performance (Yang et al. 2005).

7.3 Computational Robotics and Genome Annotation

7.3.1 Automated Sequence Analysis Pipelines

The deluge of genomic data from projects like the Multinational *Brassica* Genome Project necessitates robotic-scale computational tools. ASTRA's automated pipelines annotate *Brassica* sequences against GenBank, Swissport, and *Arabidopsis* Gene Ontology databases, while SSR and SNP markers are mined algorithmically to accelerate marker-assisted breeding (Love et al. 2004). These systems have mapped 10,204 BAC end sequences from *B. rapa* chromosome 1 onto the *Arabidopsis* genome, revealing syntenic blocks disrupted by lineage-specific transposon activity (Yang et al. 2005).

7.3.2 AI-Driven Functional Genomics

Deep learning models trained on robotic phenotyping data now predict gene functions with 89% accuracy in *B. napus*. For instance, convolutional neural networks (CNNs) analysing leaf curvature data from robotic imagers identified *BnSPATULA*, a transcription factor regulating lamina expansion, which was later validated through CRISPR-Cas9 knockout lines (Li et al. 2020; Tan et al. 2024). Such approaches bypass traditional gene cloning bottlenecks, compressing the discovery-to-application timeline from years to months.

7.4 Challenges and Future Trajectories

Despite progress, robotic systems struggle with *Brassica*'s morphological diversity, e.g. the compact rosettes of *B. oleracea* versus the sprawling architecture of *B. juncea*. Current gripper designs achieve only 76% success rates in harvesting Brussels sprouts due to occlusion by foliage (Williams 2018). Similarly, HTP systems face challenges in distinguishing overlapping leaves in dense canopies, necessitating advances in terahertz imaging and quantum dot sensors (Li et al. 2020). Emerging technologies promise to bridge these gaps. The *GummiArm* robot, with its dual flexible arms and haptic feedback, is being adapted for microsampling *Brassica* tissues for single-cell RNA sequencing (Williams 2018). Meanwhile, swarm robotics concepts envision fleets of microdrones performing real-time pollination monitoring in hybrid seed production fields, correlating floral gene expression with outcrossing rates (Anonymous 2024).

8 Future Directions in *Brassica* Crop Improvement

8.1 Integrating Omics Technologies and Biotechnology for Sustainable Crop Development

The advent of omics technologies, genomics, transcriptomics, proteomics, and metabolomics, has revolutionized our understanding of plant biology. In *Brassica* species, these tools have facilitated the identification of genes associated with desirable traits such as disease resistance, abiotic stress tolerance, and improved nutritional profiles. For instance, the development of high-quality nuclear and organellar genomes has paved the way for pan-genomic studies, enabling the exploration of genetic diversity within and across *Brassica* species (Mohd Saad et al. 2021). Biotechnological interventions, particularly genome editing techniques like CRISPR/Cas9, offer precise modifications to target genes, accelerating the breeding of climate-resilient and high-yielding *Brassica* varieties. These advancements not only enhance crop performance but also contribute to sustainable agricultural practices by reducing the reliance on chemical inputs and optimizing resource use.

8.2 Cross-Disciplinary Approaches: Bridging Genetics, AI, and Robotics

The convergence of genetics with artificial intelligence (AI) and robotics heralds a new era in crop improvement. AI algorithms can analyse vast datasets from omics studies to predict phenotypic outcomes based on genotypic information, thereby streamlining the selection process in breeding programmes. Robotic platforms equipped with advanced sensors facilitate high-throughput and high-resolution phenotyping, capturing intricate plant traits in field conditions. Such technologies enable the creation of digital twins of plants, allowing for detailed analysis and trait selection. Integrating these disciplines accelerates the breeding cycle and enhances the precision of selecting traits that confer resilience to environmental stresses, ultimately leading to the development of superior *Brassica* cultivars.

8.3 Ethical Considerations in the Genetic Modification of *Brassica* Crops

The application of genetic modification in *Brassica* crop improvement raises several ethical and societal questions. Concerns encompass environmental impacts, potential health risks, economic implications for farmers, and cultural perceptions of genetically modified organisms (GMOs). For example, debates persist regarding the safety of consuming GMOs and their potential effects on human health. Moreover, the control of GMO seeds by large biotech companies raises issues related to food sovereignty and the economic well-being of farmers. Addressing these ethical considerations necessitates transparent regulatory frameworks, inclusive dialogues with stakeholders, and public engagement to ensure that biotechnological advancements align with societal values and contribute positively to sustainable agriculture.

8.4 The Future of *Brassica* Crop Improvement in the Context of Climate Change

Climate change poses significant threats to *Brassica* crop production, including increased incidences of drought, heat stress, and pest and disease pressures. Utilizing genetic diversity within *Brassica* species and their wild relatives offers a reservoir of traits that can be harnessed to enhance climate resilience. For instance, wild relatives often exhibit increased tolerances to biotic and abiotic stresses, making them valuable resources for breeding programmes aimed at developing resilient *Brassica* varieties. Integrating traditional breeding methods with modern biotechnological tools and a comprehensive understanding of plant-environment interactions will be crucial in developing *Brassica* cultivars capable of thriving under changing climatic conditions (Quezada-Martinez et al. 2021; Idris et al. 2023).

9 Conclusion

The integration of omics technologies, biotechnology, AI, and robotics has significantly advanced *Brassica* crop improvement. These interdisciplinary approaches have enhanced our understanding of *Brassica* genetics and facilitated the development of varieties with improved traits, contributing to sustainable agricultural practices. Despite technological advancements, challenges remain, including ethical concerns, regulatory complexities, and the need for public acceptance of genetically modified crops. However, these challenges present opportunities for developing transparent policies, engaging in public education, and fostering collaborations among scientists, policymakers, and farmers to ensure that crop improvement strategies are socially responsible and widely accepted. Innovation stands at the forefront of *Brassica* crop improvement. By embracing interdisciplinary approaches and addressing ethical considerations, the agricultural community can develop resilient and productive *Brassica* varieties. These advancements will be instrumental in meeting the demands of a growing population and the challenges posed by climate change, ensuring food security and sustainability for future generations. *Brassica* species hold immense potential for global agriculture, offering nutritional, ecological, and economic benefits. Overcoming challenges such as genetic complexity, pest resistance, and environmental stress requires integrated solutions that combine biotechnology, genomics, and sustainable agronomy. Emerging technologies like AI, robotics, and CRISPR are transforming *Brassica* breeding into a data-driven, precision-oriented field. With continued innovation and cross-disciplinary collaboration, *Brassica* crops can be optimized to meet future food security and climate resilience demands.

References

Ahmad N, Fatima S, Mehmood MA, Zaman QU, Atif RM, Zhou W, Rahman M-U, Gill RA (2023) Targeted genome editing in polyploids: lessons from *Brassica*. Front Plant Sci 14:1152468. https://doi.org/10.3389/fpls.2023.1152468

Ahmar S, Zhai Y, Huang H, Yu K, Khan MHU, Shahid M, Samad RA, Khan SU, Amoo O, Fan C (2022) Development of mutants with varying flowering times by targeted editing of multiple SVP gene copies in *Brassica napus* L. Crop J 10:67–74. https://doi.org/10.1016/j.cj.2021.03.023

Ali ZA, Deng D, Shaikh MK, Hasan R, Khan MA (2024) AI-based UAV swarms for monitoring and disease identification of *Brassica* plants using machine learning: a review. CSSE 48:1–34. https://doi.org/10.32604/csse.2023.041866

Amas JC, Thomas WJW, Zhang Y, Edwards D, Batley J (2023) Key advances in the new era of genomics-assisted disease resistance improvement of *Brassica* species. Phytopathology 113:771–785. https://doi.org/10.1094/PHYTO-08-22-0289-FI

Anonymous (2024) What agricultural robotics trends you should be adopting and why. Published on March 22, 2024. Retrieved on April 9, 2025, Available via: https://www.infopulse.com/blog/agricultural-robotics-drivers-trends

Begna T (2021) Conventional breeding methods widely used to improve self-pollinated crops. Int J Res 7(1):1–16. https://doi.org/10.20431/2454-6224.0701001

Bhalla PL, Singh MB (2008) *Agrobacterium*-mediated transformation of *Brassica napus* and *Brassica oleracea*. Nat Protoc 3:181–189. https://doi.org/10.1038/nprot.2007.527

Bisht NC, Augustine R (2019) Development of *Brassica* oilseed crops with low antinutritional glucosinolates and rich in anticancer glucosinolates. In: Nutritional quality improvement in plants, pp 271–287. https://doi.org/10.1007/978-3-319-95354-0_10

Braatz J, Harloff H, Mascher M, Stein N, Himmelbach A, Jung C (2017) CRISPR-Cas9 targeted mutagenesis leads to simultaneous modification of different homoeologous gene copies in polyploid oilseed rape (*Brassica napus*). Plant Physiol 174:935–942. https://doi.org/10.1104/pp.17.00426

Branca F, Cartea E (2010) Brassica. In: Kole C (ed) Wild crop relatives: genomic and breeding resources: oilseeds. Springer, Berlin, Heidelberg, pp 17–36. https://doi.org/10.1007/978-3-642-16057-8

Cao W, Dong X, Ji J, Yang L, Fang Z, Zhuang M, Zhang Y, Lv H, Wang Y, Sun P et al (2021) BoCER1 is essential for the synthesis of cuticular wax in cabbage (*Brassica oleracea* L. var. *capitata*). Sci Hortic 277:109801. https://doi.org/10.1016/j.scienta.2020.109801

Cartea ME, Francisco M, Soengas P, Velasco P (2011) Phenolic compounds in *Brassica* vegetables. Molecules 16:251–280. https://doi.org/10.3390/molecules16010251

Cheng H, Hao M, Ding B, Mei D, Wang W, Wang H, Zhou R, Liu J, Li C, Hu Q (2021) Base editing with high efficiency in allotetraploid oilseed rape by A3A-PBE system. Plant Biotechnol J 19:87. https://doi.org/10.1111/pbi.13444

Cheng H, Cai S, Hao M, Cai Y, Wen Y, Huang W, Mei D, Hu Q (2023) Targeted mutagenesis of BnTTG1 homologues generated yellow-seeded rapeseed with increased oil content and seed germination under abiotic stress. Plant Physiol Biochem 206:108302. https://doi.org/10.1016/j.plaphy.2023.108302

Dai C, Tian X, Ma CZ (2021) CRISPR/Cas9-mediated genome editing in *Brassica*. In: Willmann MR (ed) Genome editing for precision crop breeding. Burleigh Dodds Science Publishing Limited, Cambridge, pp 403–419. https://doi.org/10.19103/AS.2020.0082.23

Dou S, Zhang T, Tu J, Shen J, Yi B, Wen J, Fu T, Dai C, Ma C (2021) Generation of novel self-incompatible *Brassica napus* by CRISPR/Cas9. Plant Biotechnol J 19:875–877. https://doi.org/10.1111/pbi.13577

Duan Z, Zhang Y, Tu J, Shen J, Yi B, Fu T, Dai C, Ma C (2020) The *Brassica napus* GATA transcription factor BnA5.ZML1 is a stigma compatibility factor. J Integr Plant Biol 62:1112–1131. https://doi.org/10.1111/jipb.12916

Duraiswamy A, Sneha ANM, Jebakani KS, Selvaraj S, Pramitha JL, Selvaraj R, Petchiammal KI, Kather Sheriff S, Thinakaran J, Rathinamoorthy S, Kumar PR (2023) Genetic manipulation of anti-nutritional factors in major crops for a sustainable diet in future. Front Plant Sci 13:1070398. https://doi.org/10.3389/fpls.2022.1070398

Duta-Cornescu G, Constantin N, Pojoga DM, Nicuta D, Simon-Gruita A (2023) Somaclonal variation advantage or disadvantage in micropropagation of the medicinal plants. Int J Mol Sci 24:838. https://doi.org/10.3390/ijms24010838

Fahey JW (2003) Brassicas. In: Caballero B (ed) Encyclopedia of food sciences and nutrition, 2nd edn. Academic Press, pp 606–615. 978-0-12-227055-0

Fan S, Zhang L, Tang M, Cai Y, Liu J, Liu H, Liu J, Terzaghi W, Wang H, Hua W (2021) CRISPR/Cas9-targeted mutagenesis of the BnaA03. BP gene confers semi-dwarf and compact architecture to rapeseed (*Brassica napus* L.). Plant Biotechnol J 19:2383. https://doi.org/10.1111/pbi.13703

Fan BL, Chen LH, Chen LL, Guo H (2025) Integrative multi-omics approaches for identifying and characterizing biological elements in crop traits: Current progress and future prospects. Int J Mole Sci 26:1466. https://doi.org/10.3390/ijms26041466

FAO (2021) FAOSTAT statistics database. http://www.fao.org/faostat

Feng Y, Cui R, Wang S, He M, Hua Y, Shi L, Ye X, Xu F (2020) Transcription factor BnaA9. WRKY47 contributes to the adaptation of *Brassica napus* to low boron stress by up-regulating the boric acid channel gene BnaA3.NIP5;1. Plant Biotechnol J 18:1241–1254. https://doi.org/10.1111/pbi.13288

Garg P, Tripathi S, Kashyap A, Arroju AK, Kumari S, Singh M, Kushwaha R, Sharma SS, Sharma J, Yadav R, Gupta NC (2025) Insights into early generation synthetic amphidiploid *Brassica juncea*: a strategy to harness maximum parental genomic diversity for improving Indian mustard. Front Plant Sci 16:1493618. https://doi.org/10.3389/fpls.2025.1493618

Gerszberg A (2018) Tissue culture and genetic transformation of cabbage (*Brassica oleracea* var. *capitata*): an overview. Planta 248:1037–1048. https://doi.org/10.1007/s00425-018-2961-3

Greer FS, Surendran A, Grant M, Lillywhite R (2023) The current status, challenges, and future perspectives for managing diseases of *Brassicas*. Front Microbiol 14:1209258. https://doi.org/10.3389/fmicb.2023.1209258

Han F, Liu Y, Fang Z, Yang L, Zhuang M, Zhang Y, Lv H, Wang Y, Ji J, Li Z (2021) Advances in genetics and molecular breeding of broccoli. Horticulturae 7:280. https://doi.org/10.3390/horticulturae7090280

Hu J, Rampitsch C, Bykova NV (2015) Advances in plant proteomics toward improvement of crop productivity and stress resistance. Front Plant Sci 6:209. https://doi.org/10.3389/fpls.2015.00209

Hu D, Jing J, Snowdon RJ, Mason AS, Shen J, Meng J, Zou J (2021) Exploring the gene pool of *Brassica napus* by genomics-based approaches. Plant Biotechnol J 19:1693–1712. https://doi.org/10.1111/pbi.13636

Idris SH, Mat Jalaluddin NS, Chang LW (2023) Ethical and legal implications of gene editing in plant breeding: a systematic literature review. J Zhejiang Univ Sci B 24:1093–1105. https://doi.org/10.1631/jzus.B2200601

Jabeen N (2020) Agricultural, economic and societal importance of *Brassicaceae* plants. In: Hasanuzzaman M (ed) The plant family *Brassicaceae*: biology and physiological responses to environmental stresses. Springer, Singapore, pp 45–128. https://doi.org/10.1007/978-981-15-6345-4_2

Jain RK, Sharma DR, Chowdhury JB (1989) High frequency regeneration and heritable somaclonal variation in *Brassica juncea*. Euphytica 40:75–81. https://doi.org/10.1007/BF00023300

Ji P, Koné D, Yin J, Jackson KL, Csinos AS (2012) Soil amendments with *Brassica* cover crops for management of Phytophthora blight on squash. Pest Manag Sci 68:639–644. https://doi.org/10.1002/ps.2308

Jiang L, Li D, Jin L, Ruan Y, Shen W, Liu C (2018) Histone lysine methyltransferases BnaSDG8.A and BnaSDG8.C are involved in the floral transition in *Brassica napus*. Plant J 95:672–685. https://doi.org/10.1111/tpj.13978

Johannessen M, Andersen B, Jørgensen R (2006) Competition affects gene flow from oilseed rape (♀) to *Brassica rapa* (♂). Heredity (Edinb) 96:360–367. https://doi.org/10.1038/sj.hdy.6800796

Jung M, Song JS, Hong S, Kim S, Go S, Lim YP, Park J, Park SG, Kim YM (2021) Deep learning algorithms correctly classify *Brassica rapa* varieties using digital images. Front Plant Sci 12:738685. https://doi.org/10.3389/fpls.2021.738685

Katche E, Quezada-Martinez D, Katche EI, Vasquez-Teuber P, Mason AS (2019) Interspecific hybridization for *Brassica* crop improvement. CBGG 1:e190007. https://doi.org/10.20900/cbgg20190007

Kehr J, Buhtz A (2011) *Brassica* proteomics and metabolomics. In: Edwards D, Batley J, Parkin I, Kole C (eds) Genetics, genomics and breeding of oilseed *Brassicas*, vol 13, 1st edn. CRC Press, Boca Raton, pp 174–193. eBook ISBN: 9780429067310

Khan SU, Saeed S, Khan MH, Fan C, Ahmar S, Arriagada O, Shahzad R, Branca F, Mora-Poblete F (2021) Advances and challenges for QTL analysis and GWAS in the plant-breeding of high-yielding: a focus on rapeseed. Biomolecules 11:1516. https://doi.org/10.3390/biom11101516

Khan MAR, Ahmed KA, Hossain MZ (2022) Machine learning to predict rapeseed traits from RNA-seq data. In: Workshops at 18th international conference on intelligent environments (IE2022), vol 31. IOS Press Ebooks, Amsterdam, pp 136–145

Khan A, Nasim N, Pudhuvai B, Koul B, Upadhyay SK, Sethi L, Dey N (2023) Plant synthetic promoters: advancement and prospective. Agriculture 13:298. https://doi.org/10.3390/agriculture13020298

Khan SU, Qian M, Li S, Fan Y, Wang H, Chang W, Alam O, Saeed S, Lu K (2024) Unveiling CRISPR/Cas in rapeseed: triumphs, trials, and tomorrow. New Crops 2:100045. https://doi.org/10.1016/j.ncrops.2024.100045

Kim H-S, Yoo JH, Park SH, Kim J-S, Chung Y, Kim JH, Kim HS (2021) Measurement of environmentally influenced variations in anthocyanin accumulations in *Brassica rapa* subsp. *Chinensis* (Bok Choy) using hyperspectral imaging. Front Plant Sci 12:1–9. https://doi.org/10.3389/fpls.2021.693854

Kioukis A, Michalopoulou VA, Briers L et al (2020) Intraspecific diversification of the crop wild relative *Brassica cretica* Lam. using demographic model selection. BMC Genomics 21:1–14. https://doi.org/10.1186/s12864-019-6439-x

Klein FB, Wilmot A, De Tejada VF, Rodriguez BL, Requena I, Busch S, Rondepierre A, Auzeeri T, Sauerwald T, Andrews WF, Rihan H (2019) Proof-of-concept modular robot platform for cauliflower harvesting. In: Precis agric'19. Wageningen Academic, pp 783–789. https://doi.org/10.3920/978-90-8686-888-9_97

Knutzon DS, Thompson GA, Radke SE, Johnson WB, Knauf VC, Kridl JC (1992) Modification of *Brassica* seed oil by antisense expression of a stearoyl-acyl carrier protein desaturase gene. PNAS 89(7):2624–2628. https://doi.org/10.1073/pnas.89.7.2624

Koul B (2022) Transgenics and crop improvement. In: Cisgenics and transgenics: strategies for sustainable crop development and food security. Springer Nature, Singapore, pp 131–347. 9789811921193

Kumar L, Bharti, Goutam E, Choudhary D (2023) Perspectives of secondary metabolites in reference to vegetable crops: a review. Int J Environ Clim Change 13:873–882. https://doi.org/10.9734/IJECC/2023/v13i92307

Laha SD, Kundu A, Podder S (2024) Impact of biotic stresses on the Brassicaceae family and opportunities for crop improvement by exploiting genotyping traits. Planta 259:97. https://doi.org/10.1007/s00425-024-04379-1

Lahari P, Tripathi V, Goutam E, Arora D, Ahlawat V, Bharti, Kumar L, Singh KK (2022) A review on integrated nutrient management in radish (*Raphanus sativus* L.). J Plant Dev Sci 14:897–904

Larkin RP, Griffin TS (2007) Control of soilborne potato diseases using *Brassica* green manures. Crop Prot 26:1067–1077. https://doi.org/10.1016/j.cropro.2006.10.004

Li H, Feng H, Guo C, Yang S, Huang W, Xiong X, Liu J, Chen G, Liu Q, Xiong L, Liu K (2020) High-throughput phenotyping accelerates the dissection of the dynamic genetic architecture of plant growth and yield improvement in rapeseed. Plant Biotechnol J 18:2345–2353. https://doi.org/10.1111/pbi.13396

Li X, Sandgrind S, Moss O, Guan R, Ivarson E, Wang ES, Kanagarajan S, Zhu L-H (2021) Efficient protoplast regeneration protocol and CRISPR/Cas9-mediated editing of glucosinolate transporter (GTR) genes in rapeseed (*Brassica napus* L.). Front Plant Sci 12:680859. https://doi.org/10.3389/fpls.2021.680859

Li J, Yu X, Zhang C, Li N, Zhao J (2022) The application of CRISPR/Cas technologies to *Brassica* crops: current progress and future perspectives. aBIOTECH 3:146–161. https://doi.org/10.1007/s42994-022-00076-3

Li X, Wang F, Ta N, Huang J (2025) The compositions, characteristics, health benefits and applications of anthocyanins in *Brassica* crops. Front Plant Sci 16:1544099. https://doi.org/10.3389/fpls.2025.1544099

Liu Y, Wei W, Ma K, Li J, Liang Y, Darmency H (2013) Consequences of gene flow between oilseed rape (*Brassica napus*) and its relatives. Plant Sci 211:42–51. https://doi.org/10.1016/j.plantsci.2013.07.002

Liu Y, Ye S, Yuan G, Ma X, Heng S, Yi B, Ma C, Shen J, Tu J, Fu T et al (2020) Gene silencing of BnaA09.ZEP and BnaC09.ZEP confers orange color in *Brassica napus* flowers. Plant J 104:932–949. https://doi.org/10.1111/tpj.14970

Love CG, Batley J, Lim G, Robinson AJ, Savage D, Singh D, Spangenberg GC, Edwards D (2004) New computational tools for *Brassica* genome research. Comp Funct Genomics 5:276–280. https://doi.org/10.1002/cfg.394

Lu L, Choi SR, Lim YP, Kang SY, Yi SY (2023) A GBS-based genetic linkage map and quantitative trait loci (QTL) associated with resistance to *Xanthomonas campestris* pv. campestris race 1 identified in *Brassica oleracea*. Front Plant Sci 14:1205681. https://doi.org/10.3389/fpls.2023.1205681

MacNish TR, Al-Mamun HA, Bayer PE, McPhan C, Fernandez CG, Upadhyaya SR, Liu S, Batley J, Parkin IA, Sharpe AG, Edwards D (2025) *Brassica Panache*: a multi-species graph pangenome representing presence absence variation across forty-one *Brassica* genomes. Plant Genome 18:e20535. https://doi.org/10.1002/tpg2.20535

Manikandan A, Muthusamy S, Wang ES, Ivarson E, Manickam S, Sivakami R, Narayanan MB, Zhu LH, Rajasekaran R, Kanagarajan S (2024) Breeding and biotechnology approaches to enhance the nutritional quality of rapeseed byproducts for sustainable alternative protein sources – a critical review. Front Plant Sci 15:1468675

Mapari AR, Mehandi S (2024) Enhancing crop resilience: advances and challenges in marker-assisted selection for disease resistance. J Adv Biol Biotechnol 27:569–580. https://doi.org/10.9734/jabb/2024/v27i71018

Mitra M, Gantait S (2020) Tissue culture-mediated biotechnological advancements in genus *Brassica*. In: Wani S, Thakur A, Jeshima Khan Y (eds) *Brassica* improvement. Springer, Cham, pp 85–107. https://doi.org/10.1007/978-3-030-34694-2_5

Mohd Saad NS, Severn-Ellis AA, Pradhan A, Edwards D, Batley J (2021) Genomics armed with diversity leads the way in *Brassica* improvement in a changing global environment. Front Genet 12:600789. https://doi.org/10.3389/fgene.2021.600789

Morris PRJ, Hepworth J, Caroline D, Wells R (2021) Machine learning identifies mechanisms of flowering in rapid-cycling *Brassica* crops. Published on 26 April, 2021. Retrieved on 9 April, 2025, Available via: https://www.jic.ac.uk/press-release/machine-learning-identifies-mechanisms-of-flowering-in-rapid-cycling-brassica-crops/

Nagraj GS, Chouksey A, Jaiswal S, Jaiswal AK (2020) Broccoli. In: Jaiswal AK (ed) Nutritional composition and antioxidant properties of fruits and vegetables. Academic Press, pp 5–17. ISBN: 978-0-12-812780-3

Neequaye M, Stavnstrup S, Harwood W, Lawrenson T, Hundleby P, Irwin J, Troncoso-Rey P, Saha S, Traka M, Mithen R et al (2021) CRISPR-Cas9-mediated gene editing of MYB28 genes impair glucoraphanin accumulation of *Brassica oleracea* in the field. CRISPR J 4:416–426. https://doi.org/10.1089/crispr.2021.0007

Park JS, Shin YH, Park YD (2021) DNA methylation level changes in transgenic chinese cabbage (*Brassica rapa* ssp. *pekinensis*) plants and their effects on corresponding gene expression patterns. Genes 12:1563. https://doi.org/10.3390/genes12101563

Paul S, Geng C-A, Yang T-H, Yang Y-P, Ji-Jun C (2019) Phytochemical and health-beneficial progress of turnip (*Brassica rapa*). J Food Sci 84:19–30. https://doi.org/10.1111/1750-3841.14417

Petruzzello M (2022) List of plants in the family Brassicaceae. Encyclopedia Britannica, pp. 1–4. Retrieved on April 10, 2025. Available via: https://ttapscenter.com/wp-content/uploads/2025/02/14N-Brassica-plants.pdf

Quezada-Martinez D, Addo Nyarko CP, Schiessl SV et al (2021) Using wild relatives and related species to build climate resilience in *Brassica* crops. Theor Appl Genet 134:1711–1728. https://doi.org/10.1007/s00122-021-03793-3

Rai KK (2022) Integrating speed breeding with artificial intelligence for developing climate-smart crops. Mol Biol Rep 49:11385–11402. https://doi.org/10.1007/s11033-022-07769-4

Rakow G (2004) Species origin and economic importance of *Brassica*. In: Pua EC, Douglas CJ (eds) *Brassica*. Biotechnology in agriculture and forestry, vol 54. Springer, Berlin, Heidelberg, pp 3–11. https://doi.org/10.1007/978-3-662-06164-0_1

Rathore SS, Shekhawat K, Kandpal BK, Premi OP, Singh SP, Singh GC, Singh D (2015) Sulphur management for increased productivity of Indian mustard: a review. Ann Plant Soil Res 17:1–12

Rauf S, Fatima S, Ortiz R (2023) Modification of fatty acid profile and oil contents using gene editing in oilseed crops for a changing climate. GM Crops Food 14:1–12. https://doi.org/10.1080/21645698.2023.2243041

Ravanfar SA, Orbovic V, Moradpour M, Abdul Aziz M, Karan R, Wallace S, Parajuli S (2017) Improvement of tissue culture, genetic transformation, and applications of biotechnology to *Brassica*. Biotechnol Genet Eng Rev 33:1–25. https://doi.org/10.1080/02648725.2017.1309821

Raza A, Hafeez MB, Zahra N, Shaukat K, Umbreen S, Tabassum J, Charagh S, Khan RS, Hasanuzzaman M (2020) The plant family *Brassicaceae*: Introduction, biology, and importance. In: The plant family *Brassicaceae*: biology and physiol responses to enviro stresses, pp 1–43. https://doi.org/10.1007/978-981-15-6345-4_1

Romero-Muñoz M, Pérez-Jiménez M (2024) Optimizing *Brassica oleracea* L. breeding through somatic hybridization using cytoplasmic male sterility (CMS) lines: from protoplast isolation to plantlet regeneration. Plants 13:3247. https://doi.org/10.3390/plants13223247

Ross IA (2024) The history of Brassicaceae plants. In: Plant-based therapeutics, vol 2: the Brassicaceae family. Springer Nature, Cham, pp 1–16. https://doi.org/10.1007/978-3-031-63681-3_1

Rubatzky VE, Yamaguchi M (2012) World vegetables: principles, production, and nutritive values, 2nd edn. Springer, New York, pp 34–41. E-ISBN: 978-1-4615-6015-9

Saban JM, Romero AJ, Ezard THG, Chapman MA (2023) Extensive crop wild hybridization during *Brassica* evolution and selection during the domestication and diversification of *Brassica* crops. Genetics 223:1–15. https://doi.org/10.1093/genetics/iyad027

Sahab S, Runa F, Ponnampalam M, Kay PT, Jaya E, Viduka K, Panter S, Tibbits J, Hayden MJ (2024) Efficient multi-allelic genome editing via CRISPR–Cas9 ribonucleoprotein-based delivery to *Brassica napus* mesophyll protoplasts. Front Plant Sci 15:1397632. https://doi.org/10.3389/fpls.2024.1397632

Sameeullah M, Kayaçetin F, Khavar KM et al (2025) Decoding genetic diversity and population structure of *Brassica* species by inter primer binding site (iPBS) retrotransposon markers. Genet Resour Crop Evol 72:417–427. https://doi.org/10.1007/s10722-024-01986-5

Sashidhar N, Harloff HJ, Potgieter L, Jung C (2020) Gene editing of three BnITPK genes in tetraploid oilseedrape leads to significant reduction of phytic acid in seeds. Plant Biotechnol J 18:2241–2250. https://doi.org/10.1111/pbi.13380

Satrio RD, Fendiyanto MH, Miftahudin M (2024) Tools and techniques used at global scale through genomics, transcriptomics, proteomics, and metabolomics to investigate plant stress responses at the molecular level. In: Shahid M, Gaur R (eds) Molecular dynamics of plant stress and its management, pp 555–607. https://doi.org/10.1007/978-981-97-1699-9

Sharma S, Rani H, Kaur G, Kumar S, Sheikh S, Samota MK (2024) Comprehensive overview of glucosinolates in crucifers: occurrence, roles, metabolism, and transport mechanisms—a review. Phytochem Rev 26:1–28. https://doi.org/10.1007/s11101-024-10021-5

Shattuck VI, Proudfoot KG (1990) Rutabaga breeding. In: Janick J (ed) Plant breeding reviews, vol 8. Wiley Online Library, pp 217–248. Online ISBN:9781118061053

Sheng X, Yu H, Wang J, Shen Y, Gu H (2022) Establishment of a stable, effective and universal genetic transformation technique in the diverse species of *Brassica oleracea*. Front Plant Sci 13:1021669. https://doi.org/10.3389/fpls.2022.1021669

Shu J, Ma X, Ma H, Huang Q, Zhang Y, Guan M, Guan C (2022) Transcriptomic, proteomic, metabolomic, and functional genomic approaches of *Brassica napus* L. during salt stress. PLoS One 17:e0262587. https://doi.org/10.1371/journal.pone.0262587

Singla P (2020) Plant synthetic biology: a paradigm shift targeting stress mitigation, reduction of ecological footprints and sustainable transformation in agriculture. In: Plant stress biology: strategies and trends, pp 435–489. https://doi.org/10.1007/978-981-15-9380-2_15

Smith B (2007) Vegetable Brassicas and related crucifers. By GR Dixon. Wallingford, UK: CABI (2007), pp. 327, £ 37.50 (paperback). ISBN 0-85199-395-8. Exp Agric 43(4):522–523. https://doi.org/10.1017/S0014479707005406

Snowdon RJ, Friedt W (2004) Molecular markers in *Brassica* oilseed breeding: current status and future possibilities. Plant Breed 123:1–8. https://doi.org/10.1111/j.1439-0523.2003.00968.x
Snowdon R, Lühs W, Friedt W (2007) Oilseed rape. In: Kole C (ed) Oilseeds. Genome mapping and molecular breeding in plants, vol 2. Springer, Berlin, Heidelberg, pp 55–114. E-ISBN: 978-3-540-34388-2
Stanic M, Hickerson N, Arunraj R, Samuel M (2021) Gene-editing of the strigolactone receptor BnD14 confers promising shoot architectural changes in *Brassica napus* (canola). Plant Biotechnol J 19:639–641. https://doi.org/10.1111/pbi.13513
Stoelen MF, de Azambuja R, López Rodríguez B, Bonsignorio F, Cangelosi A (2022) The GummiArm project: a replicable and variable-stiffness robot arm for experiments on embodied AI. Front Neurorobot 16:836772. https://doi.org/10.3389/fnbot.2022.836772
Subramanian P, Kim SH, Hahn BS (2023) *Brassica* biodiversity conservation: prevailing constraints and future avenues for sustainable distribution of plant genetic resources. Front Plant Sci 14:1–13. https://doi.org/10.3389/fpls.2023.1220134
Sujith Kumar MS, Mawlong I, Rani R (2020) *Brassica improvement*: molecular, genetics and genomic perspectives. Springer Nature. https://doi.org/10.1007/978-3-030-34694-2
Sundar IK, Sakthivel N (2008) Advances in selectable marker genes for plant transformation. J Plant Physiol 165:1698–1716. https://doi.org/10.1016/j.jplph.2008.08.002
Swain AR, Tripathi V, Verma D, Goutam E, Choudhary D, Kumar Y (2024) Effect of humic acid, boron and potassium on growth, yield and quality attributes of cauliflower. Plant Arch 24:461–468
Tan Z, Xie Z, Dai L, Zhang Y, Zhao H, Tang S, Wan L, Yao X, Guo L, Hong D (2022) Genome- and transcriptome-wide association studies reveal the genetic basis and the breeding history of seed glucosinolate content in *Brassica napus*. Plant Biotechnol J 20:211–225. https://doi.org/10.1111/pbi.13707
Tan Z, Han X, Dai C, Lu S, He H, Yao X, Chen P, Yang C, Zhao L, Yang QY, Zou J (2024) Functional genomics of *Brassica napus*: progress, challenges, and perspectives. J Integr Plant Biol 66:484–509. https://doi.org/10.1111/jipb.13635
Ton LB, Neik TX, Batley J (2020) The use of genetic and gene technologies in shaping modern rapeseed cultivars (*Brassica napus* L.). Genes 11:1161. https://doi.org/10.3390/genes11101161
Treccarichi S, Ben Ammar H, Amari M, Cali R, Tribulato A, Branca F (2023) Molecular markers for detecting inflorescence size of *Brassica oleracea* L. crops and *B. oleracea* complex species (n = 9) useful for breeding of broccoli (*B. oleracea* var. *italica*) and cauliflower (*B. oleracea* var. *botrytis*). Plants 12:407. https://doi.org/10.3390/plants12020407
Variyar PS, Gautam S, Sharma AK (2015) *Brassica* vegetables. In: Hui YH, Evranuz EÖ (eds) Handbook of vegetable preservation and processing, 2nd edn. CRC Press, Boca Raton, p 36. eBook ISBN: 9780429173073
Williams A (2018) Scientists develop harvesting robots that could revolutionise farming practices. Published on February 15, 2018, Retrieved on April 9, 2025, Available via: https://phys.org/news/2018-02-scientists-harvesting-robots-revolutionise-farming.html
Wu Z, Zhang Y, Yu H, Pan D, Wang Y, Wang Y, Li F, Liu C, Nan H, Chen W et al (2021) Programmed genome editing by a miniature CRISPR-Cas12f nuclease. Nat Chem Biol 17:1132–1138. https://doi.org/10.1038/s41589-021-00868-6
Xin Q, Wang X, Gao Y, Xu D, Xie Z, Dong F, Wan L, Yang L, Yang G, Hong D (2020) Molecular mechanisms underpinning the multiallelic inheritance of MS5 in *Brassica napus*. Plant J 103:1723–1734. https://doi.org/10.1111/tpj.14857
Xiong X, Liu W, Jiang J, Xu L, Huang L, Cao J (2019) Efficient genome editing of *Brassica campestris* based on the CRISPR/Cas9 system. Mol Gen Genomics 294:1251–1261. https://doi.org/10.1007/s00438-019-01564-w
Yang TJ, Kim JS, Lim KB, Kwon SJ, Kim JA, Jin M, Park JY, Lim MH, Kim HI, Kim SH, Lim YP (2005) The Korea *Brassica* genome project: a glimpse of the *Brassica* genome based on comparative genome analysis with *Arabidopsis*. Comp Funct Genomics 6:138–146. https://doi.org/10.1002/cfg.465

Yang H, Wu JJ, Tang T et al (2017) CRISPR/Cas9-mediated genome editing efficiently creates specific mutations at multiple loci using one sgRNA in *Brassica napus*. Sci Rep 7:7489. https://doi.org/10.1038/s41598-017-07871-9

Yang Y, Zhu K, Li H, Han S, Meng Q, Khan SU, Fan C, Xie K, Zhou Y (2018) Precise editing of *CLAVATA* genes in *Brassica napus* L. regulates multilocular silique development. Plant Biotechnol J 16:1322–1335. https://doi.org/10.1111/pbi.12872

Zaman QU, Chu W, Hao M, Shi Y, Sun M, Sang S-F, Mei D, Cheng H, Liu J, Li C (2019) CRISPR/Cas9-mediated multiplex genome editing of JAGGED gene in *Brassica napus* L. Biomolecules 9:725. https://doi.org/10.3390/biom9110725

Zandberg JD, Fernandez CT, Danilevicz MF, Thomas WJ, Edwards D, Batley J (2022) The global assessment of oilseed *Brassica* crop species yield, yield stability and the underlying genetics. Plants 11:2740. https://doi.org/10.3390/plants11202740

Zhai Y, Yu K, Cai S, Hu L, Amoo O, Xu L, Yang Y, Ma B, Jiao Y, Zhang C (2020) Targeted mutagenesis of *BnTT8* homologs controls yellow seed coat development for effective oil production in *Brassica napus* L. Plant Biotechnol J 18:1153–1168. https://doi.org/10.1111/pbi.13281

Zhang K, Nie L, Cheng Q, Yin Y, Chen K, Qi F, Zou D, Liu H, Zhao W, Wang B (2019) Effective editing for lysophosphatidic acid acyltransferase 2/5 in allotetraploid rapeseed (*Brassica napus* L.) using CRISPR-Cas9 system. Biotechnol Biofuels 12:1–18. https://doi.org/10.1186/s13068-019-1567-8

Zhang D, Yu H, Gao L, Wang J, Dong H, Guo Y, Hu S (2023) Genetic diversity in oilseed and vegetable mustard (*Brassica juncea* L.) accessions revealed by nuclear and mitochondrial molecular markers. Agronomy 13:919. https://doi.org/10.3390/agronomy13030919

Zhang X, Jia Q, Jia X, Li J, Sun X, Min L, Liu Z, Ma W, Zhao J (2025) *Brassica* vegetables an undervalued nutritional goldmine. Hortic Res 12:uhae302. https://doi.org/10.1093/hr/uhae302

Zhao C, Xie M, Liang L, Yang L, Han H, Qin X, Zhao J, Hou Y, Dai W, Du C, Xiang Y (2022) Genome-wide association analysis combined with quantitative trait loci mapping and dynamic transcriptome unveil the genetic control of seed oil content in *Brassica napus* L. Front Plant Sci 13:929197. https://doi.org/10.3389/fpls.2022.929197

Zhou XR, Li J, Wan X, Hua W, Singh S (2019) Harnessing biotechnology for the development of new seed lipid traits in *Brassica*. Plant Cell Physiol 60:1197–1204. https://doi.org/10.1093/pcp/pcz070

Zou J, Hu D, Mason AS, Shen X, Wang X, Wang N, Grandke F, Wang M, Chang S, Snowdon RJ, Meng J (2018) Genetic changes in a novel breeding population of *Brassica napus* synthesized from hundreds of crosses between *B. rapa* and *B. carinata*. Plant Biotechnol J 16:507–519. https://doi.org/10.1111/pbi.12791

Zheng Q, Wang X, Wang Z, Zhang Y, Wang H, Du K, Li T (2024) Two genes of cytochrome P450 regulate plant height via brassinosteroid biosynthesis in Brassica napus. J Integr Agric. 0 https://doi.org/10.1016/j.jia.2024.12.016

Applicability of Beneficial Microbes to Enhance Yield and Nutritional Level in *Brassica* Species Under Environmental Stress Condition

Satya Prakash, Sudhir Kumar Upadhyay, and Akshay Kumar

Abstract

Brassica species serve as hosts to intricate and specialized communities of microbes that are crucial for promoting plant growth, enhancing resilience to stress, and facilitating nutrient uptake, even in the absence of conventional mycorrhizal fungi and nodule-forming relationships. These microbial allies, often passed down through seeds, have evolved alongside *Brassicas* to carry out essential tasks such as breaking down phosphates, fixing nitrogen, producing plant hormones, assisting in iron absorption through siderophores, and reducing stress with ACC deaminase activity. Furthermore, these beneficial microbes bolster plant defence by suppressing pathogens, triggering systemic resistance, and detoxifying harmful environmental agents. The unique biochemical characteristics of *Brassicas*, particularly their production of glucosinolates, further influence the composition of microbial communities, which are predominantly made up of groups like Proteobacteria, Actinobacteria, Firmicutes, and Bacteroidetes. Various environmental pressures, both physical and biological, significantly affect the makeup and functioning of these microbiomes. Nonetheless, these microbial communities demonstrate remarkable adaptability through functional

S. Prakash (✉)
Department of Genetics and Plant Breeding, Lovely Professional University, Phagwara, Punjab, India

Research and Development Cell, Lovely Professional University, Phagwara, Punjab, India
e-mail: satya.26830@lpu.co.in

S. Kumar Upadhyay
Research and Development Cell, Lovely Professional University, Phagwara, Punjab, India

A. Kumar
Agriculture and Environment Research Unit, School of Life and Medical Sciences, University of Hertfordshire, Hatfield, UK

A. Kumar, B. Koul (eds.), *Crop Improvement Strategies in Brassica species: Applied Science*, https://doi.org/10.1007/978-981-95-3861-4_9

redundancy and shifts in genomic and physiological responses triggered by stress. Utilizing these adaptive and advantageous traits through microbe-centric management practices such as bioinoculants, seed treatments, and microbiome engineering presents sustainable alternatives to chemical fertilizers. These strategies are in harmony with integrated nutrient management and are particularly beneficial in conditions with limited inputs and high stress. Future developments in multi-omics technologies, precise microbial application, and microbiome-informed breeding will be essential for enhancing the functions of the *Brassica* microbiome and fostering resilient, climate-smart agricultural practices.

Keywords

Brassica microbiome · Plant–microbe interactions · Bioinoculants · Stress resilience · Microbiome engineering

1 Introduction

A range of environmental stresses, including drought, salt, severe temperatures, pathogenic assault, and heavy metal pollution, threatens agricultural production worldwide, therefore affecting plant growth, yield, and nutritional quality. These abiotic and biotic stresses upset cellular homeostasis, impede photosynthesis, induce oxidative damage resulting from too high reactive oxygen species (ROS) formation, and disrupt important metabolic pathways (Sachdev et al. 2021). Appreciated for their nutritional and health-promoting properties in crops, including *Brassica* species, particularly mustard (*Brassica juncea*), stresses cause insufficient biomass accumulation, delayed development, and a loss of essential phytochemicals such as glucosinolates (Singh et al. 2022). Recent studies by Singh et al. (2022) on salt stress responses in Indian mustard genotypes, particularly the salt-tolerant CS 60 and the salt-sensitive CS 245-2-80-7 during the seedling stage, employing a metabolomics approach demonstrated that CS 60 shows more active photosynthetic properties and a larger number of differently accumulated metabolites (63 DAMs) than CS 245-2-80-7 (48 DAMs); the study underscored the relevance of metabolites in salt tolerance. Important metabolic processes, mostly related to antioxidant protection, help CS 60 tolerate salty surroundings better by lowering osmotic and oxidative stress. Long-term, biologically based solutions to increase crop stress tolerance are desperately needed, considering the increasing frequency of such environmental disturbances. Using helpful microorganisms, including arbuscular mycorrhizal fungus (AMF), endophytic bacteria (Upadhyay et al. 2022; Pathak et al. 2024), and plant growth-promoting rhizobacteria (PGPR), is one promising strategy. By means of several well-documented physiological and molecular processes, these organisms establish symbiotic or associative interactions with host plants, hence improving resistance to abiotic and biotic challenges. Particularly in stress-induced growth retardation (Upadhyay et al. 2022), PGPR generates phytohormones including indole-3-acetic acid (IAA), gibberellins, and cytokinins, which support root and shoot development. By means of either induced systemic resistance (ISR) or systemic acquired resistance (SAR), they also help to produce stress-responsive genes, therefore strengthening the plant's defensive systems against abiotic damage or pathogen invasion (Khoshru et al. 2023). By spreading their extraradical hyphae into the soil matrix (Etesami et al. 2021), AMF improve water and nutrient uptake, most especially phosphate and micronutrients like Zn and Fe. In *Brassica* crops, which are non-mycotrophic or weakly mycotrophic but may benefit from microbial consortia incorporating appropriate strains, this is very helpful. By raising antioxidant enzyme activity, e.g., superoxide dismutase, catalase, and peroxidase, endophytes, which dwell within plant tissues without causing damage, help to buffer oxidative stress, hence boosting ROS scavenging under stress conditions (Zandi and Schnug 2022). By means of ionic homeostasis (e.g., Na^+/K^+ transporters under salinity stress), upregulating stress-related transcription factors (e.g., DREB, WRKY, and MYB families), and influencing osmolyte accumulation (Assaha et al. 2017; Hrmova and Hussain 2021; Trono and Pecchioni 2022), these bacteria improve plant water retention and cellular integrity. Moreover, improved root system architecture produced by microbial inoculation makes it possible to explore more soil and absorb resources, thereby boosting both production and nutrient accumulation (Bargaz et al. 2018). Though more research is being done, the efficacy and consistency of microbial consortia under multifactorial stress conditions, especially in *Brassica* spp., are still under question. Most of the existing studies are limited to single-stress scenarios or plant models. The current study is special in that it offers a thorough assessment of how certain microbial inoculants influence *Brassica* development, physiological activity, and nutritional value under many abiotic and biotic stress situations (Rad et al. 2022). It is also unique in comprehensively evaluating both primary (e.g., yield, biomass) and secondary (e.g., glucosinolates, antioxidants, mineral content) plant responses. Beneficial microbial consortia, we hypothesize, can not only lower stress-induced damage in *Brassica* spp. but also enhance plant nutritional quality by means of improved nutrient mobilization, hormonal control, ROS detoxification, and metabolic pathway modulation engaged in secondary metabolite synthesis (Tabacchioni et al. 2021). By offering a sustainable substitute for conventional chemical fertilizers and agrochemicals, this approach helps to create nutritionally rich agricultural systems with climate resilience. This chapter intends to evaluate the effects of microbial inoculation on the growth, yield, and physiological performance of *Brassica* spp. under different stress conditions; also, it aims to develop a microbial consortium-based bioformulation for sustainable *Brassica* farming under unfavourable environmental conditions by means of changes in the nutritional and biochemical profiles. Great possibilities for advancing microbial biotechnology in agriculture and matching crop output with the goals of ecological sustainability and nutritional security abound from this finding.

2 Microbial Communities Associated with *Brassica* Species

2.1 Seed-Associated

Microbial communities linked to seeds, especially endophytes, offer significant opportunities to transform sustainable farming practices. By enhancing our knowledge of their variety, methods of transmission, and roles in plant growth, we can utilize these microorganisms to boost plant vitality, increase resilience to stress, and lessen the reliance on chemical fertilizers. Acknowledging the seed microbiome as an essential element of the plant holobiont opens new avenues for creative approaches in enhancing crop quality and resilience. Ongoing research and the application of these findings in breeding and agricultural methods will be vital for establishing strong, climate-responsive farming systems that can fulfil future food needs (Samreen et al. 2021). Plant genotype, maternal environment, and seed processing conditions control these microbial assemblages; they typically survive into the next developmental stages, hence building the rhizosphere and phyllosphere microbiomes. Many studies have turned up key microbial groupings regularly linked to *Brassica* seeds. For instance, 16S rRNA gene amplicon sequencing showed that common in *B. napus* seeds include *Pseudomonas*, *Bacillus*, *Sphingomonas*, and *Paenibacillus* (Bezodis and Penfield 2024). These species are well known for their capacity to use phytohormone production, nitrogen fixation, and phosphate solubilization to drive plant growth. Particularly, *Bacillus subtilis* has been shown to produce indole-3-acetic acid, which in *Brassica oleracea* stimulates root development and seedling vigour (Patani et al. 2024). Moreover, in difficult conditions, *Sphingomonas* species might help seeds germinate and shield them from oxidative damage. The first line of protection against seed-borne and soil-borne illnesses also includes seed-associated bacteria. Enterobacter cloacae isolated from *B. oleracea* seeds, for instance, reduced *Alternaria brassicicola*, a ubiquitous fungal disease-causing seedling blight. Likewise, although boosting shoot biomass, seed inoculation with *Bacillus amyloliquefaciens* drastically reduced clubroot disease caused by *Plasmodiophora brassicae* in *B. juncea*. The utilization of seed microbiomes as a method of biological control. Still another crucial feature of seed-associated communities in *Brassica* species is the vertical transmission of microbial endophytes. Research on *Brassica rapa* revealed that seed-inhabiting taxa like *Microbacterium* and *Paenibacillus* are transferred to seedlings and survive in the developing root system, therefore helping to acquire nutrients and provide disease resistance (Truyens et al. 2015). Moreover, seeds taken from organic farming systems have shown greater microbial diversity than those taken from conventional systems, underlining the need for agricultural methods in determining the seed microbiome (Kavamura et al. 2021).

2.2 Root-Associated

Stress tolerance, nutrient absorption, and plant health all depend on root-associated microbial communities. *Brassica* plants include *Brassica napus* (canola), *Brassica oleracea* (cabbage, broccoli), and *Brassica rapa* (turnip). These microbial consortia colonise the rhizosphere (soil near roots), rhizoplane (root surface), and root endosphere. Although they lack symbiotic arbuscular mycorrhizal fungi (AMF), *Brassica* plants are non-mycorrhizal; nevertheless, they do host various bacterial and non-mycorrhizal fungal communities that offset this functional deficit via nutrient cycling, growth stimulation, and disease management (Wang et al. 2022).

Root-associated bacterial genera in *Brassica* species include *Pseudomonas*, *Bacillus*, *Rhizobium*, *Stenotrophomonas*, and *Paenibacillus*. For instance, it has been shown that *Pseudomonas fluorescens* isolated from *B. napus* rhizosphere stimulates plant growth and generates systemic resistance to root diseases such as *Rhizoctonia solani*. Another common *rhizobacterium*, *Bacillus subtilis,* produces antimicrobial compounds including lipopeptides and stimulates the manufacture of phytohormones like indole-3-acetic acid, both of which help to improve root architecture and stress tolerance in *B. juncea*.

Apart from bacteria, root-associated fungi serve purposes as well (Dimkić et al. 2022). Although mycorrhizal colonization is absent, *Brassica* roots have been proven to host endophytic fungi like *Fusarium equiseti* and *Trichoderma harzianum*. *Trichoderma* spp. are well known for their ability to secrete growth-regulating enzymes and metabolites, hence controlling root infections and boosting root biomass (Poveda et al. 2019). Studies indicate that injecting *Trichoderma atroviride* into *B. oleracea* roots increases resistance to *Pythium ultimum* and nutrient absorption.

The composition of root bacteria in *Brassica* crops is strongly influenced by environmental and soil factors. High sulfur glucosinolate levels in *Brassica* root exudates, for instance, can affect microbial populations by promoting sulfur-metabolizing bacteria such as *Desulfovirus* and *Thiobacillus*. Moreover, the kind of farming method is crucial; organically grown *B. napus* rhizospheres showed to have higher microbial diversity and richness of beneficial taxa, including *Azospirillum,* than conventionally grown fields (Kavamura et al. 2021).

2.2.1 Leaf-Associated

Diverse microbial populations found in the phyllosphere, plants' aerial surfaces, particularly leaves, play vital roles in plant health and defence. *Brassica* plants, including *Brassica oleracea*, *Brassica napus*, and *Brassica rapa,* have a leaf-associated microbiome dominated by bacteria, followed by fungi and yeasts. Among the many physiological functions these microbial assemblages control are environmental stress adaptation, pathogen defence, and host immune modulation. Like *species Pseudomonas*, *Methylobacterium*, *Sphingomonas*, *Bacillus*, and *Stenotrophomonas*, bacterial communities predominate on the phyllosphere (Bashir et al. 2022). The recognized epiphytic bacterium *Pseudomonas syringae* is often found on *Brassica* leaves, both as a commensal and, under suitable conditions, a possible pathogen. On the other hand, typically isolated from *B. napus* leaves, *Pseudomonas* fluorescens functions as a useful colonizer producing antibiotics and siderophores that inhibit dangerous fungi such as *Botrytis cinerea* (Hirano and Upper 2000). Likewise, *Methylobacterium extorquens* is well known for its ability to exploit methanol produced by leaf stomata, therefore facilitating carbon cycling and, in certain situations, host growth via phytohormone production. *Brassica* leaves host both endophytic and epiphytic fungal populations, including *Cladosporium*, *Alternaria*, and *Epicoccum*. Though *Alternaria brassicicola* is a common foliar pathogen in cabbage and mustard, non-pathogenic by means of niche competition, *Cladosporium* species might aid in excluding diseases. Often present on *Brassica phyllospheres*, *Epicoccum nigrum* has been demonstrated to generate antifungal compounds unfriendly to foliar disease. The variety of microbial communities present on leaves is influenced by leaf age, plant genotype, ambient conditions, and secondary metabolite availability including glucosinolates (Narayanasamy 2024). Glucosinolate-derived isothiocyanates produced by *Brassica* plants might provide selection pressure on microbial colonizers. *Sphingomonas* spp. have shown that they are routinely present on the phyllosphere of *Brassica oleracea* and can withstand isothiocyanate exposure, implying an adaptive relationship with the chemical defences of the host plant. Moreover, foliar microbial communities help to generate induced systemic resistance. For instance, stimulating host defence mechanisms by foliar application of *Bacillus amyloliquefaciens* to *B. juncea* leaves has been demonstrated to reduce the severity of *Xanthomonas campestris* infection (Obrien and Mary 2016).

The microbiota associated with plants is being studied in the development of sustainable farming methods. By investigating the impacts of traditional and conservation methods for establishing crops on the microbiota of winter oilseed rape (*Brassica napus*), we offer fresh perspectives on how farming practices shape microbial communities in various plant environments. Our results indicate that while the general soil and rhizosphere bacterial communities showed relative stability across different tillage approaches, the microbiota associated with roots and shoots experienced notable changes, underscoring their vulnerability to agricultural practices. The limited influence of the rhizosphere observed suggests that for winter oilseed rape, the recruitment of microbes influenced by the plant may be low in the root area under specific environmental and management conditions. The prevalent presence of Proteobacteria, Actinobacteria, and Bacteroidetes in the plant tissues further highlights their potential roles in promoting plant health (Rathore et al. 2017).

The intricate and dynamic interactions occurring at the junction of barley roots and soil reveal that both wild and cultivated varieties harbour distinct yet overlapping communities of microbes. The presence of bacterial groups like *Comamonadaceae*, *Flavobacteriaceae*, and *Rhizobiaceae* in the microbial community around the roots underscores a shared foundational microbiome shaped by the plant itself and its surrounding environment. Although the host plant's genetic characteristics do not exert an overwhelmingly strong influence on microbial diversity, their effect remains statistically notable, suggesting that domestication has left a discernible impact on the microbial populations associated with the roots. The observed enhancements in traits related to pathogen resistance, secretion processes, viral defence, and nutrient acquisition, along with evidence of positive selection affecting these protein categories, point to a mutually advantageous evolutionary connection between barley and its microbial partners. The critical roles of host–microbe and microbe–microbe interactions in determining the formation and functionality of microbial communities offer vital insights for developing crop enhancement strategies guided by microbiome studies as outlined by Bulgarelli et al. (2015).

Although historically less emphasized than rhizosphere studies, the exploration of phyllosphere microbiology is swiftly evolving thanks to techniques that do not rely on cultivation methods. These approaches reveal the variety, organization, and potential functions of microbial life in the aerial environment. Recent findings indicate that a limited number of bacterial groups predominantly inhabit the phyllosphere, influenced not only by environmental factors but also by specific characteristics of the plants, such as leaf chemistry and surface properties. The microorganisms residing on leaf surfaces exhibit unique adaptations that enable their survival and often engage in intricate relationships with both their host plants and each other. Gaining a deeper insight into the structure and functional behaviour of phyllosphere communities will enrich our foundational understanding and open new avenues for practical applications in sustainable farming, including microbial strategies aimed at enhancing plant growth and managing diseases (Vorholt 2012).

The seed microbiome is a vital but quite under-explored part of the plant holobiont. *Brassica* species seeds were reported to possess complex microbial populations, with an abundance of representatives from the *Enterobacteriaceae*, *Pseudomonadaceae*, and *Bacillaceae* families (Barret et al. 2015). Such microbes are not just contaminants but are instead vertically transferred from the parent plants to their offspring, forming a root microbial community in seedlings.

Figure 1 reveals the intricate and diverse relationships between microbial communities and *Brassica* plants, underlining their essential contributions to the health and growth of these crops. Various types of bacteria and fungi reside in the seeds, roots, and foliage, each playing a vital role through processes like producing plant hormones, acquiring nutrients, suppressing diseases, and triggering systemic resistance. These microorganisms operate through both direct and indirect methods to promote plant growth, positioning them as beneficial partners in sustainable farming practices. Gaining knowledge about these interactions provides important perspectives for improving the performance of *Brassica* crops by effectively managing their microbiomes and implementing specific microbial strategies.

3 Mechanism and Function of Associated Microbes

3.1 Direct and Indirect Mechanism

Microbial communities associated with *Brassica* species *Brassica napus*, *Brassica oleracea*, and *Brassica juncea* mustard stimulate plant growth and development in

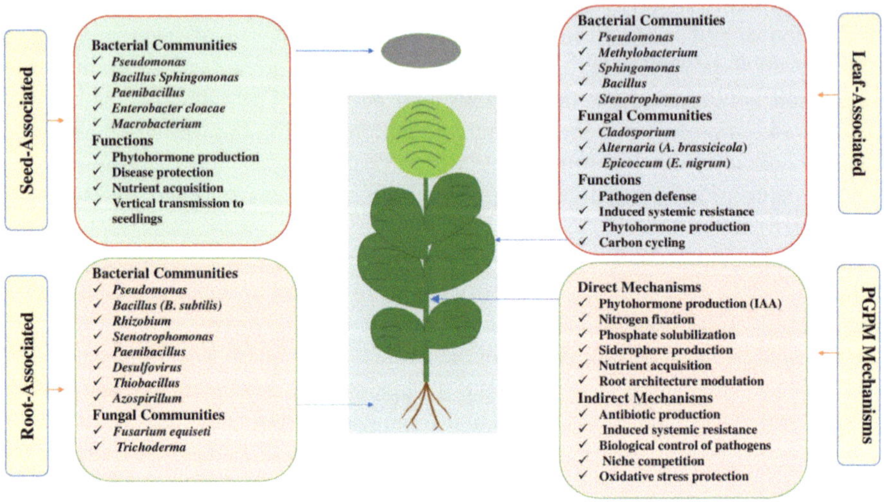

Fig. 1 Microbial communities associated with *Brassica* species and PGPM mechanisms

a variety of direct ways. These processes include physiological and biochemical activities that directly influence plant food intake, hormone synthesis, root growth, and stress reduction. *Brassica* plants, unlike legumes, do not develop symbiotic associations with arbuscular mycorrhizal fungi or rhizobia, emphasizing the need for free-living or endophytic plant growth-promoting microorganisms (PGPM) (Khoso et al. 2024).

Many times, contained in insoluble forms in soil, phosphorus is unavailable for plant uptake. Found in the rhizosphere of *B. napus* and *B. rapa* plants, *Pseudomonella fluorescens*, *Bacillus megaterium*, and *Enterobacter cloacae* are phosphate-soluble bacteria (PSB). Gluconic, lactic, and citric acids produced by these bacteria chelate cations connected to phosphate, transforming it into forms accessible for plants. For solubilizing *tricalcium phosphate*, *Bacillus amyloliquefaciens* isolated from *B. oleracea* root zones has been shown to drastically increase shoot phosphorus content and biomass (Pan and Cai 2023).

Although they lack nodulation, *Brassica* plants do have associative, free-living diazotrophs such as *Paenibacillus polymyxa*, *Klebsiella pneumoniae*, and *Azospirillum brasilense*. Fixing atmospheric nitrogen, non-symbiotic microbes convert it into ammonia (NH_3) that plants might use. Particularly associated with *Brassica* roots, *Azospirillum* spp. have been shown to enhance nitrogen absorption and chlorophyll synthesis, hence promoting higher plant development in nitrogen-limited conditions (Cruz-Hernández et al. 2022; Aasfar et al. 2021).

A wide range of bacteria and fungi generate plant hormones, which directly affect form and growth. Most studies on auxin produced by bacteria such as *Bacillus subtilis*, *Serratia marcescens*, and *Rhizobium* legumes have focused on indole-3-acetic acid. Commonly isolated from Brassica rhizospheres and endospheres, these organisms boost root elongation, lateral root growth, and root hair multiplication, therefore enhancing the plant's ability to acquire water and nutrients (Egamberdieva et al. 2017). While greatly increasing root biomass and nitrogen use efficiency, inoculating *Brassica juncea* with *Serratia plymuthica* increased IAA production.

Plant metabolism depends on iron, although its limited solubility in aerobic soils generally limits its bioavailability. Found in *Brassica* leaves and roots, microorganisms, including *Pseudomonas putida*, *Acinetobacter baumannii*, and *Stenotrophomonas maltophilia*, produce siderophores that transport Fe^{3+} into bacterial and plant cells. This increases plant iron intake while simultaneously restricting iron availability to phytopathogens, hence reducing disease burden (Rout and Sahoo 2015).

The infection of Chinese cabbage by *Plasmodiophora brassicae* reveals important genetic factors that play roles in both the initial and subsequent stages of infection. A total of 1610 differentially expressed genes (DEGs) were identified, which include 782 transcription factors from various families. This highlights the intricate interactions between the host plant and the pathogen, as well as the molecular changes that occur during the progression of clubroot disease. Notably, genes associated with alterations in the cell wall, signal transmission, and defence mechanisms were notably active during the early phases of infection. In contrast, genes involved in cell growth, auxin signalling, and metabolic functions became more prominent later. These insights provide a clearer understanding of the genetic landscape that influences tumour formation and pathogen development, presenting important targets for future research and breeding programmes focused on enhancing resistance to *P. brassicae* in *Brassica* crops (Meng et al. 2024).

Successfully extracted, analysed, and recognized zinc-solubilizing rhizobacteria from the rhizosphere of canola, emphasizing their crucial role in nutrient uptake and supporting plant growth. Out of 10 promising strains, 4 (CLS1, CLS2, CLS3, and CLS9) showed remarkable abilities to solubilize zinc and promote plant growth. Notably, *Staphylococcus succinus* (CLS1) was documented for the first time as a zinc solubilizer linked to canola, while *Priestia aryabhattai* (CLS2) and *Priestia megaterium* (CLS9) proved to be the most effective strains in boosting growth, enhancing oil content, and improving yield in both pot experiments and field trials. The specific locations where these strains were found indicate their potential for inclusion in targeted biofertilizer products. The value of native zinc-solubilizing rhizobacteria as environmentally friendly bioinoculants that can enhance zinc availability and support sustainable canola farming (Jalal-Ud-Din et al. 2024).

By means of direct mechanisms like nutrient solubilization, nitrogen fixation, phytohormone production, siderophore creation, and ethylene regulation, microbial communities in *Brassica agroecosystems* provide significant growth-promoting activities. Particularly in low-input or stress-sensitive environments, these microbial interactions offset the absence of symbiotic mycorrhizae in *Brassica* species, therefore offering long-term options to boost crop yield. Integrated nutrient management and sustainable agriculture may benefit much from the use of these helpful bacteria as bioinoculants.

3.2 Indirect Mechanism

Microbial actions protecting plants against biotic and abiotic problems, thus improving general plant health, development, and production, are known as indirect plant growth-promoting (PGP) processes. Disease prevention, modification of plant immune responses, and tolerance to environmental stresses in *Brassica* species, including *Brassica napus* (canola), *B. juncea* (mustard), and *B. oleracea* (broccoli, cabbage), depend critically on microbial communities from the rhizosphere, phyllosphere, and endosphere. Given *Brassica's* non-mycorrhizal character, disease sensitivity, and soil limitations, these indirect approaches are extremely crucial. *Brassica* roots and leaves contain many microorganisms with somewhat antagonistic action against soil and foliar diseases. This includes lytic enzyme development, resource and habitat rivalry, and antibiotic production. Separated from the rhizosphere of *B. oleracea*, *Bacillus subtilis* and *Bacillus amyloliquefaciens* generate antimicrobial lipopeptides such as surfactin and iturin, which upset fungal membranes and control diseases like *Rhizoctonia solani*, *Sclerotinia sclerotiorum*, and *Fusarium oxysporum*. Another well-researched biocontrol agent, *Pseudomonas fluorescens*, produces hydrogen cyanide (HCN) and 2,4-diacetylphloroglucinol (DAPG), both known to inhibit oomycetes and root-infecting fungi (Ludwig-Müller 2024; Martín-Cardoso et al. 2024).

Induced systemic resistance (ISR) is the mechanism by which certain beneficial bacteria can activate the immune system of the plant actively. Non-pathogenic rhizobacteria or phyllosphere microorganisms trigger ISR, which therefore produces more resistance to many kinds of illnesses and insects (Sheoran et al. 2025). Common endophytic fungi in *Brassica* roots, *Trichoderma harzianum* generates ISR by activating the jasmonic acid and ethylene signalling pathways, both of which are vital for defence against necrotrophic diseases. Likewise, *Pseudomonas putida* has been shown to induce ISR in *Brassica napus*, hence increasing resistance to *Alternaria brassicae* and *Xanthomonas campestris* by upregulating defence-related genes such as PR-1 and PDF1.2. By immobilizing or converting toxins, microorganisms also allow *Brassica* plants to survive in contaminated soils. Metal-binding proteins and efflux systems of rhizosphere bacteria like *Rhizobium legumes*, *Pseudomonas aeruginosa*, and *Arthrobacter* spp. reduce the bioavailability of heavy metals, including cadmium, lead, and arsenic (Karnwal et al. 2025). One bacterial communication method that influences pathogen colonization most typically is quorum sensing. Quorum quenching enzymes produced by certain *Brassica*-associated bacteria break down QS molecules and prevent virulence genes from expressing themselves. For instance, isolated from *B. oleracea* roots, *Rhodococcus erythropolis* produces lactonases that upset the QS system of *Erwinia carotovora*, hence lowering soft rot symptoms (Zhang et al. 2024). In the *rhizosphere* or *phyllosphere*, beneficial bacteria may compete with illnesses for vital nutrients such as iron, carbon, and amino acids. Iron is sequestered by siderophore-producing bacteria such as *Pseudomonas putida*, therefore limiting its availability to fungal infections, including *Verticillium dahliae* (Yaghoubi Khanghahi et al. 2024). Nutrient competition is a significant element restricting pathogen colonization and disease progression in

Brassica cropping systems. Not only by means of nutrient absorption but also via indirect mechanisms enhancing biotic and abiotic stress tolerance, the microbial communities linked with *Brassica* species help to build and maintain plant health. These consist of pathogen biocontrol, ISR induction, metal detoxification, interference with quorum sensing, and resource competition. Using microbial inoculants or microbiome engineering to leverage these interactions offers exciting long-term disease control and crop productivity possibilities for *Brassica*-based agriculture. Seed-associated microbial communities perform several essential functions that influence plant development and resilience.

4 Challenges in Microbiota Composition and Function Under Environmental Stress Condition

4.1 Impact of Major Abiotic Stressors on Microbiota

4.1.1 Temperature Stress

Variations in temperature significantly influence the composition and functioning of microbial communities. When exposed to high temperatures, the diversity of these microorganisms typically diminishes, leading to a prevalence of heat-resistant species. Increased heat can result in the unfolding of proteins and changes in membrane fluidity, which in turn hinders enzyme activity and overall metabolic processes. On the other hand, exposure to cold temperatures reduces microbial metabolism and activity, disrupting the cycling of nutrients. Cooler environments often favour the survival of *psychrophilic* and psychro-tolerant microbes, which produce proteins that prevent freezing and modify their membrane structures to thrive in low-temperature conditions.

The capabilities of the seed microbiome in *Brassica napus* throughout various growth phases are explored in this study. By pinpointing unique clusters of microbial communities and essential ASVs, including advantageous species like *Sphingomonas endophytica*, our research underscores the complex connections between seed maturation and microbiome activity. The link between microbial metabolic processes and seed physiology indicates a mutually developed relationship that could be vital for seed vitality and strength. These findings not only enhance our comprehension of plant–microbe dynamics during seed development but also pave the way for utilizing helpful microbes to boost crop resilience and yield (Yao et al. 2024).

4.1.2 Drought and Water Stress

Water scarcity and drought significantly influence microbial activity. When drought occurs, water films in the soil break apart, which restricts the movement of microbes and their access to nutrients. In response, microbial communities undergo physiological adaptations, including the production of osmolytes and a transition towards species that can withstand dry conditions, particularly those that can produce spores. Extended periods of drought can lead to a decrease in microbial biomass and enzymatic functions, compromising essential processes such as the decomposition of organic matter.

The reactions of plants and soil microbes to drought conditions show that while *Brassica oleracea* displayed obvious signs of stress both physiologically and biochemically, the soil microbiome stayed relatively unchanged. Notable changes in the microbial makeup and enzyme activity, like the decline in Gemmatimonadota and increased arylsulfatase activity, indicate that the microbial community is adapting functionally, which may help the ecosystem bounce back. This research underscores the potential for utilizing drought-resistant soil microbes to alleviate stress on plants and promote sustainable farming practices in the face of shifting climate patterns. Future studies should aim to discover and utilize beneficial microbial groups to create strategies that enhance crop resilience to drought (Babic et al. 2024).

4.1.3 Salinity Stress

Salt stress environments with high salinity pose two main issues: osmotic pressure and the toxicity of ions. Elevated salt concentrations can significantly diminish the variety of microbial life, while simultaneously favouring the growth of salt-tolerant and salt-loving species. While certain microorganisms can adjust their metabolic processes to cope with salty conditions, these adjustments frequently require extra energy, which can compromise the effectiveness of essential biogeochemical processes.

Salinity stress significantly changes the makeup and functional capabilities of the bacterial communities in the rhizosphere linked to *Brassica napus*. The increase in certain groups like *Sphingomonas*, *Streptomyces*, *Bacillus*, and *Pseudomonas* during stress conditions points to their potential roles in boosting plant resilience. A rise in *Chloroflexi* and *Acidobacteria*, together with diminished microbial diversity and unique clustering patterns, further highlights the effects of salinity on rhizosphere microbiota. Functional predictions that show enhanced metabolic and transport pathways under stress underscore microbial adaptation and possible support for the plant's tolerance. These insights provide crucial information for creating microbiome-based approaches aimed at improving crop performance in saline environments (Laiq et al. 2024).

4.1.4 Chemical Pollution

Chemical contamination is the introduction of harmful substances like heavy metals, pesticides, and industrial chemicals adds toxic agents to environments inhabited by microorganisms. These contaminants tend to wipe out more vulnerable species, allowing those that can withstand them to thrive. Although these resilient communities might exhibit enhanced abilities to break down pollutants, their overall range of functions is frequently diminished, restricting the adaptability of the ecosystem.

The vegetable varieties of *Brassica parachinensis* play a crucial role in boosting the breakdown of di-(2-ethylhexyl) phthalate (DEHP) within the rhizosphere, even though the specific genotype has little effect on the structure of microbial communities. The research demonstrates that the differences in ecological niches and the presence of DEHP are the main factors influencing the bacterial communities associated with roots and their interactions within the ecosystem. The increase of bacteria that can degrade phthalates and the presence of functional genes in the rhizosphere, particularly those involving important species like *Aeromicrobium* and TM7a, indicate a collaborative microbial network that aids in the breakdown of DEHP. Furthermore, the discovery of significant bacterial families through metagenomic analysis underscores the teamwork among soil microbes in degrading phthalates. These findings lay the groundwork for utilizing the rhizosphere microbiomes of vegetables to improve the bioremediation of organic pollutants in farm soils (Wang et al. 2024).

4.1.5 pH Changes

Variations in pH levels play a crucial role in shaping the behaviour of microorganisms and the availability of nutrients. When the environment becomes more acidic, it often benefits fungi at the expense of bacteria, which can significantly alter the processes of decomposition. In situations where pH levels are extremely high or low, specific microbes that possess strong cellular adaptations tend to thrive, but this can come at the cost of losing a wider range of microbial functions.

The inadequate drop in pH during the fermentation of sunki can be directly attributed to the types of microbes present and their metabolic activities in the turnip leaves undergoing fermentation. Notably, a transformation in the bacterial population marked by a decrease in *Lactobacillus* and an increase in *Limosilacto bacillus reuteri* was linked to lower levels of lactic acid and higher concentrations of ammonia, which ultimately hindered the acidification process. The build-up of ammonia, likely resulting from the breakdown of glutamine, turned out to be a significant obstacle to achieving the desired pH reduction. These observations shed light on the intricate microbial and chemical interactions at play during sunki fermentation and indicate that managing the levels of *L. reuteri* and ammonia could be essential for maintaining consistent quality in the final product (Tomita et al. 2024).

4.2 Impact of Major Biotic Stressors on Microbiota

Brassica species are susceptible to a wide range of pathogenic organisms that significantly reduce crop yield and quality. These pathogens include fungi, bacteria, and viruses, many of which are seed-borne or soil-resident and can cause recurring epidemics in *Brassica*-growing regions worldwide.

4.2.1 Fungal Pathogens

Alternaria Species

Alternaria brassicicola and *A. brassicae* are the primary fungal species responsible for Alternaria leaf spot or black spot disease in *Brassica* crops. These pathogens are mainly seed-borne but can also survive on crop debris and wild cruciferous plants (Meena et al. 2021). Infected plants exhibit dark brown to black circular lesions with concentric rings and yellow halos on leaves, stems, and siliques. Severe infections lead to premature defoliation, reduced photosynthetic area, and yield losses of up to 20–50% in susceptible cultivars (Meena et al. 2021). Additionally, seed infection by *Alternaria* significantly reduces germination and causes seedling blight. Deep and Sharma (2012) reported germination reductions of 10–80% in infected seeds compared to healthy ones.

Leptosphaeria maculans and *L. biglobosa*

Leptosphaeria maculans and *L. biglobosa* are responsible for blackleg or phoma stem canker in oilseed rape (*Brassica napus*), one of the most destructive fungal diseases globally. These pathogens are both seed-borne and residue-borne, with the ability to persist in infected debris for several years (Cornelsen et al. 2024). Initial infections appear as pale lesions containing black pycnidia on cotyledons and leaves, followed by cankers on stems that interfere with water and nutrient transport (Fitt et al. 2006). Economic losses due to blackleg can range from 50% to 80% during severe outbreaks.

Plasmodiophora brassicae

Plasmodiophora brassicae, the causative agent of clubroot disease, is an obligate soilborne pathogen affecting most *Brassica* species. Characterized by the formation of root galls, this disease disrupts water and nutrient uptake (Dixon 2009). Although not typically seed-borne, the pathogen forms durable resting spores that remain viable in soil for up to 20 years, complicating eradication (Ajithkumar et al. 2025). Yield losses due to clubroot can range from 30% to 100% in heavily infested fields (Botero-Ramirez et al. 2024).

Sclerotinia sclerotiorum

Sclerotinia sclerotiorum causes white mould or stem rot in *Brassica* crops. This fungus has a broad host range and forms sclerotia capable of surviving in soil for extended periods (Bolton et al. 2006). Disease symptoms begin as water-soaked lesions that progress to wilting, bleaching, and plant death. Although not primarily seed-borne, occasional seed contamination can occur (Guha 2025). Yield losses from Sclerotinia stem rot can range from 10% to 80% depending on environmental conditions and disease pressure.

Bacterial Pathogens

Xanthomonas campestris pv. *campestris*

Xanthomonas campestris pv. *campestris* (Xcc) causes black rot, the most significant bacterial disease of *Brassica* crops worldwide. This pathogen is predominantly seed-borne but also persists in crop debris and cruciferous weeds (Jawadagi et al. 2024). Symptoms include V-shaped chlorotic lesions starting at leaf margins and progressing to blackened veins and systemic infection (Guha 2025). Even seed infection levels as low as 0.03% can result in widespread field outbreaks under conducive conditions (Rauf et al. 2024).

Pseudomonas syringae pv. *maculicola*

In *Brassica* plants, the bacterium *Pseudomonas syringae* pv. maculicola stands out as a major leaf pathogen that results in bacterial leaf spot. This disease leads to yellowing and dead patches on the leaves, significantly hindering photosynthesis and the overall health of the plant. This pathogen poses a serious risk to the yield and market appeal of key *Brassica* crops like mustard, cabbage, and broccoli. Due to the drawbacks of chemical treatments, such as the development of resistance and environmental issues, existing management practices fall short for sustainable disease control. Although *Brassica* crops show considerable genetic variety, the resistance to *P. syringae pv. maculicola* is often inherited in a complex manner and is influenced by environmental conditions. Biological control methods present a hopeful solution, particularly using beneficial microorganisms like *Pseudomonas fluorescens*, *Bacillus subtilis*, and *Bacillus amyloliquefaciens*. These microbes can help combat the pathogen by producing antibiotics, inducing systemic resistance in the plant, and competing for space and resources. Combining these biocontrol agents with plant resistance and agricultural practices such as intercropping or crop rotation can help lower pathogen load and prolong the effectiveness of resistance. Enhancing our knowledge of the interactions between host plants, pathogens, and the microbiome, along with utilizing molecular breeding techniques, will be essential in creating resilient and disease-resistant *Brassica* farming systems (Tariq Ahmad et al. 2025).

4.2.2 Viral Pathogens

Turnip Mosaic Virus (TuMV)

Turnip mosaic virus is among the most prevalent and economically damaging viruses of *Brassica* species. Transmitted by aphids in a non-persistent manner, TuMV is occasionally seed-borne at low levels in some *Brassica* hosts (Shevchenko et al. 2024). Symptoms include mosaic patterns, stunted growth, necrosis, and leaf distortion. Yield losses depend on infection timing and host susceptibility, ranging from 30% to 90% (Tsytsiura 2023).

Cauliflower Mosaic Virus (CaMV)

Cauliflower mosaic virus infects most *Brassica* species, causing systemic mosaic, vein clearing, and plant stunting. It is transmitted by aphids and can sometimes be seed-borne (Geoffrey 2024). Cauliflower mosaic virus can reduce yields by 20–60% and frequently occurs in mixed infections, which amplify symptom severity due to synergistic interactions (Sánchez-Tovar et al. 2025).

Importance of Seed Health in *Brassica* Production

Seed health plays a critical role in the successful cultivation of *Brassica* crops, as seeds act as primary vectors for the transmission of pathogens across regions and production cycles (Geoffrey 2024). The presence of seed-borne pathogens can result in early disease establishment in the field, enabling the introduction of harmful organisms into previously disease-free environments. Additionally, such infections can facilitate long-distance dissemination of pathogens, reduce seed germination and seedling vigour, and ultimately lower crop yield and quality.

The economic impact of seed-borne diseases in *Brassica* can be significant. For example, black rot caused by *Xanthomonas campestris* pv. campestris, when seed-transmitted, has been reported to reduce marketable yields by up to 50% (Geoffrey 2024). Similarly, infections by *Alternaria* species not only affect plant health but also decrease the oil content in mustard seeds by 14–36% (Mishra et al. 2024), underscoring the importance of rigorous seed health management.

Major Seed-Borne Pathogens of Brassica

A variety of pathogens are known to be transmitted through *Brassica* seeds, with their transmission rates influenced by factors such as pathogen biology, environmental conditions, and host susceptibility. Table 1 summarizes the major seed-borne pathogens affecting *Brassica* crops, the diseases they cause, and their approximate seed transmission rates.

Mechanisms of Seed Infection and Pathogen Transmission

Pathogens can infect or contaminate *Brassica* seeds through several pathways. Systemic infection involves the colonization of the plant's vascular system, allowing pathogens such as *Xanthomonas campestris* to reach developing seeds. In direct penetration, pathogens like *Alternaria* spp. breach the seed coat during seed formation. Another route is flower infection, wherein pathogens such as *Sclerotinia sclerotiorum* infect floral tissues, leading to contaminated seeds. Additionally, many pathogens adhere externally to seed surfaces during harvesting and processing. The internal or external localization of pathogens in seeds significantly influences their persistence and capacity for transmission. Internally borne pathogens often have

Table 1 Major seed-borne pathogens of *Brassica* crops

Brassica crops	Pathogen	Disease	Seed transmission rate (%)	Reference
Cabbage, cauliflower, broccoli	*Xanthomonas campestris* pv. *campestris*	Black rot	0.03–1.5	Mao et al. (2025)
Broccoli, cauliflower, cabbage	*Alternaria brassicicola*	Alternaria leaf spot	10–90	Kaur and Dutta (2024)
Mustard, broccoli, cabbage	*Alternaria brassicae*	Alternaria blight	5–85	Kaur and Dutta (2024)
Canola, cabbage	*Leptosphaeria maculans*	Blackleg	0.3–5	Brachaczek et al. (2016)
Broccoli, cabbage, kale	*Pseudomonas syringae* pv. *maculicola*	Bacterial leaf spot	0.1–2	Noh et al. (2025)
Turnip, Chinese cabbage, broccoli	Turnip mosaic virus	Mosaic disease	0–0.5	Ton (2024)

greater longevity and pose a higher risk of transmission compared to those on seed surfaces (Geoffrey 2024).

5 Developing Microbe-Based Management Strategies

The plant microbiome comprises a diverse array of microbial communities that colonize various plant compartments, including the rhizosphere, phyllosphere, and endospheric (Enders et al. 2024). In *Brassica* species, this microbiome exhibits a complex and dynamic structure shaped by multiple factors such as plant genotype, soil composition, agricultural management practices, and environmental conditions. A distinctive feature of *Brassica* plants is their production of glucosinolates, sulfur-containing secondary metabolites that hydrolyse into bioactive compounds like isothiocyanates. These breakdown products possess antimicrobial properties, playing a critical role in modulating the composition and diversity of associated microbial communities (Martín-Cardoso et al. 2024). As a result, *Brassica* microbiomes often display unique structural profiles compared to crops (Table 2).

Core microbial taxa consistently associated with *Brassica* species include beneficial bacteria from the *phyla Proteobacteria*, *Actinobacteria*, *Firmicutes*, and *Bacteroidetes*. Among these, genera such as *Pseudomonas*, *Bacillus*, *Paenibacillus*, *Streptomyces*, and *Burkholderia* are frequently dominant, contributing to plant health through mechanisms like nutrient solubilization, hormone production, and pathogen suppression (Enders et al. 2024).

Strategies for managing diseases through microbiomes leverage helpful microbial populations to combat harmful pathogens and enhance disease resilience in *Brassica* plants. These microorganisms combat diseases via multiple pathways, including the production of antimicrobial substances, competing for essential nutrients and space on plant surfaces, activating the plant's defence mechanisms, and even preying on pathogens. Practical techniques such as biopriming seeds, managing the soil microbiome, and inoculating with endophytes have shown effectiveness in reducing disease intensity and boosting plant health. As highlighted in Table 3, these integrated methods present a sustainable and eco-friendly option compared to traditional chemical disease management approaches.

6 Microbiome-Driven Crop Improvement

6.1 Under Abiotic

The intricate relationship between soil microbes, nutrient availability, and plant growth underscores the significance of understanding the soil ecosystem. Environmental conditions, edaphic properties, and microbial communities shape plant development through direct impacts on nutrient uptake and microbial activity. Moreover, the epigenetic regulation of plants, influenced by both microbial interactions and nutrient imbalances, presents a compelling frontier in plant biology. The

Table 2 Microbe-based management strategies for *Brassica* species approaches, applications, benefits, and limitations

Strategy type	Microbial approach	Target applications	Examples in *Brassica* species	Benefits	Limitations
Single-microbe inoculants	Individual bacterial strains	Seed treatment, soil application	*Bacillus subtilis* in *B. oleracea*—IAA production and seedling vigour	Simplified production	Limited functional range
				Specific mode of action	Inconsistent field performance
				Easier regulatory approval	Environmental sensitivity
	Individual fungal strains	Root drenching, seed coating	*Trichoderma harzianum* in *B. napus*—root growth and pathogen suppression	Well-characterized mechanisms	Host specificity issues
				Targeted effects	Environmental constraints
				Long shelf life (spores)	Limited ecological persistence
	Endophytic bacteria	In-planta colonization	*Sphingomonas* sp. in *B. napus*—oxidative stress protection	Protected in-plant environment	Hard to establish in mature plants
				Access to plant metabolism	Limited dispersal
				Vertical transmission	Antagonism with native endophytes

Microbial consortia	Bacterial mixtures	Rhizosphere engineering	*Pseudomonas* + b*acillus* + *Paenibacillus* in *B. juncea*—multiple PGPR effects	Functional redundancy	Complex interactions
				Multiple action modes	Production and regulatory issues
				Synergy	
	Bacterial-fungal consortia	Plant microbiome management	*Trichoderma* + p*seudomonas* in *B. oleracea*—growth and disease suppression	Complementary niches	Antagonism risk
				Better colonization	Formulation issues
				Cross-feeding	Establishment inconsistency
	Synthetic communities (SynComs)	Targeted microbiome reconstruction	Defined community for *B. rapa*—seedling establishment	Predictable and designed effects	High R&D cost
				Reduced variability	Complex manufacturing
					Emerging approach

(continued)

Table 2 (continued)

Strategy type	Microbial approach	Target applications	Examples in *Brassica* species	Benefits	Limitations
Microbiome engineering	Plant selection for microbiome traits	Breeding programmes	*B. napus* varieties recruiting *pseudomonads*	Durable trait	Long development time
				Co-evolved interaction	Inheritance complexity
				Germplasm integration	Environmental variability
	Microbiome transfer	Soil/seedling inoculation	Suppressive soil transfer for clubroot in *B. oleracea*	Whole-community approach	Variable success
				Multiple traits	Regulatory and pathogen risks
	Precision microbiome management	Site-specific manipulation	Tailored rhizosphere in *B. napus*	Local adaptation	Needs detailed analysis
				Environmental optimization	Implementation complexity
				High efficacy potential	High cost

Formulation approaches	Encapsulation technologies	Enhanced microbial delivery	Alginate-encapsulated *B. amyloliquefaciens* in *B. juncea*	Longer shelf life	Higher cost
				Controlled release	Scaling issues
				Protected microbes	Viability concerns
	Seed coating	Pre-plant inoculation	Methylcellulose + *Enterobacter* for *B. napus*	Direct to germinating seed	Short persistence
				Less inoculant needed	Inconsistent colonization
				Simple application	Storage issues
	Biofilm-enhancing additives	Improved colonization	Exopolysaccharide-aided *pseudomonas* on *B. oleracea*	Better attachment	Variable efficacy
				Stress tolerance	Soil structure impact
				Persistence	Off-target effects
Integration with management	Integrated with reduced tillage	Conservation agriculture	Microbial consortia for strip-tillage in *B. napus*	Soil health	Slow establishment
				Low disturbance	Weed issues
				Eco-friendly	Climate sensitivity
	Combined with organic amendments	Soil health enhancement	Compost + *Trichoderma* in *B. oleracea*	Microbial support	Amendment variability
				Nutrient cycling	Complex interactions
				Soil benefits	High cost
	Precision Agri-integration	Targeted application	Variable-rate microbe application in *B. napus*	Optimized input use	Tech-heavy
				Field specific	Knowledge-intensive
				High ROI	High initial investment

Table 3 Microbiome-based disease suppression strategies in *Brassica* species

Mechanism/ strategy	Key microorganisms/ agents	Mode of action/outcomes	References
Direct antagonism	*Bacillus subtilis*	Produces lipopeptides (surfactin, iturin, fengycin) that inhibit *A. brassicicola* and *L. maculans*	Ajuna et al. (2024)
	Pseudomonas spp.	Produces DAPG, phenazines, pyrrolnitrin, and HCN that suppress soilborne pathogens	Dimkić et al. (2022)
Competition for space and nutrients	*Pseudomonas fluorescens*	Produces siderophores to sequester iron, reducing pathogen access; suppresses *Fusarium* spp.	Sah et al. (2021)
Induced systemic resistance (ISR)	*Bacillus* spp., *pseudomonas* spp.	Activates salicylic acid and jasmonic acid pathways; effective against *X. campestris pv. campestris*, *A. brassicicola*	Hossain et al. (2024)
Hyperparasitism	*Trichoderma harzianum*	Mycoparasitism of pathogens like *S. sclerotiorum*, *R. solani*; field reductions of disease by 50–70%	Natsiopoulos et al. (2024)
Seed biopriming	*Bacillus amyloliquefaciens*, *T. harzianum*, *P. fluorescens*	Reduces *Alternaria* incidence (65–85%), improves seedling vigour (15–30%)	Selim et al. (2017)
Soil microbiome management	Microbial consortia (*B. subtilis*, *B. pumilus*, *B. megaterium*)	Clubroot reduction by 75–90%; better than single strains	Zhang et al. (2022)
	Organic amendments	Enrich actinobacteria and proteobacteria; reduce *Rhizoctonia* by 40–65%	Bonanomi et al. (2022)
	Brassica cover crops, crop rotation	Restructure soil microbiota; biofumigation via isothiocyanates	Hanschen and Winkelmann (2020)
Endophyte inoculation	*Serendipita indica*	Reduces black rot by 45–60%; promotes plant growth	Saleem et al. (2023)
	Bacillus, *pseudomonas*, *Burkholderia* spp.	Induce systemic resistance; reduce *Alternaria* by 50–65%	Thokchom et al. (2020)
Black rot management	*Bacillus velezensis* B63	Biopriming + root dipping + field sprays reduce disease by 75–85%	Liu et al. (2022)

(continued)

Table 3 (continued)

Mechanism/ strategy	Key microorganisms/ agents	Mode of action/outcomes	References
Clubroot management	*Bacillus*, *Streptomyces*, *Lysobacter* (BioClub consortium)	60–78% clubroot reduction; 15–25% yield gain; commercial use	Zhang et al. (2022)
Sclerotinia stem rot control	*Coniothyriumminitans*, *Bacillus subtilis*	65–80% disease suppression via soil/foliar application and microclimate management	Kora (2008)

potential of epigenome editing, driven by insights into these molecular mechanisms, offers a novel avenue for crop improvement without altering the DNA sequence. This technology could provide stable, heritable phenotypes, presenting a sustainable and effective solution for future agricultural challenges. By integrating knowledge of soil microbiomes and their epigenetic influence, we can unlock new strategies for creating resilient crops that adapt to both biotic and abiotic stresses (Doddavarapu et al. 2024).

Soil microbial communities play a vital part in helping plants cope with drought stress, especially in *Brassica oleracea*. Although drought conditions caused noticeable stress in these plants, evidenced by stunted growth, lower photosynthesis rates, and heightened stress indicators, the soil microbiome showed remarkable resilience, maintaining a stable level of diversity despite the challenging environment. Notable changes in microbial populations were recorded, such as a decline in Gemmatimonadota and an increase in arylsulfatase activity, underscoring the complex relationships between soil microbes and how plants react to drought conditions. These results indicate that the soil microbiome is essential, yet frequently overlooked, in bolstering plant resilience during stressful times (Babic et al. 2024).

Salinity stress presents considerable obstacles to crop yields, and the plant's rhizosphere is crucial in managing stress responses. The bacterial community structure of *Brassica napus* under both typical and saline environments. Our observations indicated that changes in salinity trigger shifts in the microbial populations within the rhizosphere, resulting in differences in community diversity and makeup. Notably, there was a marked increase in the presence of genera such as *Sphingomonas*, *Streptomyces*, *Bacillus*, and *Pseudomonas*, along with a rise in *Chloroflexi* and *Acidobacteria*, pointing to a possible microbial adaptation to salt stress. The decline in diversity seen in salt-treated plants, coupled with heightened pathways related to amino acid and carbohydrate metabolism, further supports the notion that the microbial community plays an active role in lessening the harmful impacts of salinity (Laiq et al. 2024).

Severe abiotic stress conditions have created major obstacles for the health of both soil and plants. Nevertheless, the intricate biochemical processes occurring within the soil are crucial in shaping the relationships between plants and soil microbes, which are essential for boosting plant resilience against these challenges. Soil microbiomes play a key role in supporting plant nutrition and managing stress through multiple pathways, such as triggering plant defence mechanisms, producing enzymes, and regulating growth-promoting substances. The application of specially engineered soil microbes as inoculants has been shown to be advantageous in enhancing plant stability and health when faced with environmental stress by influencing oxidative stress responses and protein activities. Moreover, identifying important genes related to these functions offers potential for optimizing microbial roles in reducing plant stress. With a growing emphasis on sustainable and environmentally friendly farming practices, utilizing beneficial microbes in the rhizosphere emerges as a vital approach to enhancing plant resilience, ensuring optimal nutrition, and improving food security (Adeleke et al. 2024).

Microorganisms play a crucial role in driving biogeochemical processes and affecting plant development. However, tapping into their potential for sustainable farming presents a challenging puzzle. The complex dynamics of microbial ecosystems, coupled with our limited grasp of how plants, microbes, and their environment interact, restrict our capabilities to effectively manipulate these systems. Yet, exciting progress in the fields of microbial ecology, systems biology, and bioengineering is providing valuable tools to demystify these complexities. By studying how plants and microbes respond to non-living environmental stresses and pinpointing essential molecules and microorganisms, we have a chance to gain deeper insights into the factors that support plant health. The development of synthetic microbial communities paves the way for further exploration of plant-microbe relationships. With ongoing innovation, upcoming research could unveil novel approaches to designing ecological systems that bolster both plant and ecosystem resilience. In the end, harnessing these advancements could transform agricultural methods, creating more sustainable and effective crop production systems (Kozaeva et al. 2024).

The crucial function of seed-associated microorganisms in influencing the absorption and buildup of antibiotics in crops, especially ciprofloxacin (CIP), is noteworthy. The presence of *Bacillaceae*, particularly *Bacillus*, in the roots of varieties that accumulate high levels of antibiotics (HAV) was significantly stimulated by substances released from plant roots, playing a vital role in this accumulation process. This group of microorganisms not only promoted plant growth and root development but also improved resistance to CIP. Notably, the specific interactions between different plant varieties and their associated microbes resulted in much higher levels of CIP in the shoots of HAV compared to those with low antibiotic accumulation (LAV). These findings shed light on the intricate relationship between plant-associated microbes and antibiotic uptake, suggesting that managing these microbial communities could help reduce antibiotic levels in agricultural products. Nevertheless, additional research is essential to comprehensively grasp the wider effects of these microbial relationships on plant vitality and food safety (Wang et al. 2025).

6.2 Under Biotic

Brassica crops face increasing biotic stress from pests and pathogens, worsened by climate change and pesticide restrictions. Their associated microbiomes, diverse microbial communities in the rhizosphere, phyllosphere, and endospheric, offer promising tools for enhancing plant health and resistance. Key microbial players include plant growth-promoting rhizobacteria (PGPR), beneficial fungi, and endophytes. These organisms protect plants through competitive exclusion, antimicrobial production, induced systemic resistance (ISR), and volatile emissions. Successful applications include *Bacillus subtilis* against black rot and *Pseudomonas fluorescens* for clubroot suppression. Advanced strategies like microbiome transplantation, synthetic communities, and microbiome-aware breeding are emerging, though challenges remain in ecological understanding, delivery methods, regulation, and scalability. Future efforts should focus on high-throughput screening, multi-omics integration, precision delivery systems, and cost-benefit analysis. Microbiome-based approaches offer sustainable, effective alternatives for *Brassica* crop protection and improvement.

7 Conclusion

The microbial populations that are found in association with *Brassica* crops form a highly specialized and co-evolved ecological system that is critical to plant growth, nutrient uptake, and stress alleviation. In the absence of traditional symbiotic relationships like mycorrhizae or rhizobia, *Brassica* crops have evolved complex associations with a wide range of beneficial microorganisms that can bring about phosphate solubilization, biological nitrogen fixation, siderophore-mediated iron acquisition, and phytohormone modulation. These PGP mechanisms have direct effects on plant growth and are supported by indirect pathways such as pathogen antagonism, induced systemic resistance, abiotic stress detoxification, and modulation of plant defence signalling pathways.

Novel secondary metabolites, especially glucosinolates, additionally contribute to the assembly and functional specialization of *Brassica*-related microbiomes, which become enriched for bacteria like *Proteobacteria*, *Firmicutes*, *Actinobacteria*, and *Bacteroidetes*. Such microbes show substantial functional redundancy and flexibility, enhancing the *Brassica* holobiont's resistance under both abiotic and biotic stress situations.

Microbe-mediated management practices covering microbial inoculants, synthetic microbial communities, microbiome engineering, and seed microbiome transfer have promising potential as alternatives to traditional agrochemical practices. Integration of these approaches into crop production systems has the potential to enhance nutrient use efficiency, control pathogens, and augment stress tolerance, especially under conditions of resource constraint or environmental constraint.

To maximize the potential of *Brassica*-associated microbiomes, future research needs to focus on multi-omics-based dissection of plant–microbe–environment interactions, precision delivery systems for microbial formulations, and the integration of microbiome characteristics into breeding programmes. Ultimately, the strategic manipulation and management of microbial communities are a frontier for sustainable *Brassica* production and agroecological resilience more generally.

References

Aasfar A, Bargaz A, Yaakoubi K, Hilali A, Bennis I, Zeroual Y, Meftah Kadmiri I (2021) Nitrogen fixing Azotobacter species as potential soil biological enhancers for crop nutrition and yield stability. Front Microbiol 12:628379. https://doi.org/10.3389/fmicb.2021.628379

Adeleke BS, Chaudhary P, Ayilara MS, Ojo FM, Erinoso SM, Upadhayay VK, Akinola SA (2024) Rhizosphere microbiomes mediating abiotic stress mitigation for improved plant nutrition. Ecology 5(3):375–401. https://doi.org/10.3390/ecologies5030024

Ajithkumar K, Savitha AS, Renuka M, Mahadeva kumar S (2025) Fungal Phytopathogens and the importance of early detection through conventional methods. In: Molecular approaches for the detection of fungal Phytopathogens. CRC Press, pp 1–27

Ajuna HB, Lim HI, Moon JH, Won SJ, Choub V, Choi SI, Ahn YS (2024) The prospect of antimicrobial peptides from *bacillus* species with biological control potential against insect pests and diseases of economic importance in agriculture, forestry, and fruit tree production. Biotechnol Biotechnol Equip 38(1):2312115. https://doi.org/10.1080/13102818.2024.2312115

Assaha DVM, Ueda A, Saneoka H, Al-Yahyai R, Yaish MW (2017) The role of Na+ and K+ transporters in salt stress adaptation in Glycophytes. Front Physiol 18(8):509. https://doi.org/10.3389/fphys.2017.00509

Babic I, Šamec D, Hižak M, Huđ A, Senko H, Šangut IJ, Petrić I (2024) Evaluating drought impact on white cabbage: plant stress response and soil microbiome adaptation. Plant Stress 14:100683. https://doi.org/10.1016/j.stress.2024.100683

Bargaz A, Lyamlouli K, Chtouki M, Zeroual Y, Dhiba D (2018) Soil microbial resources for improving fertilizers efficiency in an integrated plant nutrient management system. Front Microbiol 9:1606. https://doi.org/10.3389/fmicb.2018.01606

Barret M, Briand M, Bonneau S, Préveaux A, Valière S, Bouchez O, Jacques MA (2015) Emergence shapes the structure of the seed microbiota. Appl Environ Microbiol 81(4):1257–1266. https://doi.org/10.1128/AEM.03722-14

Bashir I, War AF, Rafiq I, Reshi ZA, Rashid I, Shouche YS (2022) Phyllosphere microbiome: diversity and functions. Microbiol Res 254:126888. https://doi.org/10.1016/j.micres.2021.126888

Bezodis W, Penfield S (2024) Maternal environmental control of progeny seed physiology: a review of concepts, evidence and mechanism. Seed Sci Res:1–8. https://doi.org/10.1017/S0960258524000151

Bolton MD, Thomma BP, Nelson BD (2006) *Sclerotinia sclerotiorum* (Lib.) de Bary: biology and molecular traits of a cosmopolitan pathogen. Mol Plant Pathol 7(1):1–16. https://doi.org/10.1111/j.1364-3703.2005.00316.x

Bonanomi G, Zotti M, Idbella M, Cesarano G, Al-Rowaily SL, Abd-ElGawad AM (2022) Mixtures of organic amendments and biochar promote beneficial soil microbiota and affect *Fusarium oxysporum f.* sp. lactucae, *Rhizoctonia solani* and *Sclerotinia minor* disease suppression. Plant Pathol 71(4):818–829. https://doi.org/10.1111/ppa.13514

Botero-Ramirez A, Kirk B, Strelkov SE (2024) Optimizing Clubroot management and the role of canola cultivar mixtures. Pathogens 13(8):640. https://doi.org/10.3390/pathogens13080640

Brachaczek A, Kaczmarek J, Jedryczka M (2016) Monitoring blackleg (*Leptosphaeria spp.*) ascospore release timing and quantity enables optimal fungicide application to improved oilseed rape yield and seed quality. Eur J Plant Pathol 145:643–657. https://doi.org/10.1007/s10658-016-0922-x

Bulgarelli D, Garrido-Oter R, Münch PC, Weiman A, Dröge J, Pan Y, Schulze-Lefert P (2015) Structure and function of the bacterial root microbiota in wild and domesticated barley. Cell Host Microbe 17(3):392–403

Cornelsen JE, Ort NW, Gabert RK, Epp I, Rempel CB (2024) Current and potential pest threats for canola in the Canadian prairies. Pest Manag Sci 80(5):2220–2234. https://doi.org/10.1002/ps.7858

Cruz-Hernández MA, Mendoza-Herrera A, Bocanegra-García V, Rivera G (2022) *Azospirillum spp.* from plant growth-promoting bacteria to their use in bioremediation. Microorganisms 10(5):1057. https://doi.org/10.3390/microorganisms10051057

Deep SWATI, Sharma PRATIBHA (2012) Host age as predisposing factor for incidence of black leaf spot of cauliflower caused by *Alternaria brassicae* and *Alternaria brassicicola.* Indian Phytopathol 65(1):71–75

Dimkić I, Janakiev T, Petrović M, Degrassi G, Fira D (2022) Plant-associated *bacillus* and *pseudomonas antimicrobial* activities in plant disease suppression via biological control mechanisms-a review. Physiol Mol Plant Pathol 117:101754. https://doi.org/10.1016/j.pmpp.2021.101754

Dixon GR (2009) The occurrence and economic impact of *Plasmodiophora brassicae* and clubroot disease. J Plant Growth Regul 28(3):194–202. https://doi.org/10.1007/s00344-009-9090-y

Doddavarapu B, Lata C, Shah JM (2024) Epigenetic regulation influenced by soil microbiota and nutrients: paving road to epigenome editing in plants. Biochim Biophys Acta 1868(4):130580. https://doi.org/10.1016/j.bbagen.2024.130580

Egamberdieva D, Wirth SJ, Alqarawi AA et al (2017) Phytohormones and beneficial microbes: essential components for plants to balance stress and fitness. Front Microbiol 8:2104. https://doi.org/10.3389/fmicb.2017.02104

Enders L, French E, Kjeldgaard M, Tronson E, Kaplan I (2024) Plant-guided microbiome selection produces transient effects on insect performance and rhizosphere community assembly. Phytobiomes J 8(4):484–498. https://doi.org/10.1094/PBIOMES-01-24-0008-R

Etesami H, Jeong BR, Glick BR (2021) Contribution of arbuscular mycorrhizal fungi, phosphate–solubilizing bacteria, and silicon to P uptake by plant. Front Plant Sci 12:699618

Fitt BD, Brun H, Barbetti MJ, Rimmer SR (2006) World-wide importance of phoma stem canker (*Leptosphaeria maculans* and L. *biglobosa*) on oilseed rape (*Brassica napus*). In Sustainable strategies for managing *Brassica napus* (oilseed rape) resistance to Leptosphaeria maculans (phoma stem canker) Under the aegis of a European-Australian workshop held at INRA, Versailles, France, in association with the EU SECURE project (QLK5-CT-2002-01813), pp 3–15. https://doi.org/10.1007/1-4020-4525-5_1

Geoffrey RDV (2024) Pests and pathogens. In: Vegetable *brassicas* and related crucifers, vol 40, p 295

Guha A (2025) Common diseases of field crops and their management. Educohack Press

Hanschen FS, Winkelmann T (2020) Biofumigation for fighting replant disease-a review. Agronomy 10(3):425

Hirano SS, Upper CD (2000) Bacteria in the leaf ecosystem with emphasis on *pseudomonas syringae* a pathogen, ice nucleus, and epiphyte. Microbiol Mol Biol Rev 64(3):624–653. https://doi.org/10.1128/mmbr.64.3.624-653.2000

Hossain MT, Akhter MS, Islam MM, Faruk MI, Chung YR (2024) Cross-talks about hemi biotrophic-necrotrophic pathogens by endophytic bacillus-based EMOs. In: Plant Endophy and second Metaboli, pp 235–253. https://doi.org/10.1016/B978-0-443-13365-7.00002-6

Hrmova M, Hussain SS (2021) Plant transcription factors involved in drought and associated stresses. Int J Mol Sci 22(11):5662. https://doi.org/10.3390/ijms22115662

Jalal-Ud-Din S, Elahi NN, Mubeen F (2024) Significance of zinc-solubilizing plant growth-promoting rhizobacterial strains in nutrient acquisition, enhancement of growth, yield, and oil content of canola (*Brassica napus* L.). Front Microbiol 15:1446064. https://doi.org/10.3389/fmicb.2024.1446064

Jawadagi R, Hadimani H, Evoor S, Kishorkumar GK (2024) Vegetable Science Frontiers: Insights of Olericulture. SIP

Karnwal A, Kumar G, Mahmoud AED, Dutta J, Singh R, Al-Tawaha ARMS, Malik T (2025) Eco-engineered remediation: microbial and rhizosphere-based strategies for heavy metal detoxification. Curr Res Biotechnol 100297. https://doi.org/10.1016/j.crbiot.2025.100297

Kavamura V N, Mendes R, Bargaz A, Mauchline T H (2021) Defining the wheat microbiome: Towards microbiome-facilitated crop production. Comput Struct Biotechnol J, 19:1200–1213.https://doi.org/10.1016/j.csbj.2021.01.045

Kaur N, Dutta B (2024) Characterization of seed-to-seedling transmission of *Alternaria brassicicola* in broccoli. Plant Dis 108(7):2046–2052. https://doi.org/10.1094/PDIS-10-23-2002-RE

Khoshru B, Mitra D, Joshi K, Adhikari P et al (2023) Decrypting the multi-functional biological activators and inducers of defense responses against biotic stresses in plants. Heliyon 9(3):e13825. https://doi.org/10.1016/j.heliyon.2023.e13825

Khoso MA, Wagan S, Alam I, Hussain A, Ali Q, Saha S, Liu F (2024) Impact of plant growth-promoting rhizobacteria (PGPR) on plant nutrition and root characteristics: current perspective. Plant Stress 11:100341. https://doi.org/10.1016/j.stress.2023.100341

Kora C, McDonald MR, Boland GJ (2008) New progress in the integrated management of Sclerotinia rot of carrot. In: Integrated management of diseases caused by fungi, phytoplasma and bacteria, pp 243–270

Kozaeva E, Eida AA, Gunady EF, Dangl JL, Conway JM, Brophy JA (2024) Roots of synthetic ecology: microbes that foster plant resilience in the changing climate. Curr Opin Biotechnol 88:103172. https://doi.org/10.1016/j.copbio.2024.103172

Laiq M, Azeem MA, Begum S, Nazir N, Ahmad Z, Ullah S, Ullah A (2024) Compositional shifts in rhizobacterial communities of *Brassica napus* under salinity stress. Appl Soil Ecol 195:105258. https://doi.org/10.1016/j.apsoil.2023.105258

Liu Z, Wang H, Wang J, Lv J, Xie B, Luo S, Yu J (2022) Physical, chemical, and biological control of black rot of *brassicaceae* vegetables: a review. Front Microbiol 13:1023826. https://doi.org/10.3389/fmicb.2022.1023826

Ludwig-Müller J (2024) Clubroot disease (*Plasmodiophora brassicae*) as a threat to worldwide rapeseed or Canola (*Brassica napus*). Plant Health Cases:phcs20240021. https://doi.org/10.1079/planthealthcases.2024.002

Mao S, Kim YH, Sahu N, Kim SW, Kim HT, Watanabe M, Park JI (2025) Molecular marker development for specific amplification of *Xanthomonas campestris* pv. *campestris* race 8 causing black rot disease in *brassica* crops. JGPP 91(1):31–40. https://doi.org/10.1007/s10327-024-01210-3

Martín-Cardoso H, González-Miguel VM, Soler-López L, Campo S, San SB (2024) Association of microbiome diversity with disease symptoms in *Brassica oleracea* leaves. Horticulture 10(7):765. https://doi.org/10.3390/horticulturae10070765

Meena PD, Mehta N, Saharan GS (2021) Minor pathogens: a worldwide challenge to cultivation of *crucifers*. Agric Res J Agric 58(4). https://doi.org/10.5958/2395-146X.2021.00081.8

Meng S, Yan X, Piao Y, Li S, Wang X, Jiang J, Pang W (2024) Multiple transcription factors involved in the response of Chinese cabbage against *Plasmodiophora brassicae*. Front Plant Sci 15:1391173. https://doi.org/10.3389/fpls.2024.1391173

Mishra A, Khan NA, Jha RK, Murugesh T, Singh A (2024) Study of resistance mechanism of Alternaria blight (*Alternaria brassicicola*) by biochemical markers in Indian mustard (*Brassica juncea* L. Czern. & Coss.). Front Plant Sci 15:1420197. https://doi.org/10.3389/fpls.2024.1420197

Narayanasamy P (2024) Phyllosphere microbial plant pathogens: detection and crop disease management: volume 2 management of crop diseases. CRC Press

Natsiopoulos D, Topalidou E, Mantzoukas S, Eliopoulos PA (2024) *Endophytic Trichoderma*: potential and prospects for plant health management. Pathog 13(7):548. https://doi.org/10.3390/pathogens13070548

Noh E, Cedeno C, Bull CT, Keinath AP, Wechter WP, Wang H (2025) A quantitative PCR assay for specific detection of *Pseudomonas cannabina* pv. alisalensis. Plant Dis 109(2):461–470. https://doi.org/10.1094/PDIS-06-24-1217-RE

Obrien F, Mary J (2016) The role of the rhizosphere microbial community in plant chemistry and aphid herbivory in *Brassica oleracea*. Doctoral dissertation, University of Southampton

Pan L, Cai B (2023) Phosphate-solubilizing bacteria: advances in their physiology, molecular mechanisms and microbial community effects. Microbe 11(12):2904. https://doi.org/10.3390/microorganisms11122904

Patani A, Patel M, Islam S, Yadav VK, Prajapati D, Yadav AN, Patel A (2024) Recent advances in *bacillus*-mediated plant growth enhancement: a paradigm shift in redefining crop resilience. World J Microbiol Biotechnol 40(2):77. https://doi.org/10.1007/s11274-024-03903-5

Pathak M, Patel S, Some S (2024) Climate change mitigation and sustainable development goals: evidence and research gaps. PLOS Clim 3(3):e0000366. https://doi.org/10.1371/journal.pclm.0000366

Poveda J, Hermosa R, Monte E, Nicolás C (2019) *Trichoderma harzianum* favours the access of arbuscular mycorrhizal fungi to non-host *Brassicaceae* roots and increases plant productivity. Sci Rep 9(1):11650. https://doi.org/10.1038/s41598-019-48269-z

Rad AK, Zarei M, Astaikina A, Streletskii R, Etesami H (2022) Effects of microbial inoculants on growth, yield, and fruit quality under stress conditions. In: Sustainable horticulture. Academic Press, pp 1–38

Rathore R, Dowling DN, Forristal PD, Spink J, Cotter PD, Bulgarelli D, Germaine KJ (2017) Crop establishment practices are a driver of the plant microbiota in winter oilseed rape (*Brassica napus*). Front Microbiol 8:1489. https://doi.org/10.3389/fmicb.2017.01489

Rauf A, Mehmood MA, Ashfaq M, Kayani SB, Mushtaq I, Javeed S (2024) Effects of climate change on bacterial and viral pathogens. In: Plant quarantine challenges under climate change anxiety, pp 227–279. https://doi.org/10.1007/978-3-031-56011-8_8

Rout GR, Sahoo S (2015) Role of iron in plant growth and metabolism. Rev Agric Sci 3:1–24. https://doi.org/10.7831/ras.3.1

Sachdev S, Ansari SA, Ansari MI, Fujita M, Hasanuzzaman M (2021) Abiotic stress and reactive oxygen species: generation, signaling, and defense mechanisms. Antioxidants 10(2):277. https://doi.org/10.3390/antiox10020277

Sah S, Krishnani S, Singh R (2021) *Pseudomonas* mediated nutritional and growth promotional activities for sustainable food security. Curr Res Microb Sci 2:100084. https://doi.org/10.1016/j.crmicr.2021.100084

Saleem S, Ragasova LN, Tekielska D, Fidurski M, Sekara A, Pokluda R (2023) Serendipita indica as a plant growth promoter and biocontrol agent against black rot disease in cabbage grown in a phytotron. Agriculture (Basel) 13(11):2048

Samreen T, Naveed M, Nazir MZ, Asghar HN, Khan MI, Zahir ZA, Choudhary M (2021) Seed associated bacterial and fungal endophytes: diversity, life cycle, transmission, and application potential. Appl Soil Ecol 168:104191. https://doi.org/10.1016/j.apsoil.2021.104191

Sánchez-Tovar MR, Rivera-Bustamante RF, Saavedra-Trejo DL, Guevara-González RG, Torres-Pacheco I (2025) Mixed plant viral infections: complementation, interference and their effects, a review. Agron 15(3):620. https://doi.org/10.3390/agronomy15030620

Selim HM, Gomaa NM, Essa AM (2017) Application of endophytic bacteria for the biocontrol of *Rhizoctonia solani* (Cantharellales: ceratobasidiaceae) damping-off disease in cotton seedlings. Biocontrol Sci Tech 27(1):81–95. https://doi.org/10.1080/09583157.2016.1258452

Sheoran AR, Lakra N, Saharan BS, Luhach A, Kumar R, Seth CS, Duhan JS (2025) Enhancing plant disease resistance: insights from biocontrol agent strategies. J Plant Growth Regul 44(2):436–459. https://doi.org/10.1007/s00344-024-11480-y

Shevchenko O, Snihur H, Shevchenko T, Budzanivska I (2024) Biological and anthropogenic risk factors favoring persistence and spread of harmful plant viruses. In: Detection and management of new and emerging mystery plant virus sources, pp 275–303

Singh J, Singh V, Dutt V, Walia N, Kumawat G, Jakhar ML, Yadava DK, Sharma PC (2022) Insights into salt tolerance of mustard (Brassica juncea L. Czern & Coss): a metabolomics perspective. Environ Exp Bot 194:104760

Tabacchioni S, Passato S, Ambrosino P, Huang L, Caldara M, Cantale C, Bevivino A (2021) Identification of beneficial microbial consortia and bioactive compounds with potential as plant bio stimulants for a sustainable agriculture. Microbe 9(2):426. https://doi.org/10.3390/microorganisms9020426

Tariq Ahmad L, Quseen Mushtaq R, Jan N (2025) Bacterial pathogens of the oilseed crops: loss of yield and crop management. In: Oilseed crops, pp 91–111

Thokchom S, Malakar P, Devi RT (2020) Endophytic bacteria associated with cultivars of *Brassica juncea* and their antagonistic potentials against *Alternaria brassicae*. IJCS 8(5):1147–1150. https://doi.org/10.22271/chemi.2020.v8.i5p.10449

Tomita S, Inaoka T, Endo A, Okada S (2024) Raw material–dependent changes in bacterial and compositional profiles are involved in insufficient pH decrease in natural lactic fermentation of *Brassica rapa* leaves. Food Chem 437:137934. https://doi.org/10.1016/j.foodchem.2023.137934

Ton L (2024) Establishment of a CRISPR/CasRx system to modify Turnip mosaic virus (TuMV) resistance in *Brassica napus*

Trono D, Pecchioni N (2022) Candidate genes associated with abiotic stress response in plants as tools to engineer tolerance to drought, salinity and extreme temperatures in wheat: an overview. Plants 11(23):3358. https://doi.org/10.3390/plants11233358

Truyens S, Weyens N, Cuypers A, Vangronsveld J (2015) Bacterial seed endophytes: genera, vertical transmission and interaction with plants. Environ Microbiol Rep, 7(1):40–50. https://doi.org/10.1111/1758-2229.12181

Tsytsiura Y (2023) Viral diseases of cruciferous plants: prevalence, biology and development cycle, effective control tactics. Publishing House “Baltija Publishing

Upadhyay SK, Srivastava AK, Rajput VD, Chauhan PK, Bhojiya AA, Jain D, Minkina T (2022) Root exudates: mechanistic insight of plant growth promoting rhizobacteria for sustainable crop production. Front Microbiol 13:916488. https://doi.org/10.3389/fmicb.2022.916488

Vorholt J (2012) Microbial life in the phyllosphere. Nat Rev Microbiol 10:828–840. https://doi.org/10.1038/nrmicro2910

Wang Y, He X, Yu F (2022) Non-host plants: are they mycorrhizal networks players? Plant Divers 44(2):127–134. https://doi.org/10.1016/j.pld.2021.06.005

Wang Z, Huang YH, He M, Hu Y, Xu X, Cai QY, Lü H (2024) Root-associated bacterial communities of vegetable *Brassica* parachinensis enrich pollutant-degrading taxa and functions for enhancing phthalate dissipation. Appl Soil Ecol 202:105617. https://doi.org/10.1016/j.apsoil.2024.105617

Wang YZ, Zhao HM, Huang XP, Zhang Y, Ye JC, Feng NX, Li QX (2025) Variety-dependent seed endophytic bacteria enhance stress tolerance to and bioaccumulation of ciprofloxacin in choy sum (*Brassica parachinensis*). Microbiome 13(1):80. https://doi.org/10.1186/s40168-025-02073-2

Yaghoubi Khanghahi M, Strafella S, Filannino P, Minervini F, Crecchio C (2024) Importance of lactic acid bacteria as an emerging group of plant growth-promoting rhizobacteria in sustainable agroecosystems. Appl Sci 14(5):1798. https://doi.org/10.3390/app14051798

Yao Y, Liu C, Zhang Y, Lin Y, Chen T, Xie J, Jiang D (2024) The dynamic changes of *brassica Napus* seed microbiota across the entire seed life in the field. Plants 13(6):912. https://doi.org/10.3390/plants13060912

Zandi P, Schnug E (2022) Reactive oxygen species, antioxidant responses and implications from a microbial modulation perspective. Biol (Basel) 11(2):155. https://doi.org/10.3390/biology11020155

Zhang J, Ahmed W, Dai Z, Zhou X, He Z, Wei L, Ji G (2022) Microbial consortia: an engineering tool to suppress clubroot of Chinese cabbage by changing the rhizosphere bacterial community composition. Biol 11(6):918. https://doi.org/10.3390/biology11060918

Zhang Y, Ma N, Tan P, Ma X (2024) Quorum sensing mediates gut bacterial communication and host-microbiota interaction. Crit Rev Food Sci Nutr 64(12):3751–3763. https://doi.org/10.1080/10408398.2022.2134981

Application of Machine Learning Models for Drought Prediction with Special Reference to *Brassica* Crops

Geeta Sharma and Jyoti Gupta

Abstract

Agriculture is significantly impacted by drought because of the lack of forage and water, which reduces crop yields and impacts livestock. Due to the spatio-temporal variability, it is challenging to predict the occurrence and extent of the drought. Accurate forecasting of drought conditions is crucial for the sustainable cultivation of Brassica crops, as it empowers farmers to make well-informed decisions regarding water management and variety selection to mitigate the negative impacts of drought. Utilizing historical data and analyzing intricate datasets to uncover complex patterns, machine learning (ML) algorithms have proven to be a powerful tool in predicting droughts with high accuracy. ML algorithms have been used by various researchers in order to predict drought conditions at particular places. This is further helpful in making proactive measures for drought mitigation and resource management. This chapter performs an extensive review of popular machine learning algorithms for the effective prediction of drought situations.

Keywords

Agriculture · *Brassica* · Drought · Machine learning · Prediction

G. Sharma (✉)
School of Computer Application, Lovely Professional University, Phagwara, Punjab, India

J. Gupta
Department of Education, Lovely Professional University, Phagwara, Punjab, India

A. Kumar, B. Koul (eds.), *Crop Improvement Strategies in Brassica species: Applied Science*, https://doi.org/10.1007/978-981-95-3861-4_10

1 Introduction

Agriculture forms the backbone of every economy, ensuring food security, supplying essential raw materials, and generating employment opportunities. The growth of an economy is often closely linked to the productivity of its agricultural sector. The agricultural sector is heavily reliant on water as a fundamental resource for crop production. Water, as a vital input for agricultural production, plays a crucial role in driving economic development. It is essential for various physiological processes in plants, including germination, nutrient uptake, photosynthesis, and transpiration. Brassica crops, such as cabbage, broccoli, and cauliflower, which form an important part of the agricultural landscape, are highly water-demanding during their growth stages. These plants require an adequate water supply to maintain optimal growth rates, develop healthy leaves, and produce high-quality yields. Water scarcity can significantly limit the productivity of Brassica crops, leading to reduced yields, stunted growth, and increased susceptibility to diseases and pests. Adequate water supply is essential for their optimal growth and yield, ensuring the economic viability of Brassica cultivation and contributing to the overall prosperity of the region. Ensuring a reliable and sustainable water supply is crucial for the successful cultivation of Brassica crops and the overall health of the agricultural sector.

Droughts pose a significant obstacle within agricultural systems; droughts have both economic and environmental repercussions on agricultural systems. Droughts not only deprive crops and soils of vital precipitation but also elevate the salt content in soils and irrigation systems. Every year, approximately 55 million people are affected by the impacts of droughts. Furthermore, water scarcity affects approximately 40% of the global population, and a concerning projection indicates that by 2030, up to 700 million people could be displaced due to factors related to drought (World Health Organization n.d.). The increasing temperatures driven by climate change are exacerbating the aridity of already dry areas while intensifying the precipitation in wet regions. In regions with limited water availability, rising temperatures increase evaporation rates, making regions more prone to drought or extending dry periods. Disturbingly, floods, droughts, tropical cyclones, heat waves, and severe storms have accounted for a substantial portion (between 80% and 90%) of documented natural disasters over the past decade. These climate-related events underscore the significant impacts they have on human lives, infrastructure, and the environment (World Health Organization n.d.). Drought has significant global implications for agriculture, impacting crop production, livestock, and overall food security.

Major global implications of drought are:

- *Adverse impact on ecosystems*: Loss of biodiversity, increased forest fires, and soil degradation.
- *Adverse impact on socioeconomic factors*: Financial losses, population displacement, and social tensions.
- *Adverse impact on climate*: Higher emissions of greenhouse gases and changes in albedo.

- *Adverse impact on global trade and supply chains*: Interruptions and reliance on imported goods.
- *Adverse impact on water resource management*: Competition for water resources and strain on infrastructure.
- *Far-reaching implications of drought on agriculture*: Drought has a significant impact on agriculture, leading to lower crop yields and compromised soil quality, while also increasing the need for irrigation and raising the likelihood of pest infestations and diseases. Economic repercussions include reduced incomes for farmers and higher food prices. Additionally, livestock health deteriorates due to a scarcity of feed and water. If drought conditions persist, the long-term sustainability of farming may be threatened, necessitating changes in agricultural practices and land utilization. These issues emphasize the critical importance of developing adaptive strategies to strengthen the resilience of agricultural systems.

The spatiotemporal variability of drought makes it difficult to accurately forecast its occurrence and extent. The effective drought prediction algorithms and the integration of advanced technologies and data sources are vital for improving our understanding and ability to forecast drought events accurately. Drought prediction has witnessed a notable shift towards the utilization of ML algorithms. These algorithms play a crucial role in leveraging historical data, analyzing complex datasets with multiple dimensions, and uncovering intricate patterns. The ML algorithms, including "Support Vector Machine (SVM), Adaptive Neuro-Fuzzy Inference System (ANFIS), Random Forest (RF), Decision Trees (DT), and Multivariate Adaptive Regression Spline (MARS)," etc., have been utilized individually or in combination. These algorithms incorporate different drought indices, such as the standardized precipitation index, to enhance their predictive capabilities (Yaseen and Shahid 2021).

ML offers advanced techniques to forecast and mitigate the impacts of water scarcity. By examining a range of datasets, including meteorological information, soil moisture levels, vegetation health indicators, and historical drought patterns, these algorithms can identify subtle cues and complex connections often overlooked by traditional methods. Their capacity to analyze extensive data sets enables more precise and timely predictions, empowering decision-makers to implement proactive strategies such as water conservation measures, crop diversification, and resource management. Furthermore, machine learning algorithms contribute to enhancing our comprehension of intricate drought dynamics, aiding in the development of resilient solutions to tackle this urgent global issue.

1.1 Objective of the Study

The information contained in this chapter was obtained by conducting a comprehensive search and collection process using the Scopus source database. The search criteria included the topics "Machine Learning," "Prediction," and "Drought," with a specific focus on articles published between 2009 and 2023. This timeframe was chosen to ensure a thorough retrieval of relevant data for analysis.

1.2 Organization of the Chapter

The chapter is structured in the following manner. Section 2 discusses a review of related literature. Section 3 discusses important drought indices along with merits and limitations. Section 4 presents the discussion on popular machine learning algorithms. The key metrics used to assess the performance of drought forecasting models are discussed in Sect. 5. Section 6 presents the impact of machine learning algorithms on drought prediction. Popular machine learning algorithms used for drought prediction are discussed in Sect. 7, and their comparative analysis is presented in Sect. 8. Section 9 concludes the chapter.

2 Review of Related Literature

In Rezk et al. (2021), to predict drought occurrences and crop productivity in the farming system, researchers have explored the application of multiple classification techniques. These techniques, including random forest, artificial neural network, and k-nearest neighbor, have been employed on five datasets collected from agricultural settings. In the realm of drought classification and crop productivity prediction, the WPART method demonstrates exceptional accuracy, surpassing existing state-of-the-art algorithms. Impressively, WPART achieves accuracy levels of up to 92.51%, 96.77%, 98.04%, 96.12%, and 98.15% across the five datasets, outperforming other techniques in both drought classification and crop productivity estimation. In Belayneh and Adamowski (2013), this study examined "ANN, SVR Coupled Wavelet-ANNs," which utilize wavelet analysis (WA) as a preprocessing step for input data. The performance of these three methods was evaluated using two metrics, namely "Root Mean Square Error" and R^2 coefficient. The forecast results of the study revealed that the "Coupled Wavelet Neural Network (WA-ANN)" models outperformed the other methods in accurately predicting "SPI 3 (3-month SPI) and SPI 6 (6-month SPI)" values within lead times of 1 and 3 months in the "Awash River Basin in Ethiopia." Bacanli et al. (2009) compared ANFIS with FFNN, MLR, and ANFIS and reported that ANFIS exhibited superior accuracy and reliability in comparison to the other models evaluated.

Dastorani et al. (2010) tested RN, TLRN, and ANFIS in order to forecast precipitation values with a lead time of 12 months and revealed that both TLRN and ANFIS methods exhibited similar levels of performance and accuracy. Zhang et al. (2019) used DLNM, an "artificial neural network (ANN)" algorithm, and an XGBoost algorithm were utilized and compared for predicting the "Standardized Precipitation Evapotranspiration Index (SPEI)" with lead times ranging from 1 to 6 months in advance. The XGBoost algorithm outperformed DLNM in accurately forecasting SPEI values within these lead time intervals. Rahmati et al. (2020) used the cutting-edge machine learning algorithms, viz. "CART, BRT RF, MARS, FDA, and SVM." Among the algorithms evaluated, the RF algorithm achieved the highest level of accuracy, and the FDA algorithm demonstrated the lowest level of performance. Khan et al. (2020) tested three major ML techniques, viz. "SVM, ANN, and KNN" and reported that, as compared to "ANN and KNN," "SVM-based" algorithms exhibited superior capability in capturing the temporal and spatial characteristics of droughts across Pakistan. Another study by Malik et al. (2019) was conducted to forecast hydrological drought by employing three different approaches: "CANFIS, MLPNN, and MLR" at "Naula and Kedar stations," Uttarakhand, India. Upon analyzing the data, it was determined that the CANFIS algorithms outperformed the other algorithms in predicting hydrological drought across different time scales. Specifically, the CANFIS algorithms exhibited the best performance at "3-, 6-, and 12-month time scales for the Naula station," and at "1-, 3-, 12-, and 24-month time scales for the Kedar station."

On the other hand, the MLPNN algorithms showed superior performance at "1- and 9-month time scales for the Naula station," as well as at "6- and 9-month time scales for the Kedar station." Additionally, the "MLR algorithm" demonstrated its effectiveness solely at the "24-month time scale for the Naula station." Adnan et al. (2023) proposed that the prediction accuracy of two recently developed heuristic methods, namely "OP-ELM and DENFIS," is evaluated for drought modeling and compared with "MARS," and reported that the "DENFIS" model exhibited higher accuracy compared to both the "OP-ELM and MARS" models. Abbes et al. (2023) proposed a novel approach, named "MRA-WT-LSTM," for drought forecasting, which combines the strengths of "LSTM and MRA-WT." The proposed methodology was compared to three ML methods: "ANN, SVR, RFR," as well as "LSTM," and the results demonstrated the effectiveness of the "MRA-WT-LSTM" approach for drought forecasting. In Danandeh et al. (2023a, b), the researchers evaluated the performance of the "convolutional long short-term memory (CNN-LSTM)" hybrid model in the specific domain of short-term meteorological drought forecasting, focusing on standardized precipitation evapotranspiration data for both three-month and six-month periods in Ankara province, Turkey. The findings revealed that the CNN-LSTM model surpassed all the benchmark models in terms of performance.

In the study (Karbasi et al. 2023), the forecast of the SPEI was accomplished by employing a combination of various techniques, including the "empirical wavelet transform (EWT)," "discrete wavelet transforms (DWT)," "extended Kalman filter (EKF)," two models of "multilayer perceptrons (MLP)," and "group method of data handling (GMDH)" neural networks. The results revealed that the combination of "EWT and DWT" outperformed the standalone models. When extending the forecasting interval from 1 month to 6 months, the effectiveness of "EWT" became more prominent compared to "DWT." Additionally, the best performance in forecasting the SPEI drought index was achieved by combining the "EWT" with the "MLP-EKF" model.

In study (Acharki et al. 2023) the researchers used "random subspace, random forest, M5P, and REPTree" to predict "spatio-temporal distribution" and prediction of agricultural and meteorological drought. The findings indicated that the "REPTree algorithm" exhibited the most favorable performance during both the testing and validation phases. Conversely, the "Random Forest (RF)" algorithm was observed to have the lowest accuracy in terms of performance. In Piri et al. (2023) study, the potential of "artificial neural network (ANN)" and "support vector regression (SVR)," was examined for predicting "standardized precipitation index (SPI)" meteorological drought indices. Additionally, two selected algorithms, "particle swarm optimization (SVR-PSO)" and "response surface method (SVR-RSM)," were utilized to optimize the performance of these algorithms. The results revealed that the "SVR-RSM" algorithm exhibited superior accuracy and predictive performance compared to other models. Additionally, the "SVR-PSO" algorithm outperformed the "ANN and SVR" algorithms in terms of performance. In Vo et al. (2023), for the purpose of drought prediction, a hybrid model named "LSTM-CM" is developed in this study, combining "long short-term memory (LSTM)" with a climate model. In this study, the performance of the "LSTM-CM" hybrid model was compared with the standalone "long short-term model (LSTM-SA)" and the climate prediction model "GloSea5 (GS5)." The findings showed that by incorporating the low-bias characteristic of "LSTM-SA" and the physical process simulation capability of "GS5," the "LSTM-CM" hybrid model achieved improved drought predictions. As a result, it inherited the strengths of both models while mitigating their weaknesses.

To forecast the condition of drought, the article proposes two innovative hybrid machine learning (ML) algorithms: "Extreme Learning Machine with Water Cycle Algorithm (ELM-WCA)" and "Extreme Learning Machine with Bacterial Foraging Optimization (ELM-BFO)." These novel algorithms are specifically designed to forecast the standardized precipitation evapotranspiration index (SPEI) at the given location. The forecasting outcomes for both three- and six-month accumulation periods demonstrated that the "ELM-WCA" algorithm outperformed its counterparts, exhibiting superior performance in predicting the desired variables (Danandeh Mehr et al. 2023a, b). The primary objective of the chapter (Ham et al. 2023) is to validate the superiority of the "wavelet long short-term memory network (WLSTMN)" in comparison to "wavelet artificial neural network (WANN)" and "wavelet support vector regression (WSVR)" for drought forecasting in the western region of the Democratic People's Republic of Korea using "Standardized Precipitation Index" with timescales of "6 and 12 months (SPI-6 and SPI-12)" was utilized. The "WLSTMN" model for predicting drought at two different timescales exhibited higher values of "R^2 and LCCC," along with lower values of "RMSE and MAE," indicating its improved predictive accuracy.

In Elbeltagi et al. (2023), the objective of the study was to assess the viability and efficacy of the "Random Subspace (RSS)" algorithm, both independently and in combination with the "M5 Pruning tree (M5P)," "Random Forest (RF)," and "Random Tree (RT)" algorithms, for estimating the "Standardized Precipitation Index (SPI)" at 3, 6, and 12-month intervals between 2000 and 2019. The findings of the study revealed that the "RSS-M5P" algorithm exhibited the highest level of accuracy in predicting the "Standardized Precipitation Index (SPI)" compared to the "RSS" algorithm alone, as well as the "RSS-RF and RSS-RT" algorithms. These

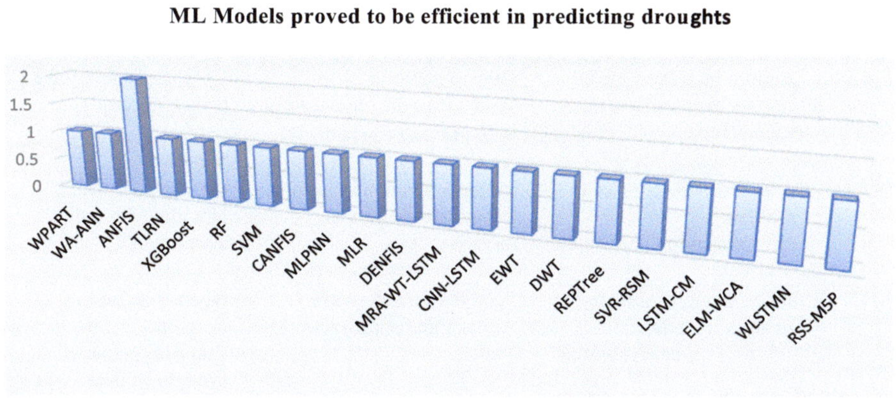

Fig. 1 Machine learning algorithms proved to be efficient in the prediction of droughts

results were particularly relevant for assessing drought conditions in the Jaisalmer region of Rajasthan. Figure 1 depicts prominent machine learning algorithms useful in the prediction of drought.

3 Important Drought Indices: An Overview

Drought indices are critical tools for quantifying and monitoring drought conditions across different regions and time scales. They provide a standardized framework to assess the severity, duration, and spatial extent of droughts, helping researchers, policymakers, and stakeholders make informed decisions. Each index is intended to address specific types of drought, such as meteorological, agricultural, hydrological, or socioeconomic, and incorporates various key climatic and environmental variables, including precipitation patterns, thermal fluctuations, soil hydrology, etc. In drought monitoring, the most widely used metrics are: Standardized Precipitation Index (SPI) for rainfall analysis, Palmer Drought Severity Index (PDSI) for long-term soil moisture, Standardized Precipitation Evapotranspiration Index (SPEI) that includes temperature effects, Crop Moisture Index (CMI) for agricultural needs, and Normalized Difference Vegetation Index (NDVI) for vegetation health assessment. Table 1 shows widely used drought indices with a detailed comparison of their purpose, advantages, and limitations.

4 Machine Learning Models for Drought Forecasting

ML algorithms now play a pivotal role in drought forecasting by detecting intricate patterns within massive climate datasets. These models process historical weather and hydrological information to identify predictive signatures for future drought events. Commonly deployed techniques range from deep learning frameworks and

Table 1 Drought indices comparison

Index	Purpose	Advantages	Limitations
SPI	Monitors meteorological droughts over various time scales	Easy to compute Works for both short and long periods	Only uses rainfall data Doesn't consider temperature or soil moisture
PDSI	Evaluates long-term water shortages Helps in farming and water management	Incorporates precipitation, temperature, and soil moisture; Ideal for extended periods	Hard to calculate Not great for quick assessments
SPEI	Studies droughts affected by temperature changes Useful for climate research	Factors in temperature Adjustable for different time frames	Needs precise evaporation data Can exaggerate drought impact
CMI	Monitors farm-related water shortages Aids crop planning	Focuses on soil moisture levels Gives timely farming insights	Only for short-term use Not for other drought types
NDVI	Checks plant health Monitors farms and ecosystems	Provides real-time, wide-area data Great for crops and nature	Indirect drought indicator Affected by pests or land changes

support vector machines to decision-tree ensembles and memory-based algorithms are commonly employed for drought prediction.

(a) *Deep Learning Models (DL):* Deep learning is a specialized branch of machine learning, employs hierarchical neural architectures to automatically learn high-level features and capture intricate non-linear dependencies within datasets. These models have demonstrated impressive results in drought prediction by capturing intricate dependencies. Convolutional Neural Networks (CNNs) are frequently used in image processing and are adapted to analyze time series data for drought forecasting. CNNs automatically identify spatial and temporal patterns, making them computationally efficient for tasks such as satellite imagery analysis. Additionally, Recurrent Neural Networks (RNNs), particularly Long Short-Term Memory (LSTM) networks, are ideal for handling time series data, capturing long-term dependencies crucial for drought forecasting. Despite their potential, the successful implementation of these models requires both substantial computational infrastructure and extensive training datasets to ensure robust pattern recognition and predictive accuracy, but they have shown superior performance in some cases compared to traditional models.

(b) *Support Vector Machines (SVM)*: Support Vector Machines (SVM) are supervised learning algorithms used for classification and regression tasks. SVMs operate by identifying a hyperplane that optimally separates data points into different classes or predicts continuous values. They have demonstrated strong accuracy in drought prediction when combined with various kernel functions, such as radial basis and polynomial kernels. SVMs have also been enhanced with preprocessing techniques like Singular Spectrum Analysis (SSA) for time series data, improving prediction accuracy. Studies in Eastern Australia have validated LSSVM's effectiveness for SPI-based drought prediction, with the algorithm demonstrating superior performance in capturing rainfall deficiency patterns compared to traditional statistical methods.

(c) *Artificial Neural Networks (ANN)*: Artificial Neural Networks (ANN) simulate human brain function to recognize patterns and make predictions. These models are highly effective for forecasting drought conditions by learning from complex relationships within time series data. One key application of ANNs in drought forecasting is the development of the Nonlinear Aggregated Drought Index (NADI), where ANN models, including Recursive Multistep Neural Networks (RMSNN) and Direct Multistep Neural Networks (DMSNN), outperformed traditional approaches like ARIMA. ANNs are well-suited for handling large datasets that encompass meteorological, hydrological, and land-use data, making them invaluable for accurate drought prediction.

(d) *Random Forest (RF)*: Random Forest algorithm enhances prediction robustness by firstly training numerous decision trees on randomly sampled data subsets, then aggregating their outputs through consensus voting (classification) or regression. It is highly effective for drought prediction as it makes minimal assumptions about the data, enabling it to model complex, non-linear relationships without needing predefined error distributions. RF is especially useful when working with large and diverse datasets, such as satellite imagery, and can handle variables with different scales without requiring adjustments. Studies have shown that RF outperforms traditional models like ARIMA in both short- and long-term drought prediction by analyzing indices such as SPI and Soil Moisture Index (SMI).

(e) *K-Nearest Neighbors (KNN)*: K-Nearest Neighbors is an instance-based learning algorithm that performs classification through majority voting among proximate observations and regression via local averaging, using distance metrics to determine neighborhood similarity. In the context of drought prediction, KNN can be used to classify drought conditions by identifying similar meteorological or hydrological patterns from historical data.

(f) *Gradient Boosting Machines (GBM)*: Gradient Boosting Machines (GBM) is an ensemble method that builds trees sequentially, each one correcting the errors of its predecessor. This technique excels in drought forecasting through its capacity to capture intricate non-linear dependencies in environmental variables. This ensemble method demonstrates superior predictive accuracy relative to many alternatives, with inherent regularization mechanisms that mitigate over-fitting risks through sequential optimization of weak decision tree learners. GBM creates a strong predictive model that can be effective for forecasting droughts across various time scales.

(g) *Extreme Gradient Boosting (XGBoost)*: XGBoost has emerged as a leading ML framework due to its computationally efficient implementation of gradient boosted decision trees, innovative handling of sparse data patterns, and consistent outperformance of alternative algorithms in benchmark studies. It has

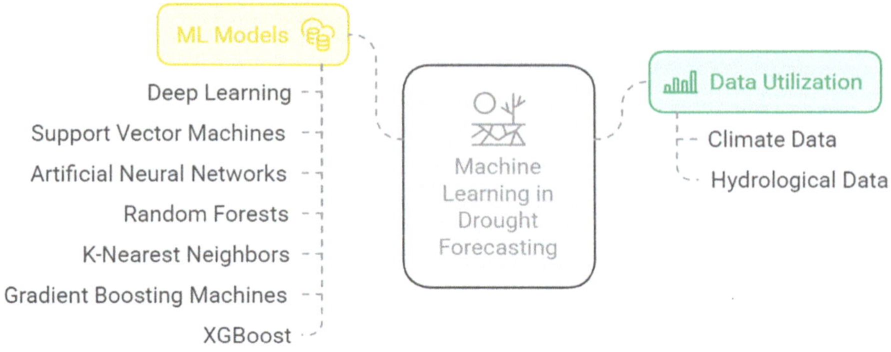

Fig. 2 Utilization of Machine Learning models in drought forecasting

proven efficient across environmental forecasting tasks, including drought prediction. XGBoost incorporates regularization techniques to prevent overfitting and utilizes advanced tree-building algorithms, making it particularly useful for large datasets with numerous features. It often outperforms traditional ML models, such as Random Forest and SVM, due to its improved accuracy and faster computation (Fig. 2).

5 Performance Metrics

In drought forecasting, various ML-based predictive modeling techniques are used to predict drought conditions with varying levels of accuracy. To determine how well these models perform, it is crucial to evaluate them using reliable metrics. These metrics help quantify the differences between predicted and actual values, providing insights into the model's ability to forecast drought events accurately. The selection of appropriate performance metrics is vital, as it directly impacts the interpretation of the model's effectiveness and the conclusions drawn from the analysis. The performance evaluation of drought forecasting models typically incorporates several critical metrics, which are discussed below. The notations used in the formula are shown in Table 2.

(a) *Root Mean Square Error (RMSE)*: RMSE is a widely used metric that calculates the average magnitude of the errors between predicted and observed values. It is particularly effective for understanding how spread out these errors are, as it places more weight on larger deviations. Lower RMSE values indicate greater predictive precision, with the magnitude directly reflecting the average discrepancy between modeled and observed drought indicators (Sundararajan et al. 2021). The formula for RMSE is:

$$\mathrm{RMSE} = \sqrt{\sum_{N}^{i=1} \frac{\left(Y_{\mathrm{FOR},i} - X_{\mathrm{OBS},i}\right)^2}{N}}$$

Table 2 Notations

Notation	Description
$Y_{\text{FOR},\,i}$	Predicted output
$X_{\text{OBS},\,i}$	Actual output
N	Number of data points
i	Individual data point

(b) *Mean Absolute Error (MAE)*: MAE quantifies the average magnitude of prediction errors by computing the absolute difference between modeled and observed values, without considering error direction. Unlike RMSE, MAE treats all errors equally, making it less sensitive to extreme outliers. This metric provides a clear and intuitive measure of the average error size. The formula for MAE is:

$$\text{MAE} = \frac{1}{N}\sum_{N}^{i=1} Y\left|_{\text{FOR},i} - X_{\text{OBS},i}\right|$$

(c) *Determination Coefficient (R^2)*: R^2 measures the proportion of variance in observed data explained by the model, where $R^2 = 1$ indicates perfect predictive capability and $R^2 = 0$ shows no improvement over mean-based prediction. In drought modeling, intermediate values (e.g., 0.6–0.8) typically indicate practically useful explanatory power (Sundararajan et al. 2021). The formula for R^2 is:

$$R^2 = \left(\frac{n\left(\sum xy\right) - \left(\sum x\right)\left(\sum y\right)}{\sqrt{\left[n\sum x^2 - \left(\sum x\right)^2\right]\left[n\sum y^2 - \left(\sum y\right)^2\right]}}\right)^2$$

(d) *Correlation Coefficient (r)*: The Pearson correlation coefficient (r) evaluates linear prediction-observation alignment, where $r = 1$ indicates perfect positive synchronization, $r = -1$ shows complete inverse correlation, and $r = 0$ reflects no linear relationship. For drought models, sustained $r > 0.6$ across validation periods suggests adequate temporal tracking capability (Sundararajan et al. (2021)). The formula for the correlation coefficient is:

$$r = \frac{\sum_{i=1}^{N}\left(X_{\text{OBS},i} - \overline{X}_{\text{OBS}}\right)\left(Y_{\text{FOR},i} - \overline{Y}_{\text{FOR}}\right)}{\sqrt{\sum_{i=1}^{N}\left(X_{\text{OBS},i} - \overline{X}_{\text{OBS}}\right)^2}\sqrt{\sum_{i=1}^{N}\left(Y_{\text{FOR},i} - \overline{Y}_{\text{FOR}}\right)^2}}$$

6 Revolutionizing Drought Prediction with Machine Learning

Drought is a complex and far-reaching natural disaster that impacts agriculture, water resources, ecosystems, and economies. While traditional drought forecasting methods, such as statistical models and climate simulations, have been valuable, they often fall short in capturing the intricate relationships between environmental factors. This limitation has spurred interest in leveraging Machine Learning (ML) techniques, which offer the ability to process vast datasets and uncover hidden patterns that conventional methods might miss. By integrating ML into drought prediction systems, researchers and policymakers can achieve more accurate and reliable forecasts, enabling better preparedness and response.

(a) *Handling Complex and Multidimensional Data*: Drought forecasting relies on analyzing diverse datasets, including weather patterns, soil moisture levels, satellite imagery, and vegetation health. Traditional approaches often struggle to integrate and interpret these high-dimensional datasets effectively. Machine learning, however, thrives in such scenarios. Algorithms like deep learning, support vector machines, and random forests are particularly adept at managing large, heterogeneous datasets. They excel at identifying complex, non-linear relationships between variables, leading to more precise predictions compared to traditional models.
(b) *Enhancing Prediction Accuracy*: One of the most compelling advantages of machine learning is its ability to deliver highly accurate predictions. Algorithms such as Random Forest, Support Vector Machines (SVM), and Artificial Neural Networks (ANN) have consistently outperformed conventional methods in forecasting drought conditions. These models continuously learn from new data, allowing them to adapt and refine their predictions over time. This adaptability is especially useful for predicting drought indices like the Standardized Precipitation Index (SPI), which depends on a wide range of influencing factors. By capturing these nuances, ML models provide a more robust understanding of drought risks.
(c) *Real-Time Monitoring and Early Warning Systems*: A standout feature of machine learning is its ability to monitor conditions in real time and deliver rapid alerts. Unlike traditional methods, which can be time-intensive, ML systems can process incoming data instantly, providing timely updates on drought conditions. This capability is critical for issuing early warnings, giving communities and policymakers more time to implement measures that minimize the impact of droughts on agriculture, water supplies, and vulnerable populations.
(d) *Preparing for Climate Change and Future Scenarios*: Climate change is making weather patterns more unpredictable, complicating drought forecasting. Machine learning models, however, can be trained to simulate various climate scenarios, offering insights into how future conditions might evolve. This flexibility allows ML-based systems to anticipate how droughts could change in response to shifting climate patterns, making them invaluable for long-term planning and resource management. By modeling potential future risks, these tools help governments and communities prepare more effectively for environmental challenges.
(e) *Streamlining Resource Use*: Traditional drought forecasting often requires significant time, expertise, and manual effort for data collection, analysis, and interpretation. Machine learning, on the other hand, automates much of this process once the models are trained. This efficiency is particularly beneficial in

regions with limited resources or technical expertise, as ML systems can deliver ongoing predictions with minimal additional cost. By reducing the manual workload, ML also frees up researchers to focus on improving model accuracy and addressing emerging challenges.

7 Machine Learning Algorithms for Drought Prediction

Machine learning algorithms are indispensable in the realm of drought prediction, offering a critical tool for leveraging extensive datasets to forecast drought occurrences more accurately and rapidly. The necessity for these algorithms stems from the escalating unpredictability of climate patterns, which profoundly affects agricultural output and water resource management. By scrutinizing historical climate records, soil moisture levels, satellite observations, and other pertinent variables, ML algorithms can discern intricate patterns and correlations that might elude human observation. Most commonly used algorithms for drought prediction are RF, SVM, ANN, and GB.

Random Forest excels in managing vast datasets and discerning complex relationships among variables, while SVM specializes in categorizing data points effectively. ANN, inspired by the human brain, excels in learning nuanced patterns from data, and GBM amalgamates decision trees' strengths to enhance predictive precision. Utilizing these algorithms empowers policymakers and farmers with early alerts and well-informed decisions to mitigate drought-related impacts on agriculture and water resources. Recurrent Neural Networks (RNN) and Long Short-Term Memory (LSTM) are well suited for sequential data but face challenges like gradient issues and high computational demands. Convolutional Neural Networks (CNNs) are effective for spatial data, such as images, but require large datasets and significant expertise to optimize.

Table 3 provides an overview of various machine learning algorithms, highlighting their key features, strengths, and limitations. Each algorithm has distinct tradeoffs, making its suitability dependent on the specific task and available resources.

8 Critical Analysis

This section presents the implementation of various machine learning algorithms. The dataset has been taken from Kaggle. The meteorological data contains 3110 rows and covers a wide range of attributes. The machine learning techniques have been applied to predict drought. The distribution of variables is shown in Fig. 3.

Drought prediction hinges on a range of factors, spanning meteorological, hydrological, and environmental domains, such as precipitation patterns, temperature fluctuations, soil moisture levels, vegetation indices, and atmospheric circulation patterns. ML offers a solution to effectively handle this diverse array of data by capturing intricate relationships and patterns. ML algorithms can seamlessly integrate these factors, enabling the development of robust predictive models. ML enhances the accuracy and reliability of drought forecasts by identifying subtle

Table 3 Machine learning algorithms

Algorithm	Description	Advantages	Disadvantages
Random Forest	This ensemble learning method constructs multiple decision trees and aggregates their predictions	Handles large and complex datasets effectively, reducing over-fitting Offers good generalization ability	Can be computationally demanding, especially with large datasets Interpretability may be challenging due to the complexity of the model
Support vector machines (SVM)	SVM is a supervised learning algorithm used for classification and regression tasks by finding the optimal hyperplane in high-dimensional space	Effective for both classification and regression tasks Performs well in high-dimensional spaces	May struggle with noisy datasets Requires careful selection of kernel and tuning of hyperparameters
Gradient boosting (GB)	GB is an ensemble learning technique that builds decision trees sequentially, refining the model with each iteration	Offers high accuracy and handles various data types well Effective for capturing non-linear relationships	Prone to over-fitting, especially with complex models Training can be time-consuming and computationally intensive
Recurrent neural networks (RNN)	Designed to analyze sequential data by incorporating memory, but sensitive to gradient issues and requires careful parameter adjustment	Well-suited for processing sequential data and capturing temporal dependencies	Vulnerable to issues like vanishing/exploding gradients, demanding meticulous parameter initialization and tuning
Long short-term memory (LSTM)	LSTM is a recurrent neural network architecture designed for sequential data, capable of capturing long-term dependencies	Well-suited for time series data and capturing temporal patterns Handles long-term dependencies effectively	Requires substantial amounts of data for training Training can be computationally demanding Vulnerable to vanishing/exploding gradient issues
Convolutional neural networks (CNN)	CNN is a deep learning architecture commonly used for image recognition tasks, adaptable to spatial data	Effective for extracting spatial patterns in data, such as satellite imagery Automatically learns relevant features	Demands large datasets for effective training Training can be resource-intensive Requires expertise in architecture design and tuning

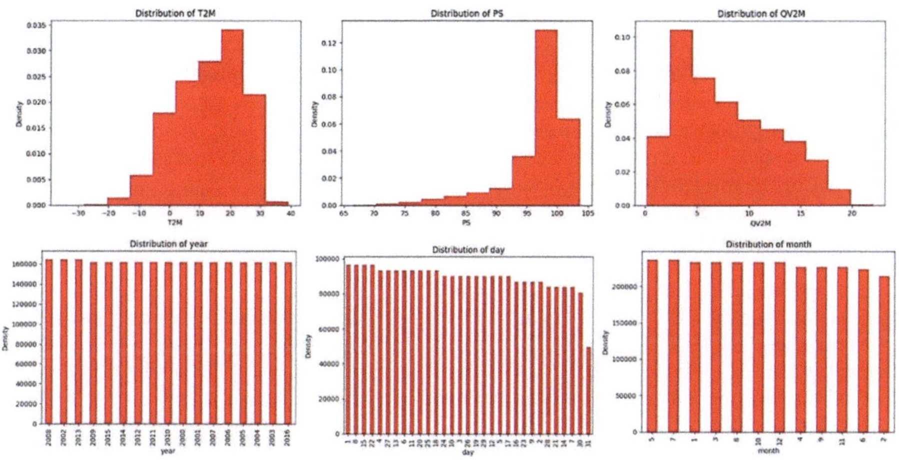

Fig. 3 Distribution of categorical variables

correlations and trends that may elude traditional statistical methods. Moreover, ML allows for the incorporation of spatial and temporal dynamics into the prediction process, facilitating more precise and timely forecasts across various scales. Consequently, ML serves as a potent tool in tackling the complexity of drought prediction by effectively addressing the multitude of factors influencing drought occurrence.

9 Conclusion

This chapter provides insights into various machine learning algorithms such as RF, ANNs, SVM, GB, and CNN for drought prediction, summarizing their theoretical foundations and applications in hydrological research. The study serves as a guide for researchers selecting models based on data availability and forecasting goals. The related work highlighted the successful application of ML algorithms in various drought prediction studies. It emphasized their role in ML in enhancing preparedness and decision-making to mitigate the widespread impacts of drought globally. By leveraging ML algorithms, stakeholders can improve their ability to anticipate and respond effectively to drought events, ultimately reducing the adverse consequences associated with water scarcity.

References

Abbes AB, Inoubli R, Rhif M, Farah IR (2023) Combining deep learning methods and multi-resolution analysis for drought forecasting modeling. Earth Sci Inf:1–10

Acharki S, Singh SK et al (2023) Spatio-temporal distribution and prediction of agricultural and meteorological drought in a Mediterranean coastal watershed via GIS and machine learning. Phys Chem Earth 131:103425

Adnan RM, Dai HL, Kuriqi A, Kisi O, Zounemat-Kermani M (2023) Improving drought modeling based on new heuristic machine learning methods. Ain Shams Eng J 14(10):102168

Bacanli UG, Firat M, Dikbas F (2009) Adaptive neuro-fuzzy inference system for drought forecasting. Stoch Env Res Risk A 23:1143–1154

Belayneh A, Adamowski J (2013) Drought forecasting using new machine learning methods. J Water Land Dev 18(9):3–12

Danandeh Mehr A, Tur R, Alee MM, Gul E, Nourani V, Shoaei S, Mohammadi B (2023a) Optimizing extreme learning machine for drought forecasting: water cycle vs. bacterial foraging. Sustainability 15(5):3923

Danandeh Mehr A, Rikhtehgar Ghiasi A, Yaseen ZM, Sorman AU, Abualigah L (2023b) A novel intelligent deep learning predictive model for meteorological drought forecasting. J Ambient Intell Humaniz Comput 14(8):10441–10455

Dastorani MT, Afkhami H, Sharifidarani H, Dastorani M (2010) Application of ANN and ANFIS models on dryland precipitation prediction (case study: Yazd in Central Iran). J Appl Sci 10(20):2387–2394

Elbeltagi A, Kumar M, Kushwaha NL, Pande CB, Ditthakit P, Vishwakarma DK, Subeesh A (2023) Drought indicator analysis and forecasting using data driven models: case study in Jaisalmer, India. Stoch Env Res Risk A 37(1):113–131

Fourth National Climate Assessment (n.d.). https://nca2018.globalchange.gov/

Ham YS, Sonu KB, Paek US, Om KC, Jong SI, Jo KR (2023) Comparison of LSTM network, neural network and support vector regression coupled with wavelet decomposition for drought forecasting in the western area of the DPRK. Nat Hazards 116(2):2619–2643

Karbasi M, Jamei M, Malik A, Kisi O, Yaseen ZM (2023) Multi-steps drought forecasting in arid and humid climate environments: development of integrative machine learning model. Agric Water Manag 281:108210

Khan N, Sachindra DA, Shahid S, Ahmed K, Shiru MS, Nawaz N (2020) Prediction of droughts over Pakistan using machine learning algorithms. Adv Water Resour 139:103562

Malik A, Kumar A, Singh RP (2019) Application of heuristic approaches for prediction of hydrological drought using multi-scalar streamflow drought index. Water Resour Manag 33: 3985–4006

Piri J, Abdolahipour M, Keshtegar B (2023) Advanced machine learning model for prediction of drought indices using hybrid SVR-RSM. Water Resour Manag 37(2):683–712

Rahmati O, Falah F, Dayal KS, Deo RC, Mohammadi F, Biggs T et al (2020) Machine learning approaches for spatial modeling of agricultural droughts in the south-east region of Queensland Australia. Sci Total Environ 699:134230

Rezk NG, Hemdan EED, Attia AF, El-Sayed A, El-Rashidy MA (2021) An efficient IoT based smart farming system using machine learning algorithms. Multimed Tools Appl 80:773–797

Sundararajan K, Garg L, Srinivasan K, Bashir AK, Kaliappan J, Ganapathy GP et al (2021) A contemporary review on drought modeling using machine learning approaches. Comput Model Eng Sci 128(2):447–487

Vo TQ, Kim SH, Nguyen DH, Bae DH (2023) LSTM-CM: a hybrid approach for natural drought prediction based on deep learning and climate models. Stoch Env Res Risk A 37(6):2035–2051

World Health Organization (n.d.) Droughts. World Health Organization. https://www.who.int/health-topics/drought#tab=tab_1

Yaseen ZM, Shahid S (2021) Drought index prediction using data intelligent analytic models: a review. In: Intelligent data analytics for decision-support Systems in Hazard Mitigation: theory and practice of Hazard mitigation, pp 1–27
Zhang R, Chen ZY, Xu LJ, Ou CQ (2019) Meteorological drought forecasting based on a statistical model with machine learning techniques in Shaanxi province, China. Sci Total Environ 665:338–346

Importance of Genetic Variability, Breeding, and Associative Transcriptomics in Rapeseed Genetic Resources

Vladislav Čurn, Eva Jozová, Ondřej Hejna, Baveesh Pudhuvai, and Bhupendra Koul

Abstract

Rapeseed (*Brassica napus* L. napus), also known as canola in Canada & Australia, oilseed rape in Europe and Asia, is a crucial oilseed crop. It ranks third among the leading oil crops with substantial production in the EU nations, Canada, China, India, and Japan. Its evolutionary lineage traces back to the Miocene period, originating from genus *Sinapis* (*B. nigra*) and Diplotaxis-Erucastrum complex (*B. oleracea*; *B. rapa*). The early records of rapeseed span from 500 to 12,000 years, influenced by the early agricultural practices and gradual evolution with two ancestral genomes (*B. rapa* and *B. oleracea*). The Mediterranean region is the centre of origin for rapeseed, which transitioned from a vegetable to an oil crop throughout the Middle Ages. The discovery of low erucic acid variants in the 1960s significantly enhanced the economic relevance of rapeseed. Its expansion in agriculture is due to the successful breeding efforts focusing on oil and meal quality improvements, notably the development of low-erucic acid and double zero varieties during the 1990s. The challenges faced in varietal development are the complex floral morphology and varied cross-pollination rate (8–80%). The hybrid development was improved after the identification of various sterile lines (CMS; GMS) to overcome the varietal crossing hindrances. Contemporary breeding techniques emphasize the creation of high-yield cultivars with enhanced resilience and stability against abiotic and biotic stresses. Advancements in biotechnology have improved breeding efficiency through the

V. Čurn · E. Jozová · O. Hejna
Faculty of Agriculture and Technology, Department of Genetics and Biotechnology, University of South Bohemia in České Budějovice, České Budějovice, Czech Republic

B. Pudhuvai
Faculty of Agriculture and Technology, Department of Plant Production, University of South Bohemia in České Budějovice, České Budějovice, Czech Republic

B. Koul (✉)
Department of Biotechnology, School of Bioengineering & Biosciences, Lovely Professional University, Phagwara, Punjab, India

A. Kumar, B. Koul (eds.), *Crop Improvement Strategies in Brassica species: Applied Science*, https://doi.org/10.1007/978-981-95-3861-4_11

application of TILLING and haploid creation. Many recent studies also used AFLP, SSR, and SNP markers to assess genetic diversity among modern cultivars and landraces and reveal the shifts in the genetic constitution. Such studies on rapeseed breeding emphasize the critical importance of genetic variability and adapting suitable breeding methods via molecular techniques will aid in maintaining genetic diversity and improve future crop traits. The chapter encapsulated key findings and insights regarding the breeding of rapeseed and the significance of genetic diversity in its cultivation, reflecting the historical context and modern advancements.

Keywords

Rapeseed · Oilseed · Markers · Haploid · Genetic diversity · AFLP

1 Origin of Rapeseed

Rapeseed, *Brassica napus* L. subsp. *napus* is an important oilseed crop across the world. It is known as oilseed rape in Europe and Asia, but known as canola in Canada and Australia. This adaptable plant is a significant source of vegetable oil, well known for its high quality. It is widely employed for its oil and protein in food/feed and various industrial applications. The evolutionary origin of *Brassica* may be traced back to the Miocene period, specifically through the *Diplotaxis-Erucastrum* complex, which is believed to have originated from the genus *Sinapidendron* (Gómez-Campo and Tortosa 1974). Further evolutionary studies demonstrate that *Brassica* species evolved in two independent pathways: *B. nigra* is likely inherited from the *Sinapis* germplasm resources, whereas *B. oleracea* and *B. rapa* originated from *Diplotaxis erucoides* (Song et al. 1990; Pradhan et al. 1992). Rapeseed, a phylogenetically young species, is nowhere to be found in wild form (Andersson and Olsson 1961). The first records about rapeseed vary widely across publications. Allender and King (2010) and Zamani-Nour et al. (2013) reported the origin of this species 500 years ago; in contrast, Cheung et al. (2009) refer to the results of phylogenetic analyses and state that the intergeneric crossing of two parental species occurred less than 10,000 years ago. Another view from Salamini et al. (2002), which dated the origin of rapeseed to 12,000 years ago, postulated the hypothesis that its origin was influenced by domestication and deliberate plant cultivation by a culture of gatherers and early farmers. It is currently believed that rapeseed evolved by a process of spontaneous hybridization around 7500 years ago, including turnip rape (*Brassica rapa, syn Brassica campestris*) and cabbage (*Brassica oleracea*) (Chalhoub et al. 2014). The centre of *Brassica napus origin* is in southern Europe and the Mediterranean, where the parental taxon is also present (Song and Osborn 1992).

The rapeseed genome $A_nA_nC_nC_n$ comprises 38 chromosomes, representing two ancestral genomes—ten chromosomes from the A_rA_r genome *Brassica rapa* L. subsp. *oleifera* (DC.) METZGER and 9 chromosomes from the C_oC_o genome

B. oleracea L. (Labana and Gupta 1993; Murphy 1994; Chalhoub et al. 2014). Cytological studies and experimental crossings have led to hypotheses about species' association with the Brassica genus and the assumption of rapeseed origin. The results from cytological studies and the behaviour of chromosomes during mitosis suggest that monogenetic species had a common, now extinct, ancestor with the basal chromosome number $x = 5$ or 6. Basal monogenetic (diploid ancestral) species originated from this ancestor through the process of autopolyploidy followed by the elimination of few chromosomes, extensive triplications of large genomic regions, and structural rearrangements (Lagercrantz and Lydiate 1996; Lagercrantz 1998). Cultivated *Brassica* species are categorized based on chromosomal number into three primary groups with genomes A, B, and C (Downey et al. 1980; Röbbelen 1985; Labana and Gupta 1993). Combining these basal species/groups by random spontaneous crossing of parental forms and subsequent amphidiploidization, other species with composite genomes have emerged (McNaughton 1976; Röbbelen 1985), and the six most important *Brassica* species are interlinked (Gupta 2007, 2009). The interrelationship between these six *Brassica* species was confirmed using comparative genome sequence analysis and shown in Fig. 1 (Schmidt and Bancroft 2011). A detailed overview of the phylogeny, evolution, and systematics of the Brassicaceae family is given by Al-Shehbaz et al. (2006).

Evidence of the origin of the cultivated rapeseed hybrid was first provided by the cytotaxonomic experiments of Morinaga (1934), supplemented and confirmed by the studies of Nagaharu (1935). Frandsen later verified/validated the conclusions of these initial studies with breeding and hybridization experiments (Frandsen 1943, 1947). Regardless of these hybridization experiments, the intrinsic process and

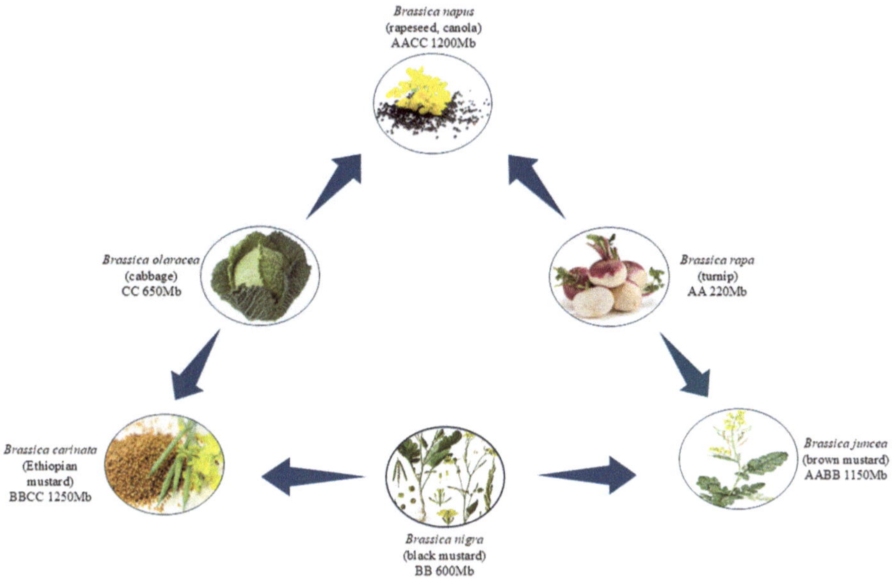

Fig. 1 Illustration of relationships among the important species of Brassica

geographic location of the initial hybridization(s) leading to rapeseed origin remain unclear. Evidence of the polyphyletic origin of rapeseed was initially confirmed by RFLP analysis of nuclear, chloroplast, and mitochondrial DNA (Song and Osborn 1992) and subsequently using nuclear and chloroplast molecular markers (Allender and King 2010) or by whole genome sequencing of *B. napus* genome (Chalhoub et al. 2014; He et al. 2021). This knowledge is also used in the resynthesis of rape from "original" parental forms, in particular as a breeding strategy for the widening of genetic variability in rapeseed breeding (Schenck and Uolf 1986; Sundberg and Glimelius 1986; Akbar 1989; Chen and Heneen 1989; Girke et al. 2012; Hilgert-Delgado et al. 2015). Parental species of *B. rapa* and *B. oleracea* are challenging to cross, mainly due to endosperm deficiency leading to embryo abortion. Spontaneous occurrence of de novo "synthesized" rape is extremely rare due to the barriers in fertilization and embryo development (Håkansson 1956).

The original occurrence of rapeseed can be found in the Mediterranean genetic centre, where there are also areas of occurrence of its original "parental" components—*Brassica oleracea* L. (cabbage) and the much more widespread *Brassica rapa* L. (turnip). Spontaneous hybridization of these two species with diploid chromosome numbers and subsequent amphidiploidization resulted in the formation of oilseed rape with its relatively complex genetic structure. Based on a study of the extent of variability in morphological forms of parental species, *B. napus* is considered to have been formed by several spontaneous hybridizations of distinct morphotypes of *B. rapa* and *B. oleracea* (McNaughton 1976; Röbbelen 1985). Because of its phylogenetic origin, rapeseed covers only part of the range of variability of parental species, and genetic variability is somewhat limited for *B. napus* (Murphy 1994).

2 Rapeseed Importance and History of Cultivation

On a global scale, rapeseed is one of the second to fourth most important oil crops (Chikkaputtaiah et al. 2017; Friedt et al. 2018). It is the only oil crop fully adapted for temperate zones, allowing cultivation in colder areas and higher altitudes (Eskin et al. 2006). The world's largest producers are the EU (European Union), Canada, China, India, Pakistan, and Japan (USDA 2023). Its massive expansion and worldwide cultivation are a result of long and successful breeding and, in particular, the breakthrough changes in the breeding focused on oil and meal quality in the 1990s—the switch to low-erucic acid varieties and double-zero varieties (Vašák and Fábry 1991).

The diverse and variable species of the *Brassica* genus were initially used as vegetables or spices. Brassica crops, including rapeseed, have an extensive association with human culture.

B. rapa is believed to be the first of the *Brassica* species to be domesticated, with the oleiferous turnip indicating the primary centre of domestication in the Mediterranean region and the foliage turnip representing the second domestication centre in China (Song et al. 1990). *B. oleraceae* was domesticated much later (Gómez-Campo 1999). Vegetable types of Brassicas were widely used in the Neolithic period. Old Indian Sanskrit writings mention that "Siddhartha" (*B. rapa*) and sarson types of oleiferous *B. napus* were already cultivated in 4000–1500 BC (Prakash 1980; Weiss 1983; Snowdon et al. 2006). Significant use of these crops was already on the European continent in the days of Ancient Greece and Rome (Fussell 1966; Labana and Gupta 1993; Murphy 1994). The use of *Brassica* species as vegetables or spicy seeds changed during the Middle Ages into the use of rapeseed and turnip for oil production for lighting or soapmaking (Matthioli 1598; Tabernaemontanum 1664). In England and the Netherlands, extended use of oilseed rape seeds for both technical purposes (lighting, lubricating machinery, production of artificial rubber–factice) and food use dates back to the early sixteenth to seventeenth centuries (Fussell 1966; Seebohm 1976). Interest in rapeseed cultivation has also led to significant diversification and creation of different morphotypes, which were helpful in oil production, forage or food (winter, spring and semi-winter oilseed rape/canola, fodder rape, swede, Siberian kale (Allender and King 2010; Wu et al. 2014). However, its use in livestock nutrition is limited due to the high glucosinolate content in the oilcake meal. This stage of rapeseed cultivation and exploitation lasted until the mid-twentieth century. After World War II, there was a shortage of animal fats (butter), and attention turned to the broader use of vegetable oils, including lower-quality rapeseed oil used to make low-quality margarine (Miller 1989; Snowdon et al. 2006). During this period, the old landraces and population varieties were under cultivation, where rapeseed breeding began to take off and the first breed varieties appeared. The main factor limiting the use of rapeseed as a full-value oil crop for food use was the high presence of erucic acid in rapeseed oil—an anti-nutritional substance that increases the cholesterol content in the blood of humans and animals and damages the heart muscle (Daun 1983, 1986). At the beginning of the 1960s, plants with almost zero erucic acid content in the oil were discovered by scientists in Canada in the German cultivar "Liho" (Stefansson et al. 1961) and later in another European variety originating from Hungary (Stefansson and Hougen 1964). By crossing these genotypes with common Canadian accessions, the first varieties without erucic acid were available already during the 1960s, and an unprecedented expansion in rapeseed breeding, cultivation and associated research, which continues to date, was started. Over the past four decades, the economic significance of rapeseed has been established through a rigorous breeding programme aimed at reducing nutritionally unfavourable constituents in the oil and meal while concurrently increasing yields and yield stability. More comprehensive information about rapeseed's economic importance, history of spreading and cultivation, and rapeseed breeding technologies is discussed by Snowdon et al. (2006).

3 Rapeseed Breeding

A significant breakthrough in breeding and rapeseed has meant the breeding of so-called "0" zero/low-erucic varieties: the French varieties Primor, the German Quinta, and the Swedish Brink and, in particular, the French variety Jet Neuf (Jönsson 1978; Stefansson 1983). Introducing zero erucic rapeseed varieties generated a keen interest in rapeseed cultivation. At the same time, this increased cultivation has also triggered new quality requirements for rapeseed breeding—a reduction in glucosinolate content so that rapeseed meal (shredded oil cakes) can be used as quality protein feed for livestock. In the late 1980s and early 1990s, the first "00" (double-zero) rapeseed varieties were introduced to the market (Downey 1990; Downey and Rakow 1987). These varieties contained traces of erucic acid in oil and diminished or negligible amounts of glucosinolates. A modified rapeseed variant was developed in Canada and named "Canola". Canola was originally a trademark name of the Rapeseed Association of Canada, and now canola varieties grown in Canada belong to three botanical *Brassica* species*: Brassica napus, B. rapa,* and *B. juncea* (Woods et al. 1991).

The removal of anti-nutritional agents, such as erucic acid, which has a negative impact on the human cardiovascular system, and liver-damaging glucosinolates, has resulted in a significant increase in rapeseed cultivation and spread across new geological areas, where the neglected plant has become a vital oil crop with a high oil evaluation (Carré and Pouzet 2014; Friedt et al. 2018; Shen et al. 2023). However, the breeding trend in rapeseed is not confined to food use and development of double-zero varieties, but it has been extended to the breeding of varieties with a specific fatty acid composition in oil usable both in food and for industrial purposes (LEAR, HEAR, HOLL, VHOA, LEHL, VGLN varieties) (Przybylski and Mag 2002; Baux et al. 2011), and as a renewable energy source (Scarisbrick et al. 1989). The apparent trend in rapeseed breeding is the development of hybrid varieties that are consistent with modern breeding approaches, as compared to the varietal lines, which perform inadequately in terms of their intended traits and yield potential. New-generation hybrid varieties are high-yielding, exhibit a strong heterosis effect even under stressful conditions, and have greater vitality, compensation ability, and improved resilience to stress conditions. The breeding strategy for hybrids relies on a stable genetic system to prevent self-pollination, the availability of elite genotypes that provide high heterosis, and the efficient transfer of pollen between desired parents under field conditions (Grant et al. 1991).

Rapeseed is a facultatively cross-pollinated plant, and considering its floral morphology, the development of hybrid varieties is challenging. The proportion of cross-pollination ranges between 8% and 80% and highly depends on the climatic and weather conditions during the flowering period. To obtain a hybrid seed, it is crucial to prevent self-pollination of the maternal component of the hybrid (Renard et al. 1998; Brandt et al. 2007). Hybrid breeding programmes use two fundamentally different approaches—pollen-sterile maternal components or self-incompatible genotypes. Pollen sterility-based breeding programmes are based on different hybrid seed production systems, namely the ogu/INRA, Polima Shaan 2A cytoplasmic male sterility, or MSL (male sterile Lembke) nuclear pollen sterility (Brown et al. 2003; Fu et al. 1995; Yan et al. 2016; Havlíčková et al. 2012). The resultants were composite hybrids, with the hybrid component of this variety being pollen sterile and pollination ensured by adding pollinators (Frauen and Baer 1996). However, this approach has switched to breeding pollen-fertile, restored hybrids using paternal lines carrying fertility restoration genes (*Rf* genes) (Delourme et al. 1998). Production of novel hybrids by employing the self-incompatibility approach is relatively complicated compared to sterility (an absolute barrier for self-pollination due to the absence of pollen). The self-incompatibility depends upon internal physiological parameters and is sometimes triggered by external factors (climate) (Koprna et al. 2005; Žaludová et al. 2013; Dou et al. 2021).

4 Further Trends in Rapeseed Breeding

Further progress in rapeseed breeding persisted due to consumer demand for oil/seed quality parameters and which highly rely on the incorporation of breeding methods (Snowdon et al. 2006; Ton et al. 2020; Cowling et al. 2023; Eckerstorfer et al. 2019). The introduction of new biotechnology techniques to breeding, such as the induction of haploid formation and regeneration of microspore haploid embryos, and the production of isogenic wholly homozygous double-haploid plants (DH plants) leads to a significant acceleration and streamlining of the breeding process (Starosta et al. 2023). A substantial limit in rapeseed breeding is the origin of this crop, the absence of "wild" rapeseed, and its minimal genetic base (Wu et al. 2014). Extension of genetic variability to overcome this limitation can be achieved by using resynthesized *B. napus* and the embryo rescue method to obtain hybrid plants after the hybridization of ancestral species (Jesske et al. 2013; Girke et al. 2012; Ihien Katche and Mason 2023). The molecular biology methods are not only confined to producing GMOs but are also much broader in their use—from identifying varieties, molecular description, and characterization to developing molecular selection markers for individual breeders' significant traits and characteristics, analysing the whole genome or transcriptome of rapeseed, and identifying genes involved in stress reaction and responsible for qualitative and yield parameters (Hejna et al. 2019; Lu et al. 2014; Huang et al. 2020; Knoch et al. 2021). The reverse genetics approach, the TILLING technique, combining classical mutagenesis and modern methods of genome analysis and identification of mutations in genes of interest for causal genes also has an irreplaceable role in contemporary breeding (Wang et al. 2008; Stephenson et al. 2010; Jhingan et al. 2023).

The following chapters will provide insight into the importance of proper management of the genetic diversity of rapeseed genetic resources, the use of techniques of plant biotechnology to get elite homozygous genotypes, the development of gene markers for the marker-assisted selection (MAS) system, and also focus on functional genomics and its use in rapeseed breeding.

5 Evaluation of Genetic Diversity in *B. napus* and Composition of Genetic Resources

Repeated and deliberate cultivation of plants leads to the development of landraces adapted to the environmental conditions in which they were grown (Friedt et al. 2007). In most crops, landraces (regional varieties) represent the first step towards creating new purposefully improved varieties with unique traits, including nutritional and cultivation aspects (Fernie et al. 2006). However, due to the selection based on specific qualitative and quantitative parameters (crop yield), there is a significant narrowing of genetic variability. The loss of several valuable genes and characteristics of the so-called "wild" types and the complete disappearance of landraces or primitive varieties resulted from the cultivation of modern varieties (Zamir 2001). There was a noticeable decline in genetic diversity throughout the twentieth century due to the enormous increase in profitable, high-yielding, high-quality varieties that are often closely related (Rao and Hodgkin 2002; van de Wouw et al. 2010). The narrowing of the genetic base then causes problems with breeding varieties more tolerant or resistant to biotic and abiotic stresses (Friedt et al. 2007). The UN's "Convention on Biological Diversity," adopted in 1992, defined the principles for the conservation and preservation of genetic resources in situ (at original sites) and ex situ (primarily by gene banks). Genetic diversity loss is considered one of today's most serious environmental problems (Ehrlich and Ehrlich 1983; Hoban et al. 2021). Unfortunately, genetic diversity cannot be enlarged to restore the status quo, so alternative conservation methods had to be developed to preserve at the level of gene reserves, populations, individuals, genomes, loci or DNA sequences (Kresovich et al. 1992; Kardos et al. 2021; Salgotra and Chauhan 2023).

Brassica napus has a relatively narrow genetic base due to its limited primary geographical distribution and intensive breeding efforts (Annisa et al. 2011; Girke et al. 2012). Although domestication of rapeseed was recently adopted (Soltis et al. 1998), breeding of *B. napus* and other oilseed brassicas has been very intensive in the last few decades and was focused only on several of the most important agronomic/economic traits. The selection of a limited number of genotypes with low erucic acid and glucosinolate contents has led to the narrowing of genetic diversity and the loss of several resistance genes found in old varieties and landraces (Hasan et al. 2006). Genotypes generated by intergeneric hybridization (Niemann et al. 2012) or resynthesized rapeseed from its ancestors—turnip (*Brassica rapa*) and cabbage (*Brassica oleracea*), using biotechnological methods (Wen et al. 2008) could be used for restoration and enhancement/widening genetic diversity. Using these approaches, prospective traits and characteristics can be introduced to the rapeseed gene pool, and these genes can enhance the biological and production potential of existing genotypes such as drought tolerance, disease and pest resistance, self-incompatibility, pollen sterility, fertility restoration, yellow seed colour, specific fatty acid composition in oil, a more robust root system, etc. (Rai et al. 2007; Rahman 2005). Research of the *Brassica* genome contributed new insight into an adequate assessment of genetic diversity. The importance of genetic diversity and the knowledge of the extent of genetic variability resulted in the establishment and management of a pangenic platform within the rapeseed and *Brassica* genomes, respectively (Bayer et al. 2021; He et al. 2021; Havlickova et al. 2018). Research on Brassica germplasm and evaluation of its genetic diversity has the potential to expedite the efficient use of genetic resources in canola breeding initiatives (Stokes et al. 2010; Hu et al. 2021).

The loss of genetic diversity limits the availability of suitable genotypes for breeding and causes a dearth of significant gene donors. The considerable degree of morphological similarity between the varieties is a further negative consequence of the narrowed genetic diversity. This raises problems in the registration and legal protection of varieties. The identification and precise description of rapeseed varieties during registration are contingent on a limited set of morphological characteristics. However, as the number of varieties increases, distinguishing the morphological variation among the varieties becomes increasingly difficult (Lombard et al. 2000). Although morphological, quality, and yield traits are under consideration for cultivar protection, the updated directions of ISTA (International Seed Testing Association) and UPOV (The International Union for the Protection of New Varieties of Plants), the molecular markers being developed will maintain the efficacy of registration and DUS (Distinctness, Uniformity, Stability) testing ensuring the quality of new varieties for farmers and merchants. Numerous markers have also been designated to characterize genetic resources, including isozymes, storage proteins, and DNA-based markers (Curn 1995; Zhao and Becker 1998; Schlötterer 2004). Molecular markers are highly advantageous in the estimation of genetic diversity within breeder populations, as well as in the evaluation of gene flow rates, genetic drift occurrences, and the extent of outbreeding (Rao and Hodgkin 2002), genetic structure of varieties and genetic resources and variety ID (Havlickova et al. 2014; Mailer et al. 1994; Sobotka et al. 2004), or controls of seed purity of hybrid varieties (Marshall et al. 1994). Several methods such as RFLP—*restriction fragment length polymorphism* (Hallden et al. 1994; dos Santos et al. 1994; Diers and Osborn 1994), RAPD—*random amplification of polymorphic DNA* (Shiran et al. 2006), AFLP—*amplified fragment length polymorphism* (Yu et al. 2007), SSR—*simple sequence repeats or microsatellites* (Hasan et al. 2006), cpSSR—*chloroplast simple sequence repeats* (Zamani-Nour et al. 2013), RAMP—*random amplified microsatellite polymorphism* (Wei et al. 2005), SNP—*single-nucleotide polymorphism* (Hayward et al. 2012) were employed to characterize and assess the *B. napus* collections. These methods can potentially enhance germplasm conservation techniques and facilitate the utilization of plant genetic resources. The following examples show the results of the molecular analyses of sets of rapeseed genetic resources regarding the assessment of the extent of genetic variability.

5.1 Analysis of AFLP Markers of a Pilot Set of Landraces and Modern Varieties

The AFLP analysis was used to assess the variability between the accessions of rapeseed genetic sources—landraces and old varieties. Varieties involved in this set were significant genotypes from the view of breeding success and cultivation. The question was how much modern high-yielding and high-quality varieties differed from landraces in terms of the extent of genetic variability. The pilot set included Czechoslovak and German landraces, Czechoslovak, Czech, and German "0" and "00" varieties. The results of molecular analyses are presented in Fig. 2. The original extent of genetic diversity of the landraces (both Czech and German) is illustrated as a yellow ellipse. Bred Czech varieties (blue ellipses) are included within this genetic pool of old gene resources, representing a narrower range of genetic variability. But unfortunately, there has been no significant differentiation from the gene pool of landraces. The German varieties (modern and older) are marked in red, and the red ellipse includes a set of the most recent varieties from the analysed group. In this case, there is a significant shift in the genetic constitution of modern

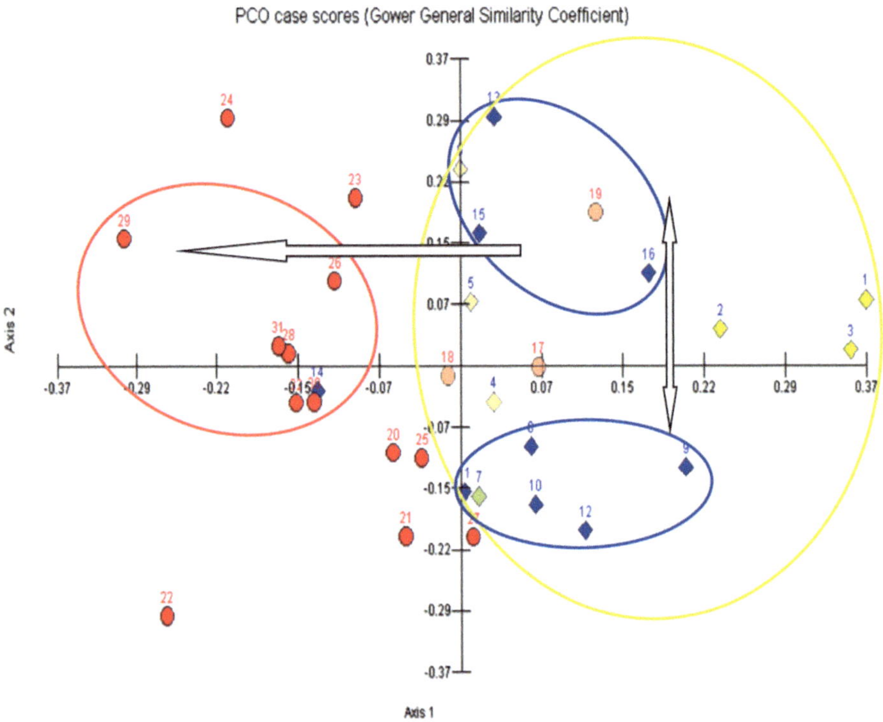

Fig. 2 PCO analysis of AFLP markers in a pilot set of landraces and modern varieties shows the shift in the genetic constitution and influence of unrelated germplasm used in breeding. The pilot set included Czechoslovak landraces—Třebíčská (*1*), Přerovská (*2*), Slapská (*3*), Česká Krajová (*4*), Valečovská (*5*), Slovenská Krajová (*6*); Czechoslovak population variety Mira (*7*); Czechoslovak and Czech "0" and "00" varieties—Solida (*8*), Silesia (*9*), Sonáta (*10*), Aglona (*11*), Omikron (*12*), Slapská Stela (*13*), Odila (*14*), Opponent (*15*), Opus (*16*), German landraces—Rechtbergs Winterraps (*17*), Olquel (*18*), WRG 19 (*19*); German varieties—Falcon (*20*), Zorro (*21*), Rasmus (*22*), Orkan (*23*), Pilot (*24*), Ramiro (*25*), Smart (*26*), Trend (*27*), Digger (*28*), Siska (*29*), Winner (*30*), Petra (*31*), Senator (*32*)

varieties and a clear distinction between them and landraces. This can result from appropriate genetic resource management and a broad base of breeding companies. Czech variety Odila (14) has exceptional status and is located in the group of German varieties. Odila originated due to CZ-German breeders' cooperation, and the first crossing of parental components from the German gene pool was done in a German station.

5.2 Analysis of AFLP Markers of the Core Collection of Rapeseed Genetic Resources

A total of 94 accessions were sourced from the core collection of genetic resources of rapeseed encompassing a diverse range of germplasm, including chosen landraces, both modern and older cultivars originating from various geographical locations, as well as current breeding materials possessing distinct qualitative features. Additionally, the set also incorporated donors of sterile cytoplasm and fertility restoration. The results from AFLP analysis of the accessions are presented in the following Fig. 3, which shows three groups: (1) A group of genetic resources marked with a yellow ellipse representing breeding material with altered fatty acid composition—genotypes with high oil content in seeds, low linoleic acid content, or genotypes with yellow seeds. Specific genetic resources or exotic trait donors were used to cross and develop these genotypes, which is also the reason for the considerable genetic difference in this group; (2) a group of genetic resources marked with a green ellipse includes varieties of different age and geographical origin; and (3) a group of genetic resources marked with a red ellipse also contains different varieties as well as new breeding materials, sterile cytoplasm donors, fertility restorers, and self-incompatible genotypes. This core collection of genetic resources illustrates proper genetic resource management, selection of appropriate parental components for crossing, and appropriate breeding and selection programmes that can ensure a broad genetic diversity in collections of breeding materials, and there is no further narrowing of genetic variability.

Similar results are presented in the study of Havlickova et al. (2014), where three molecular marker systems with varying narrative values were used. ISSR, AFLP, and SSR markers differ in their character and genome coverage capability. However, it is vital to comprehend that diverse markers possess unique characteristics, which will manifest varying facets of genetic diversity and, as a result, provide varying results (Karp and Edwards 1997; Nesbitt et al. 1995). In order to compare the extent of genetic diversity of markers of varying nature, the parameter WAM (weighted arithmetic mean of genetic distances) was used to analyse the frequency and size of genetic distance. This method facilitates a direct comparison and selection of marker systems, highlighting an alternative approach for processing genetic diversity data. The three molecular markers exhibited variations in the level of observable genetic variation and their capacity to identify genetic diversity among individual accessions. The highest genetic distances between *B. napus* accessions were recorded using ISSR markers, where WAM among all 94 accessions was 25.71%. SSR

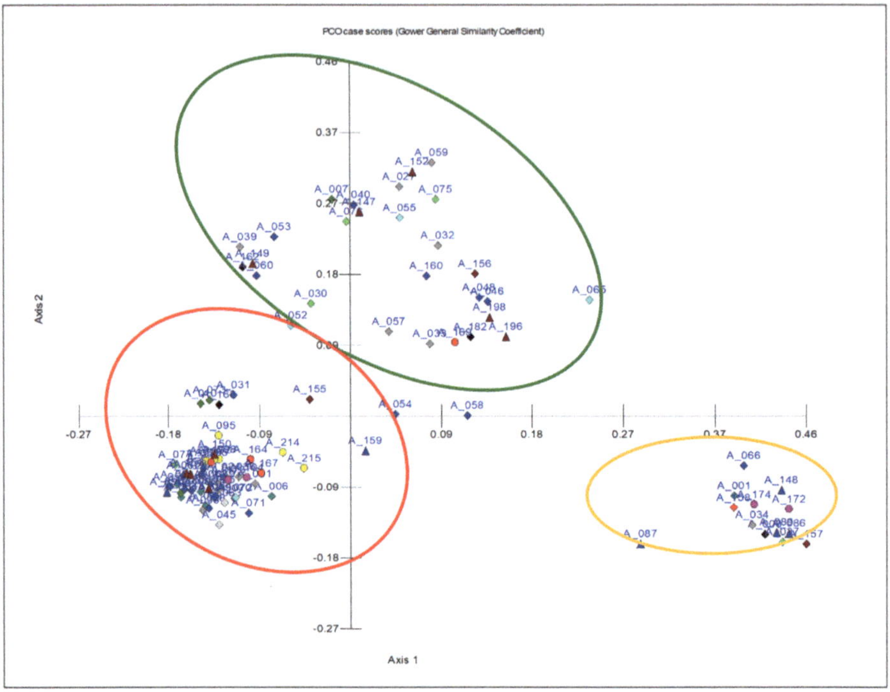

Fig. 3 PCO analysis of AFLP markers of the core collection of rapeseed genetic resources shows an unexpected range of genetic variability. Genotypes with specific traits (altered fatty acid composition and new breeding materials, sterile cytoplasm donors, fertility restorers, and self-incompatible genotypes) were significantly different from the gene pool of standard varieties, and proper management of genetic resources and appropriate selection of genotypes for breeding can preserve the range of genetic variability. PCO was performed using DARwin 6.0 software

markers reached an average WAM (22.51%); the sample's highest similarity and dense clustering were recorded in AFLP (WAM = 18.68%). However, AFLP provided the highest genome coverage and generated the highest number of markers/alleles.

Alternative approaches to standard descriptive characteristics, such as type of markers. Polymorphic bands and marker analysis employing UPGMA and PCO approaches are suitable for conducting a comprehensive evaluation of genetic variability, assessing the degree of diversity in the rapeseed gene pool, and verifying the effects of breeding efforts. These approaches are based on Bayesian statistics and may, for example, represent evaluations of the genetic structure of genetic resources. Evaluation of a population's genetic structure is a usual practice in ecological and evolutionary research (Kolář et al. 2012). According to the study by Havlickova et al. (2014), applying this methodology can facilitate the examination of cultivar classification/grouping and the expansion of genetic diversity in breeding programmes. Detailed assessment of the genetic structure of the analysed rapeseed cultivars and breeding materials can point to limited genetic flow between different groups of varieties (breeding programmes or stations), significant and sudden changes in genetic structure due to the use of new genetic resources or a change in the breeding programme, shifts in genetic structure over the course of time/years. These changes can lead to a deeper reflection on potential gene loss and appropriate management of genetic resources.

6 NGS, GWAS, and Associative Transcriptomics in Oilseed Rape

Over the past two decades, substantial progress in next-generation sequencing (NGS) and computational technologies has facilitated the development of assays for several single-nucleotide polymorphism (SNP) markers. This was also possible for plants with complex genomes or polyploid crops like wheat or oilseed rape (Trick et al. 2009; Miller et al. 2016). In polyploid organisms with duplicated genes exhibiting uneven expression, identifying correlations between genetic and phenotypic variation has demonstrated utility (Adams et al. 2003; Pires et al. 2004).

RNA sequencing is a contemporary NGS technique in plant genomics, providing a rapid and cost-effective approach to unravel novel SNP markers within gene coding areas or to measure their transcripts, which has been advantageous in studying oilseed rape (*B. napus*) (Miller et al. 2016). With the progression of mRNA sequencing and the large amount of data generated from this technique, a novel methodology known as Genome-wide Association Studies (GWAS) has emerged to investigate transcriptomics. This method has become a prominent approach for identifying quantitative trait loci (QTLs) associated with trait variation (Zhu et al. 2008).

In polyploid plants, the investigation of GWAS utilizing SNPs and GEMs (gene expression markers) obtained from mRNA-Seq analyses proved to be beneficial for polyploid crops like oilseed rape. Oilseed rape, being an allotetraploid species, exhibits a complicated genome with multiple orthologues (homologues) dispersed across different chromosomes. Investigating the genetic complexities and the relationship between gene variations and traits in oilseed rape is essential for implementing breeding initiatives. Applying mRNA-Seq data in GWAS presents a valuable opportunity to tackle these challenges and reveal the genetic mechanisms that govern desirable traits in these plants.

Associative transcriptomics (AT) is a novel bioinformatic tool that successfully integrates whole-genome association analyses and mRNAseq data from NGS. This approach utilizes transcriptome sequencing to identify SNPs and GEMs linked to trait variation to locate genomic regions controlling complex traits, making it widely applicable in plant biotechnology (Higgins et al. 2012). AT has successfully identified high-quality genetic markers associated with specific characteristics, even in polyploid crops where traditional GWAS approaches face challenges due to the size and complexity of genomes (Lu et al. 2014). The use of high-density SNP-based genetic mapping and linkage disequilibrium assessment in oilseed rape establishes a solid foundation for future genetic research (Delourme et al. 2013). This method improves understanding of the genetic basis of complex traits and aids in identifying QTLs linked to significant agronomic characteristics.

Integrating transcriptomic data with genetic information from diverse germplasm enables breeders to identify key alleles associated with heterosis and other desirable traits for developing hybrids. AT plays a crucial role in improving the nutritional quality of oilseed rape. It is employed in the genetic and molecular regulation of seed storage proteins (SSPs), which enables researchers to develop strategies for SSP synthesis and accumulation, manipulating protein content and quality in oilseed rape seeds (Gacek et al. 2018).

The investigation into the genetic foundation of flower colour variation in oilseed rape, utilizing integrated metabolome and transcriptome analyses, illustrates the capacity of associative transcriptomics to reveal new traits (Cui et al. 2024). Identifying key regulatory genes involved in anthocyanin biosynthesis enables the development of oilseed rape varieties with distinct flower colours, potentially impacting ornamental horticulture and market differentiation.

7 Development of Associative Transcriptomics

The first report on AT in *B. napus* focuses on correlating genetic variation with trait variation in oilseed, particularly on rapeseed quality-related traits. The approach involved constructing pseudomolecules with unigenes representing the *B. napus* genome by utilizing already published SNP maps and sequence data from *B. rapa* and *B. oleracea*. The mRNA-Seq data from 84 *B. napus* accessions are employed to identify and score SNP markers and quantify gene expression levels. Subsequently, transcriptome association studies were conducted using SNPs and gene expression markers (GEMs) to investigate the genetic basis of key traits related to rapeseed quality. A total of more than 62,000 SNP markers found in more than 11,000 genes were used. As a proof of concept, the erucic acid content in seed oil, a trait with known control genes, was selected for analysis. The seeds of 53 *B. napus* accessions were evaluated for erucic acid content, revealing significant variation ranging from 0% to 44.9%. These results demonstrated the abundance of polymorphisms in transcribed sequences, even in recently formed polyploid crops like oilseed rape, highlighting its broad applicability in plant biotechnology (Higgins et al. 2012). The assumption of variability of the trait is essential to AT analysis. By combining markers based on gene sequence and expression variation, AT successfully identified loci containing known genes involved in the biosynthesis and regulation of erucic acid. AT proved to be a powerful method for identifying markers linked to loci-controlling traits and deciphering causative variations (Higgins et al. 2012). Overall, the study showed that AT analysis can efficiently identify genomic regions controlling traits and provide markers for marker-assisted breeding even for crops with highly complex genomes and can be used for marker-assisted breeding.

Transcriptome association studies were conducted using over 62,000 SNPs and GEMs to investigate the genes associated with rapeseed quality traits. The evaluation of seeds from 53 *B. napus* accessions for erucic acid content demonstrated significant variation, with levels ranging from 0% to 44.9%. The results indicate a significant presence of polymorphisms in transcribed sequences, even in newly developed polyploid crops such as oilseed rape, underscoring its extensive relevance in plant biotechnology (Higgins et al. 2012). The AT has demonstrated efficacy in identifying loci-associated markers that control traits and elucidating causative variations (Higgins et al. 2012). The study revealed that AT analysis effectively identifies genomic regions associated with traits and offers markers for marker-assisted breeding, even in crops with complex genomes.

Havlickova et al. (2018) improved the AT platform by significantly increasing the genetic diversity in the panel of B. napus genotypes. The initial development of AT in B. napus involved a smaller genetic diversity panel of 84 accessions, successfully implicating orthologues of genes in connection with seed glucosinolate content, highlighting the promising potential of AT for genetic studies in this crop. Building on this early achievement, a larger genetic diversity panel comprising 383 accessions is investigated for a wide range of traits, providing valuable insights into the genetic architecture of complex traits. The platform's validity was further confirmed by analysing previously studied traits and exploring a new trait related to tocopherol content in seeds (Havlickova et al. 2018).

The study utilized an extended diversity panel of 383 *B. napus* doubled haploid (DH) or inbred accessions to represent a broad spectrum of genetic variation in the species. This comprehensive resource included various crop types of *B. napus*, making it ideal for studying genetic diversity and trait variation (Havlickova et al. 2018). In the same manner as, RNAseq data for each genotype were mapped to the Brassica AC pan-transcriptome reference, encompassing 116,098 gene models. Subsequently, the researchers discovered several times more SNP markers. In particular, 355,536 SNPs were found across the panel, with 87.0% of them being hemi-SNPs. It creates a high-density SNP map of one SNP every 0.33 kb (Havlickova et al. 2018). They also detected significant expression (RPKM >0.4) for 53,889 GEM markers, with 46.4% of all unigenes in the pan-transcriptome reference showing expression (Havlickova et al. 2018). The new AT platform's initial evaluation involved the analysis of the genetic architecture underlying seed erucic acid content. During this analysis, known loci associated with erucic acid content were successfully identified, and an additional minor-effect locus that had not been previously detected was discovered. Subsequently, the platform was utilized to investigate the variation in the relative proportions of tocopherol (vitamin E) forms present in seeds, where previously known loci were also successfully found. The new functional genotypes created for the diverse rapeseed panel are valuable resources for rapeseed researchers. Additionally, the AT approach's success in identifying linked markers and candidate genes for trait causation sets a precedent for broader applications in plant and crop science (Havlickova et al. 2018).

7.1 AT for Anion Homeostasis in *B. napus*

A further investigation employing AT was carried out by Koprivova et al. (2014) to investigate the genetic control of variation in nitrate, phosphate, and sulphate levels in the leaves of oilseed rape. The regulation of anion homeostasis is essential and significantly contributes to numerous physiological processes, ensuring the equilibrium of negatively charged ions that are crucial for cellular functions in plants (Zhu 2016). The study utilized an identical diversity panel comprising 84 *B. napus* accessions correlating quantitative trait variation with SNPs and GEMs markers. The study revealed 13 potential candidate genes that regulate sulphate, nitrate, and phosphate levels in *B. napus*, highlighting the importance of AT as an effective method for analysing complex traits (Koprivova et al. 2014).

7.2 AT for Glucosinolate in *B. napus*

Lu et al. (2014) investigated the genetic structure of various glucosinolate biosynthesis pathways in *B. napus*. Glucosinolates are essential in oilseed rape for defence mechanisms against herbivores and pests while also impacting the quality and nutritional characteristics of the oil derived from the seeds (Howell et al. 2003). The investigation utilized AT and incorporated an extensive panel of 101 *B. napus* accessions with a new reference sequence identifying 144,131 SNP markers mapped onto the specific A and C genomes. Additionally, the analysis employed RNAseq data to increase the count of GEM markers to 189,116, with 94,558 located on the A and C genomes. Using association evaluation with SNP markers, ten significant loci associated with total glucosinolate concentration in seeds were identified (Lu et al. 2014).

Furthermore, the analysis utilizing GEM markers identified seven notable associations positioned on chromosomes A2, A4, A9, C2, C4, C7, and C9. A thorough examination of the candidate genes determined that 24 were directly implicated in GSL metabolism (Li et al. 2014; Hirai et al. 2007). The investigation employed AT and enhanced the resolution of AT in comparison to a previous GWAS study (Li et al. 2014). Utilizing the candidate genes, PCR-based markers were established, and their confirmed genetic impact on GSL within the diversity panel offers a significant resource for germplasm screening and breeding selection aimed at low GSL lines (Lu et al. 2014). Kittipol et al. (2019) investigated the genetic regulation of glucosinolate (GSL) biosynthesis in the leaves and roots of *B. napus*, utilizing 288 diverse accessions from a new diversity panel, along with 355,536 SNP markers and 53,889 GEM markers. Havlickova et al. (2018). Significant associations were identified between the content of GSLs in leaves and aliphatic GSL content in regions of chromosomes A2, A9, C2, C9, and C7. The analysis identified associations between GEMs and leaf aliphatic GSL content, highlighting six genes directly involved in the biosynthesis of aliphatic GSLs. The HAG1 gene has been identified as a significant GEM marker linked to root aliphatic GSL content (Kittipol et al. 2019).

7.3 AT for Lodging Resistance in *B. napus*

A study performed by Miller et al. in 2018 focused on the lodging problem in oilseed rape to mitigate lodging risk, which is crucial for enhancing oilseed rape yields and overall crop productivity. The traits related to lodging were found to differ among a panel of 79 accessions, with lodging resistance exhibiting an approximately normal distribution within the diversity panel. Additionally, AT analysis revealed strongly associated markers to genes that regulate pectin methylesterase activity, influencing stem mechanical strength. Notable SNP association peaks were identified on chromosomes A2/C2, A3/C3, and A5/C5. Furthermore, associations of GEM with cell wall biosynthesis and mannan structure were identified. Arabidopsis T-DNA mutants were evaluated to assess their influence on stem strength, establishing that pectin methylesterase activity directly impacts absolute stem strength (Miller et al. 2018). The findings from the AT analysis corroborate the hypothesis that pectin demethylesterification influences stem strength, as previously indicated (Hongo et al. 2012). Pectins, in conjunction with cellulose and hemicelluloses, constitute fundamental elements of the plant cell wall. Two hypotheses regarding the effect of pectin demethylesterification are proposed: one suggests cell wall loosening and calcium bridge formation (Müller et al. 2013), while the other proposes that cell wall loosening facilitates cell expansion and regulates stem thickness and strength (McCann et al. 2001). Overall, AT is an effective instrument for elucidating the genetic regulation of complex traits, including stem strength (Miller et al. 2018).

7.4 AT for Ca and Mg Accumulation in *B. napus*

Alcock et al. (2017) utilized transcriptome sequences from 383 genotypes of B. napus to understand the different genes responsible for the accumulation of Ca and Mg in oilseed rape and its underlying causes. Calcium plays a crucial role in plant cell wall strengthening and signalling, supporting various physiological functions necessary for plant growth and development (Grusak et al. 2016). On the other hand, magnesium is essential for photosynthesis, and it plays a critical role in protein synthesis by bridging ribosome subunits, along with its involvement in photophosphorylation and the generation of reactive oxygen species in plants (Cakmak and Yazici 2010). The diversity panel showed variation in leaf Ca and Mg concentrations, with Ca ranging from 5838 mg/kg to 18,752 mg/kg and Mg ranging from 5118 mg/kg to 13,429 mg/kg. The frequency distribution of leaf Ca and Mg concentrations followed an approximately normal distribution (Alcock et al. 2017).

A total of 247 SNPs and 5557 GEMs showed significant associations with leaf calcium concentration. A total of 166 SNPs were significantly associated with leaf Mg concentration; of these, 86 were identical to those associated with leaf Ca concentration, suggesting a potentially similar regulatory mechanism. The findings can be utilized in marker-assisted breeding to create lines with elevated Ca and Mg concentrations.

7.5 AT for Clubroot Disease Resistance in *B. napus*

Hejna et al. (2019) conducted a significant study employing AT to investigate the genetic basis of clubroot resistance. Clubroot disease, caused by *Plasmodiophora brassicae*, forms galls or clubs on roots and hypocotyls. This condition impairs water and mineral absorption, resulting in wilting, stunted growth, chlorosis, and leaf abscission. Consequently, significant yield reductions occur, with infested regions experiencing average losses exceeding 20%, and severe instances potentially resulting in total crop failure (Pageau et al. 2006; Hejna et al. 2019). The evaluation of 245 *B. napus* accessions for clubroot resistance revealed significant variation in resistance levels. Of the accessions, 35.5% were categorized as resistant, while 19.6% displayed extreme susceptibility, with the predominant DI (disease index) range being 61–70, accounting for 24% (Hejna et al. 2019). Nine SNP association peaks or loci linked to trait variation were identified, resulting in 86 associated SNP markers. A total of 63 candidate genes were identified from the SNP analysis, regarded as potential candidates, including those that contain transcription factors with defence domains. Through GEM analysis, candidate genes were identified using a significance threshold of $-\log10P > 3.5$, identifying 21 genes (Hejna et al. 2019). This method may facilitate the development of more resilient crop varieties exhibiting improved resistance to clubroot disease.

7.6 AT for Freezing Stress Tolerance in *B. napus*

The recent publication utilizing AT analyses the genetic composition of oilseed rape and identifies genes linked to adaptation to freezing stress (Huang et al. 2020). Extended exposure to low temperatures in winter stresses semi-winter rapeseed varieties, resulting in plant damage and diminished yields (O'Neill et al. 2019; Zhang et al. 2017). The research utilized a field-grown diversity association panel subjected to prolonged low-temperature stress during winter, highlighting photosynthetic efficiency as a key factor and a dependable measure of low-temperature tolerance in plants. Gas exchange parameters related to photosynthesis were assessed in 123 rapeseed accessions, yielding insights into their photosynthetic efficiency. The application of AT has identified genomic regions that regulate photosynthetic-related traits. Of the 201 significant SNPs identified from 148 coding sequences (CDSs), most were linked to stomatal conductance to water vapour and transpiration rate. Furthermore, 145 GEMs were identified, demonstrating diverse associations with various traits. A functional analysis identified 22 candidate genes that may play a role in photosynthesis or respond to low-temperature stress, providing targets for subsequent validation. The leading candidate gene associated with low-temperature tolerance traits was identified as an orthologue of the *Arabidopsis* gene AT2G29300.2 (*BnTR1*). The findings indicate a strong correlation with the transpiration rate and a contribution to enhanced LT tolerance. The ectopic expression of BnTR1 in Arabidopsis plants led to an elevated transpiration rate and improved tolerance to low temperatures under freezing conditions. The findings demonstrate the effectiveness of AT in investigating LT tolerance in oilseed rape and emphasize the potential importance of *BnTR1* as a target gene for the genetic enhancement of LT tolerance across different plant species (Huang et al. 2020).

8 Conclusion

Rapeseed breeders may significantly benefit from precisely identifying genetic resources and varieties and detecting genetic distances between particular accessions. Precise knowledge about genetic distance can facilitate the selection of parental components based on the highest genetic diversity, not only based on morphological/qualitative characteristics. In these cases, molecular analysis results may reveal the actual diversity among the genetic resources of rapeseed. Therefore, they possess the potential to contribute significantly to the precise identification of parental components in hybridization initiatives, thereby improving the effectiveness of both line and hybrid breeding endeavours. Genome-wide association studies (GWAS) serve as a robust method for identifying candidate genes associated with traits of interest to breeders. GWAS have limitations in polyploid crops, whereas associative transcriptomics (AT) can effectively address these limitations. Associative transcriptomics is a powerful and widely adopted method for identifying genes of interest and creating molecular selection markers. Contemporary molecular biology techniques are essential to breeding practices, facilitating the efficiency and acceleration of the breeding process and the development of new, valuable varieties.

References

Adams KL, Cronn R, Percifield R, Wendel JF (2003) Genes duplicated by polyploidy show unequal contributions to the transcriptome and organ-specific reciprocal silencing. Proc Natl Acad Sci USA 100:4649–4654

Akbar MA (1989) Resynthesis of *Brassica napus* aiming for improved earliness and carried out by different approaches. Hereditas 111(3):239–246. https://doi.org/10.1111/j1601-52231990tb00402x

Al-Shehbaz IA, Beilstein MA, Kellogg EA (2006) Systematics and phylogeny of the Brassicaceae (Cruciferae): an overview. Plant Syst Evol 259(2):89–120. https://doi.org/10.1007/s00606-006-0415-z

Alcock TD, Havlickova L, He Z, Bancroft I, White PJ, Broadley MR, Graham NS (2017) Identification of candidate genes for calcium and magnesium accumulation in *Brassica napus* L by association genetics. Front Plant Sci 8. https://www.frontiersin.org/articles/10.3389/fpls.2017.01968

Allender CJ, King GJ (2010) Origins of the amphidiploid species *Brassica napus* L investigated by chloroplast and nuclear markers. BMC Plant Biol 10(1):54. https://doi.org/10.1186/1471-2229-10-54

Andersson G, Olsson G (1961) Cruciferen-0lpflanzen p61-66. In: Kappert H, Rudorf W (eds) Handbuch der Pflanzenzüchtung, vol 5, 2nd edn. Paul Parey, Berlin

Annisa YG, Sheng Ch, Wallace C (2011) Global genetic diversity of *Brassica rapa*. 17th Australian research assembly on brassicas (ARAB), Wagga Wagga NSW, August 2011, pp 17–19

Baux A, Colbach N, Pellet D (2011) Crop management for optimal low-linolenic rapeseed oil production—field experiments and modelling. Eur J Agron 35(3):144–153. https://doi.org/10.1016/j.eja.2011.05.006

Bayer PE, Scheben A, Golicz AA, Yuan Y, Faure S, Lee HT, Chawla HS, Anderson R, Bancroft I, Raman H, Lim YP, Robbens S, Jiang L, Liu S, Barker MS, Schranz E, Wang X, King GJ, Pires C, Edwards D (2021) Modelling of gene loss propensity in the pangenomes of three *Brassica* species suggests different mechanisms between polyploids and diploids. Plant Biotechnol J 19:2488–2500. https://doi.org/10.1111/pbi.13674

Brandt SA, Malhi SS, Ulrich D, Lafond GR, Kutcher HR, Johnston AM (2007) Seeding rate, fertilizer level and disease management effects on hybrid versus open pollinated canola (*Brassica napus* L). Can J Plant Sci 87(2):255–266. https://doi.org/10.4141/P05-223

Brown GG, Formanova N, Jin H, Wargachuk R, Dendy C, Patil P (2003) The radish Rfo restorer gene of Ogura cytoplasmic male sterility encodes a protein with multiple pentatricopeptide repeats. Plant J 35(2):262–272. https://doi.org/10.1046/j.1365-313X.2003.01799.x

Cakmak I, Yazici AM (2010) Magnesium: a forgotten element in crop production better. Crops 94:23–25

Carré P, Pouzet A (2014) Rapeseed market, worldwide and in Europe. OCL 21(1). https://doi.org/10.1051/ocl/2013054

Chalhoub B, Denoeud F, Liu S, Parkin IA, Tang H, Wang X, Chiquet J, Belcram H, Tong C, Samans B, Corréa M, Da Silva C, Just J, Falentin C, Koh CS, Le Clainche I, Bernard M, Bento P, Noel B, Labadie K, Alberti A, Charles M, Arnaud D, Guo H, Daviaud C, Alamery S, Jabbari K, Zhao M, Edger PP, Chelaifa H, Tack D, Lassalle G, Mestiri I, Schnel N, Le Paslier MC, Fan G, Renault V, Bayer PE, Golicz AA, Manoli S, Lee TH, Thi VH, Chalabi S, Hu Q, Fan C, Tollenaere R, Lu Y, Battail C, Shen J, Sidebottom CH, Wang X, Canaguier A, Chauveau A, Bérard A, Deniot G, Guan M, Liu Z, Sun F, Lim YP, Lyons E, Town CD, Bancroft I, Wang X, Meng J, Ma J, Pires JC, King GJ, Brunel D, Delourme R, Renard M, Aury JM, Adams KL, Batley J, Snowdon RJ, Tost J, Edwards D, Zhou Y, Hua W, Sharpe AG, Paterson AH, Guan C, Wincker P (2014) Early allopolyploid evolution in the post-Neolithic *Brassica napus* oilseed genome. Science 345(6199):950–953. https://doi.org/10.1126/science.1253435

Chen BY, Heneen WK (1989) Resynthesized *Brassica napus* L: a review of its potential in breeding and genetic analysis. Hereditas 111(3):255–263. https://doi.org/10.1111/j.1601-5223.1990.tb00404.x

Cheung F, Trick M, Drou N, Lim YP, Park J-Y, Kwon S-J, Kim J-A, Scott R, Pires JC, Paterson AH, Town C, Bancroft I (2009) Comparative analysis between homoeologous genome segments of *Brassica napus* and its progenitor species reveals extensive sequence-level divergence. Plant Cell 21(7):1912–1928. https://doi.org/10.1105/tpc.108.060376

Chikkaputtaiah C, Debbarma J, Baruah I, Havlickova L, Deka Boruah HP, Curn V (2017) Molecular genetics and functional genomics of abiotic stress-responsive genes in oilseed rape (*Brassica napus* L): a review of recent advances and future. Plant Biotechnol Rep 11(6):365–384. https://doi.org/10.1007/s11816-017-0458-3

Cowling WA, Castro-Urrea FA, Stefanova KT, Li L, Banks RG, Saradadevi R, Sass O, Kinghorn BP, Siddique KHM (2023) Optimal contribution selection improves the rate of genetic gain in grain yield and yield stability in spring canola in Australia and Canada. Plants 12(2):383. https://doi.org/10.3390/plants12020383

Cui C, Zhang K, Chai L, Zheng B, Zhang J, Jiang J, Jiang L (2024) Unraveling the mechanism of flower color variation in *Brassica napus* by integrated metabolome and transcriptome analyses. Front Plant Sci 15. https://doi.org/10.3389/fpls.2024.1419508

Curn V (1995) Acid phosphatase and leucine aminopeptidase isozymes as biochemical markers of homogeneity in oilseed rape androgenetic lines. Plant Growth Regul 16(1):59–63. https://doi.org/10.1007/BF00040508

Daun JK (1983) The introduction of low erucic acid rapeseed varieties into Canadian production. In: Kramer JKG, Sauer FD, Pigden WJ (eds) High and low erucic acid rapeseed oils production, usage, chemistry, and toxicological evaluation. Academic Press, New York, pp 161–180

Daun JK (1986) Erucic acid levels in Western Canadian canola and rapeseed. J Am Oil Chem Soc 63(3):321–324. https://doi.org/10.1007/BF02546037

Delourme R, Foisset N, Horvais R, Barret P, Champagne G, Cheung WY, Landry BS, Renard M (1998) Characterisation of the radish introgression carrying the Rfo restorer gene from the Ogu–INRA cytoplasmatic male sterility in rapeseed (*Brassica napus* L). Theor Appl Genet 97(1):129–134. https://doi.org/10.1007/s001220050876

Delourme R, Falentin C, Fomeju BF, Boillot M, Lassalle G, André I, Pauquet J (2013) High-density SNP-based genetic map development and linkage disequilibrium assessment in *Brassica napus* l. BMC Genomics 14(1):120. https://doi.org/10.1186/1471-2164-14-120

Diers BW, Osborn TC (1994) Genetic diversity of oilseed *Brassica napus* germplasm based on restriction fragment length polymorphisms. Theor Appl Genet 88(6):662–668. https://doi.org/10.1007/BF01253968

dos Santos JB, Nienhuis J, Skroch P, Tivang J, Slocum MK (1994) Comparsion of RAPD and RFLP genetic markers in determining genetic similarity among *Brassica oleracea* L genotype. Theor Appl Genet 87(8):909–915. https://doi.org/10.1007/BF00225784

Dou S, Zhang T, Tu J, Shen J, Yi B, Wen J, Fu T, Dai C, Ma C (2021) Generation of novel self-incompatible *Brassica napus* by CRISPR/Cas9. Plant Biotechnol J 19(5):875–877. https://doi.org/10.1111/pbi.13577

Downey RK (1990) Brassica oilseed breeding: achievements and opportunities. Plant Breed 60:1165–1170

Downey RK, Rakow GFW (1987) Rapeseed and mustard. In: Fehr WR (ed) Principles of cultivar development—crop species 2. Macmillan, New York, pp 437–486

Downey RK, Klassen AJ, Stringam GR (1980) Rapeseed and mustard. In: Fehr WR, Hadley HH (eds) Hybridization of crop plants. American Society of Agronomy, Inc., and the Crop Science Society of America, Inc, Madison, pp 495–509

Eckerstorfer MF, Dolezel M, Heissenberger A, Miklau M, Reichenbecher W, Steinbrecher RA, Waßmann F (2019) An EU perspective on biosafety considerations for plants developed by genome editing and other new genetic modification techniques (nGMs). Front Bioengg Biotechnol 7. https://www.frontiersin.org/articles/103389/fbioe.2019.00031

Ehrlich P, Ehrlich A (1983) Extinction: the cause and consequences of the disappearance of species. Ballantine, New York

Eskin NAM, McDonald BE, Przybylski R, Malcolmson LJ, Scarth R, Mag T, Ward K, Adolph D (2006) Aceite de canola: Historia y desarrollo Aceites Grasas. Rosário 4:594–604

Fernie AR, Tadmor Y, Zamir D (2006) Natural genetic variation for improving crop quality. Curr Opin Plant Biol 9(2):196–202. https://doi.org/10.1016/jpbi.2006.01.010

Frandsen KJ (1943) The experimental formation of *Brassica juncea* Czern et Coss. Dansk Bot Ark 11:1–17

Frandsen KJ (1947) The experimental formation of *Brassica napus* L var oleifera DC and *Brassica carinata* Braun. Dansk Bot Ark 12:1–16

Frauen M, Baer A (1996) Hybrids of winter rapeseed—varieties for the future? GCIRC Bull 12:47–51

Friedt W, Snowdon R, Ordon F, Ahlemeyer J (2007) Plant breeding: assessment of genetic diversity in crop plants and its exploitation in breeding. In: Esser K, Löttge U, Beyschlag W, Murata J (eds) Progress in botany. Springer, Berlin/Heidelberg, pp 151–178. https://doi.org/10.1007/978-3-540-36832-8_7

Friedt W, Tu J, Fu T (2018) Academic and economic importance of *Brassica napus* rapeseed. In: Liu S, Snowdon R, Chalhoub B (eds) The *Brassica napus* genome. Springer, pp 1–20. https://doi.org/10.1007/978-3-319-43694-4_1

Fu TD, Yang GS, Yang XN, Ma CZ (1995) Discovery, study and utilization of Polima cytoplasmic male sterility in *Brassica napus* L. Prog Nat Sci 2:43–51

Fussell GE (1966) Farming technique from prehistoric to modern times. Pergamon Press, Oxford/London/Edinburgh

Gacek K, Bartkowiak-Broda I, Batley J (2018) Genetic and molecular regulation of seed storage proteins (SSPS) to improve protein nutritional value of oilseed rape (*Brassica napus* l.) seeds. Front. Plant Sci 9. https://doi.org/10.3389/fpls.2018.00890

Girke A, Schierholt A, Becker HC (2012) Extending the rapeseed genepool with resynthesized *Brassica napus* L I: genetic diversity. Genet Resour Crop Evol 59(7):1441–1447. https://doi.org/10.1007/s10722-011-9772-8

Gómez-Campo C (1999) Biology of *Brassica* coenospecies, Developments in plant genetics and breeding. Elsevier, Amsterdam/New York

Gómez-Campo C, Tortosa ME (1974) The taxonomic and evolutionary significance of some juvenile characters in the Brassicaceae. Bot J Linn Soc 69:105–124. https://doi.org/10.1111/j.1095-8339.1974.tb01619.x

Grant I, Cherne DG, Patel JD (1991) Breeding strategies for the development of hybrid canola. GCIRC 1991 congress, 8th IRC Saskatoon, Saskatchewan, Canada, July 9–11, 1991, pp 82–83

Grusak MA, Broadley MR, White PJ (2016) Plant macro- and micronutrient minerals chichester. Wiley. https://doi.org/10.1002/9780470015902.a0001306.pub2

Gupta SK (2007) Rapeseed breeding advances in botanical research, vol 45. Elsevier, Amsterdam

Gupta K (2009) Biology and breeding of crucifers. CRC Press, Boca Raton. https://doi.org/10.1201/9781420086096

Håkansson A (1956) Seed development in *Brassica oleracea* and *B rapa* after certain reciprocal pollinations. Hereditas 42(3–4):373–396. https://doi.org/10.1111/j1601-52231956tb03024x

Hallden C, Nilsson NO, Rading IM, Säll T (1994) Evaluation of RFLP and RAPD markers in a comparison of *Brassica napus* breeding lines. Theor Appl Genet 88(1):123–128. https://doi.org/10.1007/BF00222404

Hasan M, Seyis F, Badani AG, Pons-Kühnemann J, Friedt W, Lühs W, Snowdon RJ (2006) Analysis of genetic diversity in the *Brassica napus* L gene pool using SSR markers. Genet Ress Crop Evol 53(4):793–802. https://doi.org/10.1007/s10722-004-5541-2

Havlíčková L, Čurn V, Jozová E, Kučera V, Klíma M (2012) Sequence analysis of the mtDNA region correlated with Shaan 2A cytoplasmic male sterility in rapeseed (*Brassica napus* L). Czech J Genet Plant Breed 48:139–142. https://doi.org/10.17221/144/2012-CJGPB

Havlickova L, Jozová E, Rychla A, Klíma M, Kucera V, Curn V (2014) Genetic diversity assessment in winter oilseed rape (*Brassica napus* L) collection using AFLP, ISSR and SSR markers. Czech J Genet Plant Breed 50:216–225. https://doi.org/10.17221/220/2013-CJGPB

Havlickova L, He Z, Wang L, Langer S, Harper AL, Kaur H, Broadley MR, Gegas V, Bancroft I (2018) Validation of an updated associative transcriptomics platform for the polyploid crop species *Brassica napus* by dissection of the genetic architecture of erucic acid and tocopherol isoform variation in seeds. Plant J 93(1):181–192. https://doi.org/10.1111/tpj.13767

Hayward A, Mason A, Dalton-Morgan J, Zander M, Edwards D, Batley J (2012) SNP discovery and applications in *Brassica napus*. J Plant Biotechnol 39:1–12

He Z, Ji R, Havlickova L, Wang L, Li Y, Lee HT, Song J, Koh C, Yang J, Zhang M, Parkin IAP, Wang X, Edwards D, King GJ, Zou J, Liu K, Snowdon RJ, Banga SS, Machackova I, Bancroft I (2021) Genome structural evolution in brassica crops. Nat Plants 7(6):757–765. https://doi.org/10.1038/s41477-021-00928-8

Hejna O, Havlickova L, He Z, Bancroft I, Curn V (2019) Analysing the genetic architecture of clubroot resistance variation in *Brassica napus* by associative transcriptomics. Mol Breed 39(8):112. https://doi.org/10.1007/s11032-019-1021-4

Higgins J, Magusin A, Trick M, Fraser F, Bancroft I (2012) Use of mRNA-seq to discriminate contributions to the transcriptome from the constituent genomes of the polyploid crop species *Brassica napus*. BMC Genomics 13(1):247. https://doi.org/10.1186/1471-2164-13-247

Hilgert-Delgado A, Klíma M, Viehmannová I, Urban MO, Fernández-Cusimamani E, Vyvadilová M (2015) Efficient resynthesis of oilseed rape (*Brassica napus* L) from crosses of winter types *B rapa* × *B oleracea* via simple ovule culture and early hybrid verification. Plant Cell Tiss Organ Cult 120(1):191–201. https://doi.org/10.1007/s11240-014-0593-2

Hirai MY, Sugiyama K, Sawada Y, Tohge T, Obayashi T, Suzuki A, Araki R, Sakurai N, Suzuki H, Aoki K, Goda H, Nishizawa OI, Shibata D, Saito K (2007) Omics-based identification of *Arabidopsis* Myb transcription factors regulating aliphatic glucosinolate biosynthesis. Proc Natl Acad Sci 104(15):6478–6483. https://doi.org/10.1073/pnas.0611629104

Hoban S, Campbell CD, da Silva JM, Ekblom R, Funk WC, Garner BA, Godoy JA, Kershaw F, MacDonald AJ, Mergeay J, Minter M, O'Brien D, Vinas IP, Pearson SK, Pérez-Espona S, Potter KM, Russo IRM, Segelbacher G, Vernesi C, Hunter ME (2021) Genetic diversity is considered important but interpreted narrowly in country reports to the convention on biological diversity: current actions and indicators are insufficient. Biol Conserv 261:109233. https://doi.org/10.1016/j.biocon.2021.109233

Hongo S, Sato K, Yokoyama R, Nishitani K (2012) Demethylesterification of the primary wall by PECTIN METHYLESTERASE35 provides mechanical support to the Arabidopsis stem. Plant Cell 24(6):2624–2634. https://doi.org/10.1105/tpc.112.099325

Howell PM, Sharpe AG, Lydiate DJ (2003) Homoeologous loci control the accumulation of seed glucosinolates in oilseed rape (*Brassica napus*). Genome 46(3):454–460. https://doi.org/10.1139/g03-028

Hu D, Jing J, Snowdon RJ, Mason AS, Shen J, Meng J, Zou J (2021) Exploring the gene pool of *Brassica napus* by genomics-based approaches. Plant Biotechnol J 19(9):1693–1712. https://doi.org/10.1111/pbi.13636

Huang Y, Hussain MA, Luo D, Xu H, Zeng C, Havlickova L, Bancroft I, Tian Z, Zhang X, Cheng Y, Zou X, Lu G, Lv Y (2020) A *Brassica napus* reductase gene dissected by associative transcriptomics enhances plant adaption to freezing stress. Front Plant Sci 11. https://www.frontiersin.org/articles/103389/fpls.2020.00971

Ihien Katche ES, Mason A (2023) Resynthesized rapeseed (*Brassica napus*): breeding and genomics. Crit Rev Plant Sci 42(2):65–92. https://doi.org/10.1080/07352689.2023.2186021

Jesske T, Olberg B, Schierholt A, Becker HC (2013) Resynthesized lines from domesticated and wild Brassica taxa and their hybrids with *B napus* L: genetic diversity and hybrid yield. Theor Appl Genet 126(4):1053–1065. https://doi.org/10.1007/s00122-012-2036-y

Jhingan S, Harloff HJ, Abbadi A, Welsch C, Blümel M, Tasdemir D, Jung C (2023) Reduced glucosinolate content in oilseed rape (*Brassica napus* L) by random mutagenesis of BnMYB28 and BnCYP79F1 genes. Sci Rep 13(1):2344. https://doi.org/10.1038/s41598-023-28661-6

Jönsson R (1978) Breeding for improved oil and meal quality in rape (*Brassica napus* L) and turnip rape (*Brassica campestris* L). Hereditas 87(2):205–218. https://doi.org/10.1111/j.1601-5223.1978.tb01264.x

Kardos M, Armstrong EE, Fitzpatrick SW, Hauser S, Hedrick PW, Miller JM, Tallmon DA, Funk WC (2021) The crucial role of genome-wide genetic variation in conservation. Proc Natil Acad Sci 118(48):e2104642118. https://doi.org/10.1073/pnas.2104642118

Karp A, Edwards KJ (1997) Molecular techniques in the analysis of the extent and distribution of genetic diversity. International Plant Genetic Resources Institute, Rome, pp 11–38

Kittipol V, He Z, Wang L, Doheny-Adams T, Langer S, Bancroft I (2019) Genetic architecture of glucosinolate variation in *Brassica napus*. J Plant Physiol 240:152988. https://doi.org/10.1016/j.jplph.2019.06.001

Knoch D, Werner CR, Meyer RC, Riewe D, Abbadi A, Lücke S, Snowdon RJ, Altmann T (2021) Multi-omics-based prediction of hybrid performance in canola. Theor Appl Genet 134(4):1147–1165. https://doi.org/10.1007/s00122-020-03759-x

Kolář F, Fér T, Štech M, Trávníček P, Dušková E, Schönswetter P, Suda J (2012) Bringing together evolution on serpentine and polyploidy: spatiotemporal history of the diploid–tetraploid complex of *Knautia arvensis* (Dipsacaceae). PLoS One 7(7):e39988. https://doi.org/10.1371/journal.pone.0039988

Koprivova A, Harper AL, Trick M, Bancroft I, Kopriva S (2014) Dissection of the control of anion homeostasis by associative transcriptomics in *Brassica napus*. Plant Physiol 166(1):442–450. https://doi.org/10.1104/pp.114.239947

Koprna R, Kučera V, Kolovrat O, Vyvadilová M, Klíma M (2005) Development of self-incompatible lines with improved seed quality in winter oilseed rape (*Brassica napus* L) for hybrid breeding. Czech J Genet Plant Breed 41:105–111. https://doi.org/10.17221/3668-CJGPB

Kresovich S, Williams JGK, McFerson JR, Routman EJ, Schaal BA (1992) Characterization of genetic identities and relationships of Brassica oleracea L via a random amplified polymorphic DNA assay. Theor Appl Genet 85(2):190–196. https://doi.org/10.1007/BF00222859

Labana KS, Gupta ML (1993) Importance and origin. In: Labana KS, Banga SS, Banga SK (eds) Breeding oilseed brassicas. Springer, Berlin/Heidelberg, pp 1–7. https://doi.org/10.1007/978-3-662-06166-4_1

Lagercrantz U (1998) Comparative mapping between *Arabidopsis thaliana* and *Brassica nigra* indicates that brassica genomes have evolved through extensive genome replication accompanied by chromosome fusions and frequent rearrangements. Genetics 150(3):1217–1228. https://doi.org/10.1093/genetics/150.3.1217

Lagercrantz U, Lydiate DJ (1996) Comparative genome mapping in brassica. Genetics 144(4):1903–1910. https://doi.org/10.1093/genetics/144.4.1903

Li F, Chen B, Xu K, Wu J, Song W, Bancroft I, Harper AL, Trick M, Liu S, Gao G, Wang N, Yan G, Qiao J, Li J, Li H, Xiao X, Zhang T, Wu X (2014) Genome-wide association study dissects the genetic architecture of seed weight and seed quality in rapeseed (*Brassica napus* L). DNA Res 21(4):355–367. https://doi.org/10.1093/dnares/dsu002

Lombard V, Baril CP, Dubreuil P, Blouet F, Zhang D (2000) Genetic relationships and fingerprinting of rapeseed cultivars by AFLP: consequences for varietal registration. Crop Sci 40(5):1417–1425. https://doi.org/10.2135/cropsci2000.4051417x

Lu G, Harper AL, Trick M, Morgan C, Fraser F, O'Neill C, Bancroft I (2014) Associative transcriptomics study dissects the genetic architecture of seed glucosinolate content in *Brassica napus*. DNA Res 21(6):613–625. https://doi.org/10.1093/dnares/dsu024

Mailer RJ, Scarth R, Fristensky B (1994) Discrimination among cultivars of rapeseed (*Brassica napus* L) using DNA polymorphisms amplified from arbitrary primers. Theor Appl Genet 87(6):697–704. https://doi.org/10.1007/BF00222895

Marshall P, Marchand MC, Lisieczko Z, Landry BS (1994) A simple method to estimate the percentage of hybridity in canola (*Brassica napus*) F1 hybrids. Theor Appl Genet 89(7):853–858. https://doi.org/10.1007/BF00224508

Matthioli PA (1598) Opera quae extant omnia Nicolai Bassaei, Francofurti

McCann MC, Roberts K, Carpita NC (2001) Plant cell growth and elongation. In: eLS. Wiley, Hoboken

McNaughton IH (1976) Swedes and rapes (*Brassica napus*). In: Simmonds NW (ed) Evolution of crop plants. Longman Inc, London/New York, pp 53–56

Miller GP (1989) Public choice at the dawn of the special interest state: the story of butter and margarine. Calif Law Rev 77:83–131

Miller CN, Harper AL, Trick M, Werner P, Waldron K, Bancroft I (2016) Elucidation of the genetic basis of variation for stem strength characteristics in bread wheat by associative transcriptomics. BMC Genomics 17(1):500. https://doi.org/10.1186/s12864-016-2775-2

Miller CN, Harper AL, Trick M, Wellner N, Werner P, Waldron KW, Bancroft I (2018) Dissecting the complex regulation of lodging resistance in *Brassica napus*. Mol Breed 38(3):30. https://doi.org/10.1007/s11032-018-0781-6

Morinaga T (1934) Interspecific hybridization in Brassica VI the cytology of F1-hybrids of *B juncea* and *B nigra*. Cytologia 6:62–67. https://doi.org/10.1508/cytologia.6.62

Müller K, Gabriel LT, Sebastian B, Karin W, Wormit A, Usadel B, Haughn G, Allison R (2013) Demethylesterification of cell wall pectins in *Arabidopsis* plays a role in seed germination. Plant Physiol 161:305–316. https://doi.org/10.1104/pp.112.205724

Murphy DJ (1994) Designer oil crops breeding, processing and biotechnology. VCH Verlagsgesellschaft mbH, Weinheim/New York. https://doi.org/10.1002/jpln.1994.1570514

Nagaharu U (1935) Genome analysis in Brassica with special reference to the experimental formation of *B. napus* and peculiar mode of fertilization. Jpn J Bot 7:389–452

Nesbitt KA, Potts BM, Vaillancourt RE, West AK, Reid JB (1995) Partitioning and distribution of RAPD variation in a forest tree species, *Eucalyptus globulus* (Myrtaceae). Heredity 74(6):628–637. https://doi.org/10.1038/hdy.1995.86

Niemann J, Wojciechowski A, Janowicz J (2012) Broadening the variability of quality traits in rapeseed through interspecific hybridization with an application of immature embryo culture. Biotechnologia 93:109–115

O'Neill CM, Lu X, Calderwood A, Tudor EH, Robinson P, Wells R, Morris R, Penfield S (2019) Vernalization and floral transition in autumn drive winter annual life history in oilseed rape. Curr Biol 29(24):4300–4306e2. https://doi.org/10.1016/j.cub.2019.10.051

Pageau K, Reisdorf-Cren M, Morot-Gaudry J-F, Masclaux-Daubresse C (2006) The two senescence-related markers, GS1 (cytosolic glutamine synthetase) and GDH (glutamate dehydrogenase), involved in nitrogen mobilization, are differentially regulated during pathogen attack and by stress hormones and reactive oxygen species in *Nicotiana tabacum* L leaves. J Exp Bot 57(3):547–557. https://doi.org/10.1093/jxb/erj035

Pires JC, Zhao J, Schranz ME, Leon EJ, Quijada PA, Lukens LN, Osborn TC (2004) Flowering time divergence and genomic rearrangements in resynthesized Brassica polyploids (Brassicaceae). Biol J Linn Soc 82(4):675–688. https://doi.org/10.1111/j.1095-8312.2004.00350.x

Pradhan AK, Prakash S, Mukhopadhyay A, Pental D (1992) Phytogeny of *Brassica* and allied genera based on variation in chloroplast and mitochondrial DNA patterns: molecular and taxonomic classifications are incongruous. Theor Appl Genet 85(2):331–340. https://doi.org/10.1007/BF00222878

Prakash S (1980) Cruciferous oilseeds in India Japan. Scientific Society Press, Tokyo

Przybylski R, Mag T (2002) Canola/rapeseed oil. In: Gunstone FD (ed) Vegetable oils in food technology: composition, properties and uses. Blackwell Publishing, Hoboken, pp 98–127

Rahman MH (2005) Resynthesis of *Brassica napus* L for self-incompatibility: self-incompatibility reaction, inheritance and breeding potential. Plant Breed 124(1):13–19. https://doi.org/10.1111/j.1439-0523.2004.01045.x

Rai B, Gupta SK, Pratap A (2007) Breeding methods. In: Incorporating advances in plant pathology rapeseed breeding, Advances in botanical research, 45. Academic Press, Elsevier, pp 21–48. https://doi.org/10.1016/S0065-2296(07)45002-9

Rao VR, Hodgkin T (2002) Genetic diversity and conservation and utilization of plant genetic resources. Plant Cell Tiss Organ Cult 68:1–19. https://doi.org/10.1023/A:1013359015812

Renard M, Delourme R, Vallée P, Pierre J (1998) Hybrid rapeseed breeding and production. Acta Hortic 459:291–298. https://doi.org/10.17660/ActaHortic.1998.459.33

Röbbelen G (1985) Raps (*Brassica napus* L). In: Hoffmann W, Mudra A, Plarre W (eds) Lehrbuch der Züchtung landwirtschaftlicher Kulturpflanzen, B2 – SpezTeil. Paul Parey, Hamburg, pp 288–302

Salamini F, Özkan H, Brandolini A, Schäfer-Pregl R, Martin W (2002) Genetics and geography of wild cereal domestication in the near east. Nat Rev Genet 3(6):429–441. https://doi.org/10.1038/nrg817

Salgotra RK, Chauhan BS (2023) Genetic diversity, conservation, and utilization of plant genetic resources. Genes 14:174. https://doi.org/10.3390/genes14010174

Scarisbrick DH, Atkinson L, Asare E (1989) Oilseed rape. Outlook Agric 18(4):152–159. https://doi.org/10.1177/003072708901800403

Schenck HR, Uolf G (1986) Characterization of somatic *Brassica napus* hybrids by polyacrylamide gel electrophoresis. Plant Breed 97(1):72–74. https://doi.org/10.1111/j1439-05231986tb01303x

Schlötterer C (2004) The evolution of molecular markers—just a matter of fashion? Nat Rev Genet 5(1):63–69. https://doi.org/10.1038/nrg1249

Schmidt R, Bancroft I (2011) Genetics and genomics of the Brassicaceae. Springer, New York

Seebohm ME (1976) The evolution of the English farm. EP Publ, Ltd, Wakefield

Shen J, Liu Y, Wang X, Bai J, Lin L, Luo F, Zhong H (2023) A comprehensive review of health-benefiting components in rapeseed oil. Nutrients 15:999. https://doi.org/10.3390/nu15040999

Shiran B, Azimkhani R, Mohammadi S, Ahmadi MR (2006) Potential use of random amplified polymorphic DNA marker in assessment of genetic diversity and identification of rapeseed (*Brassica napus* L) cultivars. Biotechnol 5:153–159

Snowdon R, Lühs W, Friedt W (2006) Oilseed rape. In: Kole C (ed) Oilseeds. Springer, Berlin/Heidelberg, pp 55–114. https://doi.org/10.1007/978-3-540-34388-2_2

Sobotka R, Dolanská L, Èurn V, Ovesná J (2004) Fluorescence-based AFLPs occur as the most suitable marker system for oilseed rape cultivar identification. J Appl Genet 45(2):161–173

Soltis DE, Soltis PS, Doyle JJ (1998) Molecular systematics of plants II DNA sequencing. Kluwer Academic Publishers, Novell

Song K, Osborn TC (1992) Polyphyletic origins of *Brassica napus*: new evidence based on organelle and nuclear RFLP analyses. Genome 35(6):992–1001. https://doi.org/10.1139/g92-152

Song K, Osborn TC, Williams PH (1990) Brassica taxonomy based on nuclear restriction fragment length polymorphisms (RFLPs). Theor Appl Genet 79(4):497–506. https://doi.org/10.1007/BF00226159

Starosta E, Szwarc J, Niemann J, Szewczyk K, Weigt D (2023) *Brassica napus* haploid and double haploid production and its latest applications. Curr Issues Mol Biol 45:4431–4450. https://doi.org/10.3390/cimb45050282

Stefansson BR (1983) The development of improved rapeseed cultivars. In: Kramer JKG, Sauer FD, Pigden WJ (eds) High and low erucic acid rapeseed oils. Academic Press, New York, pp 143–159

Stefansson BR, Hougen FW (1964) Selection of rape plants (*Brassica napus*) with seed oil practically from erucic acid. Can J Plant Sci 44(4):359–364. https://doi.org/10.4141/cjps64-069

Stefansson BR, Hougen FW, Downey RK (1961) Note on the isolation of rape plants with seed oil free from erucic acid. Can J Plant Sci 41(1):218–219. https://doi.org/10.4141/cjps61-028

Stephenson P, Baker D, Girin T, Perez A, Amoah S, King GJ, Østergaard L (2010) A rich TILLING resource for studying gene function in *Brassica rapa*. BMC Plant Biol 10(1):62. https://doi.org/10.1186/1471-2229-10-62

Stokes D, Fraser F, Morgan C, O'Neill CM, Dreos R, Magusin A, Szalma S, Bancroft I (2010) An association transcriptomics approach to the prediction of hybrid performance. Mol Breed 26(1):91–106. https://doi.org/10.1007/s11032-009-9379-3

Sundberg E, Glimelius K (1986) A method for production of interspecific hybrids within Brassiceae via somatic hybridization, using resynthesis of *Brassica napus* as a model. Plant Sci 43(2):155–162. https://doi.org/10.1016/0168-9452(86)90157-3

Tabernaemontanum JT (1664) Neu vollkommentlich Kräuterbuch Basel

Ton LB, Neik TX, Batley J (2020) The use of genetic and gene technologies in shaping modern rapeseed cultivars (*Brassica napus* L). Genes 11:1161. https://doi.org/10.3390/genes11101161

Trick M, Long Y, Meng J, Bancroft I (2009) Single nucleotide polymorphism (SNP) discovery in the polyploid *Brassica napus* using Solexa transcriptome sequencing. Plant Biotechnol J 7(4):334–346. https://doi.org/10.1111/j.1467-7652.2008.00396.x

USDA (2023) Oilseeds: world markets and trade. United States Department of Agriculture—Foreign Agricultural Service. https://apps.fas.usda.gov/psdonline/circulars/oilseeds.pdf Accessed on 21/10/2025

van de Wouw M, van Hintum T, Kik C, van Treuren R, Visser B (2010) Genetic diversity trends in twentieth century crop cultivars: a meta analysis. Theor Appl Genet 120(6):1241–1252. https://doi.org/10.1007/s00122-009-1252-6

Vašák J, Fábry A (1991) Systém výroby řepky. NPZ Hans-Georg Lembke KG Hohenlieht Thomas Mann, Bonn

Wang N, Wang Y, Tian F, King GJ, Zhang C, Long Y, Shi L, Meng J (2008) A functional genomics resource for *Brassica napus*: development of an EMS mutagenized population and discovery of FAE1 point mutations by TILLING. New Phytol 180(4):751–765. https://doi.org/10.1111/j.1469-8137.2008.02619.x

Wei W, Youliang Z, Li C, Yuming W, Zehong Y (2005) Genetic diversity among the germplasm resources of the genus *Houttuynia thunb* in China based on RAMP markers. Genet Ress Crop Evol 52(4):473–482. https://doi.org/10.1007/s10722-005-2261-1

Weiss EA (1983) Oilseed crops. Longman, London/New York

Wen J, Tu J, Li Z, Fu T, Ma C, Shen J (2008) Improving ovary and embryo culture techniques for efficient resynthesis of *Brassica napus* from reciprocal crosses between yellow-seeded diploids *B. rapa* and *B. oleracea*. Euphytica 162(1):81–89. https://doi.org/10.1007/s10681-007-9566-4

Woods DL, Capcara JJ, Downey RK (1991) The potential of mustard (*Brassica juncea* (L) Coss) as an edible oil crop on the Canadian prairies. Can J Plant Sci 71(1):195–198. https://doi.org/10.4141/cjps91-025

Wu J, Li F, Xu K, Gao G, Chen B, Yan G, Wang N, Qiao J, Li J, Li H, Zhang T, Song W, Wu X (2014) Assessing and broadening genetic diversity of a rapeseed germplasm collection. Breed Sci 64:321–330. https://doi.org/10.1270/jsbbs.64.321

Yan X, Zeng X, Wang S, Li K, Yuan R, Gao H, Luo J, Liu F, Wu Y, Li Y, Zhu L, Wu G (2016) Aberrant meiotic prophase I leads to genic male sterility in the novel TE5A mutant of *Brassica napus*. Sci Rep 6(1):33955. https://doi.org/10.1038/srep33955

Yu C, Leišová L, Kučera V, Vyvadilová M, Ovesná J, Dotlačil J, Hu S (2007) Assessment of genetic diversity of yellow-seeded rapeseed (*Brassica napus* L) accessions by AFLP markers. Czech J Genet Plant Breed 43:105–112. https://doi.org/10.17221/2071-CJGPB

Žaludová J, Havlickova L, Jozová E, Kučera V, Vyvadilová M, Klíma M, Čurn V (2013) Marker assisted selection as a tool for detection of *Brassica napus* plants carrying self-incompatibility alleles, in hybrid breeding programs. Rom Agric Res 30:13–22

Zamani-Nour S, Clemens R, Möllers C (2013) Cytoplasmic diversity of *Brassica napus* L, *Brassica oleracea* L and *Brassica rapa* L as determined by chloroplast microsatellite markers. Gen Resour Crop Evol 60(3):953–965. https://doi.org/10.1007/s10722-012-9891-x

Zamir D (2001) Improving plant breeding with exotic genetic libraries. Nat Rev Genet 2(12):983–989. https://doi.org/10.1038/35103590

Zhang Z, Li J, Pan Y, Li J, Zhou L, Shi H, Zeng Y, Guo H, Yang S, Zheng W, Yu J, Sun X, Li G, Ding Y, Ma L, Shen S, Dai L, Zhang H, Yang S, Li Z (2017) Natural variation in CTB4a enhances rice adaptation to cold habitats. Nat Commun 8(1):14788. https://doi.org/10.1038/ncomms14788

Zhao JY, Becker HC (1998) Genetic variation in Chinese and European oilseed rape (*B napus*) and turnip rape (*B campestris*) analysed with isozymes. Acta Agron Sin 24:213–220

Zhu JK (2016) Abiotic stress signaling and responses in plants. Cell 167(2):313–324. https://doi.org/10.1016/j.cell.2016.08.029

Zhu C, Gore M, Buckler ES, Yu J (2008) Status and prospects of association mapping in plants. Plant Genome 1(1). https://doi.org/10.3835/plantgenome2008.02.0089

Microspore-Derived Haploid Embryos as a Tool for Developing Double Haploid Plants

Miroslav Klíma, Vladislav Čurn, Baveesh Pudhuvai, and Bhupendra Koul

Abstract

Microspore-derived haploid embryos are a vital resource in the development of double haploid plants, substantially improving the efficacy of breeding activities by enabling the production of homozygous lines in a single generation. Conventional breeding techniques often require extensive inbreeding across multiple generations in order to attain homogeneity; however, innovations in haploid development, especially by microspore embryogenesis, have transformed this procedure. A transition from gametophyte growth to haploid development can be triggered under specific conditions, yielding structures genetically identical to haploid spores formed during meiosis. This approach has been successful, resulting in the development of over 300 commercial cultivars across more than 250 crop species, particularly benefiting oilseed rape (*Brassica napus* L.) after the identification of certain varieties with negligible erucic acid concentration. Initial achievements in haploid production in oilseed rape encompassed the identification of spontaneous haploids and the release of the first double haploid variety, "Maris Haplona". The implementation of in vitro methods, including

M. Klíma
Laboratory of Plant Stress Biology and Biotechnology, Department of Plant Genetics and Crop Breeding, Crop Research Institute, Prague, Czechia

V. Čurn
Faculty of Agriculture and Technology, Department of Genetics and Biotechnology, University of South Bohemia in České Budějovice, České Budějovice, Czech Republic

B. Pudhuvai
Faculty of Agriculture and Technology, Department of Plant Production, University of South Bohemia in České Budějovice, České Budějovice, Czech Republic

B. Koul (✉)
Department of Biotechnology, School of Bioengineering & Biosciences, Lovely Professional University, Phagwara, Punjab, India

A. Kumar, B. Koul (eds.), *Crop Improvement Strategies in Brassica species: Applied Science*, https://doi.org/10.1007/978-981-95-3861-4_12

anther cultures and isolated microspore cultivation, has significantly enhanced haploid production efficiency, with advancements such as heat-shock treatment and liquid media improving embryogenesis rates. Present investigations continue in optimizing these techniques, addressing challenges like resistant genotypes and lower diploidization rates. Recent advancements in genetic manipulation, particularly CRISPR-Cas9 methodologies, have facilitated the in vivo production of maternal haploids, offering novel opportunities for breeding. As haploid technology becomes a standard component of breeding programmes, its integration with other biotechnological methodologies is anticipated to enhance the development of novel, resilient crop varieties. However, the preference for simpler approaches may present difficulties for the traditional androgenetic methods that have been fundamental to rapeseed breeding for the past 60 years.

Keywords

Haploids · Double haploid · Microspore · Embryo · Oilseed rape · Anther culture

1 Introduction

Breeders are meticulously engaged in improving the crop yield and quality parameters. In most economically significant crops, developing new cultivars has become a highly competitive endeavour, with breeders concentrating on utilizing superior lines in crossing programmes to obtain the desired genotype rapidly. Classical techniques of line cultivar production necessitate an extended inbreeding procedure, often spanning five to eight generations, contingent upon the uniformity requirements of the resulting cultivars. Similar considerations apply to hybrid breeding; homozygous parental components are essential for producing a uniform F1 hybrid cultivar.

A remarkable approach to significantly shorten the above process is the utilization of peculiarities during micro- (mega-)-sporogenesis. Transitioning from the male (female) gametophyte developmental pathway to microspore embryogenesis (gynogenesis or haploid parthenogenesis) is feasible at a certain stage and under particular conditions. The resultant sporophytic structure resembles the zygotic embryo. However, it possesses one crucial distinction: from a genetic and cytological perspective, it is identical to haploid spores (micro- and macro-) produced during meiosis in a plant. Moreover, the ability to effectively induce chromosome doubling in such haploids, leading to a completely homozygous genotype in one generation, brings enormous opportunities to the breeding process. The success of haploid production is evident from the review written by Forster and Thomas (2005), who reported sufficient doubled haploid technology in more than 250 crop species, from which more than 300 commercial cultivars have been produced.

The interest in the prospective incorporation of haploids in oilseed rape breeding (rapeseed, canola, *Brassica napus* L.) corresponds with the period during which rapeseed transitioned from a minor crop to a significant global oilseed. Rapeseed oil of old cultivars was characterized by elevated contents of erucic acid. During the early 1960s, cultivars with almost zero erucic acid content in the oil were discovered by Canadian scientists in a German cultivar "Liho" (Stefansson et al. 1961) and later in another European variety originating from Hungary (Stefansson and Hougen 1964). By crossing these selections with common Canadian accessions, the first varieties without erucic acid were available already during the 1960s, and an unprecedented expansion in rapeseed breeding, cultivation, and associated research, which continues to date, was started.

1.1 Milestones of Haploid Production in Oilseed Rape

Initially, it was about using spontaneous haploids discovered in the stands of common cultivars. Such haploids were first reported in Sweden in the winter rapeseed cultivar "Matador", but the frequency was low—around 12 plants per hectare (Olsson and Hagberg 1955). Thompson (1969) later compared the frequency of spontaneous haploids in several Canadian spring and European winter cultivars and their inbred lines. Haploids were found in all cultivars and inbred lines. However, their occurrence was still low for routine use in breeding, from 0.1 to 6.9 haploids per 1000 plants. Based on the large differences between Canadian and European cultivars, it was also hypothesized that a higher frequency of spontaneous haploids in Canadian cultivars might be genetically based. Later, the possibility of including haploids in the breeding was confirmed when the first doubled haploid (DH) variety ever, "Maris Haplona", was released in England (Thompson 1972). The applicability of spontaneous haploids in breeding was also evident from later experiments, where DH plants of spring and winter rapeseed, surpassing the initial varieties in terms of oil yield, were obtained after the colchicine treatment (Thompson 1979). Although after the discovery and utilization of androgenetic embryogenesis, the use of spontaneous haploids no longer made sense, this approach was crucial for further rapeseed breeding and research also for the following reasons: it (i) drew attention to the existence of haploids in oilseed rape, (ii) demonstrated the feasibility of their use for the production of DH cultivars, and (iii) indicated a possible fundamental shift in rapeseed breeding—a transition from open-pollinated populations to self-pollinated line cultivars.

Further progress in terms of haploid production and utilization was brought about by the in vitro cultivation of isolated anthers with immature pollen on nutrient media. In rapeseed, the first successful regeneration of androgenetic haploids was achieved by Thomas and Wenzel (1975) in anther cultures on solid cultivation media. Surprisingly, the regeneration of a haploid plant from randomly released pollen, separated from the anther during inoculation, was recorded. It should also be mentioned that the article included experiments with isolated immature pollen grains (i.e. microspores) as well, in which the formation of multicellular structures was observed. All these results indicated promising future developments in this area. However, the frequency of embryogenesis in anther cultures at these cultivation parameters was low. For instance, Wenzel et al. (1977) reported a rate of 0.1% of responsive anthers from 1000 cultivated anthers, which was, in terms of efficiency, comparable to the occurrence of spontaneous haploids in field conditions. A substantial improvement was achieved after introducing heat-shock treatment and its positive effect on the frequency of embryogenesis (Keller and Armstrong 1978). Applying a temperature of 32.5 °C at the beginning of cultivation significantly increased the frequency of embryogenesis by hundreds of times compared to treatments without heat shock (Hoffmann et al. 1982).

For the further efficiency of the haploid production in rapeseed, a significant contribution was made by research that introduced a liquid medium based on the Nitsch and Nitsch medium (Nitsch and Nitsch 1969). Initially, it was successful in other cultures (Lichter 1981) and shortly after that, in the culture of isolated microspores (Lichter 1982). The technique of isolated microspores on liquid culture medium NLN-13 in oilseed rape prevailed wholly and quickly displaced the anther culture approach. Microspore cultures can obtain several tens of well-developed cotyledonary embryos per bud for responsible genotypes within 30 days of the culture. In the following 15 years, various optimizations of the technique took place, resulting in the publication of a detailed manual both for spring and winter oilseed rape, covering all key steps: the preparation of donor plants, isolation and cultivation of microspores, regeneration of microspore embryos, cultivation of haploids and diploidization of entire plants using colchicine solution (Coventry et al. 1998).

The last significant contribution to the routine use of androgenetic embryogenesis in rapeseed breeding was the introduction of in vitro diploidization to microspore cultures using colchicine solution (Chen et al. 1994; Möllers et al. 2000) or other antimitotic substances, e.g. trifluralin or amiprophos methyl (Hansen and Andersen 1996). This approach eliminated the occurrence of chimaeras commonly seen after colchicine application to whole plants, had a positive effect on embryogenesis (Iqbal et al. 1994), and increased the efficiency of chromosome doubling (Klíma et al. 2008) compared with in vivo diploidization. Another indisputable advantage is that a relatively small amount of these substances is needed, which makes the procedure cheaper and safer. Even though several optimization attempts have been carried out to date, making the technique more efficient, we consider the most essential milestones, such as pilot experiments, aiming at (i) utilization of heat-shock treatment, (ii) isolated microspore culture in a liquid medium, and (iii) in vitro diploidization. For detailed information on the importance of haploids in oilseed rape, we refer to some comprehensive reviews that have been published in recent decades (Forster and Thomas 2005; Möllers and Iqbal 2009; Ferrie and Möllers 2011; Starosta et al. 2023). The schematic representation of in vitro haploid production from the isolated microspore culture and the factors influencing the process are shown in Fig. 1.

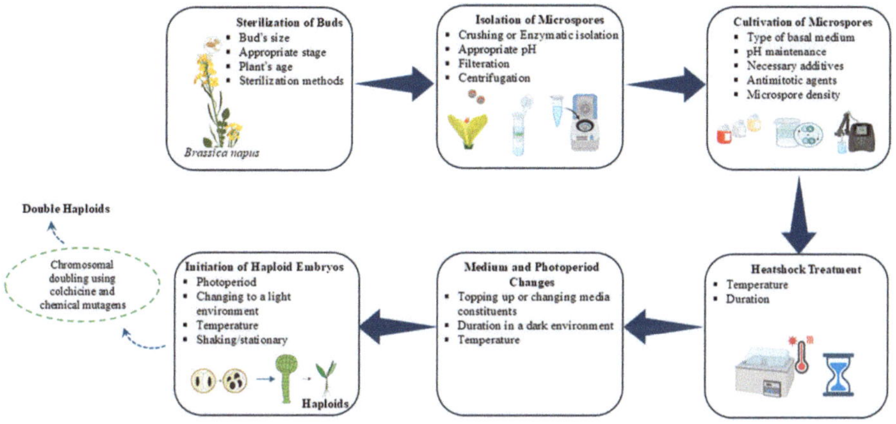

Fig. 1 Haploid induction process through microspore culture and its influencing factors

2 Improvements in Haploid Production

2.1 Microspore Embryogenesis and Plant Regeneration

Despite the adequacy of technology for the in vitro production of androgenetic haploids and doubled haploids for routine application in oilseed rape breeding initiatives by the late twentieth century (Coventry et al. 1998), ongoing research aims to improve the initial protocol. The majority of recent experiments are focused on the three main issues that hampered the efficiency of androgenetic DH technology, which were summarized in the review by Möllers and Iqbal (2009): (i) recalcitrant or poorly embryogenic genotypes, (ii) insufficient diploidization rate, and (iii) low direct plant regeneration from embryos in many cases, requiring repeated subcultivation of explants.

Tissue culture procedures are often modified to enhance embryogenic responses in microspores derived from poor embryogenic or recalcitrant varieties, cultivars, or species (Malik et al. 2008). Thus, various attempts have been made to improve embryogenic responsiveness and overall embryo production in rapeseed. A significant positive effect of lower temperatures (15/10 °C Day/night) for maintaining donor plants, cold pretreatment of flower buds on embryogenesis, and subsequent embryo development was found (Gu et al. 2004a). Experiments are also underway with the optimization of microspore isolation (Corral-Martínez et al. 2021), either selection and sorting of microspores on Percoll gradient to increase embryo production (Bhowmik et al. 2011).

Several published studies have investigated various experimental methods to initiate embryogenesis, such as manipulating the nutritive medium, adjusting inflorescence length, modifying the growth duration of the donor plant, altering heat shock conditions, and varying concentrations of activated charcoal and sucrose (Solís et al. 2015; Wan et al. 2011). However, an increase in embryogenesis was noted after the application of stress hormones: abscisic, jasmonic, and salicylic acid (Ahmadi et al. 2014b), brassinosteroids (Malik et al. 2008), or after the use of iron-free starvation media at the beginning of microspore culture (Leroux et al. 2016). Heat stress-induced embryogenesis was augmented by applying ethylene biosynthesis inhibitors (Leroux et al. 2009). There were also positive results by replacing sucrose with polyethylene glycol in the liquid NLN medium to initiate embryogenesis (Ferrie and Keller 2007). Mahasuk et al. (2017) tested different boron concentrations, which positively impacted embryogenesis. Ahmadi et al. (2014a), Ahmadi and Shariatpanahi (2015) increased embryogenesis after the application of proline, chitosan, and selected antibiotics. An interesting finding is that heat-shock treatment is not necessarily required for the initialization of embryogenesis, but a lower-than-normal cultivation temperature (around 18 °C) is also applicable. However, the frequency of embryogenesis was lower than that from heat-shock treatment (Prem et al. 2012). It has been confirmed that doubled haploids, regenerated from a poorly responsive genotype, preserve their good embryogenic responsibility, and this responsibility can even be transferred via crossing (Malik et al. 2008).

The diploidization of microspores represents a critical bottleneck in doubling haploid line production (Kampouridis et al. 2016). Therefore, experiments were carried out to increase the yield of doubled haploid regenerants in vitro by applying selected antimitotic substances in microspore culture (Klíma et al. 2008; Mohammadi et al. 2012; Szala et al. 2020). The possible genetic background of diploidization efficiency has been evaluated via inter-varietal substitution lines. Several genomic regions that may contain genetic factors governing diploidization rates have been identified. In addition, a marker with a high probability of being associated with this factor was identified (Kampouridis et al. 2016).

Experiments focused on embryo-plant conversion are also of significant importance. The positive effect of air desiccation of embryos with subsequent cultivation in the dark until the root activation phase on the subsequent regeneration of whole plants was observed (Prem et al. 2012). The cultivation of microspore embryos at 1 °C with an 8-hour light period over two weeks significantly enhanced the frequency of direct regeneration (Cegielska-Taras et al. 2002). A combination of desiccation, application of gibberellin and abscisic acid or their combinations increased the frequency of direct regeneration (Haddadi et al. 2008). Cold treatment, cotyledon excision, and subcultivation resulted in an increase in direct regeneration from embryos (Zhang et al. 2006).

Oilseed rape serves as a model system for the investigation of microspore embryogenesis (Yeung 2002; Ferrie and Möllers 2011). In recent decades, a significant quantity of research considering various genomic, transcriptomic, proteomic, and physiological approaches has been conducted, including numerous studies that investigate the influence of genes and associated physiological processes on the induction, development, and regeneration of microspore embryogenesis in plants (for instance, Tsuwamoto et al. 2007, Tsuwamoto and Takahata 2008; Joosen et al. 2007; Seguí-Simarro et al. 2011; El-Tantawy et al. 2013; Ecke et al. 2015; Rodríguez-Sanz et al. 2015; Ahmadi et al. 2016; Elahi et al. 2016; Li et al. 2016; Solís et al. 2016; Rivas-Sendra et al. 2017; Wang et al. 2020; Shahmir and Pauls 2021).

2.2 Artificial In Vivo Production of Haploids

The latest research in the haploid production field in rapeseed is also related to the development of new genetic manipulation techniques, which currently allow intraspecific crosses with haploid inducer lines for the in vivo production of maternal haploids in this crop. Zhong et al. (2021) utilized the CRISPR-Cas9 technique to generate knock-out mutants of the three BnaDMP genes. The crossing with DMP mutants resulted in the induction of in vivo maternal haploids at a frequency of 1.1–3.9%. Li et al. (2016) employed the aforementioned technique to knock out four BnaDMP genes in the Westar cultivar. Plants exhibiting DMP mutations in the T0, T1, and T2 generations demonstrated a haploid induction rate of 2.5%. Zhong et al. (2022) reported an induction rate of up to 4.4% in comparable experiments.

A new artificial in vivo induction method of doubled haploids without using an antimitotic substance was also presented. Fu et al. (2018) found a new in vivo route to produce maternal doubled haploids in rapeseed with an allo-octoploid rapeseed line as a DH inducer, which was created by interspecific crossing between *B. napus* and *B. rapa*, followed by crossing with *B. napus* and subsequent genome doubling. The induction rate via such an inducer varied between 39.9 and 98.7%, depending on the genotype of a maternal plant.

2.3 Haploids in Breeding

Researchers and rapeseed breeders have maintained close collaboration since the beginning of haploid technology development. The initial commercial doubled-haploid cultivar was introduced in oilseed rape. The preliminary idea of generating haploid and doubled haploid plants aimed to enhance plant breeding programmes. The androgenetic embryogenesis has been gradually adopted and implemented in all countries with significant canola oil production, breeding and research (Wenzel et al. 1977; Renard and Dosba 1980; Hoffmann et al. 1982; Kott 1998; Coventry et al. 1998; Cegielska-Taras 2002; Kučera et al. 2002; Gu et al. 2004b). Developing a homozygous line in one generation rather than several generations is beneficial as it accelerates breeding desirable varieties (Ferrie and Möllers 2011). However, the production of line cultivars and components of hybrids is not the only purpose of this technology. From the very beginning, it was proposed to use haploid tissues for the production of mutants (see Hoffmann 1978; Hoffmann et al. 1982) because the mutated gene is fixed in homozygous form after the doubling of the chromosome set (Szarejko et al. 1991). There were also attempts to apply in vitro selection in embryo cultures: Sacristán and Hoffmann (1979), after the treatment with *Plasmodiophora brassicae,* observed different reactions to this pathogen, and Sacristán (1982) reported the combination of mutagen application and selection pressure with the pathogen *Phoma lingam* in the culture of haploid embryos.

The microspore culture technique has emerged as the optimal tool for these purposes. According to Szarejko et al. (1991), this system demonstrates full compatibility with various biotechnological methods, including mutation and gene manipulation techniques, as it allows for the isolation and culture of millions of haploid microspores within a confined space, such as a Petri plate (Ferrie and Möllers 2011). Polsoni et al. (1988) indicated that optimizing the microspore isolation procedure by using entire inflorescences instead of individual buds enabled the production of several hundred thousand embryos within three weeks, a quantity adequate for mutation and selection studies. Anther or microspore cultures have been shown to undergo mutagenic treatment through radiation or chemical agents, resulting in variability during the haploid stage. Mutagenized populations of haploid cells facilitate rapid selection in in vitro environments. For instance, Swanson et al. (1989) obtained mutants from isolated microspore cultures with resistance to an herbicide (imidazolinone). Ahmad et al. (1991) used UV irradiation in the microspore culture to generate resistance to chlorsulfuron and *Alternaria brassicicola.* Combining a mutagen and subsequent selection using a NaCl solution in microspore embryo cultures resulted in the regeneration of plants more resistant to salinity (Rahman et al. 1995). Liu et al. (2005) published an attempt at in vitro mutation and selection of doubled haploid lines with improved resistance to *Sclerotinia sclerotiorum.* Ling et al. (2006) obtained genotypes resistant to glyphosate from a heterogeneous microspore embryo population after its application as a selection agent to the culture. In vitro selections for fatty acid content in microspore embryo cotyledons were also successful (Albrecht et al. 1994, 1995; Möllers et al. 2000). Concerning more recent information, for instance, experiments with microspore embryo transformations via *Agrobacterium* (Cegielska-Taras et al. 2008; Calabuig-Serna et al. 2023), non-destructive in vitro selection of microspore-derived embryos with the fertility restorer gene for the CMS sterility (Havlíčková et al. 2015), or investigations of the response of microspore-derived embryos to osmotic stress as a tool for in vitro selection of drought-resistant genotypes (Urban et al. 2021) have been published.

Most recent scientific publications on androgenetic haploids in rapeseed mainly focus on optimizing the method or studying the key processes determining haploid production's success (see previous chapters). At the beginning of DH implementation in breeding, lists of varieties with DH origin were regularly published, and the latest advances in implementing the method. This information proves that a modern approach, specifically doubled haploid technology, has been effectively implemented within a breeding organization. However, breeding organizations maintain cultivar development information as proprietary (Ferrie and Möllers 2011). The same probably applies to research directly applicable to breeding. Most likely, for this reason, the number of publications on this topic is gradually decreasing.

2.4 Double Haploids in Brassicaceae

Double Haploid (DH) plants possess two sets of chromosomes that are homozygous at every locus, do not segregate, and exhibit considerable phenotypic variability. The DH production method is highly effective in plant breeding, minimizing the time and costs involved in generating homozygous parents essential for hybrid seed production. The haploid embryos developed from microspores may autonomously transition to diploid status, or further chromosomal doubling methods may be required. However, DH production through chromosomal doubling may vary according to the genera and genome of various species (Segui-Simarro 2016). *Brassica napus* is considered a vital model crop for studies in microspore-derived embryogenesis due to its high innate response to stress treatments. Microspore embryogenesis also serves as an efficient tool for analysing plant morphogenesis, cell totipotency and invitro embryogenesis (Corral-Martínez et al. 2021). It gives an added advantage over the other cultures in haploid production, but the protocol should be well-optimized and set up. The methods to develop double haploids from microspore-derived embryos in Brassicaceae crops including *B. napus* (high responsive DH4079 and low responsive DH12075) (Corral-Martínez et al. 2021); *B. rapa* L ssp. chinensis (Shumilina et al. 2021), *B. rapa* L ssp. pekinensis (Adamus et al. 2021); *B. carinata* (Zatylny et al. 2021); *B. oleracea* var. italica (Alan et al. 2021); and *Raphanus sativus* L. (Kozar and Domblides 2021) have been elaborated. The recent reports contributing to the DH production in *B. napus* have been reviewed (Starosta et al. 2023; Bryzova et al. 2025).

2.5 Conclusions and Future Remarks

It is evident from the previous chapters that the androgenetic embryogenesis and the related production of haploids (and doubled haploids) have long since become a standard part of breeding programmes aimed at producing rapeseed cultivars and uniform parental components for hybrid breeding. Nevertheless, it can be expected that partial optimizations of the technology will continue and bring a not insignificant improvement of specific processes. The question is whether this information will be publicly available. Suppose we review the current status of the three bottlenecks mentioned above. In that case, Ferrie and Möllers (2011) reported more than a decade ago that possible markers of embryogenic capacity would be useful mainly for such materials. We can confirm this hypothesis based on our knowledge of the implementation of haploid technology in rapeseed breeding. In the beginning (20–30 years ago), finding sufficient genotypes and breeding materials with good embryogenic ability was quite challenging. Nowadays, the opposite is the case. Even if lots of newer materials of different provenance, origin, and quality are used in breeding, there is usually no fundamental issue in this regard after their inclusion in the breeding process. If so, it is possible to use an embryogenic material as a paternal component in crossing, and the resulting crosses will have a very high probability of being embryogenic as well.

A similar scenario currently exists with the diploidization efficiency. The optimization of in vitro chromosomal doubling has been worked out to such a stage that, in our experience, it is often difficult to find a higher number of haploid plants within a population of flowering regenerants for specific breeding material. It is probably a combination of efficient doubling and selection during in vitro cultivation; when properly regenerating, more vital and fast-growing individuals are preferentially selected for further use, doubled haploids rather than haploids. The possible genetic background of diploidization ability cannot be neglected (Kampouridis et al. 2016). We, therefore, assume that the fundamental problems with poor diploidization rates will mainly be a matter of more exotic materials.

On the contrary, there is still much room for improvement in the case of direct regeneration of whole plants from embryos. According to our experience, in many cases, this part of the androgenetic haploid technology requires repeated sub-cultivations on fresh nutrient media to achieve regeneration, which places higher demands on human power, time, space, and consumption of additional material and energy.

The development and utilization of androgenetic haploid technology have been going on for several decades, and we take it for granted as an irreplaceable part of effective rapeseed breeding. However, in the last few years, partially in connection with the development of new genome editing techniques, two "competitors" have emerged in the form of in vivo induction of maternal haploids or directly maternal doubled haploids using specific induction lines (see previous chapters). Even though the former does not yet have high efficiency, and the latter is probably strongly genetically determined and most likely patent-protected (Fu et al. 2019), it is essential to monitor their development closely. In their favour is the simplicity and unpretentiousness of the application. Thus, the adoption by breeders may be more straightforward and much faster than implementing androgenetic embryogenesis. Therefore, it cannot be ruled out that shortly, the androgenetic method will be referred to as "a traditional, gradually abandoned technique", and after almost 60 years, there will be a return to the production of haploids in the field conditions, just as it was at the beginning of haploid technology development in this important oilseed.

References

Adamus A, Szklarczyk M, Kiełkowska A (2021) Haploid and doubled haploid plant production in Brassica rapa L. subsp. pekinensis via microspore culture. In: Segui-Simarro JM (ed) Doubled haploid technology, Methods in molecular biology, vol 2288. Humana Press, New York. https://doi.org/10.1007/978-1-0716-1335-1_11

Ahmad I, Day JP, MacDonald MV, Ingram DS (1991) Haploid culture and UV mutagenesis in rapid-cycling *Brassica napus* for the generation of resistance to chlorsulfuron and *Alternaria brassicicola*. Ann Bot 67(6):519–521. https://doi.org/10.1093/oxfordjournals.aob.a088193

Ahmadi B, Shariatpanahi ME (2015) Proline and chitosan enhanced efficiency of microspore embryogenesis induction and plantlet regeneration in *Brassica napus* L. Plant Cell Tissue Organ Cult 123(1):57–65. https://doi.org/10.1007/s11240-015-0814-3

Ahmadi B, Shariatpanahi ME, Ojaghkandi MA, Heydari AA (2014a) Improved microspore embryogenesis induction and plantlet regeneration using putrescine, cefotaxime and vancomycin in *Brassica napus* L. Plant Cell Tissue Organ Cult 118(3):497–505. https://doi.org/10.1007/s11240-014-0501-9

Ahmadi B, Shariatpanahi ME, Teixeira da Silva JA (2014b) Efficient induction of microspore embryogenesis using abscisic acid, jasmonic acid and salicylic acid in *Brassica napus* L. Plant Cell, Tissue Organ Cult 116(3):343–351. https://doi.org/10.1007/s11240-013-0408-x

Ahmadi B, Masoomi-Aladizgeh F, Shariatpanahi ME, Azadi P, Keshavarz-Alizadeh M (2016) Molecular characterization and expression analysis of SERK1 and SERK2 in *Brassica napus* L: implication for microspore embryogenesis and plant regeneration. Plant Cell Rep 35(1):185–193. https://doi.org/10.1007/s00299-015-1878-6

Alan AR, Celebi-Toprak F, Lachin A, Yildiz D, Gozen V, Besirli G (2021) Doubled haploid broccoli (*Brassica olearacea* var. italica) plants from anther culture. In: Segui-Simarro JM (ed) Doubled haploid technology, Methods in molecular biology, vol 2288. Humana Press, New York. https://doi.org/10.1007/978-1-0716-1335-1_12

Albrecht S, Möllers C, Röbbelen G (1994) Selection for fatty acid composition in microspore-derived embryoids (MDE) of rapeseed, *Brassica napus* (L). J Plant Physiol 143(4):526–529. https://doi.org/10.1016/S0176-1617(11)81817-1

Albrecht S, Möllers C, Röbbelen G (1995) Selection *in vitro* for erucic acid content in segregating populations of microspore derived embryoids of *Brassica napus*. Plant Breed 114(3):210–214. https://doi.org/10.1111/j1439-05231995tb00795x

Bhowmik P, Dirpaul J, Polowick P, Ferrie AMR (2011) A high throughput *Brassica napus* microspore culture system: influence of percoll gradient separation and bud selection on embryogenesis. Plant Cell Tissue Organ Cult 106(2):359–362. https://doi.org/10.1007/s11240-010-9913-3

Bryxová P, Fernández-Cusimamani E, Klíma M (2025) Current insights into various in vitro Dihaploidization techniques used in brassica oil crops. Agronomy 15(1):179. https://doi.org/10.3390/agronomy15010179

Calabuig-Serna A, Mir R, Porcel R, Seguí-Simarro JM (2023) The highly embryogenic *Brassica napus* DH4079 line is recalcitrant to *Agrobacterium*-mediated genetic transformation. Plants 12(10):2008. https://doi.org/10.3390/plants12102008

Cegielska-Taras T (2002) The use of microspores cultures for genetic improvement of winter oilseed rape (*Brassica napus*). Monografie i Rozprawy Naukowe IHAR 18:7–107

Cegielska-Taras T, Tykarska T, Szała L, Kuraś L, Krzymański J (2002) Direct plant development from microspore-derived embryos of winter oilseed rape *Brassica napus* L ssp oleifera (DC) Metzger. Euphytica 124(3):341–347. https://doi.org/10.1023/A:1015785717106

Cegielska-Taras T, Pniewski T, Szała L (2008) Transformation of microspore-derived embryos of winter oilseed rape (*Brassica napus* L) by using *Agrobacterium tumefaciens*. J Appl Genet 49(4):343–347. https://doi.org/10.1007/BF03195632

Chen ZZ, Snyder S, Fan ZG, Loh WH (1994) Efficient production of doubled haploid plants through chromosome doubling of isolated microspores in *Brassica napus*. Plant Breed 113(3):217–221. https://doi.org/10.1111/j1439-05231994tb00725x

Corral-Martínez P, Camacho-Fernández C, Mir R, Seguí-Simarro JM (2021) Doubled haploid production in high- and low-response genotypes of rapeseed (*Brassica napus*) through isolated microspore culture. In: Segui-Simarro JM (ed) Doubled haploid technology, Methods in molecular biology, vol 2288. Humana Press, New York

Coventry J, Kott L, Beversdorf WD (1998) Manual for microspore culture techniques in *Brassica napus*. OAC Publishing, Guelph

Ecke W, Kampouridis A, Ziese-Kubon K, Hirsch AC (2015) Identification and genetic characterization by high-throughput SNP analysis of intervarietal substitution lines of rapeseed (*Brassica napus* L) with enhanced embryogenic potential. Theor App Genet 128(4):587–603. https://doi.org/10.1007/s00122-015-2455-7

Elahi N, Duncan RW, Stasolla C (2016) Effects of altered expression of LEAFY COTYLEDON1 and FUSCA3 on microspore-derived embryogenesis of *Brassica napus* L. J Genet Eng Biotechnol 14(1):19–30. https://doi.org/10.1016/jjgeb201605002

El-Tantawy AA, Solís MT, Da Costa ML, Coimbra S, Risueño MC, Testillano PS (2013) Arabinogalactan protein profiles and distribution patterns during microspore embryogenesis and pollen development in *Brassica napus*. Plant Reprod 26(3):231–243. https://doi.org/10.1007/s00497-013-0217-8

Ferrie AMR, Keller WA (2007) Optimization of methods for using polyethylene glycol as a non-permeating osmoticum for the induction of microspore embryogenesis in the Brassicaceae. In Vitro Cell Dev Biol Plant 43(4):348–355. https://doi.org/10.1007/s11627-007-9053-6

Ferrie AMR, Möllers C (2011) Haploids and doubled haploids in *Brassica* spp for genetic and genomic research. Plant Tissue and Organ Cult 104(3):375–386. https://doi.org/10.1007/s11240-010-9831-4
Forster BP, Thomas WT (2005) Doubled haploids in genetics and plant breeding. Plant Breed Rev 25:57–88. https://doi.org/10.1002/9780470650301ch3
Fu S, Yin L, Xu M, Li Y, Wang M, Yang J, Fu T, Wang J, Shen J, Ali A, Zou Q, Yi B, Wen J, Tao L, Kang Z, Tang R (2018) Maternal doubled haploid production in interploidy hybridization between *Brassica napus* and Brassica allooctaploids. Planta 247(1):113–125. https://doi.org/10.1007/s00425-017-2772-
Fu S, Li Y, Wang J, Zou Q, Tao L, Kang Z, Tang R (2019) Method for breeding *Brassica napus* varieties and materials with double haploid induction line of rapeseed. US Patent Application No 16/310,497
Gu HH, Hagberg P, Zhou WJ (2004a) Cold pretreatment enhances microspore embryogenesis in oilseed rape (*Brassica napus* L). Plant Growth Regul 42(2):137–143. https://doi.org/10.1023/B:GROW000001748829181fa
Gu HH, Hu LJ, Zhou WJ (2004b) Applications of microspore culture technology in *Brassica* breeding programs. J Agric Biotechnol 12:219–223
Haddadi P, Moieni A, Karimzadeh G, Abdollahi M (2008) Effects of gibberellin, abscisic acid and embryo desiccation on normal plantlet regeneration, secondary embryogenesis and callogenesis in microspore culture of *Brassica napus* L cv PF 704. Int J Plant Prod 2
Hansen NJP, Andersen SB (1996) In vitro chromosome doubling potential of colchicine, oryzalin, trifluralin, and APM in *Brassica napus* microspore culture. Euphytica 88(2):159–164. https://doi.org/10.1007/BF00032447
Havlíčková L, Klíma M, Přibylová M, Hilgert-Delgado AA, Kučera V, Čurn V (2015) Non-destructive in vitro selection of microspore-derived embryos with the fertility restorer gene for CMS Ogu-INRA in winter oilseed rape (*Brassica napus* L). Electron J Biotechnol 18(1):58–60. https://doi.org/10.1016/jejbt201410003
Hoffmann F (1978) Production of natural compounds by cell culture methods. In: Alfermann W, Reinhard E (eds) Proceedings of an international symposium on plant cell culture, p 319
Hoffmann F, Thomas E, Wenzel G (1982) Anther culture as a breeding tool in rape. Theor Appl Genet 61(3):225–232. https://doi.org/10.1007/BF00273779
Iqbal MCM, Möllers C, Röbbelen G (1994) Increased embryogenesis after colchicine treatment of microspore cultures of *Brassica napus* L. J Plant Physiol 143(2):222–226. https://doi.org/10.1016/S0176-1617(11)81691-3
Joosen R, Cordewener J, Supena EDJ, Vorst O, Lammers M, Maliepaard C, Zeilmaker T, Miki B, America T, Custers J, Boutilier K (2007) Combined transcriptome and proteome analysis identifies pathways and markers associated with the establishment of Rapeseed microspore-derived embryo development. Plant Physiol 144(1):155–172. https://doi.org/10.1104/pp107098723
Kampouridis A, Ziese-Kubon K, Nurhasanah EW (2016) Identification and evaluation of intervarietal substitution lines of rapeseed (*Brassica napus* L) with donor segments affecting the diploidization rate of isolated microspores. Euphytica 209:181–198. https://doi.org/10.1007/s10681-016-1656-8
Keller WA, Armstrong KC (1978) High frequency production of microspore derived plants from *Brassica napus* anther cultures. Z Pflanzenzücht 80:100–108
Klíma M, Vyvadilová M, Kučera V (2008) Chromosome doubling effects of selected antimitotic agents in *Brassica napus* microspore culture. Czech J Genet Plant Breed 44:30–36
Kott LS (1998) Application of doubled haploid technology in breeding of oilseed *Brassica napus*. Agric Biotech News Info 10:69–74
Kozar E, Domblides E (2021) Protocol of European radish (*Raphanus sativus* L.) microspore culture for doubled haploid plant production. In: Segui-Simarro JM (ed) Doubled haploid technology, Methods in molecular biology, vol 2288. Humana Press, New York. https://doi.org/10.1007/978-1-0716-1335-1_13
Kučera V, Vyvadilová M, Klíma M (2002) Utilisation of doubled haploids in winter oilseed rape (*Brassica napus* L) breeding. Czech J Genet Plant Breed 38:50–54

Leroux B, Carmoy N, Giraudet D, Potin P, Larher F, Bodin M (2009) Inhibition of ethylene biosynthesis enhances embryogenesis of cultured microspores of *Brassica napus*. Plant Biotechnol Rep 3(4):347–353. https://doi.org/10.1007/s11816-009-0109-4

Leroux BJG, Potin P, Larher FR, Bodin M (2016) Enhancement of embryo yield from isolated microspores of *Brassica napus* by early iron starvation. Plant Biotechnol Rep 10(6):483–486. https://doi.org/10.1007/s11816-016-0420-9

Li J, Huang Q, Sun M, Zhang T, Li H, Chen B, Xu K, Gao G, Li F, Yan G, Qiao J, Cai Y, Wu X (2016) Global DNA methylation variations after short-term heat shock treatment in cultured microspores of *Brassica napus* cv Topas. Sci Rep 6(1):38401. https://doi.org/10.1038/srep38401

Lichter R (1981) Anther culture of *Brassica napus* in a liquid culture medium. Z Pflanzenphysiol 103(3):229–237. https://doi.org/10.1016/S0044-328X(81)80155-9

Lichter R (1982) Induction of haploid plants from isolated pollen of *Brassica napus*. Z Pflanzenphysiol 105(5):427–434. https://doi.org/10.1016/S0044-328X(82)80040-8

Ling X, Guo-Qing Z, Hong-Hui G, Yun H, Wei-Jun Z (2006) In vitro selection of herbicide resistance in microspore-derived embryos of oilseed rape (*Brassica napus* L). Chin J Agric Biotechnol 3(1):33–38. https://doi.org/10.1079/CJB200587

Liu S, Wang H, Zhang J, Fitt BDL, Xu Z, Evans N, Liu Y, Yang W, Guo X (2005) In vitro mutation and selection of doubled-haploid *Brassica napus* lines with improved resistance to *Sclerotinia sclerotiorum*. Plant Cell Rep 24(3):133–144. https://doi.org/10.1007/s00299-005-0925-0

Mahasuk P, Kullik AS, Iqbal MC, Möllers C (2017) Effect of boron on microspore embryogenesis and direct embryo to plant conversion in *Brassica napus* (L). Plant Cell Tissue Organ Cult 130(2):443–447. https://doi.org/10.1007/s11240-017-1232-5

Malik MR, Wang F, Dirpaul JM, Zhou N, Hammerlindl J, Keller W, Abrams SR, Ferrie AMR, Krochko JE (2008) Isolation of an embryogenic line from non-embryogenic *Brassica napus* cv Westar through microspore embryogenesis. J Exp Bot 59(10):2857–2873. https://doi.org/10.1093/jxb/ern149

Mohammadi PP, Moieni A, Ebrahimi A, Javidfar F (2012) Doubled haploid plants following colchicine treatment of microspore-derived embryos of oilseed rape (*Brassica napus* L). Plant Cell, Tissue Organ Cult 108(2):251–256. https://doi.org/10.1007/s11240-011-0036-2

Möllers C, Iqbal MCM (2009) Doubled haploids in breeding winter oilseed rape. In: Touraev A, Forster BP, Jain SM (eds) Advances in haploid production in higher plants. Springer, pp 161–169. https://doi.org/10.1007/978-1-4020-8854-4_13

Möllers C, Rücker B, Stelling D, Schierholt A (2000) In vitro selection for oleic and linoleic acid content in segregating populations of microspore derived embryos of *Brassica napus*. Euphytica 112(2):195–201. https://doi.org/10.1023/A:1003884823052

Nitsch JP, Nitsch C (1969) Haploid plants from pollen grains. Science 163(3862):85–87. https://doi.org/10.1126/science163386285

Olsson O, Hagberg A (1955) Investigations on haploid rape. Hereditas 41(1–2):227–237. https://doi.org/10.1111/j1601-52231955tb02991x

Polsoni L, Kott LS, Beversdorf WD (1988) Large-scale microspore culture technique for mutation–selection studies in *Brassica napus*. Can J Bot 66(8):1681–1685. https://doi.org/10.1139/b88-230

Prem D, Solís M-T, Bárány I, Rodríguez-Sanz H, Risueño MC, Testillano PS (2012) A new microspore embryogenesis system under low temperature which mimics zygotic embryogenesis initials, expresses auxin and efficiently regenerates doubled-haploid plants in *Brassica napus*. BMC Plant Biol 12(1):127. https://doi.org/10.1186/1471-2229-12-127

Rahman MH, Krishnaraj S, Thorpe TA (1995) Selection for salt tolerance in vitro using microspore-derived embryos of *Brassica napus* cv Topas, and the characterization of putative tolerant plants. In Vitro Cell Dev Biol Plant 31:116–121. https://doi.org/10.1007/BF02632248

Renard M, Dosba F (1980) Étude de l'haploidie chez le Colza (*Brassica napus* var *oleifera*). Ann Amélior Plantes 30:191–209

Rivas-Sendra A, Calabuig-Serna A, Seguí-Simarro JM (2017) Dynamics of calcium during in vitro microspore embryogenesis and in vivo microspore development in *Brassica napus* and *Solanum melongena*. Front Plant Sci 8. https://doi.org/10.3389/fpls201701177

Rodríguez-Sanz H, Solís M-T, López M-F, Gómez-Cadenas A, Risueño MC, Testillano PS (2015) Auxin biosynthesis, accumulation, action and transport are involved in stress-induced microspore embryogenesis initiation and progression in *Brassica napus*. Plant Cell Physiol 56(7):1401–1417. https://doi.org/10.1093/pcp/pcv058

Sacristán MD (1982) Resistance responses to Phoma lingam of plants regenerated from selected cell and embryogenic cultures of haploid *Brassica napus*. Theor Appl Genet 61(3):193–200. https://doi.org/10.1007/BF00273774

Sacristán MD, Hoffmann F (1979) Direct infection of embryogenic tissue cultures of haploid Brassica *napus* with resting spores of *Plasmodiophora brassicae*. Theor Appl Genet 54(3):129–132. https://doi.org/10.1007/BF01159467

Seguı-Simarro JM (2016) Androgenesis in solanaceae. In: Germana MA, Lambardi M (eds) In vitro embryogenesis, Methods in molecular biology, vol 1359. Springer, New York, pp 209–244. https://doi.org/10.1007/978-1-4939-3061-6_9

Seguí-Simarro JM, Corral-Martínez P, Corredor E, Raska I, Testillano PS, Risueño MC (2011) A change of developmental program induces the remodeling of the interchromatin domain during microspore embryogenesis in *Brassica napus* L. J Plant Physiol 168(8):746–757. https://doi.org/10.1016/jjplph201010014

Shahmir F, Pauls KP (2021) Identification, gene structure, and expression of *BnMicEmUP*: a gene upregulated in embryogenic *Brassica napus* microspores. Front Plant Sci 11:576008. https://doi.org/10.3389/fpls2020576008

Shumilina D, Kozar E, Chichvarina O, Korottseva K, Domblides E (2021) *Brassica rapa* L. ssp. *chinensis* isolated microspore culture protocol. In: Segui-Simarro JM (ed) Doubled haploid technology, Methods in molecular biology, vol 2288. Humana Press, New York. https://doi.org/10.1007/978-1-0716-1335-1_9

Solís MT, El-Tantawy AA, Cano V, Risueño MC, Testillano PS (2015) 5-azacytidine promotes microspore embryogenesis initiation by decreasing global DNA methylation, but prevents subsequent embryo development in rapeseed and barely. Front Plant Sci 6(JUNE):10.3389/fpls201500472

Solís M-T, Berenguer E, Risueño MC, Testillano PS (2016) BnPME is progressively induced after microspore reprogramming to embryogenesis, correlating with pectin de-esterification and cell differentiation in *Brassica napus*. BMC Plant Biol 16(1):176. https://doi.org/10.1186/s12870-016-0863-8

Starosta E, Szwarc J, Niemann J, Szewczyk K, Weigt D (2023) *Brassica napus* haploid and double haploid production and its latest applications. Curr Iss Mol Biol 45(5):4431–4450. https://doi.org/10.3390/cimb45050282

Stefansson BR, Hougen FW (1964) Selection of rape plants (Brassica napus) with seed oil practically free from erucic acid. Can J Plant Sci 44(4):359–364. https://doi.org/10.4141/cjps64-069

Stefansson BR, Hougen FW, Downey RK (1961) Note on the isolation of rape plants with seed oil free from erucic acid. Can J Plant Sci 41(1):218–219. https://doi.org/10.4141/cjps61-028

Swanson EB, Herrgesell MJ, Arnoldo M, Sippell DW, Wong RSC (1989) Microspore mutagenesis and selection: Canola plants with field tolerance to the imidazolinones. Theor Appl Genet 78(4):525–530. https://doi.org/10.1007/BF00290837

Szała L, Sosnowska K, Cegielska-Taras T (2020) Induced Chromosome Doubling in Microspores and Regenerated Haploid Plants of *Brassica napus*. Acta Biol Crac Ser Bot 62:23–31. https://doi.org/10.24425/abcsb2019127750

Szarejko I, Małuszyński M, Polok K, Kilian A (1991) Doubled haploid in the mutation breeding of selected crops. In: Proceedings of an international symposium on the contribution of plant mutation breeding to crop improvement, vol. 2, IAEA, pp 355–378. https://inis.iaea.org/records/k1sqj-pap42

Thomas E, Wenzel G (1975) Embryogenesis from microspores of *Brassica napus*. Zeitschrift Fuer Pflanzenzuechtung 74:77–81

Thompson KF (1969) Frequencies of haploids in spring oil-seed rape (*Brassica napus*). Heredity 24(2):318–319. https://doi.org/10.1038/hdy196938

Thompson KF (1972) Oil-seed rape. In: Reports of the plant Breeding Institute. Cambridge University Press, Cambridge, pp 94–96

Thompson KF (1979) Superior performance of two homozygous diploid lines from naturally occurring polyhaploids in oilseed rape (*Brassica napus*). Euphytica 28(1):127–135. https://doi.org/10.1007/BF00029183

Tsuwamoto R, Takahata Y (2008) Identification of genes specifically expressed in androgenesis-derived embryo in rapeseed (*Brassica napus* L). Breed Sci 58:251–259. https://doi.org/10.1270/jsbbs58251

Tsuwamoto R, Fukuoka H, Takahata Y (2007) Identification and characterization of genes expressed in early embryogenesis from microspores of *Brassica napus*. Planta 225(3):641–652. https://doi.org/10.1007/s00425-006-0388-8

Urban MO, Planchon S, Hoštičková I, Vanková R, Dobrev P, Renaut J, Klíma M, Vítámvás P (2021) The resistance of oilseed rape microspore-derived embryos to osmotic stress is associated with the accumulation of energy metabolism proteins, redox homeostasis, higher abscisic acid, and cytokinin contents. Front Plant Sci 12. https://doi.org/10.3389/fpls2021628167

Wan G, Geng X, Xu L, Li B, Zhou W (2011) Optimization of microspore embryogenesis and plant regeneration protocols for *Brassica napus*. Int J Agric Biol 13:1560–8530

Wang F, Dirpaul JM, Zhou N, Polowick PL, Ferrie AMR, Krochko JE (2020) Transcript profiling and identification of molecular markers for early microspore embryogenesis in *Brassica napus*. Plant Physiol 144:134–154. https://doi.org/10.1104/pp106092932

Wenzel G, Hoffmann F, Thomas E (1977) Anther culture as a breeding tool in rape I Ploidy level and phenotype of androgenetic plants. Z Pflanzenzüchtg 78:149–155

Yeung EC (2002) The canola microspore-derived embryo as a model system to study developmental processes in plants. J Plant Biol 45(3):119–133. https://doi.org/10.1007/BF03030304

Zatylny A, Catinot V, Bundrock T, Wu SW, Eynck C (2021) Protocol for the production of doubled haploid plants of Brassica carinata. In: Segui-Simarro JM (ed) Doubled haploid technology, Methods in molecular biology, vol 2288. Humana Press, New York. https://doi.org/10.1007/978-1-0716-1335-1_10

Zhang GQ, Zhang DQ, Tang GX, He Y, Zhou WJ (2006) Plant development from microspore-derived embryos in oilseed rape as affected by chilling, desiccation and cotyledon excision. Biol Plant 50(2):180–186. https://doi.org/10.1007/s10535-006-0004-6

Zhong Y, Chen B, Wang D, Zhu X, Wang Y, Li M, Li Y, Liu J, Zhang J, Chen M, Wang M, Riksen T, Qi X, Cheng D, Liu Z, Li J, Chen C, Jiao Y, Liu W, Chen S (2021) A genotype independent DMP-HI system in dicot crops. BioRxiv 20210621449224. https://doi.org/10.1101/20210621449224

Zhong Y, Wang Y, Chen B, Liu J, Wang D, Li M, Qi X, Liu C, Boutilier K, Chen S (2022) Establishment of a dmp based maternal haploid induction system for polyploid *Brassica napus* and *Nicotiana tabacum*. J Integr Plant Biol 64(6):1281–1294. https://doi.org/10.1111/jipb13244

Rapeseed (*Brassica napus* L.) Breeding for Crop Improvement

Eva Jozová, Ondřej Hejna, Vladislav Čurn,
Baveesh Pudhuvai, and Bhupendra Koul

Abstract

Rapeseed (*Brassica napus*), an important oil seed crop globally, originated initially in the Mediterranean region and currently occupies third position after palm and soybean crops. The complex polyploid genome of the crop species indicates the intricacies of its origin and evolution. However, the crop has been under steady development over the past decades, and the major cultivators are Canada, the EU and China. The crop has high economic significance as it serves as a vegetable oil, technical lubricant, biofuels, and animal feed. The rapeseed oil is considered advantageous for its culinary purposes as it is characterized by a high content of unsaturated fatty acids, promoting human health. The crop promotes sustainability in agriculture by enhancing soil health, facilitating nutrient cycles, acting as a rotation crop to evade disease cycles, aligning with contemporary sustainable agricultural practices. The rapeseed crop employed for crop rotations in several counties significantly contributes to enhanced soil health and improved rural economy. The chapter pools information regarding the origin, importance, key consideration points for future improvements, and breeding approaches in rapeseed cultivation. The traditional breeding approaches for enhancing resistance, yield, and quality in rapeseed are ineffective due to genome complexity and open-pollinated varieties. Thus, a shift in the attention of breeders

E. Jozová · O. Hejna · V. Čurn
Faculty of Agriculture and Technology, Department of Genetics and Biotechnology, University of South Bohemia in České Budějovice, České Budějovice, Czech Republic

B. Pudhuvai
Faculty of Agriculture and Technology, Department of Plant Production, University of South Bohemia in České Budějovice, České Budějovice, Czech Republic

B. Koul (✉)
Department of Biotechnology, School of Bioengineering & Biosciences, Lovely Professional University, Phagwara, Punjab, India

A. Kumar, B. Koul (eds.), *Crop Improvement Strategies in Brassica species: Applied Science*, https://doi.org/10.1007/978-981-95-3861-4_13

in the generation of F1 hybrids and using male sterility has led to the development of superior varieties. The use of contemporary breeding approaches and molecular methods combined with predictive modelling of new varieties according to geographic and changing climatic conditions will aid in leveraging the crop benefits for future generations.

Keywords

Brassica napus · Erucic acid · Oilseed · Crop rotation · Glucosinolates · Breeding

1 Introduction

The family *Brassicaceae* Burnett (*Cruciferae* Juss.) represents an exceptionally large and diverse group comprising several crops with global significance, particularly oilseeds and vegetables. This family includes about 370 genera and 4600 species (Francis et al. 2021). Most of them grow from subtropics to temperate to cold zones of the northern and southern hemispheres, but the centre of occurrence is in the northern hemisphere (Al-Shehbaz et al. 2006). The oil crops of the *Brassica* genus are a collection of various plant species cultivated primarily for their oil-rich seeds due to their nutritional value, versatility/adaptability, and various industrial applications. The prominent oil crops include rapeseed (canola), mustards, and camelina, which are rich in omega-3, 6 fatty acids, vitamin E, and antioxidants (Blackshaw et al. 2011; Raboanatahiry et al. 2021). The oil derived from *Brassica* oil crops is used for human consumption (contributes to overall health and well-being) and finds applications across several industries, such as biodiesel production and technical lubricants (Ge et al. 2017; Tanner et al. 2023). Brassica crops are often employed in crop rotation to hinder the pest and disease cycles, improving overall soil health and structure and benefitting soil nutrient levels (Daniel et al. 2016; Larkin and Lynch 2018; Obermeier et al. 2022). Therefore, in the fields of agriculture, nutrition, and industry, Brassica oil crops are indispensable. Their use and cultivation have changed throughout the years, and researchers continue to investigate approaches to improve them regarding sustainability, nutritional value, and overall production. Therefore, Brassica oil crops are indispensable in agriculture, nutrition, and industry. Their cultivation and use have evolved considerably throughout the years, and researchers continue to investigate approaches to improve their crop characteristics for better productivity, nutritional content, stress resilience, and sustainability (Wittkop et al. 2009; Zandberg et al. 2022).

2 Origin and Distribution

The most important cultivated oilseed *Brassica* crop is rapeseed [*rapeseed, canola*] (*Brassica napus* L. subsp. *napus*-syn. *B. oleifera* Moench, *B. napus* var. *oleifera* (Moench) Delile, *B. napus* subsp. *pabularia* (DC.) Janch., *B. napus* var. *sahariensis* A. Chev., *Napus oleifera* (Moench) K. F. Schimp. et Spenn., *Rapa napus* (L.)

J. S. Muell., *Raphanus napus (L.) Crantz, Sinapis napus* (L.) Brot. Rapeseed has been cultivated for centuries in Europe, with evidence suggesting its use dates back to Roman times. It was primarily grown for its foliage as a vegetable rather than for oil. During the medieval and Renaissance periods, rapeseed cultivation expanded, becoming a more established crop in European agriculture (Kimber and McGregor 1995). It was used for various purposes, including oil extraction for lamps, soap making, lubrication, and as a food ingredient (Snowdon et al. 2007; Raboanatahiry et al. 2021; Nassef et al. 2023). During the 1970s, Canadian breeders B. Stefansson and K. Downey developed novel rapeseed cultivars with significantly decreased concentrations of erucic acid and glucosinolates. These new rapeseed varieties, with oil quality substantially different from the previous varieties, were named 'canola', a contraction of 'Canadian oil, low acid'. Canola was originally a trademark of the 'Rapeseed Association of Canada' where all the cultivars grown belong to *B. napus* ('Argentine canola'), *B. rapa* ('Polish canola') *or B. juncea species* (Stefansson and Hougen 1964; Stefansson and Kondra 1975; Russo et al. 2021). In contemporary, the utilization of rapeseed has significantly increased compared to the past. Thanks to breeding progress and improvement in the quality of rapeseed oil, rapeseed cultivation areas have increased many times over the last 50 years, and rapeseed has become one of the world's most essential oilseeds, producing high-quality and valuable vegetable oil having prominence in the food and chemical industries. It is generally employed for producing edible oil, but also serves as a valuable raw material for animal feed (extracted meals, green fodder, oil cake). It has emerged as a resource utilized in the chemical industry to produce biodiesel, eco-oils and eco-lubricants. It also serves as an important source of renewable energy (Snowdon et al. 2007; Friedt and Snowdon 2009; Warwick 2011).

3 Importance of Rapeseed

Canola oil ranks as the third most important oil, following palm and soybean oil, and rapeseed is considered the second most crucial oilseed crop after soybeans (Shahbandeh 2023). It is the most important oilseed crop to be cultivated in temperate zones and its importance is listed in Table 1. Due to the requirements for climatic and soil conditions, rapeseed can only be grown in some regions, unlike cereals. Over half of rapeseed production (60%) was produced in EU countries and Canada in 2020. Canada is also the world's largest grower of GM canola (genetically modified crop). Another 22% of rapeseed is produced in India and China (FAO 2023). Consequently, it is plausible to assert that despite rapeseed being the second most prevalent oilseed globally, its cultivation is limited to a few countries, and any fluctuations in production substantially affect international trade. Since 2014, rapeseed production has stagnated or declined because of EU agri-environmental measures. In 2015–20, stagnation in production was offset by stocks, but global and European rapeseed stocks continued to fall. The current increasing trend is the use of oil for non-food purposes. Finally, rapeseed also has an important position in the agricultural ecosystem; it serves as an excellent forecrop for cereals, increases soil fertility,

Table 1 Importance of rapeseed cultivation

Rapeseed holds significant importance for farmers and the agricultural industry, and its breeding history reflects efforts to enhance various characteristics	
Key benefits for farmers	
High productivity	A notable characteristic of rapeseed is its high yield of oil-rich seeds. For farmers, this makes it a crop with significant economic value and a high rate of return on investment
Versatile end uses	Rapeseed serves multiple purposes, including oil extraction for human consumption, industrial applications (biodiesel production), and use in livestock feed. Thus, the crop helps farmers to diversify their income streams
Nutrient-rich meal	The rapeseed meal or oil cake after oil extraction is used as livestock feed. Its versatility contributes to the overall sustainability of the crop by minimizing waste
Crop rotation benefits	The crop is often employed for its agronomic importance in crop rotation to break disease and pest cycles. This cultivation strategy can aid in disease avoidance and improve soil health and structure, benefiting subsequent crops in the rotation
Adaptability to different climates	The rapeseed cultivars (winter, spring, and semi-winter varieties) are grown in diverse climates, from temperate to subarctic regions. This adaptability of crops allows farmers of different geographical locations to incorporate rapeseed into their crop rotations
Economic impact	The economic impact of rapeseed extends beyond individual farms. It contributes to the overall economy by generating farmer income, supporting jobs in the agricultural supply chain, mitigating weather risks and market fluctuations, and contributing to regional and national economies

and is often used as a green fertilizer. Rapeseed straw is a source of carbon to balance the carbon/humus balance in the soil (Kirkegaard et al. 1994; Jaime et al. 2018; Jannat et al. 2022; FAO 2023).

The current occurrence of the wild *B. napus* is unknown, and the origin of this crop has remained obscure for a long period (Lu et al. 2019). Cytological studies and experimental crossbreeding have led to hypotheses about the relationship of species within the genus *Brassica* and the assumption of a hybrid origin of rapeseed (Morinaga 1929, 1934). During the 1930s, Nagaharu U (Woo Jang–choon) developed a revolutionary idea of detecting the relationships between basic and amphidiploid species of *Brassica* genera (Nagaharu 1935). The 'triangle of U' (Fig. 1) explains the basic theory and relationships among the six *Brassica* species and has been confirmed several times, even with modern methods of *Brassica* genome analysis (Kamala 1976; Town et al. 2011). Cultivated species of the genus *Brassica* are divided into three basic groups according to the number of chromosomes, with genome A, genome B, and genome C. In monogenomic species, the behaviour of chromosomes during mitosis and cytological studies suggests a common extinct ancestor with 5 or 6 chromosomes. Basic species arose from this ancestor by the process of autopolyploidy followed by the elimination of some chromosomes and structural rearrangements (Downey et al. 1980; Röbbelen 1985; Labana and Gupta 1993). Other groups were formed by spontaneous inbreeding of parental forms and

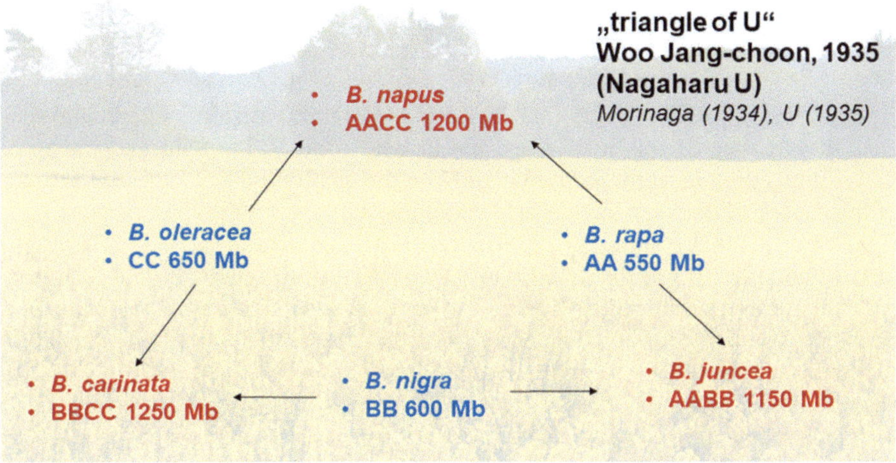

Fig. 1 The 'Triangle of U' depicts relationships between basic and amphidiploid species of *Brassica*

subsequent amphidiploidization (McNaughton 1976; Röbbelen 1985). Recent studies show that rapeseed is an allopolyploid species that probably arose relatively recently (7500 years ago) in regions where its parent species were co-cultivated (Iñiguez-Luy and Federico 2011; Chalhoub et al. 2014). Evidence of this hybrid origin of cultivated rapeseed was first provided by cytotaxonomic experiments by Morinaga (1934), confirmed by the studies of U (1935), and later supplemented by breeding and hybridization experiments (Howard 1942; Frandsen 1943, 1947; Röbbelen 1960). Evidence of polyphyletic origin of rapeseed has also been confirmed based on molecular genetic analysis, RFLP analysis of nuclear DNA, ctDNA (cytoplasmic), and mtDNA (mitochondrial) (Warwick and Black 1991; Song and Osborn 1992) and recently on the analysis of the structure of the rapeseed genome (Lysak and Koch 2011; Bancroft et al. 2011; He et al. 2015; He et al. 2021).

The hybrid origin of rapeseed is also reflected in the structure of its genome. Rapeseed behaves as a diploid species ($2n = 4\times = 38$) with AACC genomes. Genome A originates from *B. rapa* L. (syn. *B. campestris* L.) ($2n = 4\times = 20$), and genome C originates from *B. oleracea* L. ($2n = 4\times = 18$). However, the parent rapeseed species are challenging to interbreed, mainly due to the lack of endosperm, which leads to embryo abortion (Håkansson 2010). The spontaneous occurrence of de novo 'synthesized rapeseed' is extremely rare, not only because of barriers during fertilization and embryo development but also, for example, because of differences in the colour and formation of the flower of the parent components (Choudhary and Joshi 2001; FitzJohn et al. 2007; Tonosaki et al. 2016). 'Wild' rapeseed has probable origination from the Mediterranean, whereas the occurrence of *B. oleracea* and *B. rapa* overlap (Katche et al. 2019). Based on the study of the range of variability of morphological forms and comparison with parental forms, it is concluded that *B. napus* arose repeatedly several times based on spontaneous distant hybridization of different forms of *B. rapa* and *B. oleracea* (McNaughton 1976; Röbbelen 1985). Due to its phylogenetic origin, rapeseed only includes the range of variability of the subspecies or varieties *of B. oleracea* and *B. rapa*, which were involved in the original crossing, and the genetic variability of *B. napus* is very limited due to the variability of the parent species (Murphy 1994; Chalhoub et al. 2014). The hybrid origin of rapeseed, its origin from a very limited sample of genetic variability of parental species, the absence of this hybrid species in wild form in the wild and intensive (hybrid) breeding recently focused mainly on the quality traits of rapeseed are one of the causes of the very low genetic variability in the current rapeseed gene pool (Prakash et al. 2009; Rahman 2013; Friedt et al. 2018; Katche et al. 2019). Therefore, there is an effort to use the process of resynthesis of rapeseed from the 'original' parental forms as a breeding strategy to expand genetic variability in rapeseed breeding. The first studies on rapeseed resynthesis appeared as early as the 1980s (Schenck and Wolf 1986; Sundberg and Glimelius 1986; Akbar 1989; Chen and Heneen 1989), and the issue of rapeseed resynthesis and the use of resynthesized materials in breeding is being studied very intensively even now (Seyis et al. 2003; Girke et al. 2012; Hilgert-Delgado et al. 2015).

As mentioned previously, rapeseed is the world's most crucial oilseed crop, with an abundance of cultivation in the EU regions, China and Canada (Carré and Pouzet 2014). Winter, spring, and semi-winter forms of rapeseed are grown depending on the cultivation system and climatic conditions (Snowdon et al. 2007). In addition to oil, rapeseed is also cultivated for its tuberous (turnips) and deciduous (fodder rape and Siberian kale) forms. Depending on the culture, these are used as food or animal feed (Allender and King 2010). Rapeseed has gained importance only relatively recently, and the rapid rise in the cultivation of this crop is due to unprecedented successes in its breeding. Rapeseed naturally possesses higher concentrations (>50%) of erucic acid in its seeds, which is bitter and can induce cardiovascular disorders in excessive doses (Röbbelen and Rakow 1970). Consequently, rapeseed oil was exclusively used for non-edible purposes or to produce low-quality vegetable margarine. During the 1970s, breeding methods were employed and the erucic acid content in the rapeseed oil was significantly reduced. The resulting variety 'Liho' (a German variety), referred to as 'single low' or 'single zero' ('0'), contained less than 1% of erucic acid (Rakow and McGregor 1975; Downey 1990; Downey and Bell 1990; Rücker and Röbbelen 1997). This unprecedented breeding success marked the beginning of using rapeseed oil for food purposes and increased interest in rapeseed cultivation (Snowdon et al. 2007).

In the old/traditional varieties of rapeseed, a relatively large amount of sulphur secondary metabolites—glucosinolates—occurs in seeds and, thus, also in oil cakes (Bell 1993; Khajali and Slominski 2012). Glucosinolates are essential for the plant in pest control and act as natural biopesticides (Singh 2017; Liu et al. 2021). Glucosinolates do not have biological effects; they manifest only after the mechanical disruption of plant tissues and the release of the myrosinase from idioblasts (Andréasson et al. 2001; Shirakawa and Tanida 2022). The action of myrosinase enzyme leads to the hydrolysis of glucosinolates and the formation of isothiocyanates, indoles, and nitriles (Barba et al. 2016; Abdel-Massih et al. 2023). Long-term intake of such food leads to an increase in the activity of the thyroid gland, damages the liver (hepatotoxicity), kidneys, and adrenal glands (nephrotoxicity), and can also cause a decrease in fertility (Lee et al. 2020). Therefore, the production of cultivars with low glucosinolates through breeding was a subsequent achievement. In 1969, a low-glucosinolate variant of Polish spring rapeseed 'Bronowski' was released, which became the basis and starting point for further breeding (Downey et al. 1969). In 1974, these efforts resulted in the introduction of the Canadian spring rapeseed variety 'Tower,' which combines low contents of glucosinolates and erucic acid (Stefansson and Kondra 1975). Such varieties are referred to as double-low or double-zero ('00'). Starting from this historical point, there was a great boom in the cultivation of rapeseed in all growing areas (Röbbelen 1981; Daun et al. 2011; Ayton 2014).

4 Genomics and Rapeseed Breeding

In addition to the quality of oil and oil-cake (low contents of erucic acid and glucosinolates in seed), oil content, tocopherols, breeding of varieties with a specific composition of fatty acids (HEAR, LEAR, HOLL, LYCAN, HOCAN, LTCAN, GLCO), improving agronomic properties, abiotic and biotic stress tolerance is of major importance in rapeseed breeding (Baux et al. 2011; Dunford 2018; Lohani et al. 2020; Raza et al. 2021). An example is the combined resistance against *Phoma* based on the use of dominant resistance genes and polygenes ensuring horizontal resistance, resistance against *Verticillium*, *Cylindrosporium* and TuYV (Turnip Yellow Virus) and, resistance towards *Plasmodiophora brassicae*, causing clubroot disease, is gaining more importance (Neik et al. 2017; Huang et al. 2014; Mitrousia et al. 2018; Rygulla et al. 2008; Fell et al. 2023; Ibrahim et al. 2023; Hejna et al. 2019; Tonu et al. 2023). Breeders' attention is also focused on developing tolerance to abiotic stresses, like drought, resistance to cold and freezing, for increased germination and emergence, and resistance to pod shattering, which altogether contribute to the reduction of yield losses (Elferjani and Soolanayakanahally 2018; Zhu et al. 2016; Liu et al. 2022; Jian et al. 2020; Yan et al. 2019; Raman et al. 2014; Zahoor et al. 2022). Another important breeding goal is also nitrogen use efficiency (NUE). The NUE is a quantifiable trait expressed in terms of the amount of nitrogen consumed per ton of yield and can be influenced by the choice of parent components. The ideal variety will maintain superior performance even in deficit conditions and current nitrogen deficiency (Bouchet et al. 2016) (Table 2).

Since the mid-90 s, a new trend has entered the breeding and cultivation of rapeseed F1 hybrids. Hybrid varieties have become the future of rapeseed breeding. This trend was conditioned by the assumption that line varieties would not achieve the required yield potential, and this type of variety is considered obsolete. Under stress conditions, the heterosis effect is significantly manifested; hybrids are characterized by higher vitality, compensatory ability, and a higher resistance level (Brandt et al. 2007; Gehringer et al. 2007). Grant I first mentioned this trend at the GCIRC Congress, and it was pointed out that in the past decade, the development of rapeseed hybrids has attracted more interest and research activities than any other area

Table 2 Focus on rapeseed breeding

Rapeseed breeding aims to enhance the crop's agronomic traits, nutritional profile, and sustainability, thereby serving as a significant resource for farmers and the agricultural sector. Ongoing research and innovation in breeding programmes enhance the success of rapeseed as a vital global crop	
Key directions in rapeseed breeding	
Reduction of undesirable compounds	Early rapeseed varieties contained elevated concentrations of erucic acid and glucosinolates, which are potentially harmful and undesirable for both human and animal consumption. The breeding history of rapeseed includes the selection and development of varieties with low erucic acid and low glucosinolate content, thereby addressing these issues
Improved oil quality	Breeding programmes for rapeseed have concentrated on improving oil quality to comply with health and nutritional standards. This involves optimizing the fatty acid composition, decreasing saturated fats, and enhancing levels of beneficial omega-3 fatty acids
Resistance to pests and diseases	Breeding initiatives focus on creating rapeseed varieties that exhibit resistance to prevalent pests and diseases. This diminishes the need for pesticides and supports sustainable agricultural practices
Abiotic stress tolerance	Due to increasing global warming and shifting climate patterns, prioritizing the breeding of rapeseed varieties that exhibit enhanced tolerance to abiotic stresses such as drought and heat is anticipated. The development of cultivars capable of thriving in suboptimal conditions contributes to sustainable agriculture
Trait stacking for multiple resistances	The breeding of lines or cultivars for multiple resistances, including resistance to various diseases and pests, is predicted to become increasingly prevalent. Trait stacking involves the integration of several beneficial traits within a single cultivar, thereby enhancing its resilience to diverse environmental challenges
Yield improvement	Efforts to enhance rapeseed yields focus on breeding for traits including increased seed production, improved adaptation to environmental conditions, and greater stress tolerance
Genetic diversity	Maintaining genetic diversity within rapeseed varieties is crucial for resilience against evolving pests and changing environmental conditions. Breeding programmes focus on preserving and enhancing this diversity
Environmental adaptation, sustainable and eco-friendly traits	Rapeseed breeding involves the development of varieties that exhibit enhanced adaptability to diverse climates and soil types while also minimizing environmental impact. This corresponds with the growing emphasis on sustainable agriculture

(continued)

Table 2 (continued)

Genomic selection and molecular breeding	Recent developments in genomics have facilitated more accurate and efficient breeding practices. Genomic selection, which encompasses the selection of individuals based on their genomic profiles, is expected to substantially enhance breeding for traits such as yield, disease resistance, and oil quality
Precision breeding techniques	Advances in gene editing technologies, such as CRISPR-Cas9, may be applied for precise modifications in the rapeseed genome. This could expedite the development of desired traits with greater accuracy
Data analytics and machine learning	The incorporation of data analytics and machine learning into breeding programmes enhances the analysis of extensive datasets, enabling breeders to make more informed decisions. Predictive modelling and data-driven methodologies have the potential to enhance breeding strategies

of rapeseed breeding (Grant et al. 1986). Despite the initially slow, gradual increase in interest in growing hybrids, hybrid varieties are currently cultivated in more than 95% of the most important growing areas. Through hybrid breeding, yield and many quality parameters were significantly improved (Grant and Beversdorf 1985; Schmatz et al. 2017; Cuthbert et al. 2011; Wang et al. 2021). The excellent combinatorial ability of parental components used to create a hybrid variety is a prerequisite for the emergence of heterosis in the F1 generation (Xing et al. 2014). Hybrid winter rapeseed cultivars exhibit a yield advantage of 15–20% compared to conventional varieties, attributable to the expression of hybrid vigour (Brandle and McVetty 1989). The strong heterosis effect directly determines not only the seed yield potential but also the oil yield, the height of the plant, its vigour and the certainty of overwintering (Sabaghnia et al. 2010; Wolko et al. 2019; Ye et al. 2023). The production of hybrid rapeseed seeds primarily relies on cytoplasmic male sterility (CMS) or genic male sterility (GMS) and rarely on self-incompatibility (SI) (Kučera et al. 2010). Several CMS and GMS systems have been reported in rapeseed and are already used in the hybrid breeding of *B. napus* Tables 3 and 4. The *Pol*CMS was the first CMS system, discovered in 1972 (Fu 1981), applied widely in hybrid seed production (Fan et al. 1986; Downey and Röbbelen 1989).

Despite the relatively high number of different types of male sterility, only a few are employed for breeding purposes (Anisimova and Dubovskaya 2020). An optimal male sterility system must be characterized by high stability so that the parent component ('mother plant') cannot produce pollen and, therefore, only cross-pollination occurs. Another requirement is a reliable sterility maintainer needed for the propagation of the sterile maternal component and a fertility restorer necessary for the creation of restored F1 hybrid varieties (Fan and Stefansson 1986; Verma et al. 2000; Primard-Brisset et al. 2005). Apart from these requirements of the hybrid system, it is necessary to ensure a high-quality genetic background of the hybrids so that the final variety is characterized by high parameters of yield, quality,

Table 3 Sources of cytoplasmic male sterility (CMS) in rapeseed

Sterility name	Source of sterility	References
nap (napus)	Swedish varieties cv. Bronowski	Thompson (1972), Shiga and Baba (1971, 1973)
pol (Polima)	Spontaneous intraspecific sterility (cv. Polima)	Fu 1981 (discovered in 1972)
ogu (Ogura)	Sterility and fertility restorer transferred to rapeseed from Japanese radish (*Raphanus sativus*)	Heyn (1976), Rousselle and Bregeon (1982), Ogura (1968)
tour CMS	*B. tournefortii*	Rawat and Anand (1979), Stiewe and Röbbelen (1994), Landgren et al. (1996)
mur CMS	*Diplotaxis muralis* x *B. campestris* (cv. Yukina); *D. muralis* x *B. napus* (cv. Regent)	Hinata and Konno (1979), Fan et al. (1986)
Ctr CMS	From cv. Tower	Grant et al. (1986)
oxy CMS	*Brassica oxyrrhina* x *B. campestris*	Prakash and Chopra (1988)
Mori CMS	*Moricandia arvensis*	Prakash et al. (1998)
kos (Kosena) CMS	Similar to ogu-CMS, transferred to rapeseed from radish (*Raphanus sativus*) cv. Shirt	Sakai and Imamura (1992), Yamagishi and Terachi (2001)
ogu-NWSUAF	Line derived from kos-CMS	Zhao et al. (2010)
hau CMS (6–102A)	*B. juncea*	Jing et al. (2012)
Nsa CMS	*B. napus* and *Sinapis arvensis*	Hu et al. (2002), Wei et al. (2010)
Nca CMS	*B. napus*	Wei et al. (2005)
inap CMS	*B. napus* and *Isatis indigotica*	Kang et al. (2017)

Table 4 Sources of genic male sterility (GMS) in rapeseed

Sterility name	Source of sterility	References
MSL small sterility 'Lembke'	System developed by a German breeding company NPZ Lembke	Frauen (1987), Friedt and Snowdon (2009)
Syngenta 'Safecross'	MSL developed by Syngenta	Stiewe (2008)
SeedLink® hybridization system	developed by PGS, Ghent	Mariani et al. (1990)

resistance and other agronomic traits (Ji et al. 2020; Havlíčková et al. 2015; Klíma et al. 2019). The first hybrid varieties were obtained and bred by Professor Fu Ting-Dong in Wuhan (a three-line hybrid based on SI in 1975, and a hybrid based on *pol* CMS in 1985). 'Synergy', the first European winter rapeseed line-hybrid composite (LHC) variety with a mix of female hybrid plants and fertile plants from one of the parental lines in an 85:15% ratio, was registered in 1994. In 1994, the first spring Polima restored hybrid ('Hybridol') was registered in France, followed by restored F1-hybrid varieties 'Joker' and 'Pronto' based ont MSL system in Germany in 1996, and the first Ogu/INRA restored hybrid in Danemark in 1998 ('Elite') (Pelletier and Budar 2015; Pinochet and Bertrand 2000; Becker et al. 1999) (Table 5).

Table 5 Modern approaches in rapeseed breeding

Biotechnology, genomics, and transcriptomics have played crucial roles in advancing rapeseed breeding by providing tools and insights that enable breeders to select desired traits more efficiently	
Biotechnology	
Haploid production	Microspore-derived haploid embryos generated through somatic embryogenesis in pollen cultures serve as effective means to produce fully homozygous doubled haploid (DH) plants. DH plants are especially valuable in breeding programmes as they can rapidly produce homozygous lines
Embryo rescue	Embryo rescue technology and embryo cultures facilitate the production of novel rapeseed lines. These hybrids are involved in breeding programmes as sources of new genetic variation
Biotech (GM) approach and new genomic technologies (NGT)	
Genetic modification (GM)	Genetic modification techniques, including transgenic methods, have enabled the introduction of specific traits in rapeseed breeding. GM rapeseed varieties that exhibit resistance to specific herbicides have been developed, facilitating enhanced weed management
Gene editing (CRISPR-Cas9)	Gene editing tools like CRISPR-Cas9 enable precise modifications in the rapeseed genome. This technology allows for the targeted insertion, deletion, or modification of specific genes, facilitating the development of new rapeseed varieties with desired traits
Genomics	
Genome sequencing	The sequencing of the rapeseed genome provides a comprehensive understanding of its genetic makeup. This information aids in identifying genes associated with important agronomic traits, leading to more targeted breeding efforts
Marker-assisted selection (MAS)	Genetic markers linked to favourable traits are employed in marker-assisted selection. This enables breeders to more efficiently select plants with desired traits, thereby accelerating the breeding process
Genome-wide association studies (GWAS)	GWAS involves studying the entire genome to identify genetic variations associated with specific traits. This approach helps uncover the genetic basis of complex traits in rapeseed and assists in marker development
Quantitative trait loci (QTL) mapping	QTL mapping outlines genomic regions associated with quantitative traits. This information directs breeding programmes to concentrate on traits such as yield, disease resistance, and oil quality
Transcriptomics	
Gene expression profiling	Transcriptomics involves studying the complete RNA transcripts (transcriptome) set in a cell or tissue. Gene expression profiling helps identify active genes under different conditions, providing insights into how genes contribute to specific traits

(continued)

Table 5 (continued)

Associative transcriptomics	Transcriptomics allows the study of gene expression patterns across various developmental stages of rapeseed plants, as well as their responses to environmental stresses like drought and disease, facilitating the identification of genes associated with stress tolerance. This information may facilitate the development of rapeseed varieties exhibiting improved resilience
Integration with genomics and phenotypic data	Integrating transcriptomic data with genomic information and precise phenotypization provides a more comprehensive understanding of how genes function and interact. This integrated approach aids in the identification of crucial genes for targeted breeding efforts

5 Conclusions

The creation of new varieties, including lines and hybrids, is significantly accelerated by using modern biotechnological and molecular biology approaches and their integration into rapeseed breeding programmes. Biotechnological methods are mainly represented by the production of DH plants obtained from microspore embryos. DH plants are fully homozygous and allow a significant shortening of the breeding process and obtaining a new variety in less time (Starosta et al. 2023; Prem et al. 2012). The molecular methods focus on genomic selection and developing markers for the precise and timely selection of desired plants (Havlíčková et al. 2015; Klíma et al. 2019). Screening large breeding populations and selecting plants with desirable genes again allows a significant refinement of the selection process and the creation of varieties with predefined characteristics. However, to apply molecular breeding procedures, it is necessary to have deep knowledge of the rapeseed genome and the genetic basis of individual agronomically important traits. Therefore, the function of individual genes is examined, and over 100,000 genes are annotated in *B. napus* (Hu et al. 2017). Functional genomics and methodological approaches in associative transcriptomics facilitate the identification of genetic regions and candidate genes or transcription factors that regulate phenotypic traits of interest, including oil content in seeds, fertility regulation, disease and pest resistance, abiotic stress tolerance, nutrient utilization efficiency, and resistance to pod shattering (Schmidt and Bancroft 2011; Fikere et al. 2020; Abbadi and Leckband 2011; Bouchet et al. 2016; Harper et al. 2012; Havlickova et al. 2018). Significant progress in rapeseed breeding and molecular approach in breeding is also due to advances in plant genome and transcriptome analysis methods, the development of next-generation sequencing (NGS) techniques, and the use of single-nucleotide polymorphism (SNP) markers (Kittipol et al. 2019; Hu et al. 2021).

References

Abbadi A, Leckband G (2011) Rapeseed breeding for oil content, quality, and sustainability. Eur J Lipid Sci Technol 113:1198–1206. https://doi.org/10.1002/ejlt.201100063

Abdel-Massih RM, Debs E, Othman L, Attieh J, Cabrerizo FM (2023) Glucosinolates, a natural chemical arsenal: more to tell than the myrosinase story. Front Microbiol 5(14):1130208. https://doi.org/10.3389/fmicb.2023.1130208

Akbar MA (1989) Resynthesis of Brassica napus aiming for improved earliness and carried out by different approaches. Hereditas, 111(3):239–246. https://doi.org/10.1111/j1601-52231990tb00402x

Allender CJ, King GJ (2010) Origins of the amphiploid species *Brassica napus* L investigated by chloroplast and nuclear molecular markers. BMC Plant Biol 10:54. https://doi.org/10.1186/1471-2229-10-54

Al-Shehbaz IA, Beilstein MA, Kellogg EA (2006) Systematics and phylogeny of the Brassicaceae (Cruciferae): an overview. Plant Syst Evol 259:89–120

Andréasson E, Bolt Jørgensen L, Höglund AS, Rask L, Meijer J (2001) Different myrosinase and idioblast distribution in *Arabidopsis* and *Brassica napus*. Plant Physiol 127(4):1750–1763. https://doi.org/10.1104/pp.010334

Anisimova I, Dubovskaya AG (2020) CMS systems in rapeseed and their use in the breeding of domestic hybrids. Proceed App bot genet breed 181:171–180

Ayton J (2014) Variability of quality traits in canola seed, oil and meal – a review NSW Department of Primary Industries New South Wales, Australia

Bancroft I, Morgan C, Fraser F, Higgins J, Wells R, Clissold L, Baker D, Long Y, Meng J, Wang X, Liu S, Trick M (2011) Dissecting the genome of the polyploid crop oilseed rape by transcriptome sequencing. Nat Biotechnol 29(8):762–766. https://doi.org/10.1038/nbt.1926

Barba FJ, Nikmaram N, Roohinejad S, Khelfa A, Zhu Z, Koubaa M (2016) Bioavailability of glucosinolates and their breakdown products: impact of processing. Front Nutr 16:3. https://doi.org/10.3389/fnut.2016.00024

Baux A, Colbach N, Pellet D (2011) Crop management for optimal low–linolenic rapeseed oil production – field experiments and modelling. Eur J Agron 35:144–153. https://doi.org/10.1016/j.eja.2011.05.006

Becker HC, Löptien H, Röbbelen G (1999) Breeding: an overview. In: Gómez-Campo C (ed) Biology of *brassica coenospecies*. Elsevier, Amsterdam, pp 413–460

Bell JM (1993) Factors affecting the nutritional value of canola meal: a review. Can J Anim Sci 73:619–697. https://doi.org/10.4141/cjas93-075

Blackshaw RE, Johnson EN, Gan Y, May WE, McAndrew DW, Barthet V, McDonald T, Wispinski D (2011) Alternative oilseed crops for biodiesel feedstock on the Canadian prairies. Can J Plant Sci 91:889–896

Bouchet AS, Laperche A, Bissuel-Belaygue C, Snowdon R, Nesi N, Stahl A (2016) Nitrogen use efficiency in rapeseed a review. Agron Sustain Dev 36:38. https://doi.org/10.1007/s13593-016-0371-0

Brandt SA, Malhi SS, Ulrich D, Lafond GR, Kutcher HR, Johnston AM (2007) Seeding rate, fertilizer level and disease management effects on hybrid versus open pollinated canola (*Brassica napus* L). Can J Plant Sci 87:255–266. https://doi.org/10.4141/P05-223

Carré P, Pouzet A (2014) Rapeseed market, worldwide and in Europe. Ocl 21(1):D102

Chalhoub B, Denoeud F, Liu S, Parkin IA, Tang H, Wang X, Chiquet J, Belcram H, Tong C, Samans B, Corréa M, Da Silva C, Just J, Falentin C, Koh CS, Le Clainche I, Bernard M, Bento P, Noel B, Labadie K, Alberti A, Charles M, Arnaud D, Guo H, Daviaud C, Alamery S, Jabbari K, Zhao M, Edger PP, Chelaifa H, Tack D, Lassalle G, Mestiri I, Schnel N, Le Paslier MC, Fan G, Renault V, Bayer PE, Golicz AA, Manoli S, Lee TH, Thi VH, Chalabi S, Hu Q, Fan C, Tollenaere R, Lu Y, Battail C, Shen J, Sidebottom CH, Wang X, Canaguier A, Chauveau A, Bérard A, Deniot G, Guan M, Liu Z, Sun F, Lim YP, Lyons E, Town CD, Bancroft I, Wang X, Meng J, Ma J, Pires JC, King GJ, Brunel D, Delourme R, Renard M, Aury JM, Adams KL,

Batley J, Snowdon RJ, Tost J, Edwards D, Zhou Y, Hua W, Sharpe AG, Paterson AH, Guan C, Wincker P (2014) Plant genetics early allopolyploid evolution in the post–neolithic *Brassica napus* oilseed genome. Science 345(6199):950–953. https://doi.org/10.1126/science.1253435
Choudhary BR, Joshi P (2001) Genetic diversity in advanced derivatives of brassica interspecific hybrids. Euphytica 121(1):1–7. https://doi.org/10.1023/A:1012047107039
Chen BY, Heneen WK (1989) Resynthesized Brassica napus L: A review of its potential in breeding and genetic analysis. Hereditas 111(3):255–263. https://doi.org/10.1111/j1601-52231990tb00404x
Cuthbert RD, Crow G, McVetty PBE (2011) Assessment of seed quality performance and heterosis for seed quality traits in hybrid high erucic acid rapeseed (HEAR). Can J Plant Sci 91:837–846. https://doi.org/10.4141/cjps10205
Daniel C, Hommes M, Koller M (2016) Plant protection in organic production of brassica vegetables and oilseed rape. IOBC–WPRS Bull 118:110–115
Daun J, Eskin N, Hickling D (2011) Canola: chemistry, production, processing and utilization. AOCS Press, Urbana
Downey RK (1990) Brassica oil seed breeding – achievements and opportunities. Plant Breed Abstr 60(10):1165–1170
Downey RK, Bell JM (1990) New developments in canola research. In: Shahidi F (ed) Canola and rapeseed. Springer, Boston
Downey RK, Röbbelen G (1989) *Brassica* species. In: Röbbelen G, Downey RK, Ashri A (eds) oil crops of the world. McGraw-Hill, New York, pp 339–362
Downey RK, Craig BM, Youngs CG (1969) Breeding rapeseed for oil and meal quality. J Amer Oil Chem Soc 46:121–123. https://doi.org/10.1007/BF02635712
Downey RK, Klassen AJ, Stringam GR (1980) Rapeseed and mustard. In: Fehr WR, Hadley HH (eds) Hybridization of crop plants. American Society of Agronomy (ASA) and the Crop Science Society of America (CSSA), Madison, pp 495–509. https://doi.org/10.2135/1980.hybridizationofcrops.c35
Dunford N (2018) Canola oil properties food technology fact sheet, Oklahoma. https://pods.okstate.edu/fact–sheets/FAPC–134pod.pdf. Accessed on 10 Jan 2025
Elferjani R, Soolanayakanahally R (2018) Canola responses to drought, heat, and combined stress: shared and specific effects on carbon assimilation, seed yield, and oil composition. Front Plant Sci 30(9):1224. https://doi.org/10.3389/fpls.2018.01224
Fan ZG, Stefansson BR (1986) Influence of temperature on sterility of 2 cytoplasmic male–sterility systems in rape (*Brassica napus* L). Canadian J Plant Sci 66:221–227. https://doi.org/10.4141/cjps86-035
Fan Z, Stefansson BR, Sernyk JL (1986) Maintainers and restorers for three male–sterility inducing cytoplasms in rape (*Brassica napus* L). Canadian J Plant Sci 66:229–234. https://doi.org/10.4141/cjps86-036
FAO (2023) World Food and Agriculture – statistical Yearbook 2023. Rome
Fell H, Muthayil Ali A, Wells R, Mitrousia GK, Woolfenden H, Schoonbeek HJ, Fitt BDL, Ridout CJ, Stotz HU (2023) Novel gene loci associated with susceptibility or cryptic quantitative resistance to *Pyrenopeziza brassicae* in *Brassica napus*. Theor Appl Genet 136(4):71. https://doi.org/10.1007/s00122-023-04243-y
Fikere M, Barbulescu DM, Malmberg MM, Maharjan P, Salisbury PA, Kant S, Panozzo J, Norton S, Spangenberg GC, Cogan NOI et al (2020) Genomic prediction and genetic correlation of agronomic, blackleg disease, and seed quality traits in canola (*Brassica napus* L). Plants 9(6):719. https://doi.org/10.3390/plants9060719
FitzJohn RG, Armstrong TT et al (2007) Hybridisation within brassica and allied genera: evaluation of potential for transgene escape. Euphytica 158:209–230. https://doi.org/10.1007/s10681-007-9444
Francis A, Lujan-Toro BE, Warwick SI, Macklin JA, Martin SL (2021) Update on the Brassicaceae species checklist. Biodivers Data J 9:e58773
Frandsen KJ (1943) The experimental formation of *Brassica juncea*. Czern et Coss Dansk Bot Ark 11:1–17

Frandsen KJ (1947) The experimental formation of *Brassica napus* L var oleifera DC and *Brassica carinata* Braun. Dansk Bot Ark 12:1–16

Frauen M (1987) Aspects of seed production of hybrid varieties of rape. In: Feistritzer WP, Kelly F (eds) Hybrid seed production of selected cereal and vegetable crops. FAO plant production and protection papers issue 82. FAO, Rome, pp 281–300

Friedt W, Snowdon R (2009) Oilseed rape. In: Vollmann J, Rajcan I (eds) Oil crops. Handbook of plant breeding, vol 4. Springer, New York, pp 91–126

Friedt W, Tu J, Fu T (2018) Academic and economic importance of *Brassica napus* rapeseed. In: Liu S, Snowdon R, Chalhoub B (eds) The *Brassica napus* genome, 1st edn. Springer, Cham, pp 1–20

Fu TD (1981) Production and research of rape seed in the people Republic of China *Eucarpia cruciferae*. Newslett Eucarpia Cruciferae Newsl 6:6–7

Ge JC, Yoon SK, Choi NJ (2017) Using canola oil biodiesel as an alternative fuel in diesel engines: a review. App Sci 7(9):881. https://doi.org/10.3390/app7090881

Gehringer A, Spiller T, Basunanda P, Snowdon R, Friedt W (2007) New oilseed rape (*Brassica napus*) hybrids with high levels of heterosis for seed yield under nutrient–poor conditions. Breed Sci 57:315–320. https://doi.org/10.1270/jsbbs.57.315

Girke A, Schierholt A, Becker HC (2012) Extending the rapeseed gene pool with resynthesized *Brassica napus* II: Heterosis. Theor Appl Genet 124(6):1017–1026. https://doi.org/10.1007/s00122-011-1765-7

Grant I, Beversdorf WD (1985) Heterosis and combining ability in spring–planted oilseed rape (*Brassica napus* L). Can J Genet Cytol 27:472–478

Grant I, Beversdorf WD, Peterson RL (1986) A comparative light and electron microscopic study of microspore and tapetal development in male fertile and cytoplasmic male sterile oilseed rape (*Brassica napus*). Can J Bot 64(5):1055–1068

Håkansson A (2010) Seed development of *Brassica oleracea* and *B rapa* after certain reciprocal pollinations. Hereditas 42:373–396

Harper AL, Trick M, Higgins J, Fraser F, Clissold L, Wells R, Hattori C, Werner P, Bancroft I (2012) Associative transcriptomics of traits in the polyploid crop species *Brassica napus*. Nat Biotechnol 30(8):798–802. https://doi.org/10.1038/nbt.2302

Havlíčková L, Klíma M, Přibylová M, Hilgert-Delgado AA, Kučera V, Čurn V (2015) Non–destructive in vitro selection of microspore–derived embryos with the fertility restorer gene for CMS Ogu–INRA in winter oilseed rape (*Brassica napus* L). Electron J Biotechnol 18:58–60. https://doi.org/10.1016/j.ejbt.2014.10.003

Havlickova L, He Z, Wang L, Langer S, Harper AL, Kaur H, Broadley MR, Gegas V, Bancroft I (2018) Validation of an updated associative transcriptomics platform for the polyploid crop species *Brassica napus* by dissection of the genetic architecture of erucic acid and tocopherol isoform variation in seeds. Plant J 93(1):181–192. https://doi.org/10.1111/tpj.13767

He Z, Cheng F, Li Y, Wang X, Parkin IA, Chalhoub B, Liu S, Bancroft I (2015) Construction of brassica a and C genome–based ordered pan–transcriptomes for use in rapeseed genomic research. Data Brief 2(4):357–362. https://doi.org/10.1016/j.dib.2015.06.016

He Z, Ji R, Havlickova L, Wang L, Li Y, Lee HT, Song J, Koh C, Yang J, Zhang M, Parkin IAP, Wang X, Edwards D, King GJ, Zou J, Liu K, Snowdon RJ, Banga SS, Machackova I, Bancroft I (2021) Genome structural evolution in brassica crops. Nat Plants 7(6):757–765. https://doi.org/10.1038/s41477-021-00928-8

Hejna O, Havlickova L, He Z, Bancroft I, Curn V (2019) Analysing the genetic architecture of clubroot resistance variation in *Brassica napus* by associative transcriptomics. Mol Breed 39(8):112

Heyn FW (1976) Transfer of restorer genes from Raphanus to cytoplasmic male sterile *Brassica napus*. Cruciferae News 11:15–16

Hilgert-Delgado A, Klíma M, Viehmannová I (2015) Efficient resynthesis of oilseed rape (*Brassica napus* L) from crosses of winter types *B rapa* × *B oleracea* via simple ovule culture and early hybrid verification. Plant Cell Tissue Organ Cult 120:191–201. https://doi.org/10.1007/s11240-014-0593-2

Hinata K, Konno N (1979) Studies on a male–sterile strain having *brassica campestrisnucleus* and *Diplotaxis muralis* cytoplasm I breeding and some characteristics of this strain. Jpn J Breed 29:305–311. https://doi.org/10.1270/jsbbs1951.29.305

Howard HW (1942) The effect of polyploidy and hybridity on seed size in crosses between *B chinensis*, *B carinata*, amphidiploid *B chinensis–carinata* and autotetraploid *B chinensis*. J Genet 43:105–119

Hu Q, Andersen S, Dixelius C, Hansen L (2002) Production of fertile intergeneric somatic hybrids between *Brassica napus* and *Sinapis arvensis* for the enrichment of the rapeseed gene pool. Plant Cell Rep 21(2):147–152. https://doi.org/10.1007/s00299-002-0491-7

Hu Q, Hua W, Yin Y, Zhang XK, Liu LJ, Shi JQ, Zhao YG et al (2017) Rapeseed research and production in China. Crop J 5:127–135. https://doi.org/10.1016/j.cj.2016.06.005

Hu D, Jing J, Snowdon RJ, Mason AS, Shen J, Meng J, Zou J (2021) Exploring the gene pool of *Brassica napus* by genomics–based approaches. Plant Biotechnol J 19(9):1693–1712. https://doi.org/10.1111/pbi.13636

Huang YJ, Qi A, King GJ, Fitt BD (2014) Assessing quantitative resistance against *Leptosphaeria maculans* (phoma stem canker) in *Brassica napus* (oilseed rape) in young plants. PLoS One 9(1):e84924. https://doi.org/10.1371/journal.pone.0084924

Ibrahim E, Rychlá A, Alquicer G, Slavíková L, Peng Q, Klíma M, Vrbovský V, Trebicki P, Kundu JK (2023) Evaluation of resistance of oilseed rape genotypes to turnip yellows virus. Plants 12(13):2501. https://doi.org/10.3390/plants12132501

Iñiguez-Luy FL, Federico ML (2011) The genetics of *Brassica napus* L. In: Bancroft I, Schmidt R (eds) Genetics and genomics of the *Brassicaceae*. Springer, New York, pp 291–322

Jaime R, Alcántara JM, Manzaneda AJ, Rey PJ (2018) Climate change decreases suitable areas for rapeseed cultivation in Europe but provides new opportunities for white mustard as an alternative oilseed for biofuel production. PLoS One 13(11):e0207124. https://doi.org/10.1371/journal.pone.0207124

Jannat A, Ishikawa-Ishiwata Y, Furuya J (2022) Does climate change affect rapeseed production in exporting and importing countries? evidence from market dynamics syntheses. Sustainability 14(10):6051. https://doi.org/10.3390/su14106051

Ji J, Huang J, Yang L, Fang Z, Zhang Y, Zhuang M, Lv H, Wang Y, Liu Y, Li Z et al (2020) Advances in research and application of male sterility in *Brassica oleracea*. Horticulturae 6(4):101. https://doi.org/10.3390/horticulturae6040101

Jian H, Xie L, Wang Y, Cao Y, Wan M, Lv D, Li J, Lu K, Xu X, Liu L (2020) Characterization of cold stress responses in different rapeseed ecotypes based on metabolomics and transcriptomics analyses. PeerJ 31(8):e8704. https://doi.org/10.7717/peerj.8704

Jing B, Heng S, Tong D, Wan Z, Fu T, Tu J, Ma C, Yi B, Wen J, Shen J (2012) A male sterility–associated cytotoxic protein ORF288 in *Brassica juncea* causes aborted pollen development. J Exp Bot 63:1285–1295. https://doi.org/10.1093/jxb/err355

Kamala T (1976) Interspecific hybrids in brassica. Cytologia 41:407–415. https://doi.org/10.1508/cytologia.41.407

Kang L, Li P, Wang A, Ge X, Li Z (2017) A novel cytoplasmic male sterility in *Brassica napus* (inap CMS) with Carpelloid stamens via protoplast fusion with chinese woad. Front Plant Sci 6(8):529. https://doi.org/10.3389/fpls.2017.00529

Katche E, Quezada-Martinez D, Katche EI, Vasquez-Teuber P, Mason AS (2019) Interspecific hybridization for brassica crop improvement crop breeding. Genetics Genomics 1:e190007. https://doi.org/10.20900/cbgg20190007

Khajali F, Slominski BA (2012) Factors that affect the nutritive value of canola meal for poultry. Poult Sci 91(10):2564–2575. https://doi.org/10.3382/ps.2012-02332

Kimber DS, McGregor DI (1995) The species and their origin, cultivation and world production. In: Brassica oilseeds: production and utilization. CAB International, Oxford, pp 1–9

Kittipol V, He Z, Wang L, Doheny-Adams T, Langer S, Bancroft I (2019) Genetic architecture of glucosinolate variation in *Brassica napus*. J Plant Physiol 240:152988. https://doi.org/10.1016/j.jplph.2019.06.001

Klíma M, Jozová E, Jelínková I, Kučera V, Hu S, Čurn V (2019) Early *in vitro* selection of winter oilseed rape (*Brassica napus* L) plants with the fertility restorer gene for CMS *Shaan* 2A via non–destructive molecular analysis of microspore–derived embryos. Czech J Genet Plant Breed 55(4):162–165. https://doi.org/10.17221/94/2018-CJGPB

Kirkegaard JA, Gardner PA, Angus JF, Koetz E (1994) Effect of brassica break crops on the growth and yield of wheat. Aust J Agric Res 45:529–545. https://doi.org/10.1071/AR9940529

Kučera V, Vyvadilová M, Klíma M (2010) Utilization of cytoplasmic male sterility Shaan 2A in a hybrid breeding of winter oilseed rape. Agric Trop Subtrop 43:182–185

Labana KS, Gupta ML (1993) Importance and origin. In: Labana KS, Banga SS, Banga SK (eds) Breeding oilseed brassicas. Springer–Verlag, Berlin, pp 1–7

Landgren M, Zetterstrand M, Sundberg E, Glimelius K (1996) Alloplasmic male–sterile brassica lines containing *B tournefortii* mitochondria express an ORF 3′ of the atp6 gene and a 32 kDa protein. Plant Mol Biol 32:879–890. https://doi.org/10.1007/BF00020485

Larkin RP, Lynch RP (2018) Use and effects of different brassica and other rotation crops on soilborne diseases and yield of potato. Horticulturae 4(4):37. https://doi.org/10.3390/horticulturae4040037

Lee JW, Kim IH, Woyengo TA (2020) Toxicity of canola–derived glucosinolate degradation products in pigs–a review. Animals 10(12):2337. https://doi.org/10.3390/ani10122337

Liu Z, Wang H, Xie J, Lv J, Zhang G, Hu L, Luo S, Li L, Yu J (2021) The roles of cruciferae glucosinolates in disease and pest resistance. Plants 10(6):1097. https://doi.org/10.3390/plants10061097

Liu X, Wei R, Tian M, Liu J, Ruan Y, Sun C, Liu C (2022) Combined transcriptome and metabolome profiling provide insights into cold responses in rapeseed (*Brassica napus* L) genotypes with contrasting cold–stress sensitivity. Int J Mol Sci 23(21):13546. https://doi.org/10.3390/ijms232113546

Lohani N, Jain D, Singh MB, Bhalla PL (2020) Engineering multiple abiotic stress tolerance in canola, *Brassica napus*. Front Plant Sci 25(11):3. https://doi.org/10.3389/fpls.2020.00003

Lu K, Wei L, Li X, Wang Y, Wu J, Liu M, Zhang C, Chen Z, Xiao Z, Jian H, Cheng F, Zhang K, Du H, Cheng X, Qu C, Qian W, Liu L, Wang R, Zou Q, Ying J, Xu X, Mei J, Liang Y, Chai YR, Tang Z, Wan H, Ni Y, He Y, Lin N, Fan Y, Sun W, Li NN, Zhou G, Zheng H, Wang X, Paterson AH, Li J (2019) Whole–genome resequencing reveals *Brassica napus* origin and genetic loci involved in its improvement. Nat Commun 10(1):1154. https://doi.org/10.1038/s41467-019-09134-9

Lysak MA, Koch MA (2011) Phylogeny, genome, and karyotype evolution of crucifers (Brassicaceae). In: Schmidt R, Bancroft I (eds) Genetics and genomics of the Brassicaceae plant genetics and genomics: crops and models, vol 9. Springer, New York

Mariani C, De Beuckeleer M, Truettner J, Leemans J, Goldberg RB (1990) Induction of male sterility in plants by a chimaeric ribonuclease gene. Nature 347:737–741. https://doi.org/10.1038/347737a0

McNaughton IH (1976) Swedes and rapes (*Brassica napus*). In: Simmonds NW (ed) Evolution of crop plants. Longman Inc, London/New York, pp 53–56

Brandle JE, McVetty PBE (1989) Heterosis and combining ability in hybrids derived from oilseed rape cultivars and inbred lines. Crop Sci 29:1191–1195. https://doi.org/10.2135/cropsci1989.0011183X002900050020x

Mitrousia GK, Huang YJ, Qi A, Sidique SNM, Fitt BDL (2018) Effectiveness of Rlm7 resistance against *Leptosphaeria maculans* (phoma stem canker) in UK winter oilseed rape cultivars. Plant Pathol 67(6):1339–1353. https://doi.org/10.1111/ppa.12845

Morinaga T (1929) The cytology of F1 hybrids of *B napella* and various other species with ten chromosomes. Cytologia 1:16–27. https://doi.org/10.1508/cytologia.1.16

Morinaga T (1934) The cytology of F1 hybrids of *B juncea* x *B nigra*. Cytologia 6:62–67. https://doi.org/10.1508/cytologia.6.62

Nagaharu U (1935) Genome analysis in brassica with special reference to the *B napes* and peculiar mode of fertilisation. Jpn J But 7:389–452

Neik TX, Barbetti MJ, Batley J (2017) Current status and challenges in identifying disease resistance genes in *Brassica napus*. Front Plant Sci 6(8):1788. https://doi.org/10.3389/fpls.2017.01788

Obermeier C, Mason AS, Meiners T, Petschenka G, Rostás M, Will T, Wittkop B, Austel N (2022) Perspectives for integrated insect pest protection in oilseed rape breeding. Theor Appl Genet 135(11):3917–3946

Ogura H (1968) Studies on the new male–sterility in Japanese radish, with special reference to the utilization of this sterility towards the practical raising of hybrid seeds. Mem Fac Agric Kagoshima Univ 6:39–78

Nassef BG, Pape F, Poll G (2023) Enhancing the performance of rapeseed oil lubricant for machinery component applications through hybrid blends of nanoadditives. Lubricants 11(11):479. https://doi.org/10.3390/lubricants11110479

Pelletier G, Budar F (2015) Brassica Ogu–INRA cytoplasmic male sterility: an example of successful plant somatic fusion for hybrid seed production. In: Somatic genome manipulation. Springer, New York, pp 199–216

Pinochet X, Bertrand R (2000) Oilseed rape grain yield productivity increases with hybrid varietal types: a first balance sheet with post registration tests in France and Europe. OCL–Oleagineux Corps Gras Lipides 7:11–16. https://doi.org/10.1051/ocl.2000.0011

Prakash S, Chopra VL (1988) Synthesis of alloplasmic *Brassica campestris* as a new source of cytoplasmic male sterility. Plant Breed 101(3):253–255

Prakash S, Kirti P, Bhat S (1998) A *Moricandia arvensis*– based cytoplasmic male sterility and fertility restoration system in *Brassica juncea*. Theor Appl Genet 97:488–492. https://doi.org/10.1007/s001220050921

Prakash SR, Bhat SR, Quiros CF, Kirti PB, Chopra VL (2009) Brassica and its close allies: cytogenetics and evolution. Plant Breed Rev 31:21–187. https://doi.org/10.1002/9780470593783.ch2

Prem D, Solís MT, Bárány I (2012) A new microspore embryogenesis system under low temperature which mimics zygotic embryogenesis initials, expresses auxin and efficiently regenerates doubled–haploid plants in *Brassica napus*. BMC Plant Biol 12:127

Primard-Brisset C, Poupard JP, Horvais R, Eber F, Pelletier G, Renard M, Delourme R (2005) A new recombined double low restorer line for the Ogu–INRA cms in rapeseed (*Brassica napus* L). Theor Appl Genet 111(4):736–746

Raboanatahiry N, Li H, Yu L, Li M (2021) Rapeseed (*Brassica napus*): processing, utilization, and genetic improvement. Agronomy 11(9):1776. https://doi.org/10.3390/agronomy11091776

Rahman H (2013) Review: breeding spring canola (*Brassica napus* L) by the use of exotic germplasm. Can J Plant Sci 93(3):363–373

Rakow G, McGregor DI (1975) Oil, fatty acid and chlorophyll accumulation in developing seeds of two 'linolenic acid lines' of low erucic acid rapeseed. Can J Plant Sci 55:197–203

Raman H, Raman R, Kilian A, Detering F, Carling J, Coombes N, Diffey S, Kadkol G, Edwards D, McCully M, Ruperao P, Parkin IA, Batley J, Luckett DJ, Wratten N (2014) Genome–wide delineation of natural variation for pod shatter resistance in *Brassica napus*. PLoS One 9(7):e101673. https://doi.org/10.1371/journal.pone.0101673

Rawat DS, Anand IJ (1979) Male sterility in Indian mustard. Indian J Genet Plant Breed 39:412–414

Raza A, Razzaq A, Mehmood SS, Hussain MA, Wei S, He H, Zaman QU, Xuekun Z, Hasanuzzaman M (2021) Omics the way forward to enhance abiotic stress tolerance in *Brassica napus* L. GM Crops Food 12(1):251–281. https://doi.org/10.1080/21645698.2020.1859898

Röbbelen G (1960) Beitrage zur analyse des Brassica genoms. Chromosoma (Berl) 11:205–228. https://doi.org/10.1007/BF00328652

Röbbelen G (1981) Breeding for low content of glucosinolates in rapeseed. In: Bunting ES (ed) Production and utilization of protein in oilseed crops world crops: production, utilization, and description, vol 5. Springer, Dordrecht

Röbbelen G (1985) Raps (*Brassica napus* L). In: Hoffmann W, Mudra A, Plarre W (eds) Lehrbuch der Züchtung landwirtschaftlicher Kulturpflanzen, B2 – SpezTeil. Paul Parey, Hamburg, pp 288–302

Röbbelen G, Rakow, G (1970) Selection for fatty acids in rapeseed. In: Proceedings of the international conference on the science, technology and marketing of rapeseed and rapeseed products, Ste–Adble; 20–23: 47 6–490

Rousselle P, Bregeon M (1982) Premiers résultats d'un programme d'introduction de l'androstérilité " Ogura " du radis chez le colza. Agronomie 2(9):859–864

Rücker B, Röbbelen G (1997) Mutants of *Brassica napus* with altered seed lipid fatty acid composition. In: Williams JP, Khan MU, Lem NW (eds) Physiology, biochemistry and molecular biology of plant lipids springer, Dordrecht lipids. Springer, Dordrecht

Russo M, Yan F, Stier A, Klasen L, Honermeier B (2021) Erucic acid concentration of rapeseed (*Brassica napus* L) oils on the German food retail market. Food Sci Nutr 9(7):3664–3672. https://doi.org/10.1002/fsn3.2327

Rygulla W, Snowdon RJ, Friedt W, Happstadius I, Cheung WY, Chen D (2008) Identification of quantitative trait loci for resistance against *Verticillium longisporum* in oilseed rape (*Brassica napus*). Phytopathology 98(2):215–221

Sabaghnia N, Dehghan H, Alizadeh B, Moghaddam M (2010) Heterosis and combining ability analysis for oil yield and its components in rapeseed. Australian J Crop Sci 4(6):390–397

Sakai T, Imamura J (1992) Alteration of mitochondrial genomes containing atpA genes in the sexual progeny of cybrids between Raphanus sativus cms line and *Brassica napus* cv Westar. Theor Appl Genet 84:923–929. https://doi.org/10.1007/BF00227405

Schmatz AA, Alison HWP, Lucas HW, Maria EPA, Sandra RMA, de Souza NM, David JF, Ayrton BN, Augusto CA (2017) Productivity and quality of hybrid canola oil and seeding time. Int J Environ Agric Res 3(8):67–74

Schmidt R, Bancroft I (2011) Genetics and genomics of the Brassicaceae. Springer, New York

Seyis F, Snowdon RJ, Luhs W, Friedt W (2003) Molecular characterization of novel resynthesized rapeseed (*Brassica napus*) lines and analysis of their genetic diversity in comparison with spring rapeseed cultivars. Plant Breed 122(6):473–478. https://doi.org/10.1111/j.1439-0523.2003.00859.x

Shahbandeh M (2023) Global oilseed production 2022/23 Statista 2023. https://www.statista.com/statistics/263930/worldwide–production–of–rapeseed–by–country/

Shiga T, Baba S (1971) Cytoplasmic male sterility in rape plants (*Brassica napus* L). Jpn J Breed 21:16–17

Shiga T, Baba S (1973) Cytoplasmic male sterility in oilseed rape, *Brassica napus* L, and its utilization to breeding. Jpn J Breed 23:187–197. https://doi.org/10.1270/jsbbs1951.23.187

Shirakawa M, Tanida M, Ito T (2022) The cell differentiation of idioblast myrosin cells: similarities with vascular and guard cells. Front Plant Sci 10(12):829541. https://doi.org/10.3389/fpls.2021.829541

Singh A (2017) Glucosinolates and plant defense. In: Mérillon JM, Ramawat K (eds) Glucosinolates reference series in phytochemistry. Springer, Cham

Snowdon R, Lühs W, Friedt W (2007) Oilseed Rape. In: Kole C (ed) Oilseeds. Genome mapping and molecular breeding in plants 2. Springer, Berlin, Heidelberg, pp 55–114

Song K, Osborn TC (1992) Polyphyletic origins of *Brassica napus*: new evidence based on organelle and nuclear RFLP analyses. Genome 35:992–1001

Starosta E, Szwarc J, Niemann J, Szewczyk K, Weigt D (2023) *Brassica napus* haploid and double haploid production and its latest applications. Curr Issues Mol Biol 45(5):4431–4450. https://doi.org/10.3390/cimb45050282

Stefansson BR, Hougen FW (1964) Selection of rape plants (*Brassica napus*) with seed oil practically free from erucic acid. Can J Plant Sci 44:359–364

Stefansson BR, Kondra ZP (1975) Tower summer rape. Can J Plant Sci 55:343–344

Stiewe G (2008) Breeding a new hybrid family. Rape Grower 17:3

Stiewe G, Röbbelen G (1994) Establishing cytoplasmic male sterility in *Brassica napus* by mitochondrial recombination with B *tournefortii*. Plant Breed 113:294–304

Schenck HR, Uolf G (1986) Characterization of somatic Brassica napus hybrids by Polyacrylamide Gel Electrophoresis. Plant Breed 97(1):72–74. https://doi.org/10.1111/j1439-05231986tb01303x

Sundberg E, Glimelius K (1986) A method for production of interspecific hybrids within Brassiceae via somatic hybridization, using resynthesis of Brassica napus as a model. Plant Sci 43(2):155–162. https://doi.org/10.1016/0168-9452(86)90157-3

Tanner A, Baranek M, Eastlack T, Butts B, Beazley M, Hampton M (2023) Biodiesel production directly from rapeseeds. Water 15(14):2595. https://doi.org/10.3390/w15142595

Thompson KF (1972) Cytoplasmic male sterility in oilseed rape. Heredity 29:253–257. https://doi.org/10.1038/hdy.1972.89

Tonosaki K, Osabe K, Kawanabe T, Fujimoto R (2016) The importance of reproductive barriers and the effect of allopolyploidization on crop breeding. Breed Sci 66(3):333–349. https://doi.org/10.1270/jsbbs.15114

Town C, Schmidt R, Bancroft I (2011) Comparative genome analysis at the sequence level in the Brassicaceae. In: Schmidt R, Bancroft I (eds) Genetics and genomics of the Brassicaceae plant genetics and genomics: crops and models. Springer, New York

Verma JK, Sodhi YS, Mukhopadhyay A, Arumugam N, Gupta V, Pental D, Pradhan AK (2000) Identification of stable maintainer and fertility restorer lines for 'Polima' CMS in *Brassica campestris*. Plant Breed 119(1):90–92. https://doi.org/10.1046/j.1439-0523.2000.00430.x

Wang Q, Yan T, Long Z, Huang LY, Zhu Y, Xu Y, Chen X, Pak H, Li J, Wu D, Xu Y, Hua S, Jiang L (2021) Prediction of heterosis in the recent rapeseed (*Brassica napus*) polyploid by pairing parental nucleotide sequences. PLoS Genet 17(11):e1009879. https://doi.org/10.1371/journal.pgen.1009879

Warwick SI (2011) Brassicaceae in agriculture. In: Schmidt R, Bancroft I (eds) Genetics and genomics of the Brassicaceae plant genetics and genomics: crops and models, vol 9. Springer, New York, pp 33–65

Warwick SI, Black LD (1991) Molecular systematics of brassica and allied genera (subtribe Brassicinae, Brassiceae) – chloroplast genome and cytodeme congruence. Theor Appl Genet 82(1):81–92. https://doi.org/10.1007/BF00231281

Wei WL, Wang HZ, Liu GH (2005) Anatomical observations of anther development of NCa, a cytoplasmic male sterile line in rapeseed (*Brassica napus* L). Sci Agric Sin 38(6):1232–1237

Wei W, Li Y, Wang L, Liu S, Yan X, Mei D (2010) Development of a novel Sinapis arvensis disomic addition line in *Brassica napus* containing the restorer gene for Nsa CMS and improved resistance to *Sclerotinia sclerotiorum* and pod shattering. Theor Appl Genet 120(6):1089–1097

Tonu NN, Wen R, Song T, Guo X, Murphy LA, Gossen BD, Yu F, Peng G (2023) Canola with stacked genes shows moderate resistance and resilience against a field population of *Plasmodiophora brassicae* (Clubroot) pathotype X. Plants 12(4):726. https://doi.org/10.3390/plants12040726

Wittkop B, Snowdon RJ, Friedt W (2009) Status and perspectives of breeding for enhanced yield and quality of oilseed crops for Europe. Euphytica 170:131–140

Wolko J, Dobrzycka A, Bocianowski J (2019) Estimation of heterosis for yield–related traits for single cross and three–way cross hybrids of oilseed rape (*Brassica napus* L). Euphytica 215:156

Xing N, Fan C, Zhou Y (2014) Parental selection of hybrid breeding based on maternal and paternal inheritance of traits in rapeseed (*Brassica napus* L). PLoS One 9(7):e103165. https://doi.org/10.1371/journal.pone.0103165

Yamagishi H, Terachi T (2001) Intra–and inter–specific variations in the mitochondrial gene orfl38 of Ogura–type male–sterile cytoplasm from *Raphanus sativus* and *Raphanus raphanis–* trum. Theor App Gen 103:725–732. https://doi.org/10.1007/s001220100614

Yan L, Shah T, Cheng Y, Lü Y, Zhang X, Zou X (2019) Physiological and molecular responses to cold stress in rapeseed (*Brassica napus* L). J Integr Agric 18:2742–2752

Ye J, Liang H, Zhao X, Li N, Song D, Zhan J, Liu J, Wang X, Tu J, Varshney RK, Shi J, Wang H (2023) A systematic dissection in oilseed rape provides insights into the genetic architecture and molecular mechanism of yield heterosis. Plant Biotechnol J 21(7):1479–1495. https://doi.org/10.1111/pbi.14054

Zahoor A, Celaleddin B, Muhammad Z, Rana M, Muhammad A, Ejaz A (2022) Pod shattering in canola reduced by mitigating drought stress through silicon application and molecular approaches–a review. J Plant Nutr 46(1):101–128. https://doi.org/10.1080/01904167.2022.2027972

Zandberg JD, Fernandez CT, Danilevicz MF, Thomas WJW, Edwards D, Batley J (2022) The global assessment of oilseed brassica crop species yield, yield stability and the underlying genetics. Plants 11(20):2740. https://doi.org/10.3390/plants11202740

Zhao HX, Li ZJ, Hu SW, Sun GL, Chang JJ, Zhang ZH (2010) Identification of cytoplasm types in rapeseed (*Brassica napus* L) accessions by multiplex PCR assay. Theor App Gen 121(4):643–650

Zhu M, Monroe JG, Suhail Y, Villiers F, Mullen J, Pater D, Hauser F, Jeon BW, Bader JS, Kwak JM, Schroeder JI, McKay JK, Assmann SM (2016) Molecular and systems approaches towards drought-tolerant canola crops. New Phytol 210(4):1169–1189. https://doi.org/10.1111/nph.13866

Abiotic Stress Management in Cabbage

Shobna, Himanshu, Deepika Saxena,
Ramesh Kumar Sadawarti, and Rahmatullah Kazimi

Abstract

Cabbage (*Brassica oleracea*) is one of the widely cultivated vegetable crops, valued for its versatility, nutritional content, and adaptability to various climates and cuisines worldwide. Still, several abiotic factors, such as drought, high temperature stress, salinity, and nutrient toxicity affect the growth and yield of cabbage. Various stresses affect cabbage plants, leading to reduced yields and lower-quality produce. Balancing abiotic stresses is crucial for the improvement of cabbage production and the stability of its yield. This chapter presents a comprehensive discussion of abiotic stress factors affecting cabbage and the analysis of several approaches to minimize their effects. The mechanisms of stress response of cabbage at the physiological and molecular levels are discussed in detail with an understanding of how plants perceive and respond to unfavorable conditions. These are the major management practices highlighted: stress-tolerant varieties, fertility management, and water management. Moreover, for effective stress coping mechanisms, modern technologies as bio-stimulants, plant growth controllers, and integrated pest control measures, are analyzed. It also emphasizes Maine's precision agriculture practices, including the use of sensors and data, as well as climate-smart agriculture practices. Through the application of such management practices, it becomes possible for farmers to reduce the impacts of abiotic stresses on the production, enhancing yield on cabbage as well as promoting appropriate sustainable farming practices. Directions

Shobna · Himanshu · D. Saxena (✉) · R. K. Sadawarti
Department of Horticulture, School of Agriculture, Lovely Professional University, Phagwara, Punjab, India
e-mail: deepika.20662@lpu.co.in

R. Kazimi
Department of Horticulture, Daikundi of Higher Education Institution, Daikundi, Afghanistan

A. Kumar, B. Koul (eds.), *Crop Improvement Strategies in Brassica species: Applied Science*, https://doi.org/10.1007/978-981-95-3861-4_14

for future research that may help improve tolerance of abiotic stress in cabbage production are discussed.

Keywords

Abiotic stress · Climate-smart · Stress tolerance · Precision farming · Sustainable cabbage

1 Introduction

The dietary fibers found in fruits and vegetables reduce heart disease risk, alongside obesity development. Health essential vitamins A and C, in combination with important mineral content that promotes human health, are found in foods. 100 g of a serving of fruit contains 90–646 Kcal and 61–89.1 g water, and 2–14.8 g dietary fiber content (la Peña et al. 2020). Many critical stressors bring about significant work-related disappointments despite being very important in human survival. Houetohossou et al. (2023) differentiate stressors among biotic factors emanating from living agents (bacteria, fungi, viruses, and parasites), and abiotic factors that relate to nonliving things (light intensity and water amount and nutrient deficiencies, and atmospheric and climate conditions). Multiple stress factors to which plants are exposed at agricultural sites originate from living and nonliving sources. Environmental factors, like abiotic stresses, threaten the growth and reproduction of cabbage plants, while biotic stresses come from harmful living organisms such as fungi, bacteria, and insects. Climate change is increasing environmental pressures, significantly impacting horticultural crop cultivation. Fruits and vegetables are adversely affected by seasonal patterns in climate terms, exhibiting a rise in heat waves, dry periods, and flooding rains, plus prevailing high levels of solar energy intensities (Andreotti 2020). Light and heat conditions in difficult environments have led growers alongside scientists to increase vegetable yields and nutritional values. The salt stress severity in combination with drought severity depends on three major environmental factors: (i) solar radiation quantity and location and (ii) evapotranspiration needs, and (iii) soil water maintenance capacity. Highlighting the need for adequate, sustainable breeding and agricultural management strategies, to monitor these conditions, will reduce losses due to dry spells and salt stress in vegetable crops. Vegetable quality is negatively affected by both drought and salinity conditions, apart from the yield effects. Vegetables of all types demonstrate drought sensitivity at 20% volumetric water content, and salt sensitivity is recorded at 2.5 dS m^{-1} conductivity. Plant stomata respond to osmosis and ions as well as oxidative stress by prompt closure, with subsequent reduction throughout the entire plant due to drought and salt exposure. By closing stomata, two negative effects on photosynthesis are produced, i.e., the blocking of carbon fixation and a fall in the level of carbon dioxide in the chloroplasts with the encouragement of photorespiration. Following drought, as well as salt exposure on plants, these plants release more reactive oxygen species, whereas these molecules break down the organelle cellular components. Salinity and drought stress combine to affect the quality and quantity of vegetable production. Vegetables respond to drought when volumetric water content is 20%, and to salt stress when their electrical conductivity (EC) = 2.5 dS m^{-1}. Osmotic, ionic, and oxidative damage are inflicted in plants when subjected to the stressors that result from drought and salinity, which causes short-term stomatal closure before affecting the physical reduction in plant size. Such a reduction in CO_2 uptake, along with increased photorespiratory rate, arises due to the closure of stomata, which reduces carboxylation and inhibits CO_2 concentration in the cell. Plants generate more reactive oxygen species in conditions of drought and salt stress, but this generation causes major damage to cellular organelles (Khalid et al. 2023). The study of the omics sciences and the application of Arabidopsis as a model plant have revealed stress defense mechanisms, which allow scientists to identify resistant genotypes in each species. All the abiotic stress perception mechanisms in plants involve specific receptors that link to particular exocellular signals, transferring signals to cells via cyclic nucleotides (cAMP and cGMP), sugars, reactive oxygen species (ROS), calcium ions (Ca^{2+}), inositol phosphate, and nitric oxide (NO).

To promote sustainable cabbage production in such an environment, it is essential to learn how cabbage responds to abiotic stress and examine state-of-the-art management practices. Having analyzed the basic abiotic forces acting upon cabbage, this work will give a synopsis of the current management strategies and potential remedies toward enhancing output and adaptability to challenges in the environment.

2 Types of Abiotic Stresses Affecting Cabbage

Abiotic stress includes a variety of potentially harmful elements, including high and low temperatures, metal toxicity, flooding, drought, salt, and nutritional deficiencies. Additionally, it encompasses elements that are frequently erratic and extremely localized in occurrence, such as shade, UV exposure, photo-inhibition, air pollution, wind, hail, and gaseous pollutants. Abiotic stress in plants can be caused by the presence of excessive concentrations of a poisonous or antagonistic agent or by the lack of an essential resource. In certain situations, like the water supply, plants may experience stress from either too little (drought) or too much (flooding) (Rathod and Verma 2023). Various types of abiotic stresses are shown in Fig. 1.

2.1 Drought

The frequency and severity of drought have been increasing because of climate change, one of the vital factors constraining agricultural productivity. Water deficit lowers the rates of nodulation because it decreases the numbers of the existing soil rhizobia and also the ability to start root infections. The oxygen limitation problem, coupled with feedback inhibition of nitrogen fixation processes, causes the

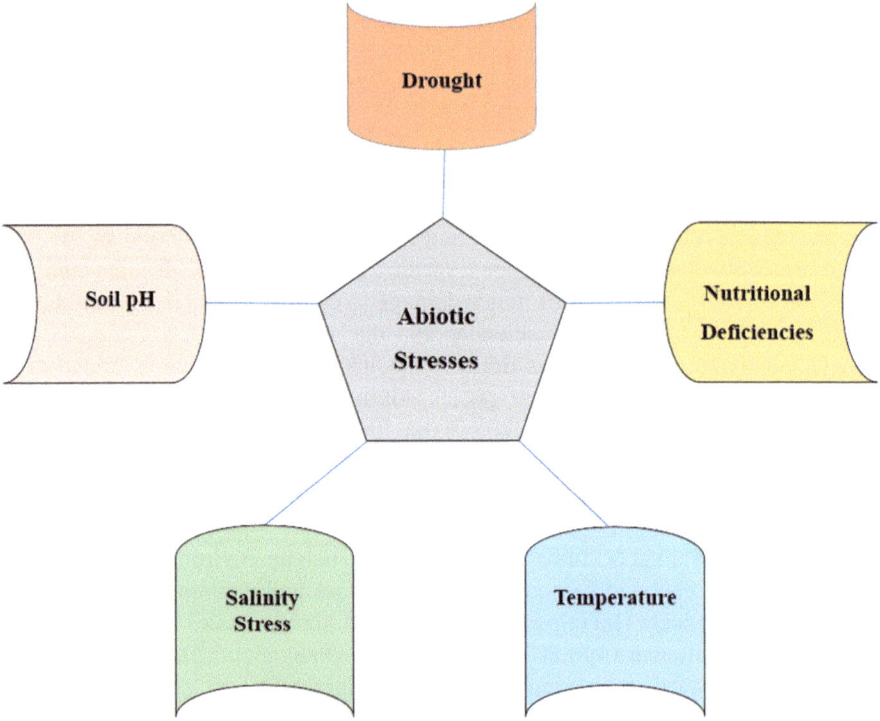

Fig. 1 Various abiotic stresses impacting cabbage growth

diminution of nitrogenase activity in Rhizobia, which inevitably causes suppression of nodule nitrogen fixation under drought conditions. Dehydration stress caused by drought disrupts stomatal functioning and impairs water and nutrient absorption, reducing water-use efficiency. It also hampers stem elongation, root growth, and leaf cell development, exacerbating the effects of dehydration. Closing of stomata and decreasing photosynthetic activities are two universal drought survival mechanisms for plants. Drought conditions force plants to elongate their roots, improving the root-to-shoot ratio and biomass in roots caused by improved segregation of biomass to roots (Dollete et al. 2024). Drought-induced stress utilizes two closely associated mechanisms of hydraulic failure as well as carbon deprivation to produce stress symptoms. Air percolation into xylem vessels is only allowed once they have become perilously low in the level of water potential, and they consequently obtain an embolism whereby water uptake to leaves is blocked. The situation resulting in air penetration through roots is called hydraulic failure because water transport has been reduced. Carbon starvation is caused in plants by a lack of movement in the phloem for nonstructural carbohydrates, since the plant processes cannot proceed despite growth and metabolic processes not occurring. The two hydraulic failure symptoms are interrelated as carbon hunger symptoms and failure to distribute water. Research indicates that under certain conditions, carbon starvation combines with hydraulic failure. The repair of xylem injury is compromised due to the lack of nonstructural carbohydrates, culminating in an even greater water stress coupled with a catastrophic loss of water status in stalling all mechanisms of carbon absorption and transportation via photosynthesis (Schwab et al. 2023). Climate change has led to an increasing frequency and severity of drought, which stands as one of the vital factors restricting agricultural productivity. Water deficit reduces nodulation rates because it diminishes the numbers of existing soil rhizobia along with their ability to initiate root infections. The oxygen limitation problem, together with feedback inhibition of nitrogen fixation processes, leads to reduced nitrogenase activity in Rhizobia, and this ultimately results in nodule nitrogen fixation suppression under drought conditions. The dehydration stress from drought complications stops the functioning of stomatal oscillations while also harming the processes of plant water and nutrient absorption and water-use efficiency, combined with stem extension and root growth and leaf cell development and growth and dehydration events. Stomatal closure and reduced photosynthetic rates stand as the principal two drought-triggered reactions in plants. Drought conditions drive plants to extend their roots, which enhances the root-to-shoot ratio and root biomass by improving biomass segregation to roots (Dollete et al. 2024). Drought-induced stress employs two related mechanisms of hydraulic failure alongside carbon deprivation to cause stress symptoms. The passage of air into xylem vessels occurs when their water potential reaches a dangerously low point, thus creating an embolism, which halts water transportation to leaves. The condition where air enters roots because water transport has declined is known as hydraulic failure. A lack of phloem movement for nonstructural carbohydrates in plants leads to carbon starvation because growth and metabolic processes cannot proceed. The two hydraulic failure symptoms occur together as carbon hunger symptoms and failure to distribute water. Research shows that hydraulic failure operates together with carbon starvation under specific conditions. The plant's ability to fix xylem damage suffers through reduced nonstructural carbohydrates, leading to worsened water stress and a catastrophic reduction of water status that blocks all photosynthetic carbon absorption and transport mechanisms (Abbas et al. 2023).

2.2 Temperature

Plants encounter direct harm from three abiotic stresses, which combine to reduce their quality results and production possibilities (Bulgari et al. 2019; Raza et al. 2022). The requirements for ideal growth conditions are progressively rising due to global warming, which endangers many vegetable crops. Climate change has two harmful effects on temperature levels: it makes them more erratic between hot and cold extremes. Different varieties of vegetable plants demonstrate diverse tolerance to freezing temperatures, prolonged exposure to cold conditions, or extreme heat based on their origin and characteristics (Bisbis et al. 2018; Raza et al. 2021a). Better vegetable varieties having tolerance to heat stress and cold stress, and general temperature changes are needed for agricultural success. Scientists study heat and cold-responsive mechanisms because this knowledge will help them develop superior vegetable varieties for breeding programs (Aleem et al. 2020; Raza et al. 2021a; Kang et al. 2022a; Chaudhary et al. 2022). Research efforts promote quick advancement of temperature-resilient crops through genome editing and mutant breeding approaches with recombinant DNA technology, along with genomic technologies as described by Aleem et al. (2020), Raza et al. (2021b), Varshney et al. (2020, 2021), and Yaqoob et al. (2023). The annual average surface temperature rises due to multiple influencing elements. Human activities that result in fast climate change serve as a primary factor behind this effect (Junaid et al. 2021). Climate variations will lead to major temperature changes that modify the plants' natural temperature range. Climate change affects vegetable crops, together with horticulture and agriculture, in their most complete form. Plants are expected to increase their water intake for growth and reproduction under the conditions of HS. Drought occurs due to severe, long-lasting weather patterns that decrease precipitation and soil moisture, thus leading to poor vegetable development and reduced growth alongside inferior quality. The phenomenon of climate change leads to increases in temperature levels, together with decreases in rainfall amounts. Drought serves as the main trigger for all undesirable heat stress impacts in plants. When drought occurs, temperatures rise in the air, thereby triggering stomata closure and reduced photosynthetic activity and damage to photosynthetic pigments. The decrease of stomatal conductance in plants due to this effect creates impairment that affects respiration and photosynthesis rates while causing growth retardation (Saeed et al. 2023).

2.2.1 Heat Stress

Heat stress (HS) is a temperature elevation lasting a specified period that generates unfixable plant damage when temperatures exceed their threshold by 10–15 °C (Raza et al. 2022; Zahra et al. 2021). The intracellular CO_2 content, together with stomatal conductance and leaf water status, are all influenced by HS. Heat-dominated cabbage led Rubio to lose essential photosynthetic rate functions because it harms the electronic component machinery (Lee et al. 2021). Multiple chlorophyll-related enzyme destruction explains why high salinity conditions lead to decreased chlorophyll levels. HS damages the first enzyme of the pyrrole biosynthesis pathway, called 5-aminolevulinate dehydratase, according to Fahad et al. (2017). Water scarcity occurs in dry conditions, making stomata instantly close down to maintain water levels. Most plants have a defensive mechanism that functions to protect them from extreme temperatures. The plant defenses suffer weakening from HS exposure, while stress causes damage to the absorption of balanced minerals and produces an excess of reactive oxygen species. ROS serves as a key controller for stomate aperture functions. The ROS production process is activated by NADPH oxidase enzymes such as AtRbohD and AtRbohF when they pair with Ca^{2+} Interacting Protein Kinases (CIPK), including CIPK26, resulting in stomatal aperture closure. The addition of flavonols as antioxidants in guard cells leads to decreased ROS production levels that result in stomatal pore opening. The ability to regulate stomatal aperture depends directly on ROS maintenance levels within cell structures (Medina et al. 2021).

2.2.2 Cold Stress

Plants developed multiple defense strategies for managing cold-related stress. The initial effect of cold exposure produces reactive oxygen species, which include OH, $O_{2,}$ and $H2O_2$. These molecules are simultaneously signaling molecules activated by stressors, yet their existence requires elimination due to their danger. An effective cold stress-defense mechanism includes the rise of alternative oxidase (AOX) activity above cytochrome c oxidase (COX) alongside antioxidant enzyme synthesis (Ding et al. 2017). Freezing conditions produce ice crystals throughout intercellular gaps, which can eventually destroy plant tissue cell membranes (Kidokoro et al. 2022). The development of discrete survival strategies exists among plants to fight against cold-related stress. When plants encounter cold, they initiate the process that creates reactive oxygen species, including OH·, O_2·, and H_2O_2 as their primary reaction. DNA damage from stress activation triggers the signaling function of these molecules, although they pose substantial harm to biological systems. Plants utilize defense mechanisms that include activating antioxidant enzyme synthesis and showing higher AOX (alternative oxidase) activity compared to COX activity to decrease oxidative stress at cold temperatures. Additionally, the production of flavonoids increases under low-temperature conditions. Various types of pigments classified as anthocyanins exist alongside flavones, together with flavonols and isoflavonoids. The pigments operate as antioxidants that protect against stress while helping to repair DNA (Sri et al. 2023).

2.3 Salinity Stress

Plant physiological functions experience negative effects because of the main abiotic factor known as salinity (Liang et al. 2018; Saddhe et al. 2019; Chen et al. 2021). Global climate change will lead to worsening salinity stress problems throughout the following years because temperature increases will coincide with reduced water availability (Etesami and Glick 2020). Soil salinity increases after inadequate irrigation results in salt accumulation throughout topsoil layers, thus negatively affecting vegetable and flower crops that receive irrigation (Baniasadi et al. 2018). Vegetable crops alongside flowers will perish when placed in areas where NaCl salt reaches between 100 and 200 mM surface saturation. The absence of damage from salinity marks salt stress in plants and directly impacts their commercial attractiveness. The signaling network produces biotic changes in osmotic potential, ions, and plant hormones, and reactive oxygen species (ROS) shape salt tolerance in different plant species. The signals reach specific receptors, which start the physiological mechanisms needed for plants to adapt to stressful environments. The mechanisms behind signal transduction in salt-stress conditions include three separate pathways, namely ion signaling pathways, along with osmolyte regulators and genetic regulation systems (Toscano et al. 2023). A decrease in growth rates becomes one of the main negative consequences for plants exposed to salt accumulation due to changes in various physiological and biochemical processes (Shahid et al. 2020). Plant adaptation to salty environments occurs through a complex signaling process involving ions besides osmotic potentials and plant regulatory substances or reactive oxygen species (ROS) signaling (Stevens et al. 2017). The signals travel to corresponding receptors to start processes that allow plants to adapt to stressful environments. Research shows that three different groups of signaling pathways involving ion signaling, osmolytes, and gene expression exist when plants face salt stress (Nxele et al. 2017). The amino acid and osmoprotectant Pro take a central role in the plant cytosol since it provides membrane stabilization and protects both proteins and enzymes, and other proteins. Protein has an essential function in salt stress because it raises membrane protein synthesis, controls ROS scavenger activities, and protects cellular solute concentrations. Numerous studies indicate that Pro enhances both antioxidant systems and water absorption within environments affected by salinity. It reduces the accumulation of destructive ions according to research by Zhang and Dai (2019) and Tang et al. (2015), and Ismail and Horie (2017). Salt stress causes global food security challenges, thus demanding sustainable approaches for improving crop growth under adverse conditions. A study in 2015 found that salt stress affected one billion hectares of world arable land, and it is expected to decrease by 50% by 2050. Plants sense environmental messages and then direct defensive functions through various regulatory monitoring systems to fight abiotic stress. Science allows researchers to break down salt tolerance complexities into separate biological mechanisms and operations. Time and increasing salt concentrations lead to two primary detrimental effects in plants called osmotic stress and particular ion toxicity. Oxidative stress, together with nutritional issues, emerges as a secondary stressor because of these initial factors. The deterioration in plant development stems from physiological and metabolic process disruptions that also lead to these specified effects. Three key disruptive effects occur in the plant cells due to these factors: membrane function failure meets diminished nutrient and water absorption abilities, and breakdown of vital cellular operations, including respiration and photosynthesis, and protein synthesis (Arzani et al. 2023).

2.4 Soil pH

The pH of the soil is crucial in controlling its characteristics and capabilities. Soil organic carbon (SOC) is influenced by enzyme activity and microbial diversity. Organic manure has been shown in several studies to be a useful tool for lowering soil pH. Soil pH can be raised by applying organic manure alone or by combining organic and inorganic fertilizers. Soil pH buffering strength increases when manure is applied, thus making Alfsol more durable against acidification. The dispersion of weakly acidic functional groups from soil organic matter into neutral molecules through buffering increases overall resistance to acidification as per Wang et al. (2023). Soil pH increased to a significant extent by 15%, 12%, 15%, 13%, 5%, and 17% through lime, charcoal, by-products, manure, straw, and their combination applications, which also enhanced crop production by 29%, 57%, 50%, 55%, 9%, and 52%. Soil pH demonstrated a direct positive relationship with crop productivity, even though the pattern changed according to different crops. The combination of extended (>6 years) treatment with amendments achieved maximum gains in soil pH as well as yield results in sandy soils exhibiting extreme acidity (pH <5.0) alongside low soil organic matter (SOM <12 g kg^{-1}) and low cation exchange capacity (CEC <100 mmolc kgs^{-1}). A combination of dung straw, charcoal, lime, and byproducts showed most effectiveness in reducing soil bulk density while simultaneously increasing soil CEC as well as SOM and base saturation; however, lime treatment specifically led to a 1% increase in soil bulk density through compaction. Soil pH, while yield increased together with CEC, SOM, and BS; however, both parameters hurt yield within compacted soil conditions. The combination of lime and straw with manure offers optimal results for acidic cultivated soil ranging from pH levels of less than 5.0 to 5.0–6.0 and 6.0–6.5, respectively, while considering costs alongside soil pH management and crop productivity values (Zhang et al. 2023).

2.5 Nutritional Deficiencies

2.5.1 Magnesium

The development and growth of crops need magnesium (Mg) as an essential nutrient because it controls protein synthesis and regulates lipid metabolism and enzyme function, and distributes carbohydrates. Mg serves as a direct component in metabolic photosynthetic compound synthesis and photosynthetic activities of shoots. A shortage of magnesium creates restrictions for plant growth that affect photosynthetic product circulation, resulting in severe declines in both crop quality and yield. The continuous deficiency of magnesium in agricultural production throughout the past 50 years has caused magnesium levels in both fruits and vegetables and cereal crops to decrease. Plants have received minimal investigation regarding their internal magnesium nutrition processes, although research on this subject extends back approximately 200 years. Absolute magnesium deficiency, together with cation competition, creates the soil conditions leading to magnesium deficit in crops. Acidic soils having a light texture and low cation exchange capacity tend to lose significant amounts of magnesium, which is the most likely cause of absolute magnesium insufficiency in plants. Long-term heavy cropping practices can deplete soil magnesium since they combine improper mineral fertilizer distribution with non-recycled agricultural residues. Excessive amounts of K^+ and Ca^{2+}, as well as NH^{4+} and Al^{3+}, reduce the availability of magnesium (Mg^{2+}) in crops. Magnesium shortage results from the competitive interaction between K^+ and Mg^{2+}, which represents the prime competition in plant mineral nutrition (Qu et al. 2023).

2.5.2 Nitrogen

The plant production sector depends on different nitrogen forms for its operations. The fertilizer ammonium nitrate ranks as one of the most popular products (NH_4NO_3, 34% N). The proper addition of ammonium nitrate leads to better growth and enhanced nutritional value of Chinese cabbage during hydroponics and pot garden cultivation (Liu et al. 2022). The findings in Zhu et al. (2021) show that optimized

N absorption and assimilation boosted Chinese cabbage bloom growth. Solid nitrogen fertilizer with the most dense concentration available today is urea at 46% N. Deficiency in nitrogen supply results in decreased chlorophyll levels and reduced photosynthetic capability, and decreased leaf N concentration. Plants suffer damage and death from overly heavy nitrogen applications because the substance lowers defensive capabilities and halts biological processes (Liu et al. 2023). When the nitrogen content increases, cabbage, as a vegetative plant, grows more leaves since it responds well to nitrogen for promoting plant growth and developing strong foliage.

2.5.3 Zinc

The absence of zinc in plants affects their growth mechanisms and development patterns negatively. Scientists have proven that this vitamin supports vital physiological activities that control DNA pathways, as well as enables enzymes and helps proteins to synthesize. The absence of zinc in plants results in reduced seed production, alongside fruit deficiencies along with leaf discoloration, and plant death, as well as impaired growth and heightened exposure to pest threats. The conditions of zinc-deficient plants can be improved using several available management methods. Soil zinc levels need to be maximized through proper maintenance practices, which is considered the most effective method. Fertilizers with zinc, composed of chelated zinc or zinc oxide, or zinc sulfate, act as suitable treatments for this purpose. The application of these fertilizers leads to toxicity alongside other problems when used in excessive amounts (Aye and Masih 2023).

2.5.4 Boron (B)

Nearly all cole crops are susceptible to deficiency. However, cauliflower's main issue, caused by a boron deficit, is its hollow stems and dark, water-soaked patches on the head. Both broccoli and cabbage can have hollow stems. As a highly soluble nutrient, boron rapidly seeps out of the soil. Sensitive crops may experience boron toxicity as a result of carryover. Peas, beans, cucumbers, and small grains are similarly impacted by an excessive amount of boron in the soil if they are grown during the same growing season as cole crops (Kundu and Chamroy 2020).

2.5.5 Copper (Cu)

Plants require copper as an essential factor to regulate metabolic process control. Many plant physiological and metabolic functions are controlled through copper, which serves as a stable cofactor for multiple proteins and enzymes. The use of iron in chlorophyll synthesis becomes possible because of this facilitation process. Research shows that copper affects the indirect formation of nodule structures. The meristem apical and younger leaf deformities, along with growth restriction and possible necrosis symptoms, appear from copper deficiency. Plants under copper deficiency develop yellowed young leaves that display tip curling and jagged edges, which leads to the top plant withering (Ahmad et al. 2024) (Table 1).

Table 1 Impact of various abiotic stresses on cabbage growth and development

Abiotic stress	Effect on cabbage	Mitigation strategies	Mechanisms involved	References
Drought stress	Reduced leaf area, stunted growth, lower head weight, reduced phytochemical content	Drought-tolerant cultivars, drip irrigation, citric acid, mulching, and antitranspirants	Osmotic adjustment, deeper rooting, enhanced antioxidant defense, ABA regulation	Kasukawa and Miyazawa (2020)
Heat stress	Premature bolting, reduced chlorophyll, curd deformation, lowered yield	Use of heat-tolerant hybrids, foliar glycine betaine, BABA, shading nets, misting systems	Heat-shock protein expression, antioxidant upregulation, membrane thermostability	Quan et al. (2022)
Salinity stress	Leaf necrosis, nutrient imbalance, osmotic stress, reduced curd size and weight	Salt-tolerant varieties, gypsum, potassium nitrate foliar sprays, AMF inoculation, biochar application	Na^+ exclusion, K^+ retention, osmolyte accumulation (proline), antioxidant enzyme activity	Kumari et al. (2023a)
Cadmium stress (Cd)	Inhibited root elongation, oxidative stress, altered canopy temperature, nutrient distortion	Soil amendments (biochar, lime), Zn/Se supplementation, phytoremediation, organic matter enrichment	Chelation by phytochelatins, reduced uptake, improved redox homeostasis	Li et al. (2022)
Waterlogging stress	Root hypoxia, wilting, leaf chlorosis, nutrient leaching, reduced yield	Raised beds, drainage improvement, tolerant genotypes, soil aeration techniques	Ethylene signaling, fermentation pathway activation, SOD/CAT activity enhancement	Shawon et al. (2020)
General climate stress	Combined effects of temperature extremes and erratic rainfall reduce head compactness and quality	Climate-smart agriculture (CSA), crop modeling, stress-resilient cultivars	Stress signaling modulation, ABA/cytokinin regulation, transcriptional reprogramming	Kumar et al. (2023)
Nutrient deficiencies	Tip burn (Ca), poor vegetative growth (N), chlorosis (Mg), reduced curd density (K, B)	Balanced fertilization, organic manure, compost, foliar nutrient sprays, biofortification	Improved root uptake, foliar absorption, microbial activity stimulation	Pavithra et al. (2022)

(continued)

Table 1 (continued)

Abiotic stress	Effect on cabbage	Mitigation strategies	Mechanisms involved	References
Cold stress	Delayed seedling emergence, leaf damage, reduced metabolic activity, bolting suppression	Cold-tolerant cultivars, priming with SA/GA_3, protected cultivation (greenhouses, tunnels)	Cold-responsive gene expression (CBF pathway), sugar accumulation, enhanced antioxidant system	Soualiou et al. (2022)
Heavy metals (Pb, As)	Reduced chlorophyll, inhibited photosynthesis, growth retardation, oxidative injury	Soil pH correction (liming), chelator use (EDTA), phytoextraction, organic amendments	Phytochelatin synthesis, antioxidant defense, metal sequestration in vacuoles	Kamal et al. (2023)
UV-B radiation	Leaf necrosis, reduced photosynthetic activity, ROS generation, DNA damage	UV-blocking greenhouse films, foliar application of flavonoids or polyphenolic elicitors	DNA repair activation, upregulation of UV-protective secondary metabolites	Yang et al. (2025)
Ozone stress	Leaf bronzing, impaired stomatal function, reduced biomass, early senescence	Ascorbic acid sprays, selection of ozone-tolerant lines, avoiding exposure peaks	ROS scavenging, enhanced catalase and peroxidase activity, stomatal closure regulation	Lee (2023)
Low atmospheric humidity	Increased transpiration, leaf desiccation, reduced stomatal conductance	Misting, mulching, humidity chambers in nurseries	Increased ABA signaling, cuticle thickening, stomatal closure	Kang et al. (2022b)

3 Physiological and Molecular Responses of Cabbage to Abiotic Stresses

3.1 Mechanisms of Stress Perception and Signal Transduction

3.1.1 Stress Perception

The role of cellular sensors: The membrane of plasma membrane contains special proteins that serve as stress perception elements for cabbage. The sensors in cabbage detect particular pressure signals, such as drought-induced osmotic stress, which causes water potential mismatches. Several types of ion channels respond to physical stress signals that include high temperatures and mechanical forces, by initiating signaling pathways. When plants experience abiotic stress, the initial vital signal that develops includes the entry of calcium ions (Ca^{2+}). Calcineurin B-like proteins (CBLs) and calmodulins (CaMs) together with calcineurin B-like proteins serve as detection elements for calcium alterations by sensing these changes and help decode the calcium signal (Abbas et al. 2023).

3.1.2 Signal Transduction

Role of reactive oxygen species and secondary messengers: during stressful situations, cells produce secondary signaling substances, including hydrogen peroxide (H_2O_2), along with other reactive oxygen species (ROS). ROS serves as both an advantage and a disadvantage in signaling due to its low quantities, enabling signals, but excessive levels cause damaging effects. The reactive oxygen species in cabbage plants regulate stomatal pore closings while activating defense antioxidants through interactions with signaling molecules such as abscisic acid (ABA) and nitric oxide (NO). The signal transduction process includes phospholipid signaling pathways as one of its key elements. The stress signal amplification function and membrane remodeling process rely on phosphatidic acid as a secondary messenger generated from lipids. The PA pathway becomes crucial for membrane integrity under salt and dry conditions (Kasukawa and Miyazawa 2020).

3.1.3 Phytohormonal Crosstalk in Stress Signaling

Positive phytohormonal mechanisms control the stress signal evaluation in cabbage plants. The majority of abiotic stress reactions in plants are controlled by abscisic acid (ABA), especially during salt stress and drought conditions. The activation of $SnRK_2$ protein kinases occurs through ABA signaling channels with PYR/PYL receptors, resulting in downstream gene activation. Through this system, the production of stress-protective proteins is enhanced while osmoprotection and stomatal control processes also become more efficient. The signaling molecules ethylene and jasmonic acid participate in stress responses, mostly when any combination of oxidative and mechanical stresses occurs. The coordination between ABA and ethylene and JA networks enables cabbage to achieve an optimal balance between its defense and growth processes (Quan et al. 2022).

3.1.4 Gene Activation and Regulatory Networks

The transduction activation of stress signals results in the activation of transcription factors such as DREB (Dehydration Responsive Element Binding) and NAC, and WRKY. The stress-responsive cis-regulatory elements in target gene promoters become activated by TFs that subsequently boost transcription of heat shock protein HSP, together with late embryogenesis abundant LEA proteins and antioxidants. Under drought conditions, DREB2A in cabbage becomes activated to manage the synthesis of osmo protectants and production of ROS-scavenging enzymes (Li et al. 2022).

3.2 Role of Stress-Responsive Genes and Hormones

Cabbage depends on stress-responsive genes and hormones for adapting to abiotic stresses. The activated genes initiate multiple pathway mechanisms that boost stress tolerance by establishing antioxidant protection and promoting osmoprotection along with structural realignments.

3.2.1 Stress-Responsive Genes

Stress-responsive genes produce proteins that are essential for shielding plant cells from harm brought on by stress. The expression of heat shock proteins (HSPs), late embryogenesis abundant (LEA) proteins, and other stress-tolerant proteins is triggered by transcription factors such as WRKY, MYB, and NAC. The expression of genes for antioxidant enzymes like as catalase (CAT) and superoxide dismutase (SOD) rises in response to cadmium stress, reducing the damage caused by reactive oxygen species (ROS). To preserve cellular osmotic balance, genes related to osmolyte biosynthesis, such as those controlling the synthesis of proline and glycine betaine, are also upregulated (Guo et al. 2023).

3.2.2 Role of Hormones

Abscisic acid (ABA), jasmonic acid (JA), and salicylic acid (SA) are three essential phytohormones that plants use to synchronize their reactions to stress. By causing stomatal closure and triggering stress-responsive genes via PYR/PYL receptors and $SnRK_2$ signaling pathways, ABA functions as a key regulator. Particularly in situations of drought and salinity stress, jasmonic acid (JA) improves defensive responses and antioxidant activity (Shawon et al. 2020). Cabbage plants can efficiently process stress signals thanks to the integration of several hormone signaling pathways, which preserve growth stability and strengthen defenses (Table 2).

3.3 Consequences on Growth, Yield, and Quality

Many external factors may negatively affect cabbage as its biology and physiology may be disrupted as a result. These stresses affect the plant's strength, head weight, and nutrition as well.

3.3.1 Impact on Growth

Drought and high temperatures limit cell expansion, which makes the plant have a lower leaf area, which in turn reduces its growth. Stresses caused by heavy metals such as cadmium affect root growth, canopy formation, and water consumption (Guo et al. 2023). This results in lowered growth rates as well as worse plant shape and structure.

3.3.2 Impact on Yield

Reduction in reproductive development carbohydrate availability may be attributed to photosynthetic inhibition, which may have causes that range from low to high stress. It has been shown that heat stress lessens both chlorophyll and stomatal development, which reduces the weight of heads and yields (Quan et al. 2022). Moreover, De Carolis et al. (2024) demonstrated that phosphorus deficiency, which results from drought, can postpone flowering and fruit production, which directly affects overall yield.

Table 2 Role of hormones in enhancing abiotic stresses in cabbage

Hormone	Function in stress tolerance	Impacts	References
Abscisic acid	Regulates stomatal closure, induces stress-responsive genes, and maintains cellular homeostasis	Enhances water-use efficiency, antioxidant activity, and drought tolerance	Abbas et al. (2023)
Jasmonic acid	Triggers antioxidant defense mechanisms and osmolyte accumulation under salinity and heavy metals	Increases SOD and POD activity; reduces oxidative stress	Raza et al. (2022)
Salicylic acid	Activates systemic acquired resistance and protects chloroplasts during stress	Improves seed germination, photosynthesis, and stress resilience	Zheng et al. (2023)
Ethylene	Modulates root growth and senescence under waterlogging and drought	Enhances aerenchyma formation and osmotic regulation under flooding and salinity	Raza et al. (2022)
Auxins	Stimulate root elongation, cell division, and differential growth under stress	Maintain shoot/root integrity and promotes stress recovery	Li et al. (2022)
Cytokinin	Delays leaf senescence and maintain chlorophyll under heat and drought stress	Sustains photosynthetic activity and cellular metabolism	Panozzo et al. (2025)
Gibberellins (GAs)	Promote growth and flowering; alleviate salt and cold-induced growth inhibition	Improve seedling vigor and reverse growth suppression	Kumari et al. (2023b)
Brassinosteroids (BRs)	Induce antioxidant enzymes and regulate stress gene expression under abiotic stress	Increase tolerance to drought, salinity, and heat by reducing ROS damage	Ahmad Lone et al. (2022)
Strigolactones (SLs)	Modulate plant architecture and interact with ABA during drought and nutrient stress	Reduce water loss and improve stress resistance mechanisms	Lastochkina et al. (2022)

3.3.3 Impact on Quality

Stress factors trigger changes in both nutritional and phytochemical content found in cabbage. Enemies perceive increased sodium levels in their tissues because of salinity stress, even though essential nutrients such as potassium and magnesium decrease in concentration. The exposure to cadmium stress leads to modification of antioxidants and glucosinolate levels, thereby diminishing both health advantages and sensory qualities of cabbage (Li et al. 2022).

4 Cultural Practices for Abiotic Stress Management

4.1 Water Management Strategies

Drought and erratic rainfall are caused by water inefficiency, which is an important resource for increasing agricultural productivity. Listed below are the best water resources management strategies:

4.1.1 Deficit Irrigation

This strategy entails supplying water below the full crop requirement during less sensitive growth stages for enhanced water-use efficiency. In semiarid regions, the Integrated Crop and Water Management (ICWM) approach, together with humates, was found to significantly increase the watermelon yield and water use efficiency (Qin and Leskovar 2020).

4.1.2 Rainwater Harvesting

Agricultural rainwater storage facilities help farmers store precipitation for agricultural needs, which reduces their dependence on seasonal rainfall and ground sources. Such measures brought success in arid and semiarid zones while they guaranteed water resource protection during dry seasons (Muralikrishnan et al. 2021).

4.1.3 Using Mulches for Moisture Saving

Straw and pumice film are examples of organic mulches that can drastically cut down on evaporation, saving up to 30% more water than areas that are left bare (Choudhary et al. 2022).

4.2 Soil Management and Cultivation Practices

Soil management and cultivation practices are the most effective techniques that increase soil resiliency to the damaging effects of drought, salinity, and other abiotic stress factors:

4.2.1 Mulching for Water Availability

Synthetic mulch also preserves soil moisture by reducing water loss, and organic mulch helps retain moisture and lowers soil evaporation. Mulching can also raise the soil's moisture content by as much as 40%, which promotes robust crop growth during the vegetative stage (Dhanyasree et al. 2022).

4.2.2 Protection from Erosion and Soil Temperature

Mulching aids in lowering soil temperature, which lessens root stress brought on by drought. It has also been demonstrated to reduce topsoil erosion in vegetable cultivation by as much as 50% (Choudhary et al. 2022).

4.2.3 Enhancing Soil Fertility through Conservation

Covering crops and no-till agriculture increase organic matter and nutrients in soils, which in turn improves the capacity of soils to infiltrate and retain water. A study conducted in sub-Saharan Africa pointed out the usefulness of these practices in alleviating drought-caused stresses. By combining mulching with soil conservation techniques, farmers can sustain productivity while becoming more resilient to abiotic stresses (Kouame et al. 2024).

4.3 Use of Protective Structures and Shade Nets

In addition to optimal resource utilization, protective structures and shade nets prevent the destruction of crops from the extreme environment.

4.3.1 Temperature Control

Shade nets help cool microclimates and lessen heat stress by lowering the amount of solar radiation that reaches crops. According to horticultural crop studies, using shade nets can boost yields by 20–30% when compared to crops grown in open fields (Singh et al. 2022).

4.3.2 Water Savings

Protective systems can assist in reducing evapotranspiration, a significant cause of water loss in agriculture, which in turn reduces the need for irrigation. It has been demonstrated that crops like pepper and tomato need a lot less water in controlled settings like greenhouses (Imran 2022).

4.3.3 UV and Pest Protection

In open fields, polysaccharide-starch nanocomposite bioinks have lessened crop losses and the requirement for chemical inputs by improving UV-ray transmittance and acting as barriers against pests and strong winds. These compounds are useful in reducing the effects of abiotic stresses on crops in certain studies, especially when it comes to protected agriculture. A more stable developing environment can be produced by combining shade nets with protective structures, which will increase crop output and resource efficiency in difficult climates (Bahudur 2020).

5 Biotechnological Interventions in Stress Management

5.1 R&D of Stress-Tolerant Crop Varieties

The creation and introduction of crop varieties with improved stress tolerance is one of the main goals of contemporary biotechnology. This entails pinpointing and altering the genes in charge of organ metabolism in harsh weather, especially in response to stressors like heat, dryness, and salinity.

5.1.1 Transcription Factors

These proteins are crucial for plant stress responses, including the regulation of osmotic adjustment, antioxidant, and metabolic pathways. They do this by binding to particular DNA sequences and controlling gene expression. Important modulators of stress response pathways in plants are transcription factors as DREB, MYB, and NAC. These transcription factors may improve crop resistance to a range of stressors, according to recent research (Khan et al. 2018).

5.1.2 CRISPR Technology

CRISPR-Cas9 technology has greatly sped up the production of plants that are more resilient to environmental problems. Using this gene-editing technology, precise changes can be made to produce plants with increased resistance to a variety of abiotic stressors (Imran et al. 2021).

5.1.3 Microbiome Interactions

Enhancing plant stress resilience has been achieved through the use of plant-microbiome interactions. Beneficial microorganisms boost antioxidant activity, osmolyte accumulation, and general plant growth when plants are stressed. A practical and sustainable method for creating crop varieties that can withstand stress is to use the plant microbiome. Among these microorganisms, plant growth-promoting rhizobacteria (PGPR) improve tolerance to abiotic stress by raising osmolyte levels, adjusting hormone balance, and enhancing the production of antioxidants. For instance, Pseudomonas fluorescens PGPR strains increase potassium uptake and proline buildup in mustard, improving salt tolerance. These developments mark a substantial step toward environmentally challenging sustainable agriculture (Ali et al. 2023).

5.2 Role of Genetic Engineering and CRISPR

Plant stress reduction has been transformed by CRISPR technology and genetic engineering. Using CRISPR-Cas9, certain stress-responsive genes can be enhanced or knocked out to control heat, salinity, and drought stress. It has been demonstrated that CRISPR-mediated disruption of ABA signaling pathways improves drought resistance. Furthermore, rice's resistance to drought was enhanced by deleting OsRR22, a negative regulator of cytokinin signaling, according to genome analysis (Imran et al. 2021).

The use of CRISPR techniques and RNA sequencing (Shi et al. 2020) to identify the important network of genes and transcriptional pathways addressing the target plant's stress response mechanisms is another important initiative of the study, as opposed to attempting to screen and modify individual stress-responsive genes. Several other components were identified by data-derived insights, and the CRISPR technology platform was then used to demonstrate that specific regulatory factors, such as transcription factors CBF1 and ICE1, highlighted cold tolerance in barley (Ahanger et al. 2020). Through metabolic pathway genetic engineering, crop stress tolerance has increased. The silicon metabolism gene is one of the integrated metabolic genes that improve the structural and biochemical responses to abiotic stress (Ahanger et al. 2020). Furthermore, by improving osmosis, proline biosynthesis manipulation in soybeans resulted in higher salt tolerance. These biotechnology instruments' accuracy enables scientists to create crops that are suited to particular environmental problems.

5.3 Application of Plant Growth Regulators

The action of plant growth regulators also assists in tissue stress by enabling various physiological and biochemical processes:

5.3.1 Melatonin (MT)

MT functions as a stress signal that shields plants from cross-stress by enhancing osmolyte storage and preventing radical lesion defenses from ROS reactions. Scientists believe that altered plants are better able to withstand the effects of salt and drought because of their enhanced melatonin biosynthesis (Hassan et al. 2022).

5.3.2 Methyl Jasmonate (MeJA)

MeJA has been shown to induce the generic stress-response pathways by cytokinin action that leads to enhanced synthesis of secondary metabolites that offer resistance to infections, drought, and oxidative stress (Yu et al. 2019).

5.3.3 Biostimulants

Organic substances that function as biostimulants can improve photosynthesis, nutrient uptake, and antioxidant activity, hence reducing yield losses brought on by abiotic stress. An economical method of stress reduction is provided by their integration into farming operations. Similarly, the use of plant growth regulators (PGRs) can increase crop productivity and resistance in harsh environments (García-García et al. 2020).

6 Role of Soil Amendments and Fertilizers

6.1 Organic Matter and Soil Functionality

Application of organic amendments such as composts and manure results in an increase in the Soil organic carbon (SOC) and the sodium adsorption ratio (SAR), which improves the structure, water holding capacity, and activities of microbes in the soils. Green manuring with sunnhemp crops promotes an increase in earthworm populations, enhancing several key aspects of soil structure and health, particularly in monoculture systems (Guo et al. 2022).

6.2 Microbial Dynamics and Nutrient Cycling

Nutrients such as nitrogen (N) comes from fresh animal manure and biochar have been shown to boost the nitrogen-cycling bacteria activity and biomass, which may help in nutrient cycling and stress mitigation. These strategies also improve soil enzyme activity that may improve crop vigor and resilience (Ouyang et al. 2022).

6.3 Balanced Nutrient Management for Stress Resistance

A fresh and stress-resistant plant may be obtained through balanced nutrient management by employing both organic and inorganic fertilizers to help build the soil fertility:

6.3.1 Integrated Nutrient Management (INM)

The use of vermicompost with other inorganic fertilizers enhanced nutrient-rich organic matter while enhancing metabolic activity in soil. In rice-wheat cropping systems, INM has been shown to promote nitrogen and phosphorus cycles and soil aggregation (Bharti et al. 2023).

6.3.2 Organic-Inorganic Interactions

Humic acids and compost, when applied alongside chemical fertilizers over extended periods, enhance nitrogen and carbon mineralization. These improvements are linked to higher crop yields and increased stress resistance in coastal soil (Sahoo et al. 2023).

6.3.3 Prevention of Soil Acidification

To counteract soil acidification, slow-maturing fertilizers and liming techniques are essential as they enhance microbial activities and nutrient availability. Integrated management systems comprising organic matter and lime enhance the soil health of acidic soils (Xie et al. 2024).

6.4 Biofertilizers and Microbial Interventions

These fertilizers are rich in microorganisms, including bacteria and algae that would either improve nutrient use in the plants or improve the condition of the soil, thus enabling the plant to withstand more stress.

6.4.1 Nitrogen Fixing and Phosphate-Solubilizing Bacteria

In the conditions of nutrient deficiency, the nitrogen-fixing bacteria (e.g., rhizobium) and phosphate-solubilizing bacteria assist in the plant's growth. Biofertilizers, which contained plants microbiota consortium, resulted in a tremendous reduction of drought stress on legume crops (Nayak and Mishra 2024).

6.4.2 Arbuscular Mycorrhizal Fungi (AMF)

They are known to increase the phosphorus uptake as they enhance the mutualism of the roots of the plant systems. AMF together with composts appeared to be beneficial to the performance and stress tolerance of plants in degraded ecosystems (Yapa et al. 2023).

6.4.3 Microbial Enzyme Activity

Use of biofertilizers enhances microbial exo-enzyme activity in soils and therefore the rate of microbial breakdown of organic matter for the purpose of serving plant nutrients increases. These activities of the microbes improve the functionality of plants in derelict soils, particularly those related to mining activities (Nayak and Mishra 2024).

6.4.4 Multifunctional Biofertilizers

In recent results, multifunctional microbial consortia were introduced, which have the capacity for nitrogen fixation, phosphate solubilization, and biosynthesis of hormones. Such consortia of stress tolerance decrease the oxidative stress damage and enhance plant antioxidant capacity, especially under drought and salinity toxicity (Shekhawat and Rawal 2024).

7 Application of Nanotechnology in Stress Management

Nano-fertilizers are steadily gaining attention as a means to increase crop productivity while alleviating "abiotic stress" because of their ability to increase nutrient use efficiency. Unlike conventional fertilizers, they incur less nutrient loss and reduce nutrient losses and deliver nutrients in a slow, highly desired fashion. A graphical representation of the application of nanotechnology in stress management in plants is shown in Fig. 2.

7.1 Enhanced Nutrient Use Efficiency (NUE)

Economically important crops like barley and maize can easily benefit from nano-urea and nano-zinc oxide fertilizers due to their increased nutrient absorption capabilities, making them an ideal choice. Kunwar et al. (2025) argued that nano-urea boosted nitrogen, enabling *Arabidopsis thaliana* to resist stress through the chlorophyll boosting, Implications, and photosynthetic efficiency. Likewise, maize ionomic and metabolic profiles under ionomic stress have been enhanced by the use of nano zinc and nano silica fertilizers (Shoukat et al. 2024).

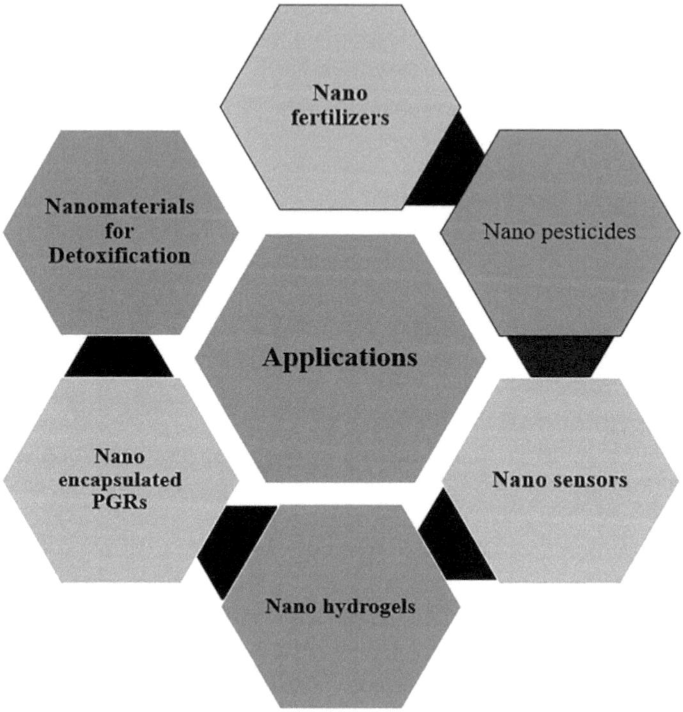

Fig. 2 Utilization of nanotechnology for stress alleviation in plants

7.2 Mitigating Abiotic Stress

Silicon and potassium nano-fertilizers can stimulate salt and drought resistance of plants by increasing plant cell and osmotic regulation strength. For example, research conducted by El-Shal et al. (2022) showed that the combination of nano K_2SO_4 and SiO_2 helped in the growth of alfalfa plants under salt stress and improved the plant stress tolerance.

7.3 Advantages over Conventional Fertilizers

Conventional fertilizers are less efficient compared to nano fertilizers, as conventional fertilizers were nearly 18 times less soluble than nano fertilizers, consequently reducing leaching. The increases in these properties resulted in an increase in nutrient uptake efficiency and increased capability to withstand environment-related stresses (Avila-Quezada et al. 2022).

7.4 Nanoparticles for Controlled Water and Nutrient Delivery

Nanoparticles (NPs), properly developed and synthesized, assist in ameliorating water and nutrients management through effective and timely delivery of resources.

7.4.1 Controlled Release of Nutrients

Nanoscale materials are made in such a way that they will enable slow and gradual release of nutrients that would be available for the crops at important times during the growth stage. As an example, zinc oxide nanoparticles (ZnO NPs) are believed to be able to offer nutrients while being environmentally friendly (Verma et al. 2024).

7.4.2 Water Retention and Irrigation Optimization

By enhancing soil water retention at the same time reducing the losscribed enhanced the aging of soil structure, nanomaterials increase the need for irrigation. When nanoscale sensors are integrated into an irrigation setup, constant monitoring can be done, enabling appropriate regulation of water use, thus managing drought conditions (Das et al. 2024).

8 Climate-Smart Practices for Sustainable Cabbage Cultivation

Cabbage varieties that are climate-smart are important because they help in reducing the adverse effects of climate variability. These types of cabbages are developed or selected to exhibit better resistance to abiotic stress, such as drought, heat, salinity, and pests and diseases, which are biotic stresses.

8.1 Breeding for Auxiliary Crop Stress Tolerance

Breeding initiatives have now shifted focus toward the identification of traits such as strong root profiles, better water management, and tolerance toward heat among crops. For example, the achievements in marker-assisted selection have helped improve cabbage lines suited to high temperatures and harsh drought conditions (Babu et al. 2024).

8.2 Pest and Disease Resistance

Due to climate change, more pests and disease problems are appearing. Novel "climate resilient" cabbage includes the addition of genes that provide resistance to well-known pests, including the diamondback moth and black rot fungus. The transfer of resistance traits from wild relatives has been effective in producing yield under such conditions (Sarma et al. 2024).

9 Precision Agriculture for Stress Monitoring and Management

In precision agriculture (PA), sensors, drones, and analytic programs are used to target farm input application at a high level of precision.

9.1 Stress Identification Monitoring in Cabbage Crop

Drones equipped with multispectral and thermal cameras illustrate real-time detection of stresses on the cabbage crop. These systems detect preconditioning of agro-climatic hazards such as water stress, lack of nutrients, or pest infestations and aid in implementing strategies. For example, thermal imaging has been used to monitor eases in canopy temperature as a drought stress indicator in the cabbage field (Ranwa et al. 2024).

9.2 Scope of Irrigation and Application of Nutrients

Precision irrigation systems, when coupled with soil moisture sensors, can deliver water specifically to the root zone, thus saving water. For instance, a combination of drip irrigation systems and soil moisture sensors has decreased the drain by 30% on cabbage farms while at the same time maintaining good yields (Das et al. 2024). On the same note, nutrients in the correct ratio can be managed at the place needed through GIS maps, which would decrease the use of fertilizer as well as the amount of nutrients released into the ecosystem (Toromade and Chiekezie 2024).

9.3 Digital Agriculture Technology

It is an approach that makes use of machine learning based on metrological and soil health conditions to know the chances of pest occurrence on a cabbage farm and prevent them by all means. This helps in minimizing the blanket application of pesticides, thus enhancing the resilience of the system (Patra et al. 2024).

9.4 Sustainability Benefits

The reduction of greenhouse gas emissions is made possible due to the use of precision agriculture that limits the application of fertilizer and water. For example, research showed that the use of precision farming techniques on various vegetables like cabbage reduces nitrogen oxide emissions by around 15% when compared to the conventional means (Singh et al. 2024).

10 Conclusion

It is evident from this chapter that better management of abiotic stress factors is an important goal for successfully increasing cabbage yields under changing environmental conditions. As global warming exacerbates the problems of drought, salinity, high temperatures, and lack of nutrients, new, improved integrated strategies have to be adopted. This chapter summarizes novel approaches such as breeding for stress-resistant cabbage varieties, precision agriculture, and the use of cutting-edge nanotechnology and biostimulants. All of these targets work toward increasing resistance to stress and increasing the efficiency of resources, as well as the consistency of crop production. Such heat-tolerant, water-use-efficient, and nutrient stress-resistant traits are among the key climate-resilient traits, and these varieties will be the mainstay of the efforts aimed at mitigating the negative climatic impacts.

References

Abbas K, Li JR, Gong B, Lu Y, Wu X, Lü G, Gao H (2023) Drought stress tolerance in vegetables: the functional role of structural features, key gene pathways, and exogenous hormones. Int J Mol Sci 24(18):13876

Ahanger MA, Bhat J, Siddiqui MH, Rinklebe J, Ahmad P (2020) Silicon and secondary metabolites integration in plants: a significant association in stress tolerance. J Exp Bot 71(21):67586774

Ahmad Lone W, Majeed N, Yaqoob U, John R (2022) Exogenous brassinosteroid and jasmonic acid improve drought tolerance in Brassica rapa L. genotypes by modulating osmolytes, antioxidants and photosynthetic system. Plant Cell Rep 41(3):603–617

Ahmad S, Munawar M, Ullah M, Khalid S, Waris I, Sher N, Eman A (2024) Crop responses and strategies for mitigating cold, salt, and drought stress in vegetables: a review. Asian J Res Crop Sci 9:680

Aleem S, Sharif I, Amin E, Tahir M, Parveen N, Aslam R, Shahid MTH (2020) Heat tolerance in vegetables in the current genomic era: an overview. Plant Growth Regul 92(3):497–516

Ali S, Tyagi A, Mir R, Rather I, Anwar Y, Mahmoudi H (2023) Plant beneficial microbiome: a boon for improving multiple stress tolerance in plants. Front Plant Sci 14:1266182

Andreotti C (2020) Management of abiotic stress in horticultural crops: spotlight on biostimulants. Agronomy 10(10):1514

Arzani A, Kumar S, Mansour MMF (2023) Salt tolerance in plants: molecular and functional adaptations. Front Plant Sci 14:1280788

Avila-Quezada GD, Ingle AP, Golińska P, Rai M (2022) Strategic applications of nano-fertilizers for sustainable agriculture: benefits and bottlenecks. Nanotechnol Rev 11(1):2123–2140

Aye HN, Masih S (2023) Role of nutrients in plants, its deficiency and management. Int J Plant Soil Sci 35(10):129–136

Babu RR, Baishya LK, Reddy AGK, Pongener A, Raghavendra HR, Omprakash Naik N (2024) Climate change effects on sustainable vegetable production in India: a review. Int J Environ Clim Chang 14(3):804–814

Bahudur AV (2020) Protected cultivation in India: meeting challenges. Int J Adv Agric Sci Technol 7(10):1–5

Baniasadi F, Saffari VR, Moud AAM (2018) Physiological and growth responses of Calendula officinalis L. plants to the interaction effects of polyamines and salt stress. Sci Hortic 234:312–317

Bharti P, Das A, Roy Choudhury S, Shambhavi S, Ganguly P, Bihari B, Rakshit R (2023) Soil aggregation controls biological activities on long-term integration of organic amend-

ments and inorganic fertilizers in rice-wheat cropping system. Commun Soil Sci Plant Anal 54(13):1865–1876
Bisbis MB, Gruda N, Blanke M (2018) Potential impacts of climate change on vegetable production and product quality – a review. J Clean Prod 170:1602–1620
Bulgari R, Franzoni G, Ferrante A (2019) Biostimulants application in horticultural crops under abiotic stress conditions. Agronomy 9(6):306
Chaudhary S, Devi P, Hanumantha RB, Jha UC, Sharma KD, Prasad PV, Nayyar H (2022) Physiological and molecular approaches for developing thermotolerance in vegetable crops: a growth, yield and sustenance perspective. Front Plant Sci 13:878498
Chen JT, Aroca R, Romano D (2021) Molecular aspects of plant salinity stress and tolerance. Int J Mol Sci 22(9):4918
Choudhary M, Kumari A, Choudhary S (2022) Effect of mulching on vegetable production: a review. Agric Rev 43(3):296–303
Das H, Singh SK, Priyadarshini E, Maurya C, Tiwari AK, Gulaiya S, Khokale SK, Panigrahi CK, Singh BV (2024) Nanotechnology facilitated real time soil monitoring for optimized crop production. Asian J Soil Sci Plant Nutr 10(2):582–607
De Carolis C, Iori V, Narciso A, Gentile D, Casentini B, Pietrini F, Grenni P, Barra Caracciolo A, Iannelli MA (2024) The effects of different combinations of cattle organic soil amendments and copper on lettuce (cv. Rufus) plant growth. Environments 11(7):134
Dhanyasree K, Rafeekher M, Premachandran A (2022) Abiotic stress management in fruit crops: a review. Agric Rev 46(3):451–456
Ding F, Liu B, Zhang S (2017) Exogenous melatonin ameliorates cold-induced damage in tomato plants. Sci Hortic 219:264–271
Dollete D, Lumactud RA, Carlyle CN, Szczyglowski K, Hill B, Thilakarathna MS (2024) Effect of drought stress on symbiotic nitrogen fixation, soil nitrogen availability and soil microbial diversity in forage legumes. Plant Soil 495(1):445–467
El-Shal RM, El-Naggar AH, El-Beshbeshy TR, Mahmoud EK, El-Kader NIA, Missaui AM, Du D, Ghoneim AM, El-Sharkawy MS (2022) Effect of nano-fertilizers on alfalfa plants grown under different salt stresses in hydroponic system. Agriculture 12(8):1113
Etesami H, Glick BR (2020) Halotolerant plant growth–promoting bacteria: prospects for alleviating salinity stress in plants. Environ Exp Bot 178:104124
Fahad S, Bajwa AA, Nazir U et al (2017) Crop production under drought and heat stress: plant responses and management options. Front Plant Sci 8:11–47
García-García AL, García-Machado FJ, Borges AA, Morales-Sierra S, Boto A, Jiménez-Arias D (2020) Pure organic active compounds against abiotic stress: a biostimulant overview. Front Plant Sci 11:575829
Giordano M, Petropoulos SA, Rouphael Y (2021) Response and defence mechanisms of vegetable crops against drought, heat and salinity stress. Agriculture 11(5):463
Guo L, Nie Z, Zhou J, Zhang S, An F, Zhang L, Wang Z (2022) Effects of different organic amendments on soil improvement, bacterial composition, and functional diversity in saline–sodic soil. Agronomy 12(10):2294
Guo BL, Feng CC, Chen Y, Lin D, Li LT (2023) Effects of polystyrene microplastics on growth, physiology, biochemistry, and canopy temperature characteristics of Chinese cabbage pakchoi (Brassica chinensis L.). Huan Jing ke Xue = Huanjing Kexue 44(9):5080–5091
Hassan M, Mahmood A, Awan M, Maqbool R, Aamer M, Alhaithloul H, Huang G, Skalický M, Brestič M, Pandey S, El Sabagh A, Qari SH (2022) Melatonin-induced protection against plant abiotic stress: mechanisms and prospects. Front Plant Sci 13:902694
Houetohossou SCA, Houndji VR, Hounmenou CG, Sikirou R, Kakaï RLG (2023) Deep learning methods for biotic and abiotic stresses detection and classification in fruits and vegetables: state of the art and perspectives. Artif Intell Agric 9:46–60
Imran (2022) Management of abiotic stresses with nano-black carbon is a tool for crop production. J Plant Nutr 46(1):145–166

Imran QM, Falak N, Hussain A, Mun BG, Yun B (2021) Abiotic stress in plants; stress perception to molecular response and role of biotechnological tools in stress resistance. Agronomy 11(8):1579

Ismail AM, Horie T (2017) Genomics, physiology, and molecular breeding approaches for improving salt tolerance. Annu Rev Plant Biol 68(1):405–434

Junaid MD, Chaudhry UK, Gökçe AF (2021) Climate change and plant growth–South Asian perspective. In: Climate change plants. Taylor and Francis, CRC Press, Boca Raton, pp 37–53

Kamal MA, Perveen K, Khan F, Sayyed RZ, Hock OG, Bhatt SC, Qamar MO (2023) Effect of different levels of EDTA on phytoextraction of heavy metal and growth of Brassica juncea L. Front Microbiol 14:1228117

Kang Y, Lee K, Hoshikawa K, Kang M, Jang S (2022a) Molecular bases of heat stress responses in vegetable crops with focusing on heat shock factors and heat shock proteins. Front Plant Sci 13:837152–837152

Kang Y, Wu Q, Qin J, Zhong M, Yang X, Chai X (2022b) High relative humidity improves leaf burn resistance in flowering Chinese cabbage seedlings cultured in a closed plant factory. PeerJ 10:e14325

Khalid MF, Huda S, Yong M, Li L, Li L, Chen ZH, Ahmed T (2023) Alleviation of drought and salt stress in vegetables: crop responses and mitigation strategies. Plant Growth Regul 99(2):177–194

Khan SA, Li MZ, Wang SM, Yin HJ (2018) Revisiting the role of plant transcription factors in the battle against abiotic stress. Int J Mol Sci 19(6):1634

Kidokoro S, Shinozaki K, Yamaguchi-Shinozaki K (2022) Transcriptional regulatory network of plant cold-stress responses. Trends Plant Sci 27(9):922–935

Kouame KP, Agrahari RK, Konjengbam NS, Koyama H, Kobayashi Y (2024) Ability of nutrient management and molecular physiology advancements to overcome abiotic stress: a study on sub-Saharan African crops. Agriculture 14(2):285

Kumar S, Kumar T, Kumar M (2023) Vegetable production under challenging conditions: abiotic stress – a review. Int J Biochem 7(2):6–14

Kumari M, Solankey SS, Sinha N (2023a) Abiotic stress management in vegetable crops. Int Res J 13(01). https://doi.org/10.51611/iars.irj.v13i01.2023.229

Kumari A, Singh SK, Mathpal B, Verma KK, Garg VK, Bhattacharyya M, Bhatt R (2023b) The biosynthesis, mechanism of action, and physiological functions of melatonin in horticultural plants: a review. Horticulturae 9(8):913

Kundu S, Chamroy T (2020) Role of micro-nutrients in Cole crops: a review. J Pharmacogn Phytochem 9(6):1017–1020

Kasukawa N, Miyazawa K (2020) Examining organic acid root exudate content and function for leafy vegetables under water-stressed conditions. J Hort Res 28(2):83-90

Kunwar UB, Manzoor N, Wen J, Pandit NR (2025) Integrating Agronomic and Molecular Advancements to Enhance Nitrogen Use Efficiency (NUE) and Promote Sustainable Rice Production. Nitrogen 6(2):34

la Peña MMD, Odriozola-Serrano I, Oms-Oliu G, Martín-Belloso O (2020) Dietary fiber in fruits and vegetables. In: Science and technology of fiber in food system. Springer, Cham, pp 123–152

Lastochkina O, Aliniaeifard S, SeifiKalhor M, Bosacchi M, Maslennikova D, Lubyanova A (2022) Novel approaches for sustainable horticultural crop production: advances and prospects. Horticulturae 8(10):910

Lee EH (2023) Early detection, mechanisms of tolerance, and amelioration of ozone stress in crop plants. In: Environmental pollution and plant responses. Routledge, London, pp 203–222

Lee J, Lee H, Wi S, Yu I, Yeo KH, An S, Jang S (2021) Enhancement of growth and antioxidant enzyme activities on kimchi cabbage by melatonin foliar application under high temperature and drought stress conditions. Hortic Sci Technol 39(5):583–592

Li L, Guo B, Feng C, Liu H, Lin D (2022) Growth, physiological, and temperature characteristics in Chinese cabbage pakchoi as affected by Cd-stressed conditions and identifying its main controlling factors using PLS model. BMC Plant Biol 22(1):3966

Liang W, Ma X, Wan P, Liu L (2018) Plant salt-tolerance mechanism: a review. Biochem Biophys Res Commun 495(1):286–291
Liu W, Muzolf-Panek M, Kleiber T (2022) The effect of various foliar treatments and nitrogen nutrition levels on the yield and physicochemical parameters of flowering Chinese cabbage. Agronomy 12(3):737
Liu W, Muzolf-Panek M, Kleiber T (2023) Effect of varied nitrogen sources and type of cultivation on the yield and physicochemical parameters of flowering Chinese cabbage (Brassica campestris L. ssp. chinensis var. utilis Tsen et Lee). Appl Sci 13(9):5691
Medina E, Kim SH, Yun M, Choi WG (2021) Recapitulation of the function and role of ROS generated in response to heat stress in plants. Plants 10(2):371
Muralikrishnan L, Padaria RN, Dass A, Choudhary AK, Kakade B, Shokralla S, Elansary HO (2021) Elucidating traditional rice varieties for consilient biotic and abiotic stress management under changing climate with landscape-level rice biodiversity. Land 10(10):1058
Nayak S, Mishra CSK (2024) Organic and biofertilizer interventions in iron mine spoil for nutrient fortification with facilitation of microbial exoenzyme activity and plant growth. Int J Eco Environ Sci 50(4):591–600
Nxele X, Klein A, Ndimba BK (2017) Drought and salinity stress alters ROS accumulation, water retention, and osmolyte content in sorghum plants. S Afr J Bot 108:261–266
Ouyang Y, Reeve JR, Norton JM (2022) The quality of organic amendments affects soil microbiome and nitrogen-cycling bacteria in an organic farming system. Front Soil Sci 2:869136
Panozzo A, Bolla PK, Barion G, Botton A, Vamerali T (2025) Phytohormonal regulation of abiotic stress tolerance, leaf senescence and yield response in field crops: a comprehensive review. Biotech 14(1):14
Patra KD, Adesh BA, Kothari DM, Nagarhalli TP (2024) DigiFarmer: precision agriculture empowered by machine learning for sustainable crop management. International conference on networking and communications (ICNWC), Chennai, India, pp. 1–5
Pavithra A, Kaleeswari R, Chitdeshwari T, Priya R, Uma D, Lourdusamy DK (2022) Agronomic biofortification of calcium in cabbage (Brassica oleracea var capitata) applied with different sources of liming in Ca deficient acidic soil of Coonoor, the Nilgiris (*Typic Dystropept*). J Appl Nat Sci 14(4):1286–1296
Qin K, Leskovar D (2020) Assessments of humic substances application and deficit irrigation in triploid watermelon. Hortic Sci 55(5):716–722
Qu S, Li H, Zhang X, Gao J, Ma R, Ma L, Ma J (2023) Effects of magnesium imbalance on root growth and nutrient absorption in different genotypes of vegetable crops. Plants 12(20):3518
Quan J, Zheng W, Wu MH, Shen Z, Tan JF, Li Z, Zhu B, Hong SB, Zhao Y, Zhu Z, Zang Y (2022) Glycine betaine and β-aminobutyric acid mitigate the detrimental effects of heat stress on Chinese cabbage (Brassica rapa L. ssp. pekinensis) seedlings with improved photosynthetic performance and antioxidant system. Plants 11(9):1213
Ranwa M, Sri KR, Khare A, Kumar V, Kumar K, Rajeshwar JR, Niharika MA (2024) A summative review of advances in sensor technology for precision agriculture. Arch Curr Res Int 24(10):257–275
Rathod A, Verma NS (2023) Impact of abiotic stress on agronomical crops. In: Frontiers of agronomy. Elite Publishing House, Rohini, p 27
Raza A, Su W, Gao A, Mehmood SS, Hussain MA, Nie W, Zhang X (2021a) Catalase (CAT) gene family in rapeseed (Brassica napus L.): genome-wide analysis, identification, and expression pattern in response to multiple hormones and abiotic stress conditions. Int J Mol Sci 22(8):4281
Raza A, Tabassum J, Kudapa H, Varshney RK (2021b) Can omics deliver temperature resilient ready-to-grow crops. Crit Rev Biotechnol 41(8):1209–1232
Raza A, Charagh S, García-Caparrós P, Rahman MA, Ogwugwa VH, Saeed F, Jin W (2022) Melatonin-mediated temperature stress tolerance in plants. GM Crops Food 13(1):196–217
Saddhe AA, Malvankar MR, Karle SB, Kumar K (2019) Reactive nitrogen species: paradigms of cellular signaling and regulation of salt stress in plants. Environ Exp Bot 161:86–97
Saeed F, Chaudhry UK, Raza A, Charagh S, Bakhsh A, Bohra A, Varshney RK (2023) Developing future heat-resilient vegetable crops. Funct Integr Genomic 23(1):47

Sahoo SK, Mishra KN, Panda N, Sethi D, Dey P (2023) Carbon and nitrogen mineralization kinetics in soil under differential nutrient management regimes in inceptisols of coastal India. Commun Soil Sci Plant Anal 54(20):2782–2799

Sarma HH, Borah SK, Dutta N, Sultana N, Nath H, Das BC (2024) Innovative approaches for climate-resilient farming: strategies against environmental shifts and climate change. Int J Environ Clim Chang 14(9):217–241

Schwab ST, Quides KW, Wendlandt CE, Trinh J, Sung M, Cardenas P, Sachs JL (2023) Effective rhizobia enhance legume growth during subsequent drought despite water costs associated with nitrogen fixation. Plant Soil 492(1):157–175

Shahid MA, Sarkhosh A, Khan N, Balal RM, Ali S, Rossi L, Garcia-Sanchez F (2020) Insights into the physiological and biochemical impacts of salt stress on plant growth and development. Agronomy 10(7):938

Shawon RA, Kang BS, Lee SG, Kim SK, Lee HJ, Katrich E, Gorinstein S, Ku Y (2020) Influence of drought stress on bioactive compounds, antioxidant enzymes, and glucosinolate contents of Chinese cabbage (Brassica rapa). Food Chem 314:125657

Shekhawat A, Rawal P (2024) Cultivating sustainability: organic treatments and soil transformation in Udaipur district. Int J Multidiscip Res 6(3). https://doi.org/10.36948/ijfmr.2024.v06i03.22786

Shi T, Wang LR, Zhang Z, Sun X, Huang H (2020) Stresses as first-line tools for enhancing lipid and carotenoid production in microalgae. Front Bioeng Biotechnol 8:610

Shoukat A, Saqib ZA, Akhtar J, Aslam Z, Pitann B, Hossain MS, Mühling KH (2024) Zinc and silicon nano-fertilizers influence ionomic and metabolite profiles in maize to overcome salt stress. Plants 13(9):1224

Singh RK, Ruiz-May E, Rajput VD, Minkina T, Gómez-Peraza RL, Verma KK, Shekhawat MS, Pinto C, Falco V, Quiroz-Figueroa FR (2022) Viewpoint of chitosan application in grapevine for abiotic stress/disease management towards more resilient viticulture practices. Agriculture 12(9):1369

Singh A, Pandey AK, Santhosh DT, Ganavi NR, Sarma A, Deori C, Das J, Shiva Kumar D (2024) A comprehensive review on greenhouse gas emissions in agriculture and evolving agricultural practices for climate resilience. Int J Environ Clim Chang 14(5):455–464

Soualiou S, Duan F, Li X, Zhou W (2022) Crop production under cold stress: An understanding of plant responses, acclimation processes, and management strategies. Plant Physiol Biochem 190:47–61

Sri BG, Anand KR, Prakash D, Lingireddy H, Gowtham KV, Akhil VS, Bhargava K (2023) A review of building community resilience for stress free future farming. Asian J Agric Ext Econ Sociol 41(9):824–833

Stevens RM, Pech JM, Grigson GJ (2017) A short non-saline sprinkling increases the tuber weights of saline sprinkler irrigated potatoes. Agronomy 7(1):4

Tang X, Mu X, Shao H, Wang H, Brestic M (2015) Global plant-responding mechanisms to salt stress: physiological and molecular levels and implications in biotechnology. Crit Rev Biotechnol 35(4):425–437

Toromade AS, Chiekezie NR (2024) GIS-driven agriculture: pioneering precision farming and promoting sustainable agricultural practices. World J Adv Sci Technol 6(1):057–072

Toscano S, Romano D, Ferrante A (2023) Molecular responses of vegetable, ornamental crops, and model plants to salinity stress. Int J Mol Sci 24(4):3190

Varshney RK, Sinha P, Singh VK, Kumar A, Zhang Q, Bennetzen JL (2020) 5Gs for crop genetic improvement. Curr Opin Plant Biol 56:190–196

Varshney RK, Bohra A, Yu J, Graner A, Zhang Q, Sorrells ME (2021) Designing future crops: genomics-assisted breeding comes of age. Trends Plant Sci 26(6):631–649

Verma N, Kaushal P, Sidhu AK (2024) Harnessing biological synthesis: zinc oxide nanoparticles for plant biotic stress management. Front Chem 12:1432469

Wang S, Hu K, Feng P, Qin W, Leghari SJ (2023) Determining the effects of organic manure substitution on soil pH in Chinese vegetable fields: a meta-analysis. J Soils Sediments 23(1):118–130

Xie Q, Xu H, Wen R, Wang L, Yang Y, Zhang H, Su B (2024) Integrated management of fruit trees and Bletilla striata: implications for soil nutrient profiles and microbial community structures. Front Microbiol 15:1307677

Yang H, Kim Y, Bae Y, Hyeon S, Choi M, Jang D (2025) Investigating growth, root development, and chlorophyll fluorescence of tomato scions and rootstocks under UV-B stress in a plant factory with artificial lighting. Sci Hortic 347:114191

Yapa N, Jayakody N, Madhushan A, Pelawatta A (2023) Effect of biofertilizers and organic amendments on germination and seedling growth of common dry zone forest species in Sri Lanka: sustainable reforestation practices in Sri Lanka. Turk J Agric Food Sci Technol 11(2):287–291

Yaqoob H, Tariq A, Bhat BA, Bhat KA, Nehvi IB, Raza A, Djalovic I, Prasad PV, Mir RA (2023) Integrating genomics and genome editing for orphan crop improvement: a bridge between orphan crops and modern agriculture system. GM Crops Food 14(1):1–20

Yu X, Zhang W, Zhang Y, Zhang X, Lang D, Zhang X (2019) The roles of methyl jasmonate to stress in plants. Funct Plant Biol 46(3):197–211

Zahra N, Shaukat K, Hafeez MB, Raza A, Hussain S, Chaudhary MT, Wahid A (2021) Physiological and molecular responses to high, chilling, and freezing temperature in plant growth and production: consequences and mitigation possibilities. In: Harsh environment and plant resilence: molecular and functional aspects. Springer, Cham, pp 235–290

Zhang Q, Dai W (2019) Plant response to salinity stress. In: Stress physiology of woody plants. CRC Press, Boca Raton, pp 155–173

Zhang S, Zhu Q, de Vries W, Ros GH, Chen X, Muneer MA, Wu L (2023) Effects of soil amendments on soil acidity and crop yields in acidic soils: a world-wide meta- analysis. J Environ Manag 345:118531

Zhu Y, Qi B, Hao Y, Liu H, Sun G, Chen R, Song S (2021) Appropriate NH4+/NO3–ratio triggers plant growth and nutrient uptake of flowering Chinese cabbage by optimizing the pH value of nutrient solution. Front Plant Sci 12:656144

Zheng Y, Wang X, Cui X, Wang K, Wang Y, He Y (2023) Phytohormones regulate the abiotic stress: An overview of physiological, biochemical, and molecular responses in horticultural crops. Front Plant Sci 13:1095363

Biofortification Strategies in Brassica Crops

Khushboo Kathayat, Mamta Sharma, and Monisha Rawat

Abstract

Vegetables are a vital source of essential nutrients and play a crucial role in enhancing human health. The Brassicaceae family, also known as crucifer, includes various crops such as cabbage, mustard, turnip, and radish, which provide a sufficient amount of nutrition (vitamins, minerals, and antioxidants). Despite their nutritional benefit, malnutrition and micronutrient deficiency remain serious global issues, particularly in developing countries. To combat this serious concern, biofortification has proven to be a suitable solution. Various techniques like agronomical practices, breeding, and transgenic approaches have been applied to crops to increase their nutrient content. Agronomical biofortification involves the application of nutrients at the soil and foliar levels, seed priming, and soilless cultivation. Conventional breeding enhances genetic characters for nutritional improvement. Different transgenic methods like CRISPR/Cas9 and genome editing further increase the nutritional content of crops. The integration of these technologies can be used to improve human health and ensure the global availability of nutritious food.

Keywords

Biofortification · Brassicaceae · Micronutrient deficiency · Agronomic practices · Conventional breeding · Transgenic methods · Zero hunger

K. Kathayat · M. Rawat
School of Agriculture, Lovely Professional University, Phagwara, Punjab, India

M. Sharma (✉)
School of Bioengineering and Biosciences, Lovely Professional University, Phagwara, Punjab, India
e-mail: mamta.18431@lpu.co.in

A. Kumar, B. Koul (eds.), *Crop Improvement Strategies in Brassica species: Applied Science*, https://doi.org/10.1007/978-981-95-3861-4_15

1 Introduction

Vegetables are one of the most important sources of nutrients in human food and are known as protective food because of the sufficient amount of vitamins and minerals that are able to fight against different diseases, They are an important part of growth and development (Lal et al. 2020). The Brassicaceae group of vegetables contains various vitamins, including A, C, and E, along with antioxidants that contribute to improving human health (Table 1). Brassicaceae or cruciferae is an economically important family also known as mustard, crucifers, or cabbage family. Mostly, they are herbaceous plants. There are different members present in this family, like cole crops, mustard, turnip, and radish (Fig. 1). Mostly, they are cultivated in cool season in sandy loamy soil with a pH of 5.5–7.

The United Nations' Sustainable Development Goals (SDGs) emphasize the eradication of hunger by 2030. However, one in three countries struggles with inadequate food supply, and nearly two billion people worldwide experience "hidden hunger" which refers to poor intake of essential micronutrients in their daily diet. Even after meeting basic nutritional needs, over 40% of people worldwide suffer from some form of micronutrient deficiency (Sujith Kumar et al. 2020). Approximately 800 million people worldwide suffer from malnutrition due to limited access to diets and essential micronutrient interventions. This problem is more common in developing countries (Malik and Maqbool 2020). As the world population increases day by day, arable land is shrinking, and due to changes in climate, there is an urgent need for high-yielding, nutritionally enriched crop varieties that can withstand diverse climatic and biotic stresses (Kumar et al. 2020). Biofortification, a technique aimed at enriching the nutritional value of crops, offers a promising solution to mitigate hidden hunger by enhancing the nutrient composition of staple

Table 1 Main phytonutrients present in *Brassica* species

Category	Compounds/nutrients	Health benefits
Glucosinolates	Aliphatic GSLs: sinigrin, glucoraphanin → SFN, glucoiberin → iberverin, glucoerucin → erucin	Anti-cancer, antimicrobial
	Indolic GSLs: glucobrassicin → I3C, Neoglucobrassicin → NI3C	Anti-neurodegenerative diseases
	Aromatic GSLs: gluconasturtiin → PEITC, Glucotropaeolin → BITC	Anti-cardiovascular diseases
Vitamins	Vitamin C, B vitamins, folate, tocopherol, phylloquinone	Skin health, antioxidant, normal blood coagulation
Carotenoids	Lutein, β-carotene, α-carotene, lycopene, zeaxanthin, β-cryptoxanthin, violaxanthin, neoxanthin	Eye health, anti-type 2 diabetes
Phenols	Polyphenol (flavonoids, anthocyanins)	Antioxidant, anti-obesogenic
Minerals	Calcium, potassium, iron, magnesium, sulfur, manganese, phosphorus, zinc, selenium	Bone quality, anti-obesogenic

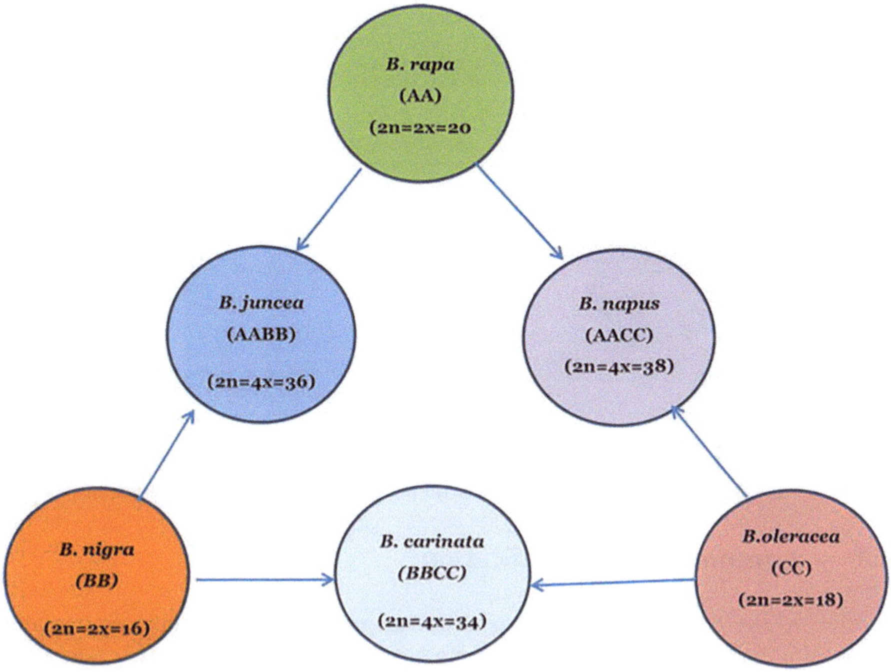

Fig. 1 Genetic evolution of *Brassica* species (Chand et al. 2021)

foods (Prasad et al. 2015). Innovative approaches like nutrition-sensitive agriculture and targeted rural micronutrient programs (Ansari and Thapa 2019) can effectively enhance nutrient availability. Biofortified crops, by enriching diets with essential nutrients, hold significant promise for improving public health and addressing global malnutrition (Singh et al. 2016).

2 Methods of Biofortification

The ultimate goal of biofortification is to produce healthy, nutritious, and safe foods (Saltzman et al. 2014). It can be achieved by three main ways, namely, agronomic, conventional, and transgenic (Garg et al. 2018; Connorton and Balk 2019; Jha and Warkentin 2020; Van Der Straeten et al. 2020) (Fig. 2), which involve the application of fertilization, crop breeding, and biotechnological application, respectively. As an innovative and promising strategy, biofortification plays a vital role in addressing nutrient deficiencies and promoting overall human health. This approach, referred to as multi-nutrient or combinatorial biofortification, focuses on developing crops with a well-balanced nutritional profile.

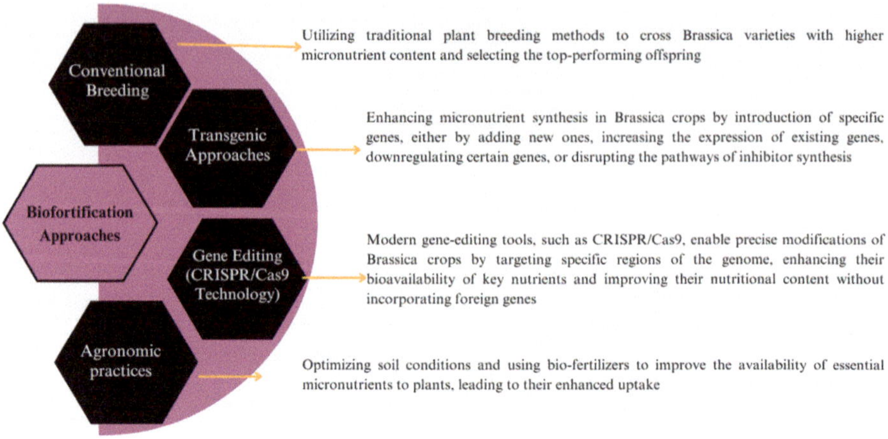

Fig. 2 Various biofortification approaches used for developing improved crop varieties

2.1 Agronomical Biofortification

It refers to the process of increasing micro micronutrient properties of vegetable crops with the help of different agronomic practices (Prasad et al. 2015). The most important micronutrients are Zn, iodine, and selenium. There are different methods of agronomic bio-fortification out of which the foliar method is a quick and easy method. Agronomic biofortification has been explored extensively in *Brassica* crops, achieving notable success. It basically depends on fertilizer application methods, solubilization, and mobilization of elements from source to sink. Successful examples of agronomical methods are mentioned in Table 2.

2.1.1 Techniques for Applying Agronomic Biofortification

Different types of agronomical bio-fortification techniques are listed in Fig. 3.

Soil Application

Applying mineral fertilizers to the soil enhances the nutrient content in edible plant parts and is the easiest method of micronutrient supplementation for crops. Soil application of zinc sulfate ($ZnSO_4$) has been shown to significantly increase zinc (Zn) content in *Brassica napus* (canola), enhancing seed Zn concentration and oil quality. A study demonstrated that soil-applied Zn improved branching, pod number, and seed production in oilseed rape (Safdar et al. 2023). Another research indicated that Zn fertilization positively affected yield attributes and seed oil quality in canola (Manaf et al. 2019). Similarly, soil application of sodium selenate (Na_2SeO_4) or sodium selenite (Na_2SeO_3) increases selenium (Se) accumulation in *Brassica juncea* (Indian mustard), enhancing its nutritional benefits. de Sousa Lima et al. (2015) reported that applying 300 mg Zn kg^{-1} soil resulted in up to a 28-fold increase in zinc concentration in kale.

Table 2 Examples of agronomical biofortification in *Brassica* crops

Crop	Targeted nutrients	Treatments	Main findings	References
Brassicaceae microgreens	Fe	Fe conc. 0, 10, 20, 40 mgL^{-1} in Soilless media	Concentration of Fe shoot increases 64% in arugula and up to 278% in red cabbage	Di Gioia et al. (2019)
Brassicaceae microgreens	Zn	$ZnSO_4$ (5–10 mg/L) in Soilless media	75–281% more zinc content in crop	Di Gioia et al. (2019)
Broccoli	Zn	0.25% $ZnSO_4$. 7 H_2O @ 15 mL per pot	Foliar application of zinc increases Zn concentration in florets, stems, and leaves by 78 mg/kg and 23 mg/kg, respectively, while also enhancing bioavailability by influencing phytic acid levels	Rivera Martin et al. (2020)
Mustard	Zn	Application of Zn in hydroponics system	More Zn content is found in mustard pods	Miliauskienė et al. (2022)
Arugula	Zn	Three doses (0.5, 1, and 1.5 kg/ha Zn	Higher Zn content, i.e., 279% more, was found in leaves	Rugeles-Reyes et al. (2019)
Brassica oleracea cv. Bronco	Zn	80–100 μM Zn	Increases the amino acid content and phytochemical content	Barrameda-Medina et al. (2017)
Cabbage	Iodine	Potassium iodide @ 15 kg I /ha	Iodine concentration (109.1 mg/kg) was increased by foliar application of KI and KIO_3 in cabbage heads as compared to sole application of KIO_3	Ojok et al. (2019)
Broccoli	Selenium	Application of sodium selenite	Higher amounts of phenolic compounds and antioxidant activity	Bachiega et al. (2016)
Mustard sprout	Selenium	Selenate and selenite @ 20 μmol L^{-1}	Increased concentration of anthocyanin content	Woch and Hawrylak-Nowak (2019)
Cabbage	Selenium	8 mg kg^{-1} and 16 mg kg^{-1} Se yeast	Increased characters like free amino acid (81.6%), soluble sugar (46.5%), ascorbic acid (34.9%), phenolic acid (12.3%), glucosinolates (44.8%), and SOD activity (25.2%) were found by application of 8 mg/kg Se yeast	Liao et al. (2021)

(continued)

Table 2 (continued)

Crop	Targeted nutrients	Treatments	Main findings	References
Kohl rabi	Selenium	10 mgSe/l	Increased chlorophyll and carotene content	Golob et al. (2020)
Mustard	Selenium	Enrichment with Na_2SeO_4	Found more selenium content in Indian mustard	Montes-Bayón et al. (2002)
Canola	Selenium	Enrichment with Na_2SeO_4	Leaf material holds as much as 3.5 mg Se/kg	Bañuelos et al. (2005)
Cabbage	Calcium	Limestone application	Lowe Fe and Mg content, and high Ca content	Pavithra et al. (2022)
Kohl rabi	Iodine	1 g Iodine/l	Anthocyanin content was increased	Golob et al. (2020)
Cabbage	Iodine	Iodine	More plant growth is observed	Weng et al. (2008)
Radish	Iodine	10–150 mg kg^{-1}	Iodine content is found up to 100–1300 μg	Weng et al. (2008)
Cabbage	Iodine	150 mg kg^{-1}	The biomass of cabbage increases by 34.8%	Hong et al. (2008)

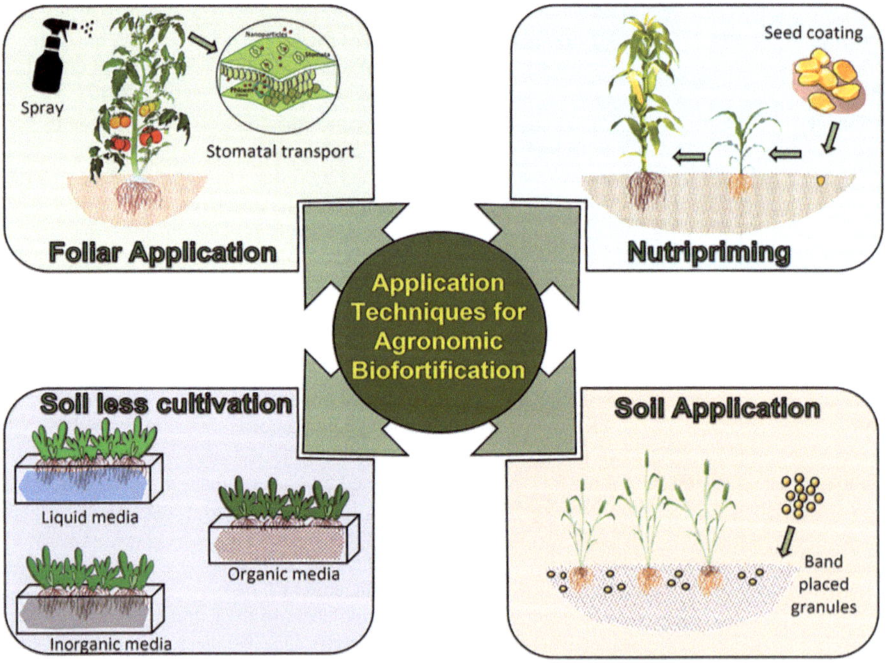

Fig. 3 Different types of agronomical biofortification. (Source: Bhardwaj et al. 2022)

Foliar Application

Foliar application is recognized as one of the most effective methods for enhancing the nutritional content of crops (Smoleń 2012). It enables the direct supply of nutrients at various growth stages, particularly through leaves (Poblaciones and Rengel 2017; Doolette et al. 2018; Kandil et al. 2022). This method facilitates nutrient absorption, promoting efficient translocation to edible plant parts such as seeds, while also reducing nutrient loss and supporting plant growth under unfavorable environmental conditions. The application of boron and nitrogen through foliar spraying has been shown to increase crop yield (Choudhary and Bhogal 2017; Chauhan et al. 2020). Due to its efficiency in fertilizer utilization, minimal infrastructure requirements, and ease of application requiring less technical expertise, foliar application is one of the most widely adopted biofortification techniques. For instance, the combined foliar application of borax (1.0%) at the flowering and capsule formation stages significantly improved crude fiber (3.77%), total soluble solids (TSS) (24.9%), and protein content (27.53%) in Indian mustard (*Brassica juncea)* (Dhaliwal et al. 2022). Additionally, foliar application of sodium selenite was found to enhance mineral content in Savoy cabbage (*Brassica oleracea* var. *sabauda*) (Golubkina et al. 2018).

Nutri Priming/Seed Priming

It is defined as a soaking of seed before planting in different solutions. The primary objectives of seed priming are to enhance germination, promote a healthy root system, improve seedling establishment, and ultimately increase yield (Lutts et al. 2016; Farooq et al. 2019; Raj and Raj 2019). Improved chlorophyll content (high value of SPAD) and faster germination were observed by bio-nutri priming with compost leachate of Bak choy seed (Phooi et al. 2023). In turnip stress generated due to oxidation was reduced by the help of seed priming with Selenium. It also increases antioxidant content and uptake of nutrients, which ultimately results in better growth under saline conditions (Hussain et al. 2023). Cysteine- and ascorbate-treated seeds have been used to increase seed yield and oil content in canola (Molnar et al. 2023). Seeds treated with magnesium oxide at different concentrations increase different characters, namely, leaf area and biomass, antioxidant activities, and chlorophyll content, ultimately leading to increased yield and growth (Gautam et al. 2023).

Soilless Cultivation

It is one of the latest methods of crop cultivation that utilizes organic, inorganic, or liquid growing media instead of soil. Different systems like hydroponics, aeroponics, and vertical farming are included in soilless cultivation. Microgreen production using soilless media serves as one of the best examples of latest cultivation method. Plant productivity can be increased by optimum nutrient solutions with controlled temperature and light (Treftz and Omaye 2016). Direct root exposure to fertilizer solutions improves the nutritional quality of crops (Wiesner-Reinhold et al. 2017). Use of soil-less cultures promotes year-round production by reducing soil-related problems like poor soil fertility and soilborne pathogens (Tomasi et al.

2015). Additional benefits of this technique include reduced labor requirements, lower weed populations, easier harvesting, and simplified plant maintenance (Tomasi et al. 2015). In white cabbage, zinc (Zn) content was found to increase (Barrameda-Medina et al. 2017). Notably, high micronutrient levels have been observed in Brassicaceae microgreens fortified with Zn and iron (Fe), as well as in tatsoi enriched with selenium (Se) (Puccinelli et al. 2019; Pannico et al. 2020). Additionally, 100 g of mustard biofortified with 300 ppm calcium (Ca) can meet 41.6% of the body's daily calcium requirement (Gustiar et al. 2020).

2.2 Conventional Breeding

It is regarded as the most widely accepted biofortification method, offering greater cost-effectiveness and sustainability compared to other approaches (Bouis and Saltzman 2017). Sufficient genotypic variation of a particular trait is a must requirement for breeding approaches. This sufficient variation is utilized in breeding approaches to improve the nutritional content of the crop. Crossing between parents with high nutrient content and recipient lines that are good in agronomic values was done for a larger generations, and the objective of this crossing was to produce healthy plants with high nutrient and good agronomic characters. In biofortification by breeding, important steps are to identify target populations in which a set level of nutrients has to be incorporated. For this purpose, screening of germplasm has to be done, and after screening biofortified crop is to be produced. Later on, its performance is to be evaluated and if its nutrient retention is crops is found to be suitable, then development of different strategies to disseminate seeds and promotion of crops to the market is done. Finally, it will improve the nutritional status of the target population. This breeding method is sometimes affected by limited genetic variation in conventional biofortification methods have been successfully attempted in numerous crops, including rice, wheat, maize, sorghum, pearl millet, lentils, cowpea, beans, potato, cassava, sweet potato, cauliflower, banana, mango, and grapes, leading to the release of several improved varieties. In cauliflower, Pusa Beta Kesari released by IARI in 2015–2016, New Delhi, and some other varieties like Purple Graffiti and Orange Cheddar released by Cornell University, New York found to be rich in beta carotene (Garg et al. 2018).

2.3 Transgenic Methods of Biofortification

Various approaches, ranging from conventional selective breeding to advanced genetic engineering, have been employed for biofortification of crops (Fig. 2). Compared to other nonspecific procedures like traditional breeding and other agronomic practices, genetic engineering is a specialized procedure to produce new crop varieties with desirable attributes. A key advantage of genetic engineering is its ability to transfer and express desirable traits from a donor organism to a host organism across evolutionary and taxonomic boundaries (Malik and Maqbool 2020).

Additionally, genetic engineering technologies are more precise, efficient, reliable, and time-saving compared to traditional methods (Kiran 2020). In cases when a specific micronutrient is naturally not produced in a crop, a transgenic approach is the only viable option to introduce a foreign gene into the target crop (Malik and Maqbool 2020).

Transgenic crops are created through genetic modifications, such as the incorporation of novel genes using recombinant DNA technology. For instance, some crops have been engineered to express *Bacillus thuringiensis* toxin, providing resistance against specific pests. Additionally, the introduction and overexpression of phytoene synthase and phytoene desaturase genes have successfully enabled crops to synthesize beta-carotene, a precursor to vitamin A (Kumar et al. 2020). Another approach involves the overexpression of genes already present in the crop to enhance specific traits. Additionally, the downregulation of genes responsible for undesirable characteristics, such as those causing browning and bruising, has been applied to extend the shelf life of certain crops. Finally, interrupting the synthesis pathway of inhibitor genes is also a strategy used in the development of transgenic crops (Malik and Maqbool 2020).

Various transgenic approaches have been utilized across different crops to develop iron-biofortified varieties. For example, the insertion of an iron-binding protein gene, such as lactoferrin, has been employed to enhance Fe accumulation in seeds. Studies have shown that iron biofortification can be achieved through various transgenic strategies, including the insertion of iron-chelator and transporter genes, the overexpression of iron reductase genes, and the reduction of Fe-inhibitors or antinutrients. Similarly, transgenic approaches have been used to enhance zinc content in plants by introducing genes encoding zinc-binding proteins, increasing the expression of proteins involved in zinc uptake, and overexpressing zinc-storage proteins. Genetic modifications have also been used to improve the vitamin profiles of diverse crops, including the enhancement of vitamin A, B, and C levels. For example, the bacterial phytoene synthase (crtB) gene has been used in mustard to improve vitamin A content (Malik and Maqbool 2020). However, research on iodine biofortification in crops using recombinant DNA technology remains limited, largely due to the insufficient understanding of iodine physiology in plants (Gonnella et al. 2019).

Advancements in genome editing technologies have also enabled more efficient and effective gene editing for enhanced traits such as yield, nutritional value, and disease resistance. The first-generation genome editing tools, like zinc-finger nucleases and transcription activator-like effector nucleases, have shown promise for biofortification. However, the second-generation clustered regularly interspaced short palindromic repeats (CRISPR/Cas9)-based genome editing tool has emerged as a breakthrough technique, rapidly evolving to facilitate the development of improved crop varieties. Altering endogenous pathways that involve multiple genes using genome editing tools can enhance the accumulation of micronutrients, as demonstrated in the modification of the carotenoid pathway in *Brassica napus* (Kiran 2020). Gene editing has shown promising results in Brassica and various other crops like Arabidopsis, potato, soybean, rice, and maize (Kumar et al. 2020).

Biofortification in Brassica crops has been explored across diverse aspects, including selenium enrichment (Wiesner-Reinhold et al. 2017; Li et al. 2018), enhancement of bioactive compounds in Brassicaceae vegetables (Raiola et al. 2018), and zinc and iron biofortification of Brassicaceae microgreens (Di Gioia et al. 2019) and *Brassica oleracea* cv. Bronco (Barrameda-Medina et al. 2017). Additionally, biofortification has been used to increase micronutrient concentrations (Billard et al. 2014), improve the antioxidant properties of broccoli (Bachiega et al. 2016) and Indian mustard (*Brassica juncea* L.) (Golubkina et al. 2018), and enhance the iron tolerance of *Brassica napus* (Zhu et al. 2016). Examples include the biofortification of oilseed Brassica juncea with the anticancer compound glucoraphanin through the suppression of the GSL-ALK gene family (Augustine and Bisht 2015) and the improvement of oil quality, yield, and nutrient uptake in *Brassica juncea* L. (Dhaliwal et al. 2022). Furthermore, transgenic approaches have resulted in Provitamin A biofortified crop varieties. The transgenic approach has also been utilized for metabolic engineering of oilseed crops, i.e., for altering the endogenous fatty acid composition (Kumar et al. 2020). Biofortification of broccoli with selenium has been shown to increase phenolic compound content and enhance antioxidant activity (Bachiega et al. 2016). Table 3 summarizes transgenic *Brassica* species, detailing introduced transgenes, resulting traits (e.g., salt tolerance, carotenoid enhancement), and their mechanisms.

3 Benefits and Limitation

This section presents a critical comparison of various biofortification methods developed to combat malnutrition in plant-based foods (Carvalho and Vasconcelos 2013) (Table 4).

4 Challenges in the Case of Biofortification

Poor soil conditions, like low organic matter and pH imbalance, can reduce plant nutrient absorption. Sometimes inhibitor like phytates reduces iron absorption by humans (Frossard et al. 2000). Poor storage and processing practices and cooking techniques reduce the effectiveness of biofortified crops. Sometimes farmers are not interested in adopting these crops due to higher input costs, lack of awareness, and unfamiliar agronomic practices. People also feel that biofortified crops differ in taste, texture, and appearance, so they are not willing to use these varieties. The development and commercialization of transgenic crops took approximately 13 years and the average cost of this whole process is $35 million USD (Kumar et al. 2020). Also, the patenting of these transgenic crops by private companies causes the monopoly of these crops. There has been an ongoing debate about whether the development of GM crops is driven by privatization and profit-making or aimed at enhancing food security (Sumanth et al. 2022). Regulations and legislations governing GM crops should be adjusted, especially in developing countries, to

Table 3 Transgenic modifications in *Brassica* species

Transgenic plant	Transgene	Trait	Mechanism	References
B. napus, B. juncea, B. compestris L. spp. *Chinensis*	CodA	Salt tolerance	Oxidation of choline to betaine through constitutive expression of a bacterial choline oxidase gene imparts moderate resistance to salt stress	Huang et al. (2000), Shah et al. (2018)
Brassicaceae oilseeds, Brassica napus	Ld-LPAAT + Bn-fae1	Erucic acid production	Overexpression of the rapeseed fatty acid elongase gene (fae1) in combination with expression of the lysophosphatidic acid acyltransferase gene from *Limnanthes douglasii* (Ld-LPAAT)	Nath et al. (2009), Wang et al. (2022)
Brassica napus, Brassica oleracea	crtB, crtI, crtY	Carotenoid enhancement	Expression of bacterial genes using the seed-specific napin promoter increased total carotenoid levels	Bai et al. (2011)
Brassica juncea	γ-TMT	Increased α-tocopherol content (vitamin E)	Overexpression of reframe-γ-tocopherol methyltransferase cDNA from *Arabidopsis thaliana*, which encodes the enzyme responsible for converting the large γ-tocopherol pool to α-tocopherol	Yusuf and Sarin (2007)
Brassica oleracea var. capitata	betA	Enhanced salt tolerance	The bacterial betA gene, responsible for the biosynthesis of glycinebetaine, was transferred to the cabbage cultivar "Golden Acre"" via Agrobacterium-mediated transformation of hypocotyl explants	Bhattacharya et al. (2004)

(continued)

Table 3 (continued)

Transgenic plant	Transgene	Trait	Mechanism	References
Brassica juncea	Lectin gene	Resistance to fungal diseases	Introduction and expression of chickpea lectin gene into a widely grown cultivar of Indian mustard	Kumar et al. (2015)
B. napus var. Wester	DREB	Expression of many stress-inducible genes	Incorporation of DREB factor into *Brassica napus var.* wester genome regulated the expression of stress-inducible genes	Qamarunnisa et al. (2015)
Brassica juncea	Gly I and Gly II	Abiotic stress tolerance, detoxification of methylglyoxal	Expression of glyoxalase I (gly I) and glyoxalase II (gly II) genes under the control of either a stress-inducible rd29A promoter or a constitutive CaMV 35S promoter, showed enhanced tolerance to salinity, heavy metal, and drought stress	Rajwanshi et al. (2016)
B. napus	–	Enriched in glucoraphanin	RNA interference (RNAi) of the GSL-ALK gene family was used to downregulate the expression of GSL-ALK genes	Liu et al. (2012)
Brassica juncea	Genes encoding for APS, ECS, and GS	Enhanced selenium (Se) accumulation and tolerance	Overexpression of genes involved in (GS)sulfur (S)/Se metabolism enhanced Se accumulation and tolerance. The study also reported successful field performance	Bañuelos et al. (2005)
Brassica juncea L.	Bacterial glutathione reductase	Enhanced cadmium (Cd) tolerance	Overexpression of glutathione reductase in *Brassica juncea* showed enhanced cadmium tolerance at the chloroplast level	Pilon-Smits et al. (2000)

(continued)

Table 3 (continued)

Transgenic plant	Transgene	Trait	Mechanism	References
Brassica napus L., Nicotiana tabacum L.	Human metallothionein-II (MT-II) processed gene	Heavy metal tolerance	Gene introduced via *Agrobacterium tumefaciens*; demonstrated potential for phytoremediation (heavy metal tolerance/ sequestration)	Misra and Gedamu (1989)
Brassica napus	DsRed (red fluorescence protein)	Established transgenic lines	Five vectors carrying the red fluorescence protein encoding gene from Discosoma sp. (DsRed) were constructed and integrated into rapeseed via *Agrobacterium*-mediated hypocotyl transformation	Zhang et al. (2020)
Brassica juncea	Wheat germ agglutinin (WGA) cDNA	Increased resistance to mustard aphids	A cDNA encoding wheat germ agglutinin (WGA), which is antimetabolic, antifeedant, and insecticidal, was transferred to Indian mustard (*Brassica juncea* cv. RLM-198) through *Agrobacterium*-mediated transformation	Kanrar et al. (2002)
B. juncea cv. Pusajaikisan	Snowdrop lectin gene (gna)	Insect resistant	*Agrobacterium*-mediated transformation of stem segments with the gna gene for insect resistance	Sharma et al. (2004)
Brassica juncea L.	Overexpression under promoter 35S CaMV	Enhanced defense response against *Alternaria brassicae*	Overexpression of chitinase gene through 35S promoter in Brassica resulted in the activation defense response against the fungal pathogen *Alternaria brassicae*	Mondal et al. (2003)

(continued)

Table 3 (continued)

Transgenic plant	Transgene	Trait	Mechanism	References
Brassica juncea	MsrA1 (antimicrobial peptide)	Resistance to fungal phytopathogens	Transgenic expression of an antimicrobial peptide MsrA1, a synthetic chimera of cecropin A and melittin CAPs with antimicrobial properties, conferred fungal resistance	Rustagi et al. (2014)
Brassica juncea	AtPCS1 (phytochelatin synthase)	Enhanced arsenic (As) and cadmium (Cd) tolerance and accumulation	Transgenic plants demonstrated increased tolerance and accumulation of heavy metals and metalloids from soil by expressing an *Arabidopsis thaliana AtPCS1* gene, encoding phytochelatin synthase	Gasic and Korban (2007)

Table 4 SWOT analysis of biofortification approaches

Types	S (strength)	W (weakness)	O (opportunity)	T (threats)
Agronomical	Simple method, suitable for immediate results	Needs regular application of nutrients, expensive and difficult to distribute	Often used as a compliment to other types	Negative environmental impact, reverse exhaustion
Conventional plant breeding	Successful for minerals and vitamins, easier distribution	Time taking process	Wide public acceptance, simple legal framework	Requires genetic variation
Transgenic	Successful for minerals and vitamins, easier distribution	A long-time taking process	Fast omics development	Low public acceptance, high sociopolitical problems, environmental impact

be less rigorous, and cost and time-effective to promote the adoption of GM crops. Approval of these crops at global level takes a longer time due to strict rules. These rules should be less rigorous and less time taking (Sumanth et al. 2022). Government and food industries may not interest to prioritize these crops due to lack of knowledge or competing food security initiatives (Kant et al. 2025). Intensive farming can reduce micronutrient levels in the soil, making agronomical biofortification less sustainable over time. Increase in temperature, extreme weather, and degradation of soil can affect crop nutrient and yield (Ismail et al. 2007). Fertilizer and water accumulation in soil is also a reason for less acceptance of these crops (Waters and Sankaran 2011). For biofortification of crops, significant research is required, which needs a large funding and it may not always be available. Export and import of biofortified crops are also a barrier to the adoption of these crops by people (Watanabe et al. 2005).

5 Conclusion

Biofortification presents a promising and viable strategy for mitigating malnutrition and nutrient deficiency globally. Various techniques, including soil and foliar applications, nutrient priming, soilless cultivation, conventional breeding, and transgenic approaches, have been employed to enhance the nutritional content of crops. While conventional breeding offers a cost-effective method, advancements in genetic engineering and genome editing provide precise tools for increasing crop nutrient levels. Despite its significance, several challenges, such as poor soil conditions, nutrient loss, high input costs, regulatory barriers, and consumer perceptions, impede the widespread adoption of biofortified crops. Furthermore, environmental factors like climate change and soil degradation contribute to the limited uptake of these crops. To promote their adoption, concerted efforts are necessary, encompassing increased investment, streamlined regulatory frameworks, and comprehensive training programs for farmers and public awareness campaigns. Incorporating biofortified crops into our diets can significantly improve global food security and enhance human health.

References

Ansari SA, Thapa S (2019) Biofortification of food crops: an approach towards improving nutritional security in South Asia. Int J Adv Agric Sci Technol 6(12):23–33

Augustine R, Bisht NC (2015) Biofortification of oilseed Brassica Juncea with the anti-cancer compound glucoraphanin by suppressing GSL-ALK gene family. Sci Rep 5:18005

Bachiega P, Salgado JM, de Carvalho JE, Ruiz AL, Schwarz K, Tezotto T, Morzelle MC (2016) Antioxidant and antiproliferative activities in different maturation stages of broccoli (*Brassica oleracea* var. *italica*) biofortified with selenium. Food Chem 190:771–776

Bai C, Twyman RM, Farré G, Sanahuja G, Christou P, Capell T, Zhu C (2011) A golden era – provitamin A enhancement in diverse crops. In Vitro Cell Dev Biol Plant 47:205–221

Bañuelos G, Terry N, LeDuc DL, Pilon-Smits EA, Mackey B (2005) Field trial of transgenic Indian mustard plants shows enhanced phytoremediation of selenium-contaminated sediment. Environ Sci Technol 39(6):1771–1777

Barrameda-Medina Y, Blasco B, Lentini M, Esposito S, Baenas N, Moreno DA, Ruiz JM (2017) Zinc biofortification improves phytochemicals and amino-acidic profile in *Brassica oleracea* cv. bronco. Plant Sci 258:45–51

Bhardwaj AK, Chejara S, Malik K, Kumar R, Kumar A, Yadav RK (2022) Agronomic biofortification of food crops: an emerging opportunity for global food and nutritional security. Front Plant Sci 9(13):1055278

Bhattacharya RC, Maheswari M, Dineshkumar V, Kirti PB, Bhat SR, Chopra VL (2004) Transformation of *Brassica oleracea var. capitata* with bacterial betA gene enhances tolerance to salt stress. Sci Hortic 100(1–4):215–227

Billard V, Etienne P, Jannin L, Garnica M, Cruz F, Garcia-Mina JM, Yvin JC, Ourry A (2014) Two biostimulants derived from algae or humic acid induce similar responses in the mineral content and gene expression of winter oilseed rape (*Brassica napus* L.). J Plant Growth Regul 33(2):305–316

Bouis HE, Saltzman A (2017) Improving nutrition through biofortification: a review of evidence from HarvestPlus, 2003 through 2016. Glob Food Secur 12:49–58

Carvalho SM, Vasconcelos MW (2013) Producing more with less: strategies and novel technologies for plant-based food biofortification. Food Res Int 54(1):961–971

Chand S, Patidar OP, Chaudhary R, Saroj R, Chandra K, Meena VK (2021) Rapeseed-mustard breeding in India: scenario, achievements and Brassica breeding and biotechnology, 19

Chauhan ZY, Patel DK, Bhabhor K (2020) Effect of nitrogen, phosphorus and sulphur on yield, nutrient uptake and soil fertility after harvest of mustard (*Brassica juncea* L.). Int J Curr Microbiol Appl Sci 9(6):3506–3512

Choudhary S, Bhogal NS (2017) Effect of boron on yield, quality and its uptake in Indian mustard (*Brassica juncea* L) Genotypes. Ann Plant Soil Res 19:394–397

Connorton JM, Balk J (2019) Iron biofortification of staple crops: lessons and challenges in plant genetics. Plant Cell Physiol 60(7):1447–1456

De Sousa Lima F, Nascimento CW, da Silva SC (2015) Zinc fertilization as an alternative to increase the concentration of micronutrients in edible parts of vegetables. Revista Brasileira de Ciências Agrárias 10(3):403–408

Dhaliwal SS, Sharma V, Shukla AK, Kaur M, Verma V, Sandhu PS, Alsuhaibani AM, Gaber A, Hossain A (2022) Biofortification of oil quality, yield, and nutrient uptake in Indian mustard (*Brassica Juncea* L.) by foliar application of boron and nitrogen. Front Plant Sci 13:976391

Di Gioia F, Petropoulos SA, Ozores-Hampton M, Morgan K, Rosskopf EN (2019) Zinc and iron agronomic biofortification of brassicaceae microgreens. Agronomy 9(11):677

Doolette CL, Read TL, Li C, Scheckel KG, Donner E, Kopittke PM, Schjoerring JK, Lombi E (2018) Foliar application of zinc sulphate and zinc EDTA to wheat leaves: differences in mobility, distribution, and speciation. J Exp Bot 69(18):4469–4481

Farooq M, Usman M, Nadeem F, ur Rehman H, Wahid A, Basra SM, Siddique KH (2019) Seed priming in field crops: potential benefits, adoption and challenges. Crop Pasture Sci 70(9):731–771

Frossard E, Bucher M, Machler F, Mozafar A, Hurrell R (2000) Potential for increasing the content and bioavailability of Fe, Zn and Ca in plants for human nutrition. J Sci Food Agric 80:861–879

Garg M, Sharma N, Sharma S, Kapoor P, Kumar A, Chunduri V, Arora P (2018) Biofortified crops generated by breeding, agronomy, and transgenic approaches are improving lives of millions of people around the world. Front Nutr 5(12):1–33

Gasic K, Korban SS (2007) Transgenic Indian mustard (*Brassica juncea*) plants expressing an Arabidopsis phytochelatin synthase (AtPCS1) exhibit enhanced As and Cd tolerance. Plant Mol Biol 64:361–369

Gautam A, Sharma P, Ashokhan S, Yaacob JS, Kumar V, Guleria P (2023) Magnesium oxide nanoparticles improved vegetative growth and enhanced productivity, biochemical potency and storage stability of harvested mustard seeds. Environ Res 229:116023

Golubkina N, Kekina H, Caruso G (2018) Yield, quality and antioxidant properties of Indian mustard (*Brassica Juncea* L.) in response to foliar biofortification with selenium and iodine. Plants 7(4):80

Golob A, Novak T, Maršić NK, Šircelj H, Stibilj V, Jerše A, Kroflič A, Germ M (2020) Biofortification with selenium and iodine changes morphological properties of Brassica oleracea L. var. gongylodes and increases their contents in tubers. Plant Physiol Biochem, 150: 234–243

Gonnella M, Renna M, D'Imperio M, Santamaria P, Serio F (2019) Iodine biofortification of four brassica genotypes is effective already at low rates of potassium iodate. Nutrients 11(2):451

Gustiar F, Munandar M, Ningsih SW, Ammar M (2020) Biofortification of calcium on mustard (*Brassica juncea* L.) and lettuce (*Lactuca sativa*) cultivated in floating hydroponic system. Buletin Agroteknologi 1(1):27–36

Hong CL, Weng HX, Qin YC, Yan AL, Xie LL (2008) Transfer of iodine from soil to vegetables by applying exogenous iodine. Agron Sustain Dev 28:575–583

Huang J, Hirji R, Adam L, Rozwadowski KL, Hammerlindl JK, Keller WA, Selvaraj G (2000) Genetic engineering of glycinebetaine production toward enhancing stress tolerance in plants: metabolic limitations. Plant Physiol 122(3):747–756

Hussain S, Ahmed S, Akram W, Li G, Yasin NA (2023) Selenium seed priming enhanced the growth of salt-stressed Brassica rapa L. through improving plant nutrition and the antioxidant system. Front Plant Sci 13:1050359

Ismail AM, Heuer S, Thomson MJ, Wissuwa M (2007) Genetic and genomic approaches to develop rice germplasm for problem soils. Plant Soil 65:547–570

Jha AB, Warkentin TD (2020) Biofortification of pulse crops: status and future perspectives. Plants 9(1):73

Kandil EE, El-Banna AA, Tabl DM, Mackled MI, Ghareeb RY, Al-Huqail AA, Ali HM, Jebril J, Abdelsalam NR (2022) Zinc nutrition responses to agronomic and yield traits, kernel quality, and pollen viability in rice (*Oryza sativa* L.). Front Plant Sci 13:1–1

Kanrar S, Venkateswari J, Kirti P, Chopra V (2002) Transgenic Indian mustard (*Brassica juncea*) with resistance to the mustard aphid (*Lipaphis erysimi Kalt.*). Plant Cell Rep 20:976–981

Kant S, Patel KK, Maurya SK (2025) Biofortification challenges and opportunities. Eco Farm 5(1):4–6

Kiran K (2020) Advanced approaches for biofortification. In: Advances in agri-food biotechnology. Springer, pp 29–55

Kumar D, Shekhar S, Bisht S, Kumar V, Varma A, Kumar M (2015) Ectopic overexpression of lectin in transgenic *Brassica juncea* plants exhibit resistance to fungal phytopathogen and showed alleviation to salt and drought stress. J Bioeng Biomed Sci 5(1):147

Kumar K, Gambhir G, Dass A, Tripathi AK, Singh A, Jha AK, Yadava P, Choudhary M, Rakshit S (2020) Genetically modified crops: current status and future prospects. Planta 251(4):1–27

Lal MK, Kumar A, Kardile HB, Raigond P, Changan SS, Thakur N, Dutt S, Tiwari RK, Chourasia KN, Kumar D, Singh B (2020) Biofortification of vegetables. In: Advances in agri-food biotechnology, pp 105–129

Li X, Wu Y, Li B, Yang Y, Yang Y (2018) Selenium accumulation characteristics and biofortification potentiality in turnip (*Brassica rapa* var. rapa) supplied with selenite or selenate. Front Plant Sci 8:2207

Liao X, Rao S, Yu T, Zhu Z, Yang X, Xue H, Gou Y, Cheng S, Xu F (2021) Selenium yeast promoted the Se accumulation, nutrient quality and antioxidant system of cabbage (*Brassica oleracea* var. *capitata* L.). Plant Signal Behav 16(6):1907042

Liu Z, Hirani AH, McVetty PB, Daayf F, Quiros CF, Li G (2012) Reducing progoitrin and enriching glucoraphanin in *Braasica napus* seeds through silencing of the GSL-ALK gene family. Plant Mol Biol 79:179–189

Lutts S, Benincasa P, Wojtyla L, Kubala S, Pace R, Lechowska K, Quinet M, Garnczarska M (2016) Seed priming: new comprehensive approaches for an old empirical technique. In: New challenges in seed biology-basic and translational research driving seed technology, vol 46, p 64420

Malik KA, Maqbool A (2020) Transgenic crops for biofortification. Front Sustain food Syst 4:571402

Manaf A, Raheel M, Sher A, Sattar A, Ul-Allah S, Qayyum A, Hussain Q (2019) Interactive effect of zinc fertilization and cultivar on yield and nutritional attributes of canola (*Brassica napus* L.). J Soil Sci Plant Nutr 19:671–667

Miliauskienė J, Brazaitytė A, Vaštakaitė-Kairienė V, Sutulienė R, Laužikė K (2022) Zinc biofortification of hydroponic mustard microgreens grown under different red and blue LED lighting ratios. Biol Life Sci Forum 2:2–7

Misra S, Gedamu L (1989) Heavy metal tolerant transgenic *Brassica napus* L. and *Nicotiana tabacum* L. Theor Appl Genet 78:161–168

Molnár K, Biró-Janka B, Domokos E, Nyárádi II, Fodorpataki L, Stoie A, Duda MM (2023) Effects of seed priming and foliar treatment with ascorbate, cysteine, and triacontanol on canola (*Brassica napus* L.) under field conditions. Horticulturae 9(2):207

Mondal KK, Chatterjee SC, Viswakarma N, Bhattacharya RC, Grover A (2003) Chitinase-mediated inhibitory activity of Brassica transgenic on growth of *Alternaria brassicae*. Curr Microbiol 47:171–173

Montes-Bayón M, LeDuc DL, Terry N, Caruso JA (2002) Selenium speciation in wild type and genetically modified Se accumulating plants with HPLC separation and ICP-MS/ES-MS detection. J Anal At Spectrom 17:872–879

Nath UK, Wilmer JA, Wallington EJ, Becker HC, Möllers C (2009) Increasing erucic acid content through combination of endogenous low polyunsaturated fatty acids alleles with Ld-LPAAT+ Bn-fae 1 transgenes in rapeseed (*Brassica napus L.*). Theor Appl Genet 118:765–773

Ojok J, Omara P, Opolot E, Odongo W, Olum S, Gijs DL, Gellynck X, De Steur H, Ongeng D (2019) Iodine agronomic biofortification of cabbage (*Brassica oleracea* var. *capitata*) and cowpea (*Vigna unguiculata* L.) is effective under farmer field conditions. Agronomy 9(12):797

Pannico A, El-Nakhel C, Graziani G, Kyriacou MC, Giordano M, Soteriou GA, Zarrelli A, Ritieni A, De Pascale S, Rouphael Y (2020) Selenium biofortification impacts the nutritive value, polyphenolic content, and bioactive constitution of variable microgreens genotypes. Antioxidants 9(4):272

Pavithra A, Kaleeswari RK, Chitdeshwari T, Priya RS, Uma D, Lourdusamy DK (2022) Agronomic biofortification of calcium in cabbage (*Brassica oleracea* var capitata) applied with different sources of liming in Ca deficient acidic soil of Coonoor, The Nilgiris (Typic Dystropept). J Appl Nat Sci 14(4):1–1

Phooi CL, Azman EA, Ismail R (2023) Effect of priming on *Brassica rapa* subsp. *chinensis* (Bok Choy) seeds germination. J Agritech 43(4):288–296

Pilon-Smits EA, Zhu YL, Sears T, Terry N (2000) Overexpression of glutathione reductase in *Brassica juncea*: effects on cadmium accumulation and tolerance. Physiol Plant 110(4):455–460

Poblaciones MJ, Rengel Z (2017) Combined foliar selenium and zinc biofortification in field pea (*Pisum sativum*): accumulation and bioavailability in raw and cooked grains. Crop Pasture Sci 68(3):265–271

Prasad BV, Mohanta S, Rahaman S, Bareily P (2015) Bio-fortification in horticultural crops. J Agric Eng Food Technol 2(2):95–99

Puccinelli M, Malorgio F, Rosellini I, Pezzarossa B (2019) Production of selenium-biofortified microgreens from selenium-enriched seeds of basil. J Sci Food Agric 99(12):5601–5605

Qamarunnisa S, Jamil I, Raza S, Azhar A, Naqvi SM (2015) Genetic improvement of canola against abiotic stress through incorporation of DREB gene. Asian J Agric Biol 3(3):77–104

Raiola A, Errico A, Petruk G, Monti DM, Barone A, Rigano MM (2018) Bioactive compounds in Brassicaceae vegetables with a role in the prevention of chronic diseases. Molecules 23(1):15

Raj AB, Raj SK (2019) Seed priming: an approach towards agricultural sustainability. J Appl Nat Sci 11(1):227–234

Rajwanshi R, Kumar D, Yusuf MA, DebRoy S, Sarin NB (2016) Stress-inducible overexpression of glyoxalase I is preferable to its constitutive overexpression for abiotic stress tolerance in transgenic *Brassica juncea*. Mol Breed 36:1–5

Rivera Martín AM, Broadley MR, Poblaciones Suárez-Bárcena MJ (2020) Soil and foliar zinc application to biofortify broccoli (*Brassica oleracea* var. *italica* L.): effects on the zinc concentration and bioavailability. Plant Soil Environ 66(3):113–118

Rugeles-Reyes SM, Cecilio Filho AB, Lopez Aguilar MA, Silva PH (2019) Foliar application of zinc in the agronomic biofortification of arugula. Food Sci Technol 39(4):1011–1017

Rustagi A, Kumar D, Shekhar S, Yusuf MA, Misra S, Sarin NB (2014) Transgenic *Brassica juncea* plants expressing MsrA1, a synthetic cationic antimicrobial peptide, exhibit resistance to fungal phytopathogens. Mol Biotechnol 56:535–545

Safdar ME, Qamar R, Javed A, Nadeem MA, Javeed HM, Farooq S, Głowacka A, Michałek S, Alwahibi MS, Elshikh MS, Ahmed MA (2023) Combined application of boron and zinc improves seed and oil yields and oil quality of oilseed rape (*Brassica napus* L.). Agronomy 13(8):2–18

Saltzman A, Birol E, Bouis HE, Boy E, De Moura FF, Islam Y (2014) Biofortification: progress toward a more nourishing future. Glob Food Secur 2(1):9–17

Shah N, Anwar S, Xu J, Hou Z, Salah A, Khan S, Gong J, Shang Z, Qian L, Zhang C (2018) The response of transgenic *Brassica* species to salt stress: a review. Biotechnol Lett 40:1159–1165

Sharma M, Sahni R, Kansal R, Koundal KR (2004) Transformation of oilseed mustard *Brassica juncea (L.) Czern* and *Coss cv. Pusajaikisan* with snowdrop lectin gene. Indian J Biotechnol 3:97–102

Singh U, Praharaj CS, Singh SS, Singh NP (2016) Biofortification of food crops. Springer, pp 51–60

Smoleń S (2012) Foliar nutrition: current state of knowledge and opportunities. In: Advances in citrus nutrition, pp 41–58

Sujith Kumar MS, Mawlong I, Rani R (2020) Biofortification of Brassicas for quality improvement. In: *Brassica* improvement: molecular, genetics and genomic perspectives. Springer, pp 127–145

Sumanth KS, Anusha T, Prakash S (2022) Transgenic crops: present status, problems and future prospects. J Pharm Innov 11:753–758

Tomasi N, Pinton R, Dalla Costa L, Cortella G, Terzano R, Mimmo T, Scampicchio M, Cesco S (2015) New ‘solutions’ for floating cultivation system of ready-to-eat salad: a review. Trends Food Sci Technol 46(2):267–276

Treftz C, Omaye ST (2016) Hydroponics: potential for augmenting sustainable food production in non-arable regions. Nutr Food Sci 46(5):672–684

Van Der Straeten D, Bhullar NK, De Steur H, Gruissem W, MacKenzie D, Pfeiffer W, Qaim M, Slamet-Loedin I, Strobbe S, Tohme J, Trijatmiko KR (2020) Multiplying the efficiency and impact of biofortification through metabolic engineering. Nat Commun 11(1):5203

Wang P, Xiong X, Zhang X, Wu G, Liu F (2022) A review of erucic acid production in Brassicaceae oilseeds: progress and prospects for the genetic engineering of high and low-erucic acid rapeseeds (*Brassica napus*). Front Plant Sci 11(13):899076

Watanabe KN, Sassa Y, Suda E, Chen CH, Inaba M, Kikuchi A (2005) Global political, economic, social and technological issues on transgenic crops – review. Plant Biotechnol 22(5):515–522

Waters BM, Sankaran RP (2011) Moving micronutrients from the soil to the seeds: genes and physiological processes from a biofortification perspective. Plant Sci 180(4):562–574

Weng HX, Weng JK, Yan AL, Hong CL, Yong WB, Qin YC (2008) Increment of iodine content in vegetable plants by applying iodized fertilizer and the residual characteristics of iodine in soil. Biol Trace Elem Res 123:218–228

Wiesner-Reinhold M, Schreiner M, Baldermann S, Schwarz D, Hanschen FS, Kipp AP, Rowan DD, Bentley-Hewitt KL, McKenzie MJ (2017) Mechanisms of selenium enrichment and measurement in brassicaceous vegetables, and their application to human health. Front Plant Sci 8:1365

Woch W, Hawrylak-Nowak B (2019) Selected antioxidant properties of alfalfa, radish, and white mustard sprouts biofortified with selenium. Acta Agrobot 72(2):1–11

Yusuf MA, Sarin NB (2007) Antioxidant value addition in human diets: genetic transformation of *Brassica juncea* with γ-TMT gene for increased α-tocopherol content. Transgenic Res 16:109–113

Zhang K, He J, Liu L, Xie R, Qiu L, Li X, Yuan W, Chen K, Yin Y, Kyaw MM, San AA (2020) A convenient, rapid and efficient method for establishing transgenic lines of *Brassica napus*. Plant Methods 16:1–4

Zhu W, Zuo R, Zhou R, Huang J, Tang M, Cheng X, Liu Y, Tong C, Xiang Y, Dong C, Liu S (2016) Vacuolar iron transporter BnMEB2 is involved in enhancing iron ftolerance of *Brassica napus*. Front Plant Sci 7:1353

Role of Brassica Crops in Climate-Smart Agriculture: Mitigating Environmental Challenges

Navjot Rana, Swati Mehta, Pankaj Dahiya,
Shivanshu Ladohia, and Mohamed A. Mattar

Abstract

Brassica crops such as mustard, rapeseed, cabbage, cauliflower, and broccoli are vital for climate-smart agriculture because they enhance carbon sequestration, soil health, water use efficiency, and resilience to climate extremes. Brassicas naturally manage soilborne pests and pathogens with their biofumigation potential, while their deep root systems improve soil aeration and nutrient cycling. When incorporated into conservation agriculture, crop rotation, and intercropping, brassicas enhance soil fertility, weed control, and sustainability. Brassica crops also help in the reduction of soil erosion, improve water retention, and reduce risk by acting as a cover crop along with improving soil fertility. The brassica crops enhance carbon sequestration and add organic carbon to the soil, also enhancing water use efficiency. Brassicas lead to a reduction in greenhouse gas emissions, such as methane and nitrous oxide, and increase food security by drought-tolerant cultivars that maintain steady yields under limited moisture conditions. Despite the benefits of growing Brassica, issues such as soil erosion, pest outbreaks, and climate variability threaten Brassica production. Creating stress-tolerant cultivars, implementing precision farming practices, and using

N. Rana (✉) · S. Mehta · P. Dahiya
Department of Agronomy, School of Agriculture, Lovely Professional University, Phagwara, India

S. Ladohia
Department of Agronomy, Chaudhary Sarwan Kumar Himachal Pradesh Krishi Vishvavidyalaya (CSK HPKV), Palampur, HP, India

M. A. Mattar
Department of Agricultural Engineering, College of Food and Agricultural Sciences, King Saud University, Riyadh, Saudi Arabia

Agricultural Engineering Research Institute (AEnRI), Agricultural Research Centre, Giza, Egypt

A. Kumar, B. Koul (eds.), *Crop Improvement Strategies in Brassica species: Applied Science*, https://doi.org/10.1007/978-981-95-3861-4_16

Integrated Pest Management (IPM) are needed to overcome these challenges. Regulatory changes such as carbon credit programs, committed research money, and grower subsidies are also important to facilitate greater adoption of brassicas in CSA. Scientific progress, global cooperation, and farmer initiatives are very necessary in order to realize the full potential of Brassica crops in climate-smart agriculture for the future. The dependence on vegetable oils is reduced by the development of high-yielding and climate-tolerant varieties and has enhanced economic resilience because most of the vegetable oil is imported from abroad. The sustainability efforts are also enhanced by the climate-resilient strategies, digital farming technologies and agroecological approaches.

Keywords

Climate-smart agriculture · *Brassica* · Greenhouse gases · Carbon sequestration · Sustainability · Organic carbon · Water use efficiency · Soil fertility

1 Introduction

Climate-smart agriculture (CSA) is an essential approach to adapting agricultural systems to climate change while maintaining food security, resilience, and sustainability (Matteoli et al. 2020). As climate change results in extreme temperatures, prolonged droughts, and erratic weather conditions, climate-smart agriculture seeks to enhance productivity, enhance adaptability, and reduce greenhouse gas emissions (Raihan 2024). Conservation tillage, agroforestry, climate-resilient crops, and water use efficiency are some of the sustainable practices that lead to this (Kizito et al. 2022). Among various crops that sustain CSA, Brassica crops are important because of their economic, nutritional, and environmental significance (Kandegama et al. 2022). The Brassica crops, which are cultivated extensively in various agroclimatic regions, are mustard, rapeseed, cabbage, cauliflower, broccoli, kale, and Brussels sprouts (Akhatar et al. 2025). These crops are critical to global agricultural systems due to their utility for numerous purposes, including the generation of food, edible oil, animal feed, and biofuel (Friedt et al. 2018). Brassica crops are an attractive choice for farmers seeking hardy and sustainable farming practices due to their quick growth periods and capability to thrive under different climatic conditions (Gillespie 2024). Brassica crops play a very important role in climate-smart agriculture Fig. 1.

Brassica crops support the immune system, digestion, and bone health because these crops are nutritionally rich in vitamins such as vitamin A, C, and K, along with vital minerals like calcium, potassium, and iron (Kumar et al. 2022). Oilseed brassicas such as mustard and rapeseed are key producers of biofuels, edible oil, and animal feed and a great source of protein (Sachan et al. 2024). These crops have high market value; both small and commercial farmers benefit from growing them. These Brassica crops play a crucial role in ensuring economic development and food security (Mpumi et al. 2020). In CSA, one of the key benefits of Brassica is its capability to withstand changing weather patterns. Because of its tolerance to heat,

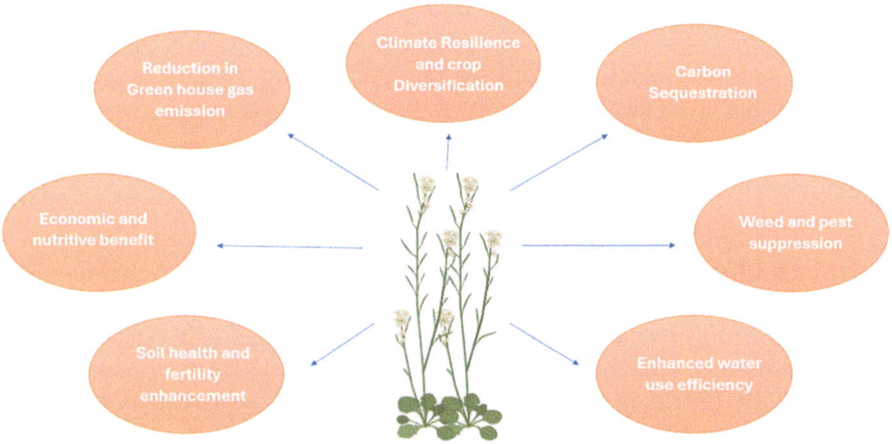

Fig. 1 Role of Brassica crops in climate-smart agriculture

salinity, drought, and waterlogging, most varieties can thrive in environments where other crops may struggle (Panjabi et al. 2019). These crops enhance water retention, boost organic matter content in soil, and reduce soil erosion due to their deep root system.

Some species of Brassica crops help with integrated pest management (IPM) by releasing biofumigant chemicals, such as glucosinolates, which naturally suppress soil-dwelling pests and diseases, reduce the need for chemical pesticides, and encourage sustainable insect control and higher ecological balance (Choudhary et al. 2021). Brassica crops absorb atmospheric carbon and store it in soil and plant biomass, hence, they help in mitigating climate change (Kwiatkowski et al. 2023). Brassica crops lead to enhancing the soil structure and aid in the sequestration of carbon when integrated into conservation agriculture and cover crops (Thapa et al. 2023). Oilseed brassicas provide a renewable substitute for fossil fuels when used to produce biofuel, further lowering greenhouse gas emissions from agriculture (Taheripour et al. 2022).

Integrating Brassica crops into climate-smart agricultural practices such as agroforestry, crop diversification, and intercropping enhances resistance to climate variability while ensuring sustainability and long-term yields (Tripathi et al. 2022). Farmers employing these strategies experience increased ecosystem health, reduced risks, and more consistent yields. In CSA, there are numerous advantages to the promotion of Brassica crops, which include enhanced food security, economic sustainability, and environmental conservation (Sileshi et al. 2023).

As global agriculture continues to be threatened by climate change, incorporating robust and resilient crops like Brassicas into sustainable agricultural systems will be essential to tackling upcoming issues (Shahzad et al. 2021). Future research and policy initiatives should focus on promoting sustainable agriculture practices, developing climate-resilient Brassica varieties, and expanding farmer market opportunities (Singh and Rao 2023). Farming communities can continue to produce food sustainably while leveraging Brassica crops' potential to help them adapt to the effects of climate change by adopting these initiatives (Sileshi et al. 2023). By utilizing Brassica crops in CSA systems, agricultural systems can become more resilient, productive, and environmentally friendly, thereby promoting food security and economic resilience (Goswami et al. 2023). In a world that is changing quickly, this will protect economic resilience and food security.

2 Contribution of Brassica Crops to Climate-Smart Agriculture

Brassica crops improve soil health and water use efficiency and reduce pest and disease infestation with less inputs and match the characteristics of climate-smart agriculture. By providing healthy food and income, they also assist in mitigating carbon sequestration and improving climate change resilience as addressed in Table 1.

Table 1 Key contributions of Brassica crops in climate-smart agriculture

Category	Key contributions	Impact on agriculture	Implementation strategies	References
Carbon sequestration	Absorb atmospheric carbon, enrich soil organic carbon	Reduces greenhouse gas emissions, enhances soil fertility	Use as cover crops, conservation tillage, organic amendments	Fasolo (2024), Singh (2022)
Soil health improvement	Deep-root systems enhance aeration and nutrient cycling	Prevents soil erosion, improves structure and biodiversity	Crop rotation, intercropping, biofumigation techniques	Akchaya et al. (2025), (Goud et al. (2022)
Water use efficiency	Improve moisture retention, reduce water stress	Ensure yield stability in drought-prone regions	Drought-tolerant Brassica varieties, precision irrigation	Tyagi and Haritash (2024), Hamann et al. (2018)
Pest and weed suppression	Biofumigation properties reduce soil pathogens and weed growth	Lowers pesticide use, promotes eco-friendly pest management	Companion planting, integrated pest management (IPM)	Reicosky et al. (2021), Remolif et al. (2023)
Climate resilience	Adaptability to various climates, withstand environmental stress	Maintains yield under extreme weather fluctuations	Breeding stress-resistant cultivars, genetic improvement	Singh and Rao (2023)
Economic and nutritional benefits	High market value, rich in vitamins, minerals, antioxidants	Improves farmer income, reduces dependence on imports	Promoting domestic oilseed production, market support policies	Shahzad et al. (2021), Sileshi et al. (2023)

3 Carbon Sequestration and Greenhouse Gas Mitigation

Brassica crops play a fundamental role in climate-smart agriculture as they reduce greenhouse gas emissions and assist with carbon sequestration. These Brassica crops play a great role in increasing soil organic carbon and decreasing the emission of nitrous oxide (N_2O) and methane (CH_4) making them to becoming imperative elements in agro-ecosystems for sustainable agricultural practices (Shakoor et al. 2022).

The Brassica crops build up soil organic carbon, which is one of the main benefits of these crops. Brassica crops have a deep root system, which stimulates microbial activity, raises soil organic matter levels, and enhances soil structure (Tyagi and Haritash 2024). With the passage of time, the decomposition of Brassica residues goes for decomposition which enhances carbon storage capacity and fertility of soil by adding the organic matter, which is high in carbon (Meena et al. 2020). The Brassica crop leftovers enhance the process of humus development, which stabilizes carbon in the soil and also lowers CO_2 in the atmosphere (Singh 2022).

Brassica crops create the aerobic soil conditions that avoid methane production and lower overall emissions; on the other hand, cattle and flooded rice are the major sources of CH_4 emissions (Beniche et al. 2021). Brassicas have also been discovered in decreasing emissions of N_2O by affecting nitrogen cycling. Brassica root systems enhance their nitrogen use efficiency by minimizing the chances of volatilization and leaching of nitrogen fertilizers (Otchere 2022). Utilizing Brassica crops as part of an integrated nutrient plan can greatly reduce the carbon footprint of farm produce (Mahto et al. 2024). Using Brassica crops as cover crops is well-established method to reduce soil carbon loss. Growing Brassicas as cover crop (like mustard, radish) enhances soil organic matter retention, reduces soil erosion, which ultimately reduces soil degradation and controls the weed infestation (Reicosky et al. 2021). These crops protect the soil from degradation, reducing carbon loss due to erosion and oxidation (Tyagi and Haritash 2024). Brassicas minimize soil nitrogen loss due to their ability to scavenge the nitrogen from the residues in the soil which thereby reduces the N_2O emission (Otchere 2022). In Conservation agriculture, such as no-till or reduced tillage farming, Brassica crops can be integrated, which enhances carbon sequestration and maintains soil health (Thapa et al. 2023). By using Brassica crops in climate-smart agriculture, farmers can increase soil carbon storage, reduce greenhouse gas emissions, and improve sustainability (Cao 2024). Brassicas are vital allies in halting climate change and guaranteeing long-term agricultural resilience because of their contributions to soil conservation, carbon sequestration, and effective nitrogen use (Fasolo 2024).

4 Soil Health and Fertility Management

Deep-root systems found in Brassica crops, such as mustard and rapeseed, are essential for enhancing soil aeration and nutrient cycling (Akchaya et al. 2025). By penetrating compacted soil layers, these roots facilitate water circulation through the soil profile. Better aeration facilitates better gas exchange, which in turn creates an environment that is conducive to the growth of beneficial bacteria that help break down organic matter and release nutrients (Mohamed et al. 2021). When plant residue decomposes, the deep roots also retrieve nutrients from the lower soil layers and transfer them to the upper layer of the soil (Callesen et al. 2016). This lowers nutrient loss through leaching and increases the availability of vital elements like nitrogen, phosphate, and potassium for succeeding crops, and the remaining organic residues enhance soil aggregation, fortifying its structure and halting deterioration (Goud et al. 2022). Crop rotations that include Brassica crops improve soil quality, which eventually raises agricultural production and sustainability.

Brassica crops have a reputation as biofumigation crops because of their high glucosinolate content. As this plant waste breaks down in the soil, glucosinolates are turned into substances beneficial to our plants like isothiocyanates, which naturally suppress weeds, nematodes, and soilborne diseases (Ntalli and Caboni 2017). By helping to control diseases, such as clubroot, Verticillium wilt, and Rhizoctonia root rot, these substances offer a sustainable alternative to synthetic pesticides (Bangarwa and Norsworthy 2017). Biofumigation enhances soil health without leaving toxic residues and builds beneficial microbial communities that suppress plant pathogens as compared to chemical fumigants (Remolif et al. 2023).

Brassica crops help maintain soil by reducing erosion and increasing levels of organic matter. Its heavy cover acts as a shield and helps to prevent soil displacement by limiting surface runoff and rain force (Taylor 2013). The root systems of these crops strengthen soil structure and reduce susceptibility to soil erosion by trapping soil particles, particularly in dry and windy regions. This is because the breakdown of Brassica leftovers after harvest significantly increases the organic matter content of the soil, which increases the soil fertility and the water holding capacity of the soil (Hussain et al. 2024). The impact of Brassica crops on soil organic carbon over time is graphically shown in Fig. 2. Cation exchange capacity

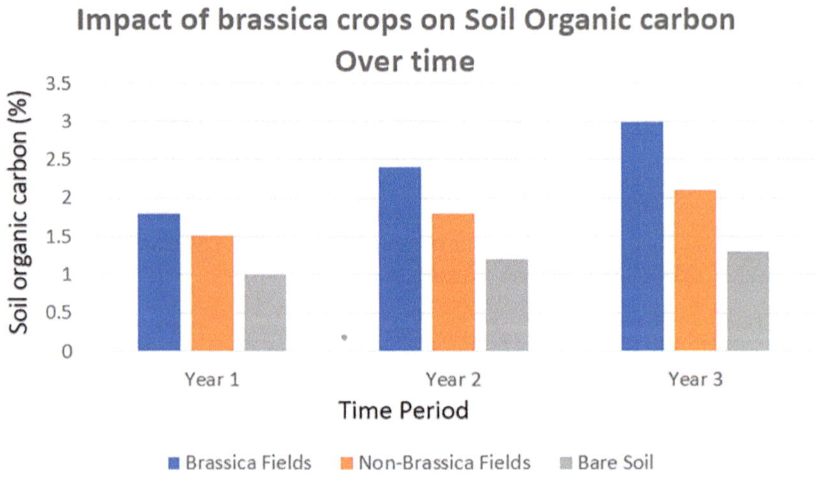

Fig. 2 Impact of Brassica crops on soil organic carbon over time. (Hussain et al. 2024)

of soil is also increased by the higher organic matter in the soil which helps retain essential nutrients for plant development (Callesen et al. 2016). Soil fortified with organic remnants helps it resist environmental stress such as drought and temperature variations for longer periods.

5 Water Use Efficiency and Drought Resilience

Brassica crops, including mustard, rapeseed, and cabbage, are extremely resilient in drought situations and, as such, have developed unique traits that allow them to survive and thrive in dry climates (Sheoran et al. 2024). The deep root system of brassicas is one of their most beneficial features. Unlike many other crops, Brassica crops can withstand drought as they have developed strong, deep root systems that allow them to access water bodies deep into the soil (Tyagi and Haritash 2024). These crops also store compounds, including proline and carbohydrates, to keep their cells hydrated in drought, thus keeping an optimal internal water balance in the cell (Barickman et al. 2020). Another smart survival mechanism to reduce excessive water loss through transpiration is the regulation of the stomata (Sheoran et al. 2024). Some *Brassica* species have even shorter growing seasons, enabling them to reach maturity before drought conditions become more severe (Hamann et al. 2018). Over the years, plant breeders have developed even more drought-resistant Brassica varieties, making Brassicas an excellent alternative for farmers in arid and semi-arid regions (Dar 2022). Due to their natural adaptations, Brassica is a reliable food and oilseed crop type to grow under variable rainfall conditions (Panjabi et al. 2019).

Brassica crops not only provide a drought-tolerant crop to growers but they also improve the soil's ability to store water (Sheoran et al. 2024). Their deep roots will loosen the soil and create natural passages to allow the rainwater to travel downward rather than run off the surface. This reduces erosion and increases the length of time that water is retained in the soil (Tyagi and Haritash 2024). The Brassica plants will also act as a cover crop, protecting the soil from direct sunlight will reduce evaporation, as well as reduce the chance for the soil to dry out too quickly (Kaye and Quemada 2017). The biomass from the broccoli or kale plants that are left in the field from a previous harvest can add organic matter back into the soil, improving porosity and water retention (Brazel 2023). During subsequent growing seasons, the roots of the plants that have grown in the past season will have made it easier to get water to the roots (Hunter et al. 2014). The organic materials released from plants can mitigate soil disease, creating a welcoming environment for other crops. Brassica crops can enhance the farming experience and improve yields in dry environments by holding moisture in the soil if they are introduced as a crop rotation (Brazel 2023).

Brassica crops are revolutionizing rainfed and drought-affected agricultural systems. They are perfect for low-input farming systems when water available for irrigation is scarce or not stable, as they have low water needs (Zhang et al. 2023). The comparative analysis of Brassica crops with non-Brassica crops on soil moisture retention is described in Fig. 3. Brassica crops are frequently grown by farmers

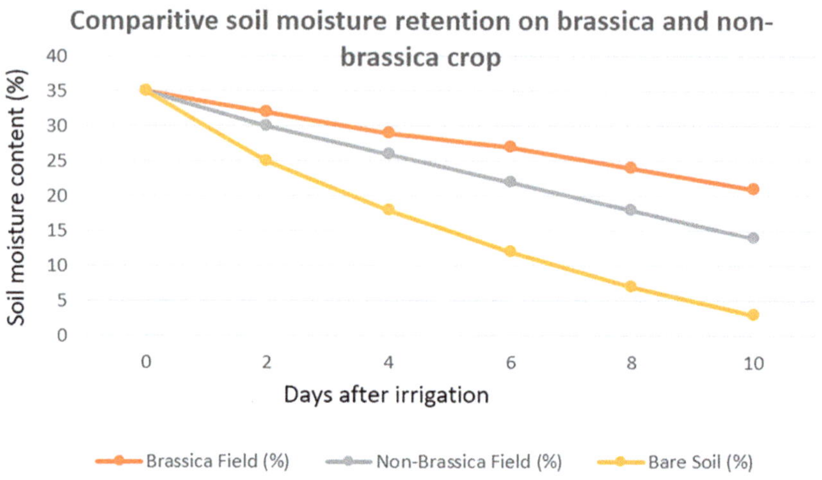

Fig. 3 Comparative soil moisture retention on Brassica and non-Brassica crops. (Zhang et al. 2023)

using conservation agriculture practices that help improve the moisture status of the soil, for instance, mulch, minimal tillage (Thapa et al. 2023). Brassicas complement other crops like cereals and legumes, overall improving soil health, reducing pest and disease pressure, and improving water uptake (Bell et al. 2020). Brassicas also promote biodiversity, reduce irrigation frequency, and aid agriculture's resilience to the effects of climate change, all to the benefit of the environment (Hunter et al. 2014). Farmers in arid areas prefer drought-resilient Brassica cultivars because they consistently produce economic yields despite erratic rainfall. Incorporating these crops into rainfed farming can help producers improve long-term food security and maximize land and water resources (Barickman et al. 2020).

6 Climate-Resilient Crop Diversification

Besides serving as essential food crops, Brassica crops like mustard, rapeseed, and cabbage contribute to sustainable cropping systems and climate resilience (Panjabi et al. 2019). In intercropping systems, brassicas help maximize the use of available resources when grown with other crops (Lal et al. 2019). They have taproots that break up compacted soil, allowing air and water to penetrate the roots of neighboring crops (Tyagi and Haritash 2024). They are self-suppressing weeds and soilborne diseases because brassicas produce their biofumigants that stimulate microbial communities that are effective against weeds and pathogens, which in turn can reduce the need for chemical applications (Van Niekerk 2018). Brassicas also disrupt pest and disease cycles, which can build-up when the same crop is planted in successive seasons, as nature's soil cleaners when used during crop rotations (Bell et al. 2020). The plant material will decompose and help nourish and replenish the soil, and as such, rotating brassicas with grains, legumes, or roots can enhance soil fertility (Callesen et al. 2016). Brassica crops will improve the health of the soil, help with moisture retention, and, when integrated into the cropping system, will provide the opportunity for farmers to manage irregular weather patterns and promote greater overall yields (Jat et al. 2021).

Brassicas are a unique plant family that produces glucosinolates that decompose into biofumigants that help manage soil pathogens such as bacteria, fungi, and nematodes (Van Niekerk 2018). The inclusion of brassicas in your farm system acts to interfere with the habitat of various insect pests, limiting their dispersal and creating a reduced dispersal dynamic in an intercropping system (Lal et al. 2019). Some brassicas also attract beneficial insects to prey on crop damaging pests as an added pest and disease management strategy (Bell et al. 2020). By incorporating brassicas in a crop rotation, farmers can help decrease pesticide reliance, save money, and contribute to environmental sustainability and the health of the crop as well (Lal et al. 2019). Brassica integration into cropping systems highlights the beauty of diversified farming systems, which continue to keep farmers farming in an economically productive way while also creating healthy, resilient fields (Panjabi et al. 2019).

Brassica crops represent reliable agricultural options in a time of increased growing condition volatility due to climate change (Singh and Rao 2023). Their extreme versatility enables production in a variety of growing regions, including cold or temperate climates and dry growing regions; hence, these crops can give farmers reliable crop options during inclement weather (Shahzad et al. 2021). A number of brassicas grow quickly and produce multiple harvests throughout the year, which can provide a constant, reliable food source. *Brassicas* not only work as reliable crops, but they also provide important components of the diet as they are a source of vitamins, minerals, and healthy oils (Kumar et al. 2022). Farmers will have more success in achieving food availability security and can further improve their resilience to uncertain climate impacts while maintaining the ongoing productivity of the farm for future generations when they practice sustainable agriculture that includes Brassicas (Gillespie 2024).

7 Weed and Pest Suppression

Brassica crops occupy a valuable niche in sustainable agriculture because they can act as a natural means of weed suppression (Reicosky et al. 2021). Using the term "allelopathy," we can indicate the plant chemicals that they release in the soil via decaying plant matter and roots, as well as decomposing plant materials, and these chemicals specifically inhibit germination and growth of weeds as a means of limited competition for light, water, and nutrients (Ntalli and Caboni 2017). This capacity for suppression through allelopathy is especially useful in conservation agriculture systems, where the goal is to minimize herbicidal use (Bell et al. 2020). When included as an under-sown cover crop in a cropping system or grown as a companion crop, brassicas such as mustard and rapeseed can help suppress weeds while simultaneously providing soil cover (Kaye and Quemada 2017). The dense leaf structure of these cover crops interrupts light availability, therefore preventing weed germination when combined with soil biomass (Reicosky et al. 2021). When brassica cover crops are incorporated into the soil as a result of decomposition, they will suppress weed germination in the soil (Kaye and Quemada 2017). Using brassica crops for weed management reduces farmers' reliance on synthetic herbicides and therefore economic inputs, while promoting soil health and more sustainable systems.

Brassica crops are not only natural weed suppressants. They also suppress harmful pathogens and nematodes in the soil (Remolif et al. 2023). Upon decomposition, the plant tissue will release bioactive compounds, namely, glucosinolates that will convert to natural fumigants, such as isothiocyanate nematodes (Van Niekerk 2018). This suppresses soilborne pathogens, fungi, bacteria, and nematodes that damage roots and crops (Van Niekerk 2018). Many producers incorporate Brassica waste for green manure purposes to improve biofumigation. Using Brassica crops for biofumigation reduces dependence of farmers on chemicals for soil treatment and is a key practice in organic and regenerative agriculture systems (Choudhary et al. 2021).

Brassicas contribute to a reduction in reliance on chemical pesticides by naturally mitigating weeds, pests, and soilborne diseases (Bell et al. 2020). They pull beneficial insects to some degree, which may contribute to the natural control of pests. The allelopathic characteristics of many brassicas, as well as the biofumigation effects of brassicas, help limit herbicide and soil fumigant applications (Remolif et al. 2023). When used as trap crops, some *Brassica* species limit the application of insecticides that would otherwise be needed to prevent insect pests from overconsuming more susceptible crops (Abbas et al. 2022). Likewise, planting brassicas with other crops promotes biodiversity and contributes to a healthy ecology of natural predators (Choudhary et al. 2021). The inclusion of Brassica crops into a grower's operation allows producers to adopt more sustainable pest management practices without sacrificing productivity (Choudhary et al. 2024). Moving to farming practices that are more sustainable, with fewer chemical inputs and a healthier food system, benefits not only farmers, but also health and the land they farm (Remolif et al. 2023).

8 Economic and Nutritional Benefits

Brassica crops, which encompass mustard, rapeseed, and other green vegetables, have great economic significance for their numerous industrial, food, and oil purposes. Farmers have a constant income from the consistent demand of these crops in the regional and global market (Shahzad et al. 2021). The oil-rich seeds of mustard and rapeseed are particularly valuable because they are a key ingredient in edible oil production (Raboanatahiry et al. 2021). Profitability of the crop is increased further from the general use of the by-products of oil extraction, mustard cake, as an animal feed (Vichare and Morya 2024). These Brassica crops are a great choice for farmers seeking to maximize land use and increase output because of their adaptability to a variety of climatic conditions and relatively short growing seasons (Akhatar et al. 2025). The incorporation of Brassica crops in farming systems reduces pest pressure, improves soil fertility, and increases total productivity, and all these kinds of factors lead to an increase in profits (Remolif et al. 2023). Consumer interest in healthier foods has fueled demand for Brassica vegetables such as kale, cauliflower, and broccoli, Brassica vegetables are available and marketed in both fresh and processed forms to health-conscious populations (Shinali et al. 2024). The organic Brassica crops have more demand in the market and can fetch a greater price in the market. Farmers can get more profit from Brassica crop production by adopting effective storage methods and post-harvest handling (Jat et al. 2021).

Brassicas provide a good source of essential nutrients as they represent an important aspect of a balanced diet. They are rich in vitamins, minerals, and healthy bioactive compounds that contribute to overall health. These are high in fiber, which helps gut health and promotes good digestion (Kumar et al. 2022). Cruciferous vegetables, such as broccoli, kale, and Brussels sprouts, are particularly high in vitamins A, C, and K, which are all important for immunity, skin health, and blood clotting (Zhang et al. 2025). Brassica crops have one of the most impressive nutritional advantages is the glucosinolate concentration (Li et al. 2024). Glucosinolate is a group of sulfur-containing chemicals that have been well-studied for their potential cancer-preventive properties (Choudhary et al. 2021). The consumption of glucosinolates breaks down the one type of bioactive compound known as isothiocyanates, which exhibit anti-inflammatory, detoxifying, and antioxidant properties, which can protect cells from oxidative damage and decrease the risk of chronic disease (Li et al. 2024).

Brassica crops are rich in iron, calcium, magnesium, potassium, and other minerals. Potassium helps regulate blood pressure, while calcium and magnesium promote bone health (Kumar et al. 2022). Leafy Brassicas like kale and mustard greens improve hemoglobin levels and prevent anemia as they contain iron. Some Brassica seeds (mustard and rapeseed) provide omega-3 fatty acids, which are good for heart health and relieving inflammation (Sachan et al. 2024). Brassica vegetables have become fundamental to many healthy eating practices, as plant-based diets are becoming a global trend; their integration into a regular eating pattern can help increase overall health and wellbeing of humans, and contribute positively to a healthy diet (Kumar et al. 2022).

Incorporating Brassica crops into sustainable farming practices provides a boost to local agriculture economies (Shahzad et al. 2021). Jobs and better livelihoods are provided by facilitating direct links with farm market to markets and by the production of value-added products (Sileshi et al. 2023). Investing in Brassica crops under sustainable agriculture practices will improve local economies, promote healthier eating by consumers, and reduce dependence on imported seed sources (Kizito et al. 2022).

Brassica crops hold great value for public health and sustainable agriculture due to their high nutrition and economic benefits (Jat et al. 2021). Brassica crops have high market value, nutrient content, and a role in enhancing food security; hence, the climate-smart agriculture is highly dependent on these Brassica crops (Biondi et al. 2021). Increase in the production and consumption of Brassica crops will lead to support the long-term profitable outcome and improve public health (Kizito et al. 2022).

9 Challenges and Future Prospectives of Brassica Crops in Climate-Smart agriculture

9.1 Addressing Climate-Related Challenges: Extreme Weather, Pests, and Diseases

Factors related to climate change elevated temperatures, unpredictable rainfall, periods of drought, and more pests and diseases make Brassica crops more vulnerable. Heat stress can negatively impact seed germination, blooming, and pod development, which reduces yield (Siomos et al. 2022). Climate variability disrupts the growth cycle of crops, which impacts the market window and overall production (Matteoli et al. 2020). The complications caused by the pests and diseases exacerbate these complications. Warmer temperatures paired with increased humidity create the ideal conditions for diseases caused by bacteria, fungi (e.g., Alternaria blight, blackleg), and insects (e.g., aphids, flea beetles) to flourish (Van Niekerk, 2018). These biotic stresses significantly impact crop quality and production, and can put growers' economics at risk (Das Laha et al. 2024). In order to tackle these hurdles and ensure the crop productivity, mercantile farmers must utilize biological control agents, integrated pest management (IPM), and climate-adaptive agricultural practices, which include mulching, stress-tolerant crop types, and precision irrigation (Choudhary et al. 2021). Different Brassica crops have different traits and characteristics to tackle the crop change and support climate resilience. The different traits of different Brassica crops are given in Table 2.

10 Breeding and Genetic Advancements for Stress-Resilient Brassica Crops

Through advances in genome editing, biotechnology, and plant breeding, Brassica crops with climate resilience can be developed (Meena et al. 2025). Traditional breeding, as well as genome editing with CRISPR and marker-assisted selection (MAS), are particularly important to develop Brassica cultivars with high yield potential and tolerating abiotic and biotic stress (Saha et al. 2022). The desire for climate-smart Brassica crops is of utmost priority in the breeding goals development of drought tolerant cultivars able to effectively uptake soil moisture with deep root systems, improvement in heat tolerance to ensure good pod and flowering development through hot weather, incorporation of disease resistance genes to minimize the use of chemical pesticides and increasing the efficiency in phosphorus and nitrogen uptake and metabolism to improve soil health sustainably (Meena et al. 2025). Research from private seed companies and universities continues to improve the resilience of Brassica cultivars; this is published to ensure a viable yield during a deteriorating climate (Daniel et al. 2023).

Table 2 Climate resilience traits of major Brassica crops

Brassica crop	Drought tolerance	Heat tolerance	Cold tolerance	Soil improvement ability	Pest and disease resistance	References
Mustard (*B. juncea*)	High	Moderate	Low	Enhances organic matter and aeration	Resistant to aphids, moderately resistant to fungal diseases	Quezada-Martinez et al. (2021), Ram et al. (2021)
Rapeseed (*B. napus*)	Moderate	Low	High	Improves soil carbon sequestration	Susceptible to clubroot, resistant to *Alternaria* blight	Quezada-Martinez et al. (2021)
Cabbage (*B. oleracea* var. *capitata*)	Low	Low	High	Enhances microbial activity and nitrogen cycling	Moderate resistance to black rot and cabbage worms	Moreb et al. (2020)
Cauliflower (*B. oleracea* var. *botrytis*)	Low	Moderate	High	Helps suppress soil nematodes	Susceptible to clubroot, resistant to downy mildew	Singh, et al. (2018)
Broccoli (*B. oleracea* var. *italica*)	Moderate	Low	High	Improves nitrogen efficiency	Resistant to some fungal pathogens, moderate aphid resistance	Nagraj et al. (2020)

11 Policies Interventions to Encourage Climate-Smart Agriculture's Use of Brassica Crops

To improve the contribution of Brassica crops to climate-smart agriculture, we will require effective national and international policy support (Singh & Rao, 2023). The role of Brassica and related crops in farming systems, the government support that would be most relevant to this goal is defined as subsidies and cash payments to directly encourage farmers to adopt these adaptive Brassica cultivars to increase funding for research aimed at developing more sustainable practices and improved stress resistant cultivars and to develop irrigation and soil conservation programs which would enhance the productivity and moisture retention on the soil and also develop policies that support price stability and market access which are essential for the long-term economic viability of Brassica crop production (Sileshi et al. 2023). Through their diverse uses, Brassica crops can also enhance their contributions to climate resilience and sustainable agriculture by being integrated into agroforestry systems and crop rotation systems, as well as through carbon sequestration practices.

12 Future Outlook of the Use of Brassica Crops in Climate-Smart Agriculture

Brassica crops are perfectly positioned for their adaptability, their potential economic significance, and their environmental benefits; however, realizing the full potential of Brassica crops is a complex endeavor that requires synergies among technology, science, policy, and farmer engagement (Raboanatahiry et al. 2021). Farmers' ability to apply inputs will be enhanced by precision agriculture with remote sensing technology, soil moisture probes, and AI-driven insect trapping (Aziz et al. 2025). Real-time crop monitoring via drones or satellite images can also limit potential yield losses caused by crop diseases and nutrient deficiencies (Ali et al. 2024). Automated and precision irrigation systems will reduce the water footprint of crops and encourage Brassica cropping in rainfed and semi-arid regions (Ali et al. 2024). Each of these developments enhances crop resilience under variable climate conditions while also driving the efficiency of production inputs.

Brassica crops have an advantage in climate change mitigation because they have a high potential for carbon sequestration (Tyagi and Haritash 2024). By building organic matter in soils, preventing degradation of soils, and sequestering carbon dioxide from the atmosphere, they contribute to improving soil health when used in cover crop systems (Meena et al. 2025). The deep rooting system they generate increases soil health through the enhanced activity of soil microbes (Tyagi and Haritash 2024). If the federal government develops a carbon credit market for farmers who implement regenerative agriculture with sustainable soil management, the role of Brassica as a crop in regenerative agriculture or climate-smart agriculture would become more mainstream and part of the broader "solution" of climate change (Wilson et al. 2023). Brassica crops allow producers to reduce greenhouse gas emissions while maintaining productivity. Brassica crops, when included in multiple cropping systems, can contribute to agricultural resiliency in the face of climate extremes and potentially reduce pest and disease pressures and reliance on synthetic pesticides when using crop rotation and intercropping practices (Thapa et al. 2023). Research should be directed at assessing the best companion crops for improving soil structure, suppressing weed populations, and cycling nutrients (Kommula, 2023). By incorporating Brassica crops alongside other companion crops such as legumes, cereals, and root crops, the agriculture industry can contribute to a more sustainable (and less variable) soil fertility (Lal et al. 2019).

Collaboration among seed companies, research institutions, universities, and policymakers needs to be strengthened in order to enable innovation (Singh and Rao 2023). International collaborations can help enhance plant germplasm, technology transfer, and genetics to ensure that farmers have access to climate-adaptive Brassica varieties globally (Martins et al. 2024). Research and development investment in stress-tolerant breeding, sustainable agronomic practices, and pest-resistant Brassica varieties must be increased to address long-term agricultural sustainability and global food security (Kommula 2023). There lies the potential for Brassica crops supporting climate-smart agriculture as long as science, regulations, and farmer engagement align on progress (Singh and Rao 2023). Brassica crops still have significant potential to contribute positively to food security, reduce impacts of climate change, and contribute to the overall resilience of agricultural systems as we consider new technologies, sustainable practices, and financial motivation (Raihan 2024). The next phase of agricultural issues relating to climate change, agricultural sustainability, will depend on the capacity of brassicas (Kwiatkowski et al. 2023).

13 Conclusion

Brassica crops play a significant role in climate-smart agriculture (CSA), directly contributing to soil health, water productivity, climate resiliency, and carbon sequestration. Planting Brassica crops as part of conservation agriculture, crop rotation, and other cropping systems is important for soil fertility improvement, pest management, and productivity all vital aspects related to moral-based agricultural systems given numerous potential outcomes for soil health. Brassica crops possess biofumigation capabilities, which assist in the natural management of soilborne pathogens. As a deep-rooted crop with unique nutrient cycling, Brassica crops promote soil aeration. Brassica crops also aid in mitigating climate change risk by serving as cover crops, mitigating soil erosion, increasing soil water infiltration, and improving soil carbon levels. They assist in reducing greenhouse gases by sequestering carbon from the atmosphere and decreasing nitrous oxide and methane emissions. Drought-tolerant cultivars also produce stable harvests during water-limited growing seasons. Breeding or growing stress-tolerant cultivars, using precision agriculture practices, and working with today's policies that enable practices like tax incentive structures, research funding, or even carbon credits for agriculture will help move us along towards addressing these issues. To continue maximizing the potential of Brassica crops to contribute to food security, climate.

References

Abbas M, Saleem M, Hussain D, Ramzan M, Saleem MJ, Abbas S, Hussain N et al (2022) Review on integrated disease and pest management of field crops. Int J Trop Insect Sci 42(5):3235–3243

Akchaya K, Parasuraman P, Pandian K, Vijayakumar S, Thirukumaran K, Mustaffa MRAF, Choudhary AK (2025) Boosting resource use efficiency, soil fertility, food security, ecosystem services, and climate resilience with legume intercropping: a review. Front Sustain Food Syst 9:1527256

Akhatar J, Upadhyay P, Kumar H (2025) Crop cultivation and hybrid seed production strategies in rapeseed-mustard. In: Kumar H, Upadhyay P (eds) Hybrid seed production for boosting crop yields: applications, challenges and opportunities. Springer, Singapore, pp 177–224

Ali ZA, Deng D, Shaikh MK, Hasan R, Khan MA (2024) AI-based UAV swarms for monitoring and disease identification of Brassica plants using machine learning: a review. Comput Syst Sci Eng 48(1)

Aziz D, Rafiq S, Saini P, Ahad I, Gonal B, Rehman SA, Iliya MN (2025) Remote sensing and artificial intelligence: revolutionizing pest management in agriculture. Front Sustain Food Syst 9:1551460

Bangarwa SK, Norsworthy JK (2017) Glucosinolate and isothiocyanate production for weed control in plasticulture production system. In: Mérillon JM (ed) Glucosinolates, pp 201–235

Barickman TC, Ku KM, Sams CE (2020) Differing precision irrigation thresholds for kale (Brassica oleracea L. var. acephala) induces changes in physiological performance, metabolites, and yield. Environ Exp Bot 180:104253

Bell LW, Watt LJ, Stutz RS (2020) Forage brassicas have potential for wider use in drier, mixed crop–livestock farming systems across Australia. Crop Pasture Sci 71(10):924–943

Beniche I, El Bari H, Siles JA, Chica AF, Martín MÁ (2021) Methane production by anaerobic co-digestion of mixed agricultural waste: cabbage and cauliflower. Environ Technol 42(28):4550–4558

Biondi F, Balducci F, Capocasa F, Visciglio M, Mei E, Vagnoni M, Mazzoni L (2021) Environmental conditions and agronomical factors influencing the levels of phytochemicals in Brassica vegetables responsible for nutritional and sensorial properties. Appl Sci 11(4):1927

Brazel SR (2023) The influence of overwatering, underwatering, and waterlogging on the growth of kale (*Brassica oleracea* var. *acephala*). MS thesis, Mississippi State University

Callesen I, Harrison R, Stupak I, Hatten J, Raulund-Rasmussen K, Boyle J, Zabowski D (2016) Carbon storage and nutrient mobilization from soil minerals by deep roots and rhizospheres. For Ecol Manag 359:322–331

Cao M (2024) Soil matters: evaluating soil water dynamics and soil greenhouse gas emissions under climate-smart agriculture. PhD thesis, University of British Columbia

Choudhary V, Kumar KK, Sahu BH, Muthappa SK, Das AD (2021) Agronomic innovations in biotic stress management and its combined effect with abiotic stresses in crop production. Indian J Agron 66(5):237–257

Choudhary RL, Jat RS, Singh HV, Dotaniya ML, Meena MK, Meena VD, Rai PK (2024) Low budget natural farming practices for cultivation of rapeseed and mustard. In: Meena RK, Ram H, Saxena A, Kumar R (eds) Natural farming: basics and application. Venture Graphics, pp 122–133

Daniel KAM, Muindi EMD, Muti SMD (2023) Cabbage (*Brassica oleracea*) production in Kenya: a review of its economic importance, ecological requirement and production constraints. Int J Plant Soil Sci 35(18):245–254

Dar ZA (2022) Genetic analysis of drought tolerance in Indian mustard [*Brassica juncea* (L.) Czern & Coss.]. PhD thesis, Sher-e-Kashmir University of Agricultural Sciences and Technology of Jammu

Das Laha S, Kundu A, Podder S (2024) Impact of biotic stresses on the Brassicaceae family and opportunities for crop improvement by exploiting genotyping traits. Planta 259(5):97

Fasolo A (2024) Regenerating soil fertility through cover crops and soil conservation approach: experiences in adoption and evaluation of these innovative techniques. PhD thesis, University of Padua, Italy

Friedt W, Tu J, Fu T (2018) Academic and economic importance of *Brassica napus* rapeseed. In: Snowdon R, Stein N (eds) The *Brassica napus* Genome. Springer, Cham, pp 1–20

Gillespie S (2024) Practical regeneration: realistic strategies for climate smart agriculture. Friesen Press

Goswami M, Gupta AK, Kishan R, Baidya S, Khan YI, Prakash S, Nautiyal S (2023) An evaluation of climate resilient agricultural practices in India: a narrative synthesis of literature. Environ Sustain 6(1):7–23

Goud BR, Raghavendra M, Prasad PS, Hatti V, Halli HM, Nayaka GV, Rajpoot SK (2022) Sustainable management and restoration of the fertility of damaged soils. Agric Issues Policies 113

Hamann E, Weis AE, Franks SJ (2018) Two decades of evolutionary changes in *Brassica rapa* in response to fluctuations in precipitation and severe drought. Evolution 72(12):2682–2696

Hunter PJ, Teakle GR, Bending GD (2014) Root traits and microbial community interactions in relation to phosphorus availability and acquisition, with particular reference to *Brassica*. Front Plant Sci 5:27

Hussain A, Elkarmout AF, Mansour EZ, Awais M, Usman M, Ahmad H, Ahmad T (2024) An environment friendly practice, the climate smart agriculture crop production and soil management systems: a review. J Sustain Agric Environ Sci 3(3):101–124
Jat RS, Choudhary RL, Singh HV, Meena MK, Singh VV, Rai PK (2021) Sustainability, productivity, profitability and soil health with conservation agriculture based sustainable intensification of oilseed brassica production system. Sci Rep 11(1):13366
Kandegama WWW, Rathnayake RMPJ, Baig MB, Behnassi M (2022) Impacts of climate change on horticultural crop production in Sri Lanka and the potential of climate-smart agriculture in enhancing food security and resilience. In: Behnassi M, Barak G (eds) Food security and climate-smart food systems: building resilience for the global South. Springer, Cham, pp 67–97
Kaye JP, Quemada M (2017) Using cover crops to mitigate and adapt to climate change. A review. Agron Sustain Dev 37:4
Kizito F, Chikowo R, Kimaro A, Swai E (2022) Soil and water conservation for climate-resilient agriculture. In: Sustainable agricultural intensification: a handbook for practitioners in East and Southern Africa. CABI, London, pp 62–80
Kommula D (2023) Plant–plant interactions in *Brassica* and legumes intercropping. MSc thesis, Wageningen University and Research, Department of Plant Breeding, Wageningen
Kumar S, Sharma S, Kumar V, Sharma R, Minhas A, Boddu R (2022) Cruciferous vegetables: a mine of phytonutrients for functional and nutraceutical enrichment. In: Gupta VK, Treichel H, Pandey A (eds) Current advances for development of functional foods modulating inflammation and oxidative stress. Academic Press, pp 401–426
Kwiatkowski CA, Pawłowska M, Harasim E, Pawłowski L (2023) Strategies of climate change mitigation in agriculture plant production – a critical review. Energies 16(10):4225
Lal BL, Rana KS, Rana DS, Shivay YS, Sharma DK, Meena BP, Gautam P (2019) Biomass, yield, quality and moisture use of *Brassica carinata* as influenced by intercropping with chickpea under semiarid tropics. J Saudi Soc Agric Sci 18(1):61–71
Li L, Ma P, Nirasawa S, Liu H (2024) Formation, immunomodulatory activities, and enhancement of glucosinolates and sulforaphane in broccoli sprouts: a review for maximizing the health benefits to human. Crit Rev Food Sci Nutr 64(20):7118–7148
Mahto R, Singh RK, Singh JP, Tiwari RK, Vishwakarma DK, Obaidullah AJ et al (2024) Evaluation of *Brassica* species for growth, yield and heat use efficiency under nitrogen nutrition and iron sulphide nanoparticles application. Sci Hortic 333:113278
Martins S, Pereira S, Dinis LT, Brito C (2024) Enhancing olive cultivation resilience: Sustainable long-term and short-term adaptation strategies to alleviate climate change impacts. Horticulturae 10(10):1066
Matteoli F, Schnetzer J, Jacobs H (2020) Climate-smart agriculture (CSA): an integrated approach for climate change management in the agriculture sector. In: Handbook of climate change management: research, leadership, transformation. Springer, Cham, pp 1–29
Meena RS, Kumar S, Yadav GS (2020) Soil carbon sequestration in crop production. In: Nutrient dynamics for sustainable crop production. Springer, Singapore, pp 1–39
Meena BL, Meena HS, Singh VV, Meena MD, Sharma HK, Rai PK et al (2025) Breeding climate resilient rapeseed–mustard in climate change era: current breeding strategies and prospects. In: Breeding climate resilient and future ready oilseed crops. Singapore, Springer, pp 303–332
Mohamed HI, Sofy MR, Almoneafy AA, Abdelhamid MT, Basit A, Sofy AR et al (2021) Role of microorganisms in managing soil fertility and plant nutrition in sustainable agriculture. In: Plant growth-promoting microbes for sustainable biotic and abiotic stress management. Springer, Singapore, pp 93–114
Moreb N, Murphy A, Jaiswal S, Jaiswal AK (2020) Cabbage. In: Nutritional composition and antioxidant properties of fruits and vegetables. Academic Press, Cambridge, pp 33–54
Mpumi N, Machunda RS, Mtei KM, Ndakidemi PA (2020) Selected insect pests of economic importance to *Brassica oleracea*, their control strategies and the potential threat to environmental pollution in Africa. Sustainability 12(9):3824
Nagraj GS, Chouksey A, Jaiswal S, Jaiswal AK (2020) Broccoli. In: Nutritional composition and antioxidant properties of fruits and vegetables. Academic Press, Cambridge, pp 5–17

Ntalli N, Caboni P (2017) A review of isothiocyanates bio-fumigation activity on plant parasitic nematodes. Phytochem Rev 16:827–834

Otchere O (2022) The impact of cover cropping on soil nitrogen availability, crop nitrogen use, and nitrous oxide emissions in a prairie potato-grain crop rotation. Ph.D. dissertation, University of Saskatchewan

Panjabi P, Yadava SK, Kumar N, Bangkim R, Ramchiary N (2019) Breeding *Brassica juncea* and *B. rapa* for sustainable oilseed production in the changing climate: progress and prospects. In: Genomic designing of climate-smart oilseed crops. Springer, Cham, pp 275–369

Quezada-Martinez D, Addo Nyarko CP, Schiessl SV, Mason AS (2021) Using wild relatives and related species to build climate resilience in *Brassica* crops. Theor Appl Genet 134(6):1711–1728

Raboanatahiry N, Li H, Yu L, Li M (2021) Rapeseed (*Brassica napus*): Processing, utilization, and genetic improvement. Agronomy 11(9):1776

Raihan A (2024) A review of recent advances, challenges, and potential future directions of climate-smart agriculture. In: Proceedings of the international conference on climate-smart agriculture

Ram B, Priyamedha MS, Sharma HK, Rani R, Singh KH, Singh VV et al (2021) Development and evaluation of early maturing thermo-tolerant Indian mustard (*Brassica juncea* L. Czern & Coss) genotypes for cultivation in semi-arid region of India. Electron J Plant Breed 12(1):200–206

Reicosky DC, Calegari A, dos Santos DR, Tiecher T (2021) Cover crop mixes for diversity, carbon and conservation agriculture. In: Cover crops and sustainable agriculture. CRC Press, Boca Raton, pp 169–208

Remolif G, Maria E, Schiavon G, Garello M, Buonsenso F, Spadaro DC (2023) Efficacy of biofumigation with essential oils in the control of postharvest rots of nectarines. In: Book of abstracts of the 12th international congress of plant pathology, pp 874–875

Sachan DS, Naimuddin SK, Patra D, Subha L, Senthilkumar T, Chittibomma K, Prasad SV (2024) Advancements in enhancing oil quality in rapeseed and mustard: a comprehensive review. J Exp Agric Int 46(5):181–193

Saha P, Singh S, Aditika BR, Dey SS, Das Saha N et al (2022) Genomic designing for abiotic stress resistant Brassica vegetable crops. In: Genomic designing for abiotic stress resistant vegetable crops. Springer, Cham, pp 153–185

Shahzad A, Ullah S, Dar AA, Sardar MF, Mehmood T, Tufail MA et al (2021) Nexus on climate change: agriculture and possible solution to cope future climate change stresses. Environ Sci Pollut Res 28:14211–14232

Shakoor A, Dar AA, Arif MS, Farooq TH, Yasmeen T, Shahzad SM et al (2022) Do soil conservation practices exceed their relevance as a countermeasure to greenhouse gases emissions and increase crop productivity in agriculture? Sci Total Environ 805:150337

Sheoran AR, Lakra N, Saharan BS, Luhach A, Mandal NK, Seth CS et al (2024) Enduring drought: effects and strategies for Brassica crop resilience. Acad Biol 2(3)

Shinali TS, Zhang Y, Altaf M, Nsabiyeze A, Han Z, Shi S et al (2024) The valorization of wastes and byproducts from cruciferous vegetables: a review on the potential utilization of cabbage, cauliflower, and broccoli byproducts. Foods 13(8):1163

Sileshi GW, Dagar JC, Nath AJ, Kuntashula E (2023) Agroforestry as a climate-smart agriculture: strategic interventions, current practices and policies. In: Agroforestry for sustainable intensification of agriculture in Asia and Africa. Singapore, Springer Nature Singapore, pp 589–640

Singh S (2022) Decomposition of wide C:N ratio crop residues on the release of active carbon pool and available nutrients in Inceptisols of Varana SI. Ph.D. dissertation, Banaras Hindu University, Varanasi

Singh HP, Rao NK (2023) Climate resilient and sustainable development of horticulture – options and opportunities – a review. Int J Innov Hortic 12(1):1–14

Singh BK, Singh B, Singh PM (2018) Breeding cauliflower: a review. Int J Veg Sci 24(1):58–84

Siomos AS, Koularmanis K, Tsouvaltzis P (2022) The impacts of the emerging climate change on broccoli (Brassica oleracea L. var. italica Plenck.) crop. Horticulturae 8(11):1032

Taheripour F, Sajedinia E, Karami O (2022) Oilseed cover crops for sustainable aviation fuels production and reduction in greenhouse gas emissions through land use savings. Front Energy Res 9:790421

Taylor FI (2013) Control of soil borne potato pathogens using Brassica spp. mediated biofumigation. Ph.D. dissertation, University of Glasgow

Thapa VR, Ghimire R, Adhikari KP, Lamichhane S (2023) Soil organic carbon sequestration potential of conservation agriculture in arid and semi-arid regions: a review. J Arid Environ 217:105028

Tripathi HG, Kunin WE, Smith HE, Sallu SM, Maurice S, Machera SD et al (2022) Climate-smart agriculture and trade-offs with biodiversity and crop yield. Front Sustain Food Syst 6:868870

Tyagi A, Haritash AK (2024) Climate-smart agriculture, enhanced agroproduction, and carbon sequestration potential of agroecosystems in India: a meta-analysis. J Environ Stud Sci:1–19

Van Niekerk AJ (2018) A comparison of different strategies to control pests and diseases in Brassica spp. production in the Western Cape. M.Sc. thesis, Nelson Mandela University, George, South Africa

Vichare SA, Morya S (2024) Exploring waste utilization potential: nutritional, functional and medicinal properties of oilseed cakes. Front Food Sci Technol 4:1441029

Wilson K, Mercier S, Myers R (2023) Encouraging farmer adoption of regenerative agriculture practices in the United States. Chicago Council Glob Aff:1–20

Zhang H, Ghahramani A, Ali A, Erbacher A (2023) Cover cropping impacts on soil water and carbon in dryland cropping system. PLoS One 18(6):e0286748

Zhang X, Jia Q, Jia X, Li J, Sun X, Min L et al (2025) Brassica vegetables – an undervalued nutritional goldmine. Hortic Res 12(2):uhae302

Crop Improvement Approaches for the Mitigation of Biotic Stresses in Mustard Crops

Rakesh Roshan Satapathy

Abstract

Globally, mustard is considered to be the most significant vegetable oil source. Due to their great susceptibility to biotic stresses, *brassica* crops substantially impact crop quality and yield. Although different pathogens quickly overcame that resistance by altering their metabolic pathways, breeding approaches were previously employed to build resistance in mustard to minimize biotic stresses. It is imperative to use cutting-edge methods to create biotic stress-tolerant plants to meet the challenge of comprehending the physiological pathways of both pathogens and plants. Various measures are necessary to prevent stress, including selecting genes, metabolites, enzymes, and microbes to survive it. Several strategies, including using DNA markers in breeding, identifying quantitative trait loci, tissue culture, etc., are practical methods to increase mustard's resistance to biotic stress. These methods must be combined to create high-yield, healthy cultivars. Methods such as in vitro tissue culture and identifying quantitative trait loci (QTL) may be more beneficial to increasing mustard's resistance to biotic stress. These methods must be combined to create high-yield, healthy cultivars.

Keywords

Mustard · Biotic stress · Molecular techniques

R. R. Satapathy (✉)
Faculty of Agricultural Sciences, Siksha O Anusandhan University, Bhubaneswar, Odisha, India
e-mail: rakeshsatapathy@soa.ac.in

A. Kumar, B. Koul (eds.), *Crop Improvement Strategies in Brassica species: Applied Science*, https://doi.org/10.1007/978-981-95-3861-4_17

1 Introduction

One significant crop grown in India is *Brassica juncea*, accounting for 90% of the area (Source: www.drmr.res.in). Initially, its cultivation was confined to the Northern belt, but it has spread to non-traditional areas in the country's eastern, Western, and Southern regions. It is an amphidiploid (2n = 36, genome AABB) derived from an interspecific cross of *Brassica nigra* (2n = 16, Genome BB) and *Brassica rapa* (2n = 20, Genome AA) as shown in Fig. 1.

2 Production Scenario

Figures 2, 3, and 4 present that in 2018–2019, the world's rapeseed-mustard output, yield, and area were expected to be 36.5 mha, 72.3 mt, and 1980 kg/ha, respectively. India accounts for 19.80% and 9.80% of the world's total production and acreage,

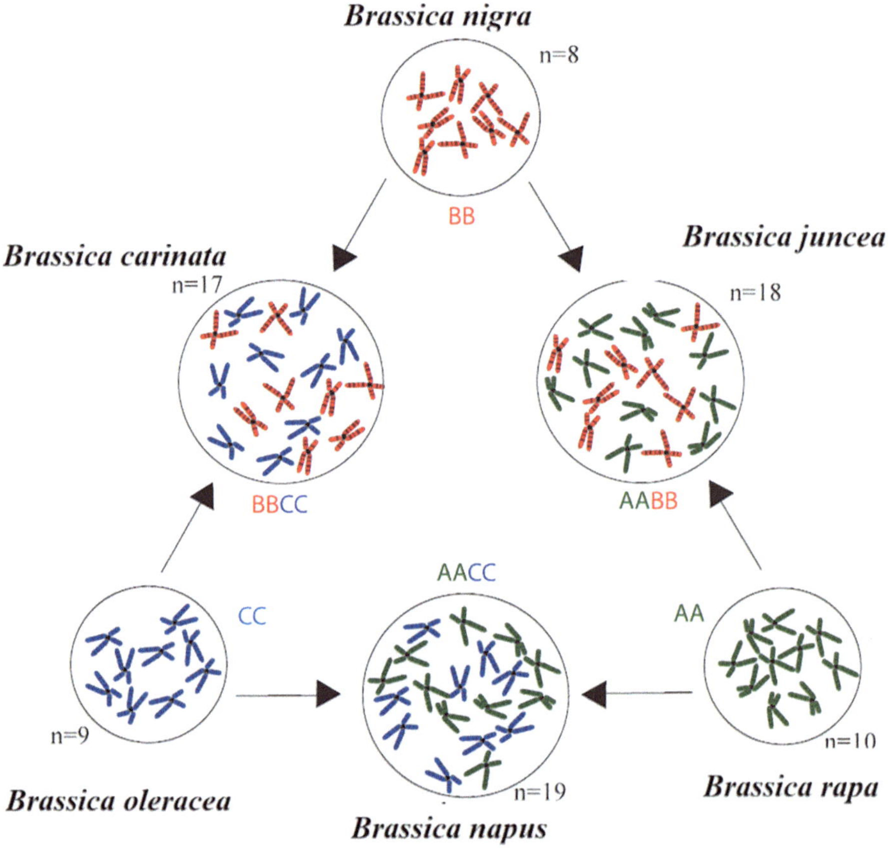

Fig. 1 A triangle showing the genetic relationships between different species of the genus *Brassica*. (Source: https://en.wikipedia.org/wiki/Triangle_of_U)

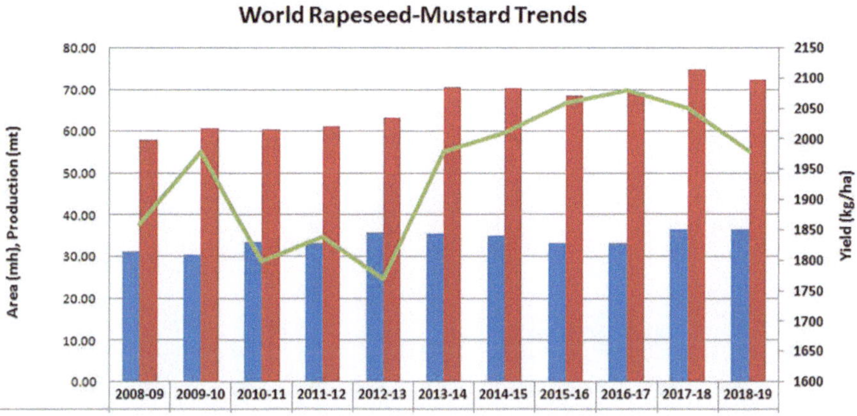

Fig. 2 World rapeseed—mustard trend

Fig. 3 Contribution of nations to the average mustard acreage (2013–2014 to 2017–2018)

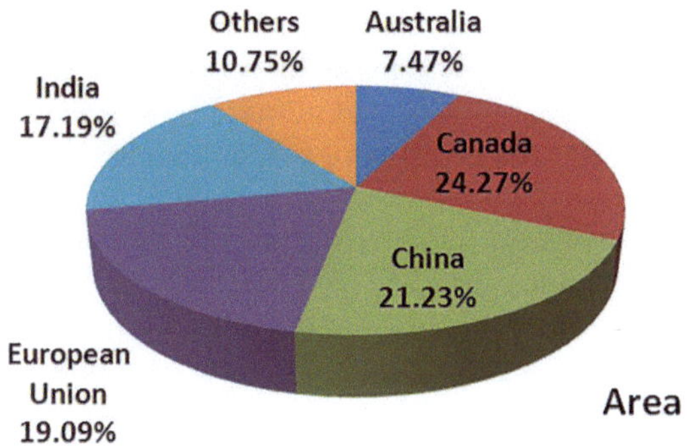

Fig. 4 Contribution of nations to the mustard production average (2013–2014 to 2017–2018)

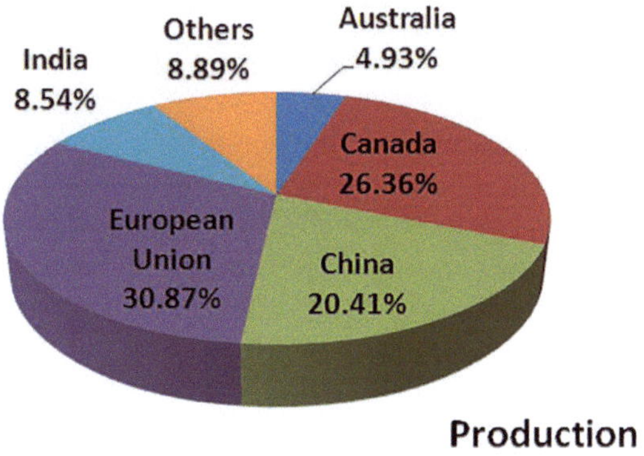

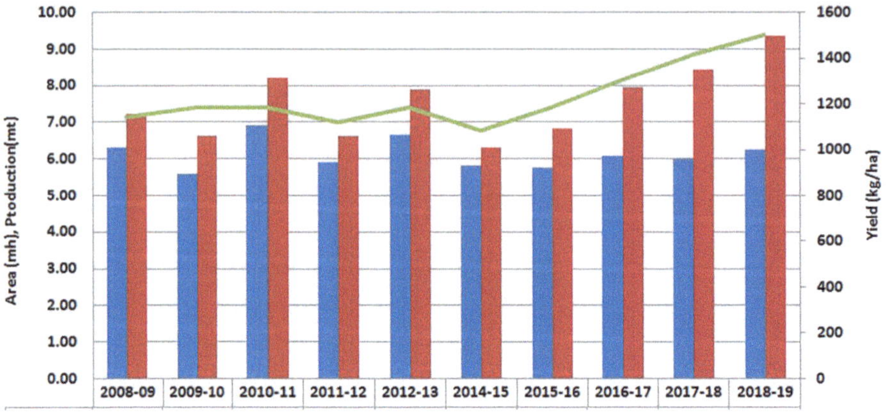

Fig. 5 Rapeseed-mustard production scenario in India

respectively (USDA). Productivity has improved significantly over the past 8 years, rising from 1840 kg/ha in 2010–2011 to 1980 kg/ha in 2018–2019. Production increased from 61.6 mt in 2010–2011 to 72.4 mt in 2018–2019.

In India, rapeseed and mustard crops are grown in various agroclimatic settings, from the hills in the northeast and northwest to the southern regions, using saline soils, early or late sowing, irrigation, and mixed cropping. Figure 5 shows that approximately 75–80% of the 6.23 million hectares planted to these crops in the nation during the 2018–2019 planting season are Indian mustard.

The main oilseed crops in India are soybeans, groundnuts, and rapeseed-mustard, which account for around 88% and 84% of the country's total production and acreage, respectively (Average of 2014–2015 to 2018–2019), as shown in Figs. 6, 7 and 8.

3 Directorate of Rapeseed—Mustard Research

3.1 Biotic Stresses in the Mustard Crop

Pests and diseases are the two main categories of biotic stresses. Among diseases, fungal diseases have significant biotic restrictions in mustard, causing substantial crop output losses globally. In India, *Brassica* crops are susceptible to over 30 diseases (Saharan et al. 2005). However, based on their distribution throughout the nation and the economic yield losses they cause, only a small number are considered severe diseases. Leaf blight (*Alternaria brassicae*), White rust and head stag (*Albugo candida*), Downy mildew (*Peronospora parasitica*), Powdery mildew (*Erysiphe cruciferarum*), and Stem rot (*Sclerotinia sclerotiorum*) are the main biotic stresses of rapeseed-mustard in India. These diseases affect seed quantity and quality (Saha and Singh 1988; Kolte 2005; Saharan et al. 2005). White rust and head

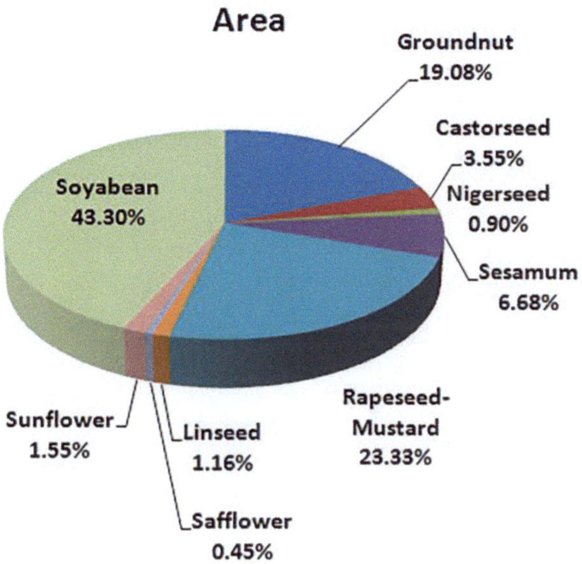

Fig. 6 Average acreage of contribution of various oilseed crops (2014–2015 to 2018–2019)

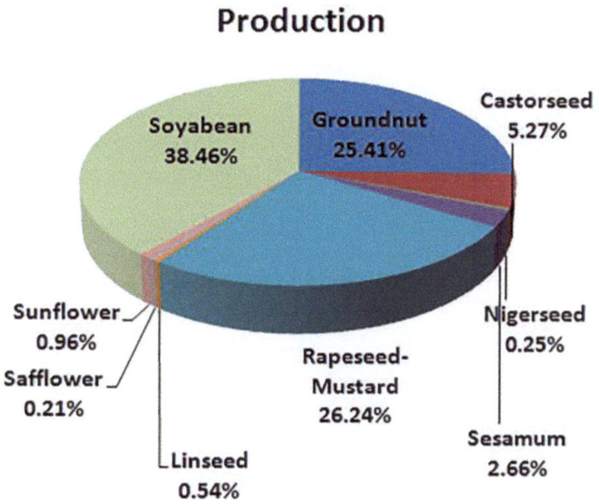

Fig. 7 Production average of contribution of various crops (2014–2015 to 2018–2019). (Source: https://www.drmr.res.in/about_rmcrop.php)

stag (*A. candida*) and *Alternaria* blight (*A. brassicae*) have been reported as the most prevalent and harmful fungal infections affecting mustard worldwide (Kotle 1985). The details of these diseases with their Yield losses are presented in Table 1.

Among insect pests, mustard aphids, and painted bugs, mustard sawfly, and diamondback moths are major pests of mustard.

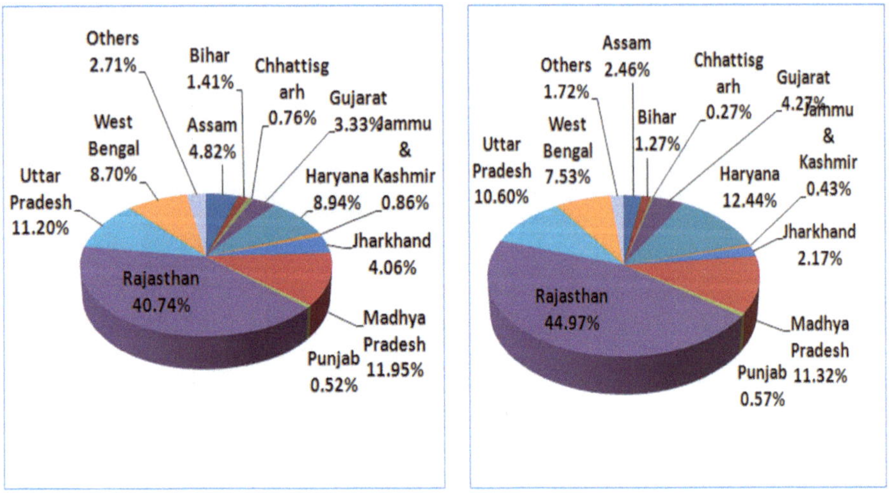

Fig. 8 Contribution of various oilseed-producing states in total area and production (average of 2014–2015 to 2018–2019)

Table 1 Yield loss in mustard crop due to diseases

Diseases	Causative organism	Yield reduction (%)
Leaf blight	*Alternaria brassicae*	15–72
White rust and head stag	*Albugo candida*	Up to 46
Stem rot	*Sclerotinia sclerotiorum*	Up to 34
Powdery mildew	*Erysiphe cruciferarum*	Up to 17
Downy mildew	*Peronospora parasitica*	15–36

1. **Mustard Aphid** (*Lipaphis erysimi*): This is a serious mustard pest. The crop suffers losses from December to March. Losses from this insect might range from 25% to 40%. Plants remain stunted, have fewer pods and grains, their pods shrivel up, and their seeds do not mature as a result of nymphs and adults sucking cell sap from leaves, stems, inflorescence, and developing pods.
2. **Painted Bug** [(*Bagrada hilaris (cruciferarum)*)]: The crop is attacked by this insect twice throughout the growing season: once in October or November and again in March or April when the crop reaches maturity. When leaves and pods wilt and dry out, the nymphs and adults consume the cell sap from them. By sucking pods, which shrivel up and prevent seed development, this bug also decreases the amount of oil.
3. **Diamondback Moth** (*Plutella xylostella*): These small greyish brown moths lay eggs on leaves and the larvae feed on the leaves and pods, causing extensive damage.
4. **Mustard saw fly** (*Athalia lugens proxima)*: The larvae feed from the margin of the leaf towards the midrib and the grubs cause numerous shot holes which cause an extensive damage in the crop.

4 Conventional Methods of Disease Management

The below-mentioned strategies are beneficial for managing diseases.

- Since sowing timing significantly impacts disease incidence and severity, it is crucial to optimise it.
- Using approved seeds, pesticides, and tolerant and disease-resistant cultivars.
- Applying different biocontrol agents, such as *Trichoderma viride* and *Pseudomonas*, or botanicals, such as *Allium sativum*, to seeds. Because biocontrol agents (BCA) are frequently effective against a variety of soil-borne diseases, their use is beneficial. Additionally, they are economical and environmentally safe, and their use reduces the possibility that the virus may become resistant to the control agent.
- Maintain the ideal plant population with the suggested crop spacing using the necessary N, P, and K fertiliser amounts.

Increased pesticide use harms biodiversity and causes the target pests to become resistant. More significantly, the consequences of climate change will make plants more vulnerable to pests. In these circumstances, improving our understanding of plant physiology may result in managing biotic stresses on a long-term basis. To develop biotic stress tolerance in mustard, there has been made considerable use of various plant breeding approaches. With connections between stress pathways and overlaps between the roles of stress metabolites and stress proteins under various stress conditions, plants' stress responses show great redundancy and complexity in their expression, sensitivity, and reaction. There are restrictions on the number of genes with known functions that can be used in the case of stress proteins, but probably more crucially, there is the question of whether one or many gene changes will result in persistent resistance. To create better solutions for biotic stresses, technology must be updated on a regular basis. These days, a number of molecular approaches are being explored for improved crop disease control. A major issue and challenging task for agricultural development is the lack of disease-resistant sources in plant breeding, which can be resolved by applying biotechnology techniques. This is one of the greatest options and opportunities to create crop-stress tolerance in mustard (Singh et al. 2020). Various methods are deployed to develop biotic stress tolerance, which are cited below.

5 Objectives of Mustard Breeding for Biotic Stress Resilience

1. **Resistance to biotic stresses:** Developing plants can tolerate or resist stresses caused by pests and diseases.
2. **Genetic diversity:** Developing plants with a diverse genetic pool increases resilience against a wide range of biotic stresses.
3. **Adaptability:** Developing plants adapt to the biotic stresses that are caused by climate changes.

4. **Yield stability:** Developing plants can maintain high yields in the face of biotic stresses.
5. **Improved nutritional quality:** Developing plants with improved nutritional quality can better withstand biotic stress conditions.

6 Traditional Breeding Approaches for Biotic Stress Resilience

1. **Mass selection:** It is the oldest method of breeding technique where a large number of desirable phenotypes are selected from a population, and their seeds are mixed to create a new variety. It utilizes the existing genetic variations. It is very simple and quick, and varieties are widely adapted. The mustard varieties having biotic stress tolerance are described in Table 2.
2. **Pedigree selection:** In this method, individual plants are selected from the F2 generation, their progenies are grown, and a record of parent-progeny relationship (pedigree) is maintained. It was first outlined by Lowe in 1927. The mustard varieties having biotic stress tolerance are described in Table 3.
3. **Recurrent selection**: Cross-pollinated crops are improved by the use of this breeding technique. In this case, single plants are chosen either by progeny testing or by their phenotype. Self-seeding is done by the chosen single plants. They are crossed in every feasible way to create a population for the following selection cycle in the following generation. The mustard varieties having biotic stress tolerance are described in Table 4.
4. **Backcross breeding:** Backcrossing is a technique used to produce offspring with a genetic identity more similar to the parent by crossing a hybrid with one of its parents or one that shares genetic traits with the parent. The mustard varieties having biotic stress tolerance are described in Table 5.

Table 2 Varieties released through mass selection

Sl. no.	Variety name	Resistance to the disease
1	Pusa Bold	Aphid, White rust, and *Alternaria* blight
2	Pusa Basanti	White rust
3	Pusa mustard-21	White rust
4	Pusa mustard-29	*Alternaria* blight
5	Pusa mustard-28	White rust
6	Pusa mustard-30	*Alternaria* blight

Table 3 Varieties released through pedigree selection

Sl. no.	Variety name	Resistance to the disease
1	Varun	*Alternaria* blight and white rust
2	Sarson shakti	*Alternaria* blight and white rust
3	Varun-25	*Alternaria* blight and white rust
4	Varun-26	*Alternaria* blight and white rust

Table 4 Varieties released through recurrent selection

Sl. no.	Variety name	Resistance to the disease
1	RH-819	White rust
2	NRCHB-506	*Alternaria* blight and white rust
3	RH-2525	*Alternaria* blight and white rust
4	RH-881	*Alternaria* blight and white rust
5	RH-0481	White rust

Table 5 Varieties released through backcross breeding

Sl. no.	Variety name	Resistance to the disease
1	RH-0745	*Alternaria* blight
2	Coral-432	White rust
3	Coral-419	*Alternaria* blight and white rust

7 Modern Breeding Approaches for Biotic Stress Resilience

1. **CRISPR-Cas9 Technology:** It can cleave DNA and is a part of bacterial immune systems. It has been modified to be used as a gene-editing tool. It is an accurate set of molecular scissors that may cut a particular DNA sequence when guided by a customizable guide. Assou et al. (2022) removed the major allergen Bra j I, a seed storage protein of the 2S albumin class, in the allotetraploid brown mustard (*Brassica juncea*) using genome editing via CRISPR/Cas9.
2. **High throughput sequencing:** Recent and constantly evolving technology known as High Throughput Sequencing (HTS) allows for the parallel sequencing of numerous DNA molecules, producing massive data sets faster than ever before. used high-throughput genotyping by sequencing technology to identify molecular markers associated with resistance to Fusarium wilt disease in mustard. They identified several candidate genes and markers that can be used for marker-assisted breeding in developing *Fusarium* wilt-resistant mustard varieties.
3. **Gene stacking:** Combining two or more desired genes into a single plant is called gene cloning. Stacked traits are the combined traits that emerge from this procedure. A biotech stack is a genetically modified crop type that possesses stacked features. Scientists from Indian Agricultural Research created the hybrid cultivar NRCHB-101 by stacking genes to make it resistant to disease.
4. **Omics technology:** The term "omics" describes a branch of the biological sciences that includes terms like "genomics", "transcriptomics", "proteomics", and "metabolomics". Different types of omics technology have been used for biotic

Table 6 Different disease resistance QTLs identified in *Brassica juncea*

Population used	Markers used	Mapped gene/QTLs	References	Disease
Donskaja × Jubilejnaja F1-derived DH	RFLP	Acr = Ac2a1, monogenic dominant gene (race 2A)	Cheung et al. (1998)	White rust
BEC144 × Varuna (F7: F2) RIL Populations	RAPD, CAPS, AFLP	AC2 (t)	Mukherjee et al. (2001)	White rust
Heera × Varuna (F1-derived DH)	AFLP, SSR	AcB1-A4.1	Panjabi-Massand et al. (2010)	White rust
Donskaja IV × TM-4	AFLP, SSR	AcB1-A5.1	Panjabi-Mass and et al. (2010)	White rust
Tumida × Varuna F1-derived DH	SSR, SNP	BjuWRR2	Bhayana et al. (2020)	White rust
Brassica juncea-*Brassica fruticulose* introgression lines	SSR, SNP	A-01, A-03, A-04, A-05, A-08, A-09 And B-05	Rana et al. (2017)	*Sclerotinia* stem rot
Brassica juncea-*Brassica fruticulose* introgression lines	GBS-GWAS	A-01, A-03, A-04, A-05, A08, A-09 And B-05 and 20 candidate genes	Atri et al. (2019)	*Sclerotinia* stem rot
Brassica juncea-*Elletaria cardamonides* introgression lines	GBS-GWAS	A03, B03	Rana et al. (2019)	*Sclerotinia* stem rot
Recombinant lines (*Brassica napus* and *Brassica juncea*)	RAPD, RFLP	B8 (Rlm6)	Chevre et al. (1997)	Blackleg
AC Vulcan/UM3132 (F2)	RFLP, SSR	LMJR1	Christianson et al. (2006)	Blackleg

stress resilience of mustard association mapping, introgressive breeding, NGS-based BSA, transcriptomics, and proteomics. The varieties released through this technology are described in Table 6.

- **Association mapping:** Quantitative trait loci mapping that links genotypes by taking the help of linkage disequilibrium. Raman et al. (2016) used a *Brassica napus* diversity panel with 179 lines to map Rlm-12 on ChrA01. They also discovered several novel SNPs strongly associated with the blackleg resistance trait and the previously mapped Rlm4 on A07.
- **Introgressive breeding:** The method uses transmission of disease resistance gene from exotic and wild relatives into popular cultivated varieties from wild species. During the collection of introgression lines of *Brassica juncea—Brassica fructiculosa*, Atri et al. (2019) used GWAS to map stem rot resistance responses on seven *Brassica juncea*: A-01, A-03, A-04, A-05, A-08, A-09, and B-05.

- **NGS: NGS-based bulked segregant analysis** involves bulking deoxyribonucleic acid or ribonucleic acid from separated phenotypes to create pods. After the genotyping of pods was done, QTLs were found using SNP calling. The blackleg resistance gene has been fine-mapped and cloned using NGS-based BSA. Rlm1 of *Brassica napus*, which encodes a serine/threonine protein kinase implicated in inducing the systemic immune response through the generation of reactive oxygen species, is homologous to STN& (*Brassica rapa*, *Brassica oleracea*, and *Arabidopsis*).

8 Molecular Techniques for Disease Management

8.1 Tissue Culture Technique

Traditional breeding approaches are used to integrate genes of interest by intercrossing species to improve disease and pest tolerance in crops; however, these strategies have proven to be less effective and have produced unfavourable outcomes (Manoj et al. 2010). Furthermore, the creation of stress-tolerant plants can benefit from applying biotechnological approaches. Genetic transformation involves transferring disease and pest tolerance genes from gene pools in several plant species to produce crops that can withstand stress. Genetic engineering may be most useful for improving crop varieties, but its primary drawbacks are poor gene expression, transgene silencing, and limited transformation competence (Rai et al. 2011).

The tissue culture method is found to be an appropriate and cost-efficient method for producing disease and pest-resistant plants. These plants are raised under regulated conditions, allowing for the growth of disease and pest-tolerant plants. Additionally, this method advances our knowledge of plants' metabolic and biochemical mechanisms that flourish in harsh conditions (Perez-Clemente and Gomez-Cadenas 2012). While *Brassica napus* resistant to *Phoma lingam* was obtained using embryonic culture (Sacristan 1982), *Brassica napus* resistant to *Alternaria brassicicola* was obtained using partially filtered culture filtrates (MacDonald and Ingram 1985).

In vitro selection by increasing the expression of PR proteins, anti-fungal peptides, or phytoalexin synthesis is essential for choosing suitable plants (Ganesan and Jayabalan 2006; Kumar et al. 2008). When creating better disease-tolerant crops, this technology is better than transgenic approach (Jayasankar et al. 2000). Organogenic or embryogenic callus, shoots, somatic embryos, or cell suspensions can be deployed to develop pathogen resistance by in vitro selection. It is possible to produce resistant plants by subjecting these cultures to diverse poisons from different plant diseases (Kumar et al. 2008). The first mustard variety with a high yield through tissue culture appropriate for unconventional locations is "Pusa Jaikisan".

9 Plant Defence Using RNA Interference to Build Tolerance to Biotic Stress

A potent technique for identifying functional genetic sequences and utilizing the downregulation of gene(s) expression is RNA interference (RNAi). Co-suppression, post-transcriptional gene silencing, virus-induced gene silencing, and other gene silencing strategies can be used to change the expression of a particular gene for a particular trait. Globally, current plant biology research is now centred on this molecular phenomenon. Consequently, it has benefited crop development and emerged as a useful tool for *Brassica* (*Brassica* sp.) functional genomics. RNA interference (RNAi) has quickly supplanted earlier antisense technology. RNA interference (RNAi) and qualitative improvements in oilseed have helped identify genes in mustard involved in fertility, somatic embryogenesis, and tolerance to biotic and abiotic stresses. It also plays a big part in maturity and yield characteristics.

10 Microorganisms' Application in Biotic Stress Tolerance

Symbiotic interactions between microorganisms and plants can lead to various outcomes, depending on the fitness benefits each partner receives (Lewis 1985). Interactions with host plants can be either favourable, negative, or neutral. Changes in the external environment cause the plant's metabolism to become unbalanced, necessitating complex genetic and metabolic systems within its cellular system (Gill and Tuteja 2010). The process by which bacteria associated with plants that are not harmful cause the host to react negatively to a disease is called "induced systemic resistance". By triggering plant defence mechanisms, the ISR protects newly exposed plants from subsequent attacks by dangerous microorganisms and insects. The phyto-hormone like ethylene and jasmonic acid control signalling pathways in induced systemic resistance (Pieterse et al. 2014). The induction of systemic resistance (SR) caused by free-living rhizobacterial strains has been the subject of numerous studies (Choudhary and Johri 2009). Using a virulent isolate of *Alternaria brassicae*, mustard (*Brassica juncea*) established resistance to *Alternaria* Black Spot, while *Alternaria alternata* strains could not do so (Vishwanath and Kolte 1999). Later, ISR mediated by several additional *Pseudomonas* species came to light, and the impact was described in several plant-pathogen systems. It has been demonstrated that *Pseudomonas* sp. strain-1 inhibits *Sclerotinia* stem rot on mustard stems triggered by *Sclerotinia sclerotiorum*. When endophytes' roles are configured, their contribution to stress tolerance significantly increases. Endophytic bacteria can improve plant health by preventing disease and herbivory and increasing plant growth and development. The endophytic bacterium *Pseudomonas syringae* had the ability to develop disease resistance using defensive priming (Pavlo et al. 2011).

11 Mustard Varieties Developed Using Different Genetic Engineering Methods and Marker-Assisted Selection for Biotic Stress Resistance

1. Braj mustard (*Brassica juncea*): Developed using genetic engineering to express a modified **cry1Ac** gene from *Bacillus thuringiensis* (Bt) for resistance against the diamondback moth (*Plutella xylostella*).
2. Coral mustard (*Brassica juncea*): Developed using genetic engineering to express the Bt cry1Ac gene for resistance against the diamondback moth.
3. Parinay mustard (*Brassica juncea*): Developed using marker-assisted selection to introgress the Bt cry1Ab gene of *Brassica napus* for resistance against the diamondback moth.
4. Surya mustard (*Brassica juncea*): Developed using genetic engineering to express the Bt cry1Ab gene for resistance against diamondback moth.
5. Pro agro mustard (*Brassica juncea*): Developed using genetic engineering to express the Bt cry1Ac and cry1Ba genes for resistance against the diamondback moth and cabbage looper (*Trichoplusia ni*).
6. Pusa mustard 30 (*Brassica juncea*): Developed using marker-assisted selection to introgress the sclerotinia resistance gene (Rlm 2) from *Brassica rapa* for resistance against *Sclerotinia* stem rot.
7. RH-30 (*Brassica juncea*): Developed using marker-assisted selection to introduce the clubroot resistance gene (CRb) from *Brassica rapa* for resistance against club root.
8. Pusa mustard hybrid-26 (*Brassica juncea*): Developed using marker-assisted selection to introgress the Sclerotinia resistance gene (Rlm2) from *Brassica rapa* for resistance against *Sclerotinia* stem rot.
9. NUDHHS-1 (*Brassica napus*): Developed using genetic engineering to express the Bt cry1Ac gene for resistance against the diamondback moth.
10. RWG1 (*Brassica juncea*): Developed using marker-assisted selection to introduce the clubroot resistance gene (CRa) from *Brassica rapa* for resistance against clubroot.

Various transgenic approaches, like signalling pathway, PR Protein, R genes, etc., that are used to improve *Brassica juncea* for biotic stress tolerance are mentioned in Table 7.

12 Conclusion

According to the foregoing chapter, biotic and abiotic stressors are the main obstacles to increasing the rapeseed-mustard crop's yield. Until now, traditional plant breeding methods have not given us complete defence against biotic and abiotic stressors. Therefore, reducing the negative impacts of these pressures on *Brassica* crops is imperative to increase production and productivity and satisfy the nation's constantly rising oil demand. The identification of DNA regions closely associated

Table 7 Transgenic approaches to improve *Brassica juncea* biotic stress tolerance

Category	Genes used	Sources	Transgenic plants	Resistance against plant pathogens	References
Signalling pathways	BjNPR1	*Brassica juncea*	*B. juncea*	*Alternaria brassicae, Erysiphe cruciferarum*	Ali et al. (2017)
Germin and Germin-like protein	HvOxO	*Hordeum vulgare*	*B. juncea*	*Sclerotinia sclerotiorum*	Verma and Kaur (2021)
PR protein	Chitinase glucanase	*Solanum lycopersicum*	*B. juncea*	*Alternaria brassicae*	Mondal et al. (2007), Chhikara et al. (2012)
R genes	WRR4	*Arabidopsis thaliana*	*B. juncea, B. napus*	*Albugo candida*	Borhan et al. (2010)
	WRR1	*Brassica juncea*	*Brassica juncea*	*Albugo candida*	Arora et al. (2019)
	Hevein	Rubber tree	*Brassica juncea*	*Alternaria brassicae*	Kanrar et al. (2002)
Lectins	Agglutinin	*Colocasia esculenta*	*Brassica juncea*	*Lipaphis erysimi*	Das et al. (2018)
	Agglutinin ACA	*Allium cepa*	*Brassica juncea*	*Lipaphis erysimi*	Hossain et al. (2006)

with mustard agronomical features through the application of genetic and genomic analysis. By eliminating the time-consuming process of direct screening in the screen house and field settings, molecular markers for the indirect selection of improved crops expedite the selection process. Because PGPRs can produce helpful enzymes, proteins, hormones, and other microbes that help plants cope with biotic and abiotic stress, their use in mustard can also be very successful. Since mustard is a crop that is particularly vulnerable to stress in its early stages, tissue culture is another helpful option. Therefore, applying a specific technology or a combination of various techniques may enhance rapeseed mustard's ability to withstand stress.

References

Ali S, Mir ZA, Tyagi A, Mehari H, Meena RP, Bhat JA, Grover A (2017) Overexpression of NPR1 in *Brassica juncea* confers broad-spectrum resistance to fungal pathogens. Front Plant Sci 8:1693

Arora H, Padmaja KL, Paritosh K, Mukhi N, Tewari AK, Mukhopadhyay A, Pental D (2019) BjuWRR1, a CC-NB-LRR gene identified in *Brassica juncea*, confers resistance to white rust caused by *Albugo candida*. Theor Appl Genet 132:2223–2236

Assou J, Zhang D, Roth KD, Steinke S, Hust M, Reinard T, Boch J (2022) Removing the major allergen Bra j I from brown mustard (*Brassica juncea*) by CRISPR/Cas9. Plant J 109(3):649–663

Atri C, Akhatar J, Gupta M, Gupta N, Goyal A, Rana K, Banga SS (2019) Molecular and genetic analysis of defensive responses of *Brassica juncea– B. fruticulosa* introgression lines to *Sclerotinia* infection. Sci Rep 9(1):17089

Borhan MH, Holub EB, Kindrachuk C, Omidi M, Bozorgmanesh-Frad GHZ, Rimmer SR (2010) WRR4, a broad-spectrum TIR-NB-LRR gene from *Arabidopsis thaliana* confers white rust resistance in transgenic oilseed *brassica* crops. Mol Plant Pathol 11(2):283–291

Cheung WY, Gugel RK, Landry BS (1998) Identification of RFLP markers linked to the white rust resistance gene (Acr) in mustard (*Brassica juncea* (L.) Czern. and Coss.). Genome 41(4):626–628

Chèvre AM, Barret P, Eber F, Dupuy P, Brun H, Tanguy X, Renard M (1997) Selection of stable *Brassica napus-B. juncea* recombinant lines resistant to blackleg (*Leptosphaeria maculans*) Identification of molecular markers, chromosomal and genomic origin of the introgression. Theor Appl Genet 95:1104–1111

Choudhary DK, Johri BN (2009) Interactions of *Bacillus* spp. and plants–specifically to induced systemic resistance (ISR). Microbiol Res 164(5):493–513

Christianson JA, Rimmer SR, Good AG, Lydiate DJ (2006) Mapping genes for resistance to *Leptosphaeria maculans* in *Brassica juncea*. Genome 49(1):30–41

Das A, Ghosh P, Das S (2018) Expression of Colocasia esculenta tuber agglutinin in Indian mustard provides resistance against *Lipaphis erysimi,* and the expressed protein is non-allergenic. Plant Cell Rep 37:849–863

Ganesan M, Jayabalan N (2006) Isolation of disease-tolerant cotton (*Gossypium hirsutum* L. cv. SVPR 2) plants by screening somatic embryos with fungal culture filtrate. Plant Cell Tissue Organ Cult 87:273–284

Gill SS, Tuteja N (2010) Reactive oxygen species and antioxidant machinery in abiotic stress tolerance in crop plants. Plant Physiol Biochem 48(12):909–930

Hossain MA, Maiti MK, Basu A, Sen S, Ghosh AK, Sen SK (2006) Transgenic expression of onion leaf lectin gene in Indian mustard offers protection against aphid colonization. Crop Sci 46(5):2022–2032

Jayasankar S, Li Z, Gray DJ (2000) In-vitro selection of *Vitis vinifera* Chardonnay with *Elsinoe ampelina* culture filtrate is accompanied by fungal resistance and enhanced chitinase secretion. Planta 211:200–208

Kanrar S, Venkateswari JC, Kirti PB, Chopra VL (2002) Transgenic expression of hevein, the rubber tree lectin, in Indian mustard confers protection against *Alternaria brassicae*. Plant Sci 162(3):441–448

Kolte SJ (2005) Tackling fungal diseases of oilseed brassicas in India

Kotle SJ (1985) Diseases of annual edible oilseed crop. In: Rapeseed mustard and sesame diseases, vol II

Kumar VJ, Ranjitha Kumari BD, Sujatha G, Castano E (2008) Production of plants resistant to *Alternaria carthami* via organogenesis and somatic embryogenesis of safflower cv. NARI-6 treated with fungal culture filtrates. Plant Cell Tissue Organ Cult 93:85–96

Lewis DH (1985) Symbiosis and mutualism: crisp concepts and soggy semantics. In: The biology of mutualism, pp 29–39

MacDonald MV, Ingram DS (1985) In vitro selection for resistance to *Alternaria brassicicola* in *Brassica napus* spp. *oleifera* (winter oilseed rape) using partially purified culture filtrates. Cruciferae Newslett 10:97–100

Mondal KK, Bhattacharya RC, Koundal KR, Chatterjee SC (2007) Transgenic Indian mustard (*Brassica juncea*) expressing tomato glucanase leads to arrested growth of *Alternaria brassicae*. Plant Cell Rep 26:247–252

Mukherjee AK, Mohapatra T, Varshney A, Sharma R, Sharma RP (2001) Molecular mapping of a locus controlling resistance to *Albugo candida* in Indian mustard. Plant Breed 120(6):483–497

Panjabi-Massand P, Yadava SK, Sharma P, Kaur A, Kumar A, Arumugam N, Pental D (2010) Molecular mapping reveals two independent loci conferring resistance to *Albugo candida* in the east European germplasm of oilseed mustard *Brassica juncea*. Theor Appl Genet 121:137–145

Pavlo A, Leonid O, Iryna Z, Natalia K, Maria PA (2011) Endophytic bacteria enhance potato growth and disease resistance (*Solanum tuberosum* L.). Biol Control 56(1):43–49

Pérez-Clemente RM, Gómez-Cadenas A (2012) In vitro tissue culture, a tool for studying and breeding plants subjected to abiotic stress conditions. In: Recent advances in plant in vitro culture, pp 91–108

Pieterse CM, Zamioudis C, Berendsen RL, Weller DM, Van Wees SC, Bakker PA (2014) Induced systemic resistance by beneficial microbes. Annu Rev Phytopathol 52(1):347–375

Rai MK, Kalia RK, Singh R, Gangola MP, Dhawan AK (2011) Developing stress tolerant plants through in vitro selection – an overview of the recent progress. Environ Exp Bot 71(1):89–98

Raman H, Raman R, Coombes N, Song J, Diffey S, Kilian A, Marcroft S (2016) Genome-wide association study identifies new loci for resistance to *Leptosphaeria maculans* in canola. Front Plant Sci 7:1513

Rana K, Atri C, Gupta M, Akhatar J, Sandhu PS, Kumar N, Banga SS (2017) Mapping resistance responses to *Sclerotinia* infestation in introgression lines of Brassica juncea carrying genomic segments from wild Brassicaceae *B. fruticulosa*. Sci Rep 7(1):5904

Rana K, Atri C, Akhatar J, Kaur R, Goyal A, Singh MP, Banga SS (2019) Detection of first marker-trait associations for resistance against *Sclerotinia sclerotiorum* in *Brassica juncea–Erucastrum cardaminoides* introgression lines. Front Plant Sci 10:1015

Sacristan MD (1982) Resistance responses to *Phoma lingam* of plants regenerated from selected cell and embryogenic cultures of haploid *Brassica napus*. Theor Appl Genet 61:193–200

Saha LR, Singh HB (1988) Diseases of rapeseed and mustard and their management

Saharan GS, Mehta N, Sangwan MS (2005) Diseases of oilseed crops. Indus Publishing

Singh RK, Prasad A, Muthamilarasan M, Parida SK, Prasad M (2020) Breeding and biotechnological interventions for trait improvement: status and prospects. Planta 252:1–18

Verma R, Kaur J (2021) Expression of barley oxalate oxidase confers resistance against *Sclerotinia sclerotiorum* in transgenic *Brassica juncea* cv Varuna. Transgenic Res 30:143–154

Vishwanath V, Kolte SJ (1999) Methods of inoculation for resistance to *Alternaria* blight of rapeseed and mustard. J Mycol Plant Pathol 29(1):96–99

MIX
Papier aus verantwortungsvollen Quellen
Paper from responsible sources
FSC® C105338

If you have any concerns about our products,
you can contact us on
ProductSafety@springernature.com

In case Publisher is established outside the EU,
the EU authorized representative is:
Springer Nature Customer Service Center GmbH
Europaplatz 3, 69115 Heidelberg, Germany

Printed by Libri Plureos GmbH
in Hamburg, Germany